THE
ras SUPERFAMILY
of
GTPases

THE
ras SUPERFAMILY
of
GTPases

Juan Carlos Lacal
Frank McCormick

CRC Press
Taylor & Francis Group
Boca Raton London New York

CRC Press is an imprint of the
Taylor & Francis Group, an **informa** business

First published 1993 by CRC Press
Taylor & Francis Group
6000 Broken Sound Parkway NW, Suite 300
Boca Raton, FL 33487-2742

Reissued 2018 by CRC Press

A Library of Congress record exists under LC control number: 92044952

Publisher's Note
The publisher has gone to great lengths to ensure the quality of this reprint but points out that some imperfections in the original copies may be apparent.

Disclaimer
The publisher has made every effort to trace copyright holders and welcomes correspondence from those they have been unable to contact.

ISBN 13: 978-1-138-10512-6 (hbk)
ISBN 13: 978-1-138-56230-1 (pbk)
ISBN 13: 978-0-203-70993-1 (ebk)

Visit the Taylor & Francis Web site at http://www.taylorandfrancis.com and the CRC Press Web site at http://www.crcpress.com

PREFACE

During the early years of molecular oncology, in which RNA and DNA tumor viruses led the charge towards understanding cellular transformation, *ras* proteins stood apart from their many contemporaries by virtue of their unusual biochemical properties: they failed to show signs of kinase activity, and they bound to GTP. During recent years, in which molecular analysis of tumor cells has set the pace in cancer research, *ras* oncogenes have emerged as being of exceptional interest by virtue of the high frequency at which they occur in human tumors. Indeed, H-*ras* was the first human oncogene to be identified, putting *ras* firmly in the spotlight. This leading role was maintained by the intriguing discovery that oncogenic forms of *ras* differed from their normal counterparts by single amino acid substitutions, usually at the same codon. This made it likely that an understanding of the basis of oncogenic activation was within reach. In the mid-1980s molecular cloning and expression technology facilitated the production of recombinant *ras* proteins in abundant quantities, opening the door for biochemical studies that had been virtually impossible using native material. These studies revealed that *ras* activation was caused by failure in the GTPase mechanisms, leaving oncogenic proteins locked in the active GTP-state constitutively. The GTPase mechanism was found to depend on another protein, GAP, which has since been implicated in a number of complex cell signaling pathways that have yet to be unraveled. Meanwhile, the abundance of recombinant *ras* protein made in bacteria made it possible to perform elegant structural analysis culminating in high resolution X-ray structures of active and inactive forms of *ras* p21, and of normal and mutant proteins. In spite of these biochemical and biophysical approaches, we are not yet certain how *ras* proteins hydrolyse GTP, how GAP participates in this process, and more important, we do not know how *ras* proteins function to bring about cellular transformation.

In parallel with these advances, several other approaches were being pursued toward the goal of understanding *ras* function. One involved the use of *S. cerevisiae* as a model, taking advantage of yeast genetics to identify pathways in which *ras* proteins are involved. This approach led relatively quickly to the first candidate for a *ras* effector, adenylyl cyclase. However, it was realized just as quickly that *ras* proteins do not regulate adenylyl cyclase in mammalian cells, so that the relevance of studying *ras* in yeast seemed questionable. However, this system has provided a wealth of information about *ras* regulation and processing, most of which is probably relevant to mammalian cells. Furthermore, the mechanism by which *ras* activates cyclase, which has not yet been elucidated, is likely to resemble *ras* action on mammalian effectors when scrutinized at the molecular level. Currently, it seems that the yeast *S. pombe* may be more analogous to mammalian cells with respect to *ras* function, so that much of the technology developed in *S. cerevisiae* can now be applied in the *pombe* system. More recently, other genetic approaches toward *ras* function using *Drosophila* and *C. elegans* have appeared very promising, and may well lead to new insights into *ras* function.

Substantial progress has been made in identifying and understanding *ras* family members, such as proteins of the *rho* and *rab* sub-groups. Again, a combination of biochemistry and genetics have been applied, and we now have a wealth of information concerning proteins that regulate *ras*-like GTPases (GAPs and exchange factors) and biological systems in which they are involved (from microorganisms and higher eukaryotes). We are not yet at the point where an effector of any single *ras*-like protein can be named, but the existence of cell-free systems for analyzing their function should accelerate this search.

While we have witnessed an explosion of information relating to function and regulation of *ras* and its relatives, and have become entangled in a web of complex pathways, impressive progress has been made toward understanding the function of tumor suppressor genes such as Rb and p53, and in unraveling complexities of the cell cycle. Progress has also been made in understanding many aspects of signal transduction, including the realization that many signaling pathways involve protein–protein interactions such as those using SH2/tyrosine phosphate interactions. Looking forward, our major challenge will be the collection of this wealth of information into a coherent and comprehensible scheme. Before this happens, we expect the fog to thicken rather than clear, but ultimately, clarification is assured.

Juan Carlos Lacal
Frank McCormick

THE EDITORS

Juan Carlos Lacal, Ph.D., is a Group Leader, Institute for Biomedical Research, Spanish Research Council, Madrid, Spain, and Honorary Professor of the Universidad Autónoma of Madrid, Spain.

Dr. Lacal graduated in 1979 from the Universidad Autónoma of Madrid, with a B.S. degree in Biology (summa cum laude). In 1979 he obtained the M.S. degree in Biochemistry and Molecular Biology, with Special Class Honors, from the Universidad Autónoma of Madrid. In 1982, he obtained his Ph.D. in Biochemistry and Molecular Biology (summa cum laude) from the Universidad Autónoma of Madrid.

Dr. Lacal has been the recipient of the European Molecular Biology Organization Fellowship and the Fogarty International Center Fellowship. He was appointed as Visiting Associate at the National Cancer Institute, Bethesda, Maryland.

Dr. Lacal is a member of the American Association for the Advancement of Science, American Association for Microbiology, Spanish Biochemistry Society, Spanish Cell Biology Society, Spanish Association for Cancer Research, and Spanish Federation of Cancer Research Societies. He is Secretary of the Spanish Association for Cancer Research and Vocal of the Spanish Federation of Cancer Research Societies.

Dr. Lacal has been the recipient of research grants from the Spanish Government and private industry. He is author of over 50 international publications, has edited three books, and has co-authored several other national and international books. Finally, he has participated in the organization of several national and international meetings on cell biology and cancer. His major research interest relates to investigation of the mechanism of regulation of proliferation and differentiation by oncogenes, especially those related to small GTPases of the *ras* superfamily.

Frank McCormick, Ph.D., is Vice President of Research and Scientific Founder of Onyx Pharmaceuticals. He is author of over 100 papers on the subject of *ras* and *ras* regulators. Prior to Onyx's formation, Dr. McCormick was Vice President, Therapeutics and Biological Research at Chiron Corporation. He began his career at Cetus Corporation as a Postdoctoral Researcher in 1981, becoming Vice President, Discovery Research in 1990. Dr. McCormick conducted post-doctoral research at the State University of New York at Stony Brook and the Imperial Cancer Research Fund in London. He received his Ph.D. in Biochemistry from Cambridge University in England.

CONTRIBUTORS

Arthur S. Alberts, B.A.
Graduate Student
Departments of Medicine
University of California
La Jolla, California

Rafael P. Ballestero, B.S.
Graduate Student
Instituto de Investigaciones Biomedicas
Madrid, Spain

Johannes L. Bos, Ph.D.
Professor
Laboratory for Physiological Chemistry
Utrecht University
Utrecht, Netherlands

Boudewijn M. Th. Burgering, Ph.D.
Assistant Professor
Laboratory for Physiological Chemistry
Utrecht University
Utrecht, Netherlands

Carmela Calés, Ph.D.
Profesor Ayudante
Departamento de Bioquímica
Universidad Autónoma
Madrid, Spain

James Cardelli, Ph.D.
Associate Professor
Department of Microbiology and
 Immunology
Louisiana State University Medical
 Center
Shreveport, Louisiana

Amancio Carnero, B.S.
Graduate Student
Instituto de Investigaciones Biomedicas
Madrid, Spain

Richard A. Cerione, Ph.D.
Associate Professor
Department of Pharmacology
Schurman Hall
Cornell University
Ithaca, New York

Pierre Chardin, Ph.D.
Senior Scientist
Institut de Pharmacologie Moleculaire et
 Cellulaire du CNRS
Valbonne, France

Jean-Bernard Créchet, Ph.D.
Scientist
Laboratoire de Biochimie
Ecole Polytechnique
Palaiseau, France

Antonio Cuadrado, Ph.D.
Investigador Contratado, CSIC
Instituto de Investigaciones Biomedicas
Madrid, Spain

Peter D'Eustachio, Ph.D.
Associate Professor
Department of Biochemistry
New York University School of Medicine
New York, New York

Julian Downward, Ph.D., M.A.
Head, Signal Transduction Laboratory
Imperial Cancer Research Fund
London, England

George T. Drivas, B.S., M.S.
Graduate Assistant
Department of Biochemistry
New York University School of Medicine
New York, New York

Renee Emkey, B.S.
Graduate Student
Department of Biochemistry
Tufts University School of Medicine
Boston, Massachusetts

R. Keith Esch, Ph.D.
Postdoctoral Fellow
Department of Chemistry
University of North Carolina
Chapel Hill, North Carolina

Tony Evans, Ph.D.
Scientist
Department of Cell Biology
Genentech, Inc.
San Francisco, California

Larry A. Feig, Ph.D.
Assistant Professor
Department of Biochemistry
Tufts University School of Medicine
Boston, Massachusetts

James R. Feramisco, Ph.D.
Professor
Department of Medicine and
  Pharmacology
UCSD Cancer Center
University of California
La Jolla, California

Richard A. Firtel, Ph.D.
Professor
Department of Biology
Center for Molecular Genetics
University of California
San Diego, California

Dieter Gallwitz, M.D.
Professor, Director
Department of Molecular Genetics
Max-Planck-Institute for Biophysical
  Chemistry
Göttingen, Germany

John F. Hancock, M.A., M.B., Ph.D.
Senior Scientist
Onyx Pharmaceuticals
Richmond, California

Matthew J. Hart, Ph.D.
Department of Biochemistry, Cellular and
  Molecular Biology
Department of Pharmacology
Schurman Hall
Cornell University
Ithaca, New York

Ludger Hengst, Ph.D.
Postdoctoral Fellow
Department of Molecular Genetics
Max-Planck-Institute for Biophysical
  Chemistry
Göttingen, Germany

Eric Jacquet, Ph.D.
Scientist
Laboratoire de Biochimie
Ecole Polytechnique
Palaiseau, France

Douglas I. Johnson, Ph.D.
Assistant Professor
Department of Microbiology and
  Molecular Genetics
University of Vermont
Burlington, Vermont

Kozo Kaibuchi, M.D., Ph.D.
Associate Professor
Department of Biochemistry
Kobe University School of Medicine
Kobe, Japan

Masahito Kawata, M.D.
Research Associate
Department of Biochemistry
Kobe University School of Medicine
Kobe, Japan

Warren A. Kibbe, Ph.D.
Postdoctoral Fellow
Department of Molecular Genetics
Max-Planck-Institute for Biophysical
  Chemistry
Göttingen, Germany

Akira Kikuchi, M.D., Ph.D.
Assistant Professor
Department of Biochemistry
Kobe University School of Medicine
Kobe, Japan

Alan R. Kimmel, Ph.D.
Senior Investigator
Laboratory of Cellular and
 Developmental Biology
National Institutes of Health
Bethesda, Maryland

Hitoshi Kitayama, Ph.D.
Member
Laboratory of Molecular Oncology
Tsukuba Life Science Center
RIKEN
Ibaraki, Japan

Juan Carlos Lacal, Ph.D.
Investigador Científico
Instituto de Investigaciones Biomedicas
Madrid, Spain

Javier León, Ph.D
Assistant Professor
Department of Molecular Biology
Faculty of Medicine
University of Cantabria
Santander, Spain

David Leonard, B.S.
Graduate Student
Department of Biochemistry Cellular and
 Molecular Biology
Department of Pharmacology
Cornell University
Ithaca, New York

Zeev Lev, D. Sc.
Senior Lecturer
Department of Biology, Technion
Israel Institute of Technology
Haifa, Israel

**Christopher J. Marshall, M.A.,
 D.Phil.**
Chairman
Section of Cell and Molecular Biology
Institute of Cancer Research
Chester Beatty Laboratories
London, England

Kunihiro Matsumoto, Ph.D.
Professor
Department of Molecular Biology
Faculty of Science
Nagoya University
Nagoya, Japan

Frank McCormick, Ph.D.
Vice President — Research
Onyx Pharmaceuticals
Richmond, California

Judy L. Meinkoth, Ph.D.
Assistant Professor
Department of Medicine and Cancer
 Center
University of California, San Diego
La Jolla, California

Michel-Yves Mistou, Ph.D.
Postdoctoral Fellow
Laboratoire de Biochimie
Ecole Polytechnique
Palaiseau, France

Narito Morii, M.D., Ph.D.
Associate Professor
Department of Pharmacology
Kyoto University Faculty of Medicine
Kyoto, Japan

Shuh Narumiya, M.D., Ph.D.
Professor
Department of Pharmacology
Kyoto University Faculty of Medicine
Kyoto, Japan

Yasuo Nemoto, M.D., Ph.D.
Scientist
Department of Pharmacology
Kyoto University Faculty of Medicine
Kyoto, Japan

Makoto Noda, Ph.D.
Member and Chief
Department of Viral Oncology
Cancer Institute
Tokyo, Japan
and
Visiting Scientist
Molecular Oncology
Tsukuba Life Sciences Center
RIKEN
Ibaraki, Japan

Andrea Parmeggiani, M.D.
Senior Scientist
Laboratoire de Biochimie
Ecole Polytechnique
Palaiseau, France

Angel Pellicer, M.D., Ph.D.
Professor
Department of Pathology
New York University Medical Center
New York, New York

Rosario Perona, Ph.D.
Titulado Superior, CSIC
Instituto de Investigaciones Biomedicas
Madrid, Spain

Paul Polakis, Ph.D.
Senior Scientist
Onyx Pharmaceuticals
Richmond, California

Patrick Poullet, Ph.D.
Postdoctoral Fellow
Laboratoire de Biochimie
Ecole Polytechnique
Palaiseau, France

Tracy Ruscetti, B.S.
Doctoral Student
Department of Microbiology and
  Immunology
Louisiana State University Medical
  Center
Shreveport, Louisiana

Mark G. Rush, B.S., M.S., Ph.D.
Associate Professor
Department of Biochemistry
New York University School of Medicine
New York, New York

Takuya Sasaki, M.D., Ph.D.
Research Associate
Department of Biochemistry
Kobe University School of Medicine
Kobe, Japan

Axel Schönthal, Ph.D.
Assistant Professor
Department of Microbiology
University of Southern California
Los Angeles, California

Katsuhiro Shinjo, D.V.M.
Staff Scientist
Pfizer Research
Nagoya, Japan

Steven Sorscher, M.D.
Clinical Instructor
Department of Medicine and Cancer
  Center
University of California, San Diego
La Jolla, California

Yoshimi Takai, M.D., Ph.D.
Professor
Department of Biochemistry
Kobe University School of Medicine
Kobe, Japan

Kazuma Tanaka, Ph.D.
Assistant Professor
Department of Biochemistry
Kobe University School of Medicine
Kobe, Japan

Armand Tavitian, M.D.
Director
INSERM
Paris, France

Akio Toh-e, Ph.D.
Professor
Department of Biology
Faculty of Science
University of Tokyo
Tokyo, Japan

Gerald Weeks, Ph.D.
Professor
Department of Microbiology
University of British Columbia
Vancouver, Canada

Michael H. Wigler, Ph.D.
Senior Scientist, Head Mammalian Cell
  Genetics
Department of Mammalian Cell Genetics
Cold Spring Harbor Laboratory
Cold Spring Harbor, New York

Fred Wittinghofer, Ph.D.
Research Assistant
Max-Planck-Institut fur Medizinische
  Forschung
Abteilung Biophysik
Heidelberg, Germany

Takeshi Yamamoto, M.D.
Graduate Student
Department of Biochemistry
Kobe University School of Medicine
Kobe, Japan

Ahmed Zahraoui, Ph.D.
Associate Researcher
INSERM
Paris, France

TABLE OF CONTENTS

I. THE WORLD OF *ras* GENES AND THEIR PRODUCTS

Oncogenic Potential of *ras* Genes

Structure and Properties of *ras* Proteins

Genetic Approaches to *ras* Function

II. THE *ras* SUPERFAMILY

General Overview

The *ras* Branch

The *rho* Branch

The *rab* Branch

III. FUNCTIONAL REGULATION OF *ras* AND *ras*-RELATED GTPases

GAPs and Related Factors

Exchange Factors for *ras* and *ras*-Related Proteins

SECTION I
The World of *ras* Genes and Their Products

Chapter 1

ras GENES INVOLVEMENT IN CARCINOGENESIS: LESSONS FROM ANIMAL MODEL SYSTEMS

Javier León and Angel Pellicer

TABLE OF CONTENTS

0-8493-5214-2/93/$0.00 + $.50
© 1993 by CRC Press, Inc.

I. INTRODUCTION AND HISTORICAL PERSPECTIVE

ras oncogenes were originally discovered as part of the genomes of highly oncogenic murine retroviruses. At least five sarcoma viruses containing activated versions of *ras* genes have been isolated (reviewed by Lacal and Tronick[1]). All these tumors cause sarcomas in rats or mice, the Harvey sarcoma virus and the Kirsten sarcoma virus being the representative isolates carrying the activated forms of H- and K-*ras* genes, respectively. Thus, the story of *ras* genes and *ras*-related genes is linked to animal carcinogenesis. Years later (1982) several groups showed that genes isolated from human tumor cell lines by transfection methods were homologs to the oncogene from Harvey sarcoma virus. This gene was a mutated version from the normal H-*ras* found in nontumor cells.[2-5] Since then, activated *ras* genes have been found in numerous human and animal tumors.

The mammalian *ras* gene subfamily belongs to the superfamily of small GTPases (reviewed in References 6 and 7 and Section II of this book). The *ras* subfamily has three members: H-, K-, and N-*ras*. *ras* proteins (termed p21 for their molecular mass, 21 kDa) contain 188 (K-*ras*-B) or 189 (H- N-, and K-*ras*-A) amino acids. The first 80 amino acids are almost completely conserved among the four proteins. The protein segment encompassing amino acids 80 to 164 show slightly less similarity among *ras* proteins (70 to 80%). The rest of the protein, the hypervariable region, is specific for each *ras* gene, except for the last four amino acids.[8,9]

ras protooncogenes, as well as *neu*, are activated by point mutations in their coding sequences, although overexpression of *ras* genes can also transform cells *in vitro.*[10-14] Activated *ras* oncogenes contain point mutations affecting either codons 12 (Gly), 13 (Gly), or 61 (Gln). Rarely, mutations on other codons (e.g., in codon 117[15] or 146[16]) have also been reported.

The fact that *ras* genes are activated by point mutations opens the way for the search of the molecular mechanisms relating the genetic injury by the carcinogenic agent with the uncontrolled cell proliferation leading to tumor formation.

II. ANIMAL MODEL SYSTEMS AND *ras* ACTIVATION

A large body of data has been published over the last decade on *ras* activation in human tumors, and significant *ras* activation frequencies have been linked to several types of human tumors. Despite that, few general conclusions can be drawn on the involvement of *ras* genes in tumor generation based on human studies. In human carcinogenesis, key parameters as to the carcinogenic agent and genetic background are frequently unknown. Therefore, a considerable number of animal model systems have been established to study carcinogenesis at the molecular level. Some model systems are based on selected animal strains showing high frequencies of certain spontaneous tumors. Nevertheless, most models are based on protocols that use a wide variety of carcinogenic agents to treat particular animal species or strains with increased sensitivity to the carcinogen.

Although a large variety of model systems have been designed for carcinogenetic studies, only a few have been analyzed for oncogene detection. Previous reviews[17-19] dealt with oncogene activation in animal carcinogenesis. Here we will briefly review the results obtained on *ras* oncogene activation in animal model systems, albeit the list of models cited is not exhaustive. *ras* genes carrying point mutations are by far the most prevalent oncogenes found both in human cancer and in experimental carcinogenesis. *ras* gene amplification or increased expression have occasionally been reported in some animal models,[20-22] as well as some human tumors (reviewed by Field and Spandidos[23]). However, here we will consider *ras* activation as synonymous with point mutation. Apart from *ras* activation, few activated oncogenes have been consistently found in tumors of animal model systems. These oncogenes include mutated *neu* genes[24] and rearranged or amplified c-*myc*.[25,26] Detection of other oncogenes has rarely been reported. The prevalence of *ras* oncogenes found in experimental carcinogenesis correlates with the situation in human cancer, where activated *ras* genes are again the oncogenes most commonly found (reviewed by Bos[27]). Although there are some correlations between the finding of particular *ras* genes in human and animal induced tumors (i.e., squamous cell carcinomas, keratoacanthomas, and bladder and lung carcinomas) this is not the general rule.

The high frequency of *ras* activation could be partly explained because the methodology originally used (NIH 3T3 transformation assays, see below) was biased in favor of *ras* oncogene detection. As a consequence, mutations in *ras* genes are more actively sought than alterations on other genes. However, as discussed below, the high prevalence of mutated *ras* genes in induced and spontaneous tumors speaks in favor of important roles for these genes in the cellular proliferation and differentiation control systems.

A. DETECTION OF *ras*-MUTATED GENES

In model systems, *ras* activation has been detected by one or more of the following methods.

1. Biological Assays

ras-activated genes introduced in rodent fibroblast cell lines, such as mouse NIH 3T3, induce cellular transformation in a dominant fashion. Therefore, activated *ras* oncogenes can be detected either by a focus-forming assay *in vitro* or by tumor formation in nude mice after injection of the transfected cells. As noted above, the first assay led to the discovery of cellular *ras* oncogenes. However, the focus-forming assay is frequently replaced by the more sensitive nude mice assay.[28,29]

2. Molecular Assays
a. *Gene Cloning*

After the focus-forming assay mentioned above, the first method used in *ras* oncogene detection, was the cloning and sequencing of the genes out of the transformed cells in order to identify the activating mutation of the *ras* gene. Today, this identification can be more easily done by the application of the more recent techniques referred to below, particularly those using the polymerase chain reaction (PCR) technology.

b. *RNase Mismatch Cleavage Analysis*

Some RNases are very specific for single-strand RNA and can cut an RNA:RNA or RNA:DNA hybrid at single mismatch sites. Usually, a mixture of RNase A and T_1 that cut most (but not all) mismatches is used.[30] These features were originally exploited by Perucho's group for detecting K-*ras* mutations in human colonic tumors.[30] Regions of the *ras* genes containing the 12th or 61st codons are cloned downstream from SP6 or T7 phage promoters. The *ras* antisense RNA is produced *in vitro* by using the corresponding RNA polymerase, hybridized to the tumor RNA, and digested with RNases. The presence of mutations is detected by analysis of the digestion products. This method is sensitive and has the advantage of detecting the expression of the mutation since RNA from the tumor is analyzed. However, it does not reveal the type of particular base substitution. A modification is to use PCR-amplified DNA instead of the tumor RNA.[31]

c. *Oligonucleotide Differential Hybridization in Gel*

DNA from tumor or transformed cells is restricted, separated in a gel, and the gel is hybridized to an allele-specific oligonucleotide (usually 19-mer) encompassing the codon where mutations are to be detected. Several oligonucleotides, each carrying a different base change, are used as probes to detect any possible mutation. Washing conditions can be worked out so oligonucleotides containing only one mismatch would not hybridize. Therefore, mutations can be directly detected using tumor DNA, providing that the fraction of cells carrying the mutated gene is not too low.[32-34] Today, this method has been rendered obsolete by those based on PCR described below.

d. Allele-Specific Oligonucleotide Hybridization of PCR Products

Today this is likely the most popular method for detecting *ras* mutations. Selected fragments of *ras* genes encompassing the 12th or 61st codons are amplified, transferred to nylon, and hybridized to different oligonucleotides carrying the possible mutations. The PCR technique requires minute amounts of DNA and therefore, can be applied to detect *ras* mutations in small preneoplastic lesions[35,36] or in formalin or paraffin-fixed tissues.[31,37] This latter application has been very useful in the retrospective analysis of human tumors.

e. Restriction Fragment Length Polymorphism (RFLP)

In some cases, a *ras*-mutated codon can be detected by RFLP. For example an A → T transversion in the second base of mouse H-*ras* creates a new XbaI site.[22,38,39] A refinement of this technique, described by Barbacid and Kumar, combines PCR with RFLP by using specifically modified amplimers to create new restriction sites only when *ras* mutations are present.[40]

f. Direct Sequencing of Amplified DNA

By this methodology selected fragments of the *ras* genes are amplified by PCR and the amplified product is sequenced. This allows us to directly visualize the activating mutation and therefore, it is increasingly used despite its technical difficulties. The technique can also be applied to paraffin-fixed tissue.[37]

We will briefly review the model systems where *ras* mutations have been detected following the chronological order in which the oncogene activation was described.

B. SKIN TUMORS

Skin tumors constitute an attractive model system because it is a tissue easily accessible, where carcinogen delivery can be well controlled, and because tumor development and progression is easily monitored. Two general protocols have been used: a "complete carcinogenesis" protocol where repeated doses of the carcinogenic agent are applied, and the "initiation-promotion" protocols, where a single dose of the initiating carcinogenic agent is followed by repeated applications of a tumor promotor, usually a phorbol ester (for a recent review, see Drinkwater[41]). In this protocol, the treatment with either the initiating agent or the promoter alone is not sufficient for tumor induction.

TABLE 1
ras Activation in Chemically Induced Skin Tumors

Animal and type of tumor	Carcinogen	Activated *ras* genes[a]	Frequency[b] (%)	Ref.
Mouse carcinomas	DMBA	H-*ras*-61	75 (3/4)	38
Mouse carcinomas	DBACR	H-*ras*-61	80 (4/5)	38
Mouse carcinomas	DMBA	H-*ras*-61	50 (4/8)	142
Mouse carcinomas	NQO + TPA	H-*ras*-61	50 (2/4)	142
Mouse papillomas	DMBA + TPA	H-*ras*-61	100 (48/48)	45
Mouse carcinomas	DMBA + TPA	H-*ras*-61	77 (10/13)	45
Mouse papillomas	MNNG + TPA	H-*ras*-12	73 (11/15)	45
Mouse carcinomas	MMNG + TPA	H-*ras*-12	15 (2/13)	45
Mouse papillomas	MNU + TPA	H-*ras*12	42 (5/12)	45
Mouse papillomas	MCA + TPA	H-*ras*-61 H-*ras*-13	16 (4/20) 16(4/20)	45
Mouse carcinomas	MCA + TPA	H-*ras*-13	57 (4/7)	45
Mouse papillomas	DMBA[c] + TPA	H-*ras*-61	75 (9/12)	83
Mouse carcinomas	DMBA[c] + TPA	H-*ras*-61	50 (2/4)	83
Mouse tumors[d]	NCM + TPA	H-*ras*-61	44 (4/9)	143
Mouse tumors[d]	Promoters[e]	H-*ras*-61	78 (7/9)	21
Rabbit keratoacanthomas	DMBA	H-*ras*-61	82 (18/22)	42
Rabbit carcinomas	DMBA	H-*ras*61 H-*ras*-12 N-*ras*	40 (4/10) 10 (1/10) 10 (1/10)	42
Rat carcinomas	β-propiolactone	H-*ras*-61	17 (1/6)	144
Rat carcinomas	Radiation	K-*ras*	83 (5/6)	26
Rat carcinomas	Dimethoxibenzidine	H-*ras* N-*ras*	77 (10/13) 8 (1/13)	145
Rat carcinomas	Dimethylbenzidine	H-*ras* N-*ras*	78 (7/9) 11 (1/9)	145

Note: DMBA, 7,12-dimethylbenz[a]anthracene; DBACR, dibenz[c,h]acridine; MCA, 3-methylcholanthrene; MNNG, *N*-methyl-*N'*-nitro-*N*-nitrosoguanidine; MNU, *N*-methyl-*N*-nitrosourea; NCM, *N*-nitrosocimetidine; NQO, 4-nitroquinoline-1-oxide; and TPA, 12-*O*-tetradecanoyl-phorbol-13-acetate.

[a] Mutated codon is indicated when known.
[b] Tumors with *ras* mutated genes/tumors analyzed.
[c] Transplacental exposure to DMBA.
[d] Papillomas and carcinomas.
[e] TPA, mezerein.

ras activation has been detected in skin tumors induced by both types of protocols. Table 1 includes the frequencies of *ras* activation found in mouse, rat, and rabbit skin tumors. Two important conclusions can be deduced from Table 1: (1) there is a high frequency of *ras* activation and (2) the activated gene is H-*ras* in most cases, regardless of the animal, carcinogenic agent used, and type of tumor induced.

Most of the experiments carried out with complete carcinogenesis protocols use 7,12-dimethylbenz(a)anthracene (DMBA)[38,42,43] although other agents have also been used including dibenz(c,h)acridine (DBACR),[38] and ionizing radiation.[26] In this latter case, activated K-*ras* genes were found, and constitute the only example thus far reported where activation of a *ras* gene other than H-*ras* has been found in induced skin tumors. In the initiation-promotion protocols, DMBA and 12-*O*-tetradecanoyl-13-phorbol acetate (TPA) is the combination most frequently used in rodent skin carcinogenesis. Balmain's group was the first to detect *ras* activation using this carcinogenesis protocol.[44] The same group also demonstrated that infection by Harvey-*ras* virus substitutes for DMBA treatment[45] and found similar frequencies of H-*ras* activation using different initiating agents, such as DMBA, MNU, *N*-methyl-*N'*-nitro-*N*-nitrosoguanidine (MNNG), and 3-methylcholanthrene (MCA).[46] However, the type of mutation varied with the carcinogen, an effect that will be discussed below. Basically, the tumors induced in mice are papillomas and squamous cell carcinomas. DMBA treatment of rabbit skin induced a significant fraction of keratoacanthomas, together with carcinomas and papillomas.[42,43] Keratoacanthomas are benign tumors that undergo spontaneous and complete regression, unlinked to an immunological reaction.[47] Thus, this model system provides self-regressing tumors (keratoacanthomas), benign tumors (papillomas), and malignant tumors (squamous cell carcinomas). As shown in Table 1, H-*ras* activation is found in all kinds of skin tumors induced in mice and rabbits. It is noteworthy that a significant incidence of H-*ras* activation is detected in human keratoacanthomas and less in squamous cell carcinomas.[37]

C. MAMMARY TUMORS

Rat and mouse mammary carcinomas are induced at high frequencies by systemic treatment with MNU or DMBA.[34,39,48,49] Intraperitoneal single injections of MNU (50 mg/kg weight) or DMBA are sufficient to obtain high incidence of tumors in female rats. As shown in Table 3, frequencies of *ras* activation in these tumors are always high, close to 90% in various reports. More interesting is the finding that most of the activated oncogenes found are H-*ras* with mutations on codon 12 when MNU was used as a carcinogen, while mutations on codon 61 are more common in DMBA-induced tumors. Moreover, as seen below, there is a close correlation between the base changed in either codon and the carcinogen used. Conversely to the liver model system (see below), similar percentages of tumor induction and *ras* activation in the tumors does not significantly vary among the different rat strains assayed.[34] In this model, tumor development is estrogen-dependent and therefore, tumors only appear once the rat has reached sexual maturity, and many of them partially regress upon ovariectomy or antiestrogen treatment.[50,51] However, there is no relationship between the presence of activated H-*ras*, the mutated codon (12th or 61st), and the status of estrogen dependence[50] Interestingly, if MNU is given to neonate rats or transplacentally stillborn rats, H-*ras*-activated genes can be detected in mammary tissue two

TABLE 2
ras **Activation in Chemically Induced**
Mammary Tumors

Animal and type of tumor	Carcinogen	Activated ras genes[a]	Frequency[b] (%)	Ref.
Rat carcinomas	MNU	H-*ras*-12	86 (61/71)	34
Rat carcinomas	DMBA	H-*ras*-61	20 (5/25)	34
Rat carcinomas	MNU	H-*ras*-12	54 (15/28)	51
		K-*ras*-12	18 (5/25)	
Rat fibroadenomas	MNU	H-*ras*-12	43 (3/7)	51
		K-*ras*-12	14 (1/7)	
Rat preneoplastic lesions[c]	MNU	H-*ras*-12	16 (8/50)	51
		K-*ras*-12	15 (3/20)	
Rat carcinomas	MNU	H-*ras*-12	88 (23/26)	50
Rat carcinomas	DMBA	H-*ras*	21 (6/29)	50
Rat carcinomas	MNU[d]	H-*ras*-12	100 (3/3)	52
Rat carcinomas	MNU[e]	H-*ras*-12	45 (78/173)	108
Mouse carcinomas[f]	DMBA	H-*ras*-61	100 (4/4)	39
Mouse carcinomas[f]	DMBA	H-*ras*-61	67 (12/18)	146
Mouse carcinomas[f]	DMBA	H-*ras*-61	92 (11/12)	36
Mouse hyperplastic lesions[f]	DMBA	H-*ras*-61	33 (2/6)	36
Mouse carcinomas	MNU[g]	K-*ras*-12	90 (19/21)	81
Mouse hyperplastic lesions	MNU[g]	K-*ras*-12	90 (9/10)	81

Note: DMBA, 7,12-dimethylbenz[a]anthracene and MNU, *N*-methyl-*N*-nitrosourea.

[a]   Mutated codon is indicated when known.
[b]   Tumor with *ras* mutated genes/tumors analyzed.
[c]   Detected 2 weeks after exposure to MNU.
[d]   Transplacental exposure to MNU.
[e]   Different MNU doses were used.
[f]   Tumors derived from hyperplastic mammary cell lines transplanted into mammary fat pads.
[g]   Cells were exposed *in vitro* to MNU and reimplanted in mammary glands.

weeks after carcinogen exposure and subsequently, in tumors which developed after puberty. Therefore, the H-*ras* gene remains mutated in the target cells long before the tumor onset.[51,52] However, conversely to the situation in rats exposed to MNU at puberty where H-*ras* is the only *ras* oncogene detected, in rats treated neonatally activated K-*ras* is found in tumors and preneoplastic lesions at significant percentages (Table 2).[51]

D. LYMPHOMAS

ras activation has been detected in mouse thymic lymphomas after exposure to MNU (intraperitoneal injections) or γ-radiation. These treatments result in

TABLE 3
ras Activation in Mouse Lymphomas

Mouse strain	Carcinogen	Activated ras genes[a]	Frequency[b] (%)	Ref.
RF/J mouse[c]	MNU	K-*ras*-12	82 (23/28)	32
		N-*ras*-12	3 (1/28)	
		N-*ras*-13	3 (1/28)	
		N-*ras*-61	11 (3/28)	
RF/J mouse[c]	Gamma radiation	K-*ras*-12	78 (7/9)	32
		N-*ras*-12	11 (1/9)	
RF/J mouse[c]	Neutron radiation	K-*ras*-12	8 (2/24)	16
		K-*ras*-146	4 (1/24	
		N-*ras*-61	4 (1/24	
C57BL/6J mouse[c]	MNU	K-*ras*	80 (8/10)	56
C56BL/6J mouse	MNU[d]	K-*ras*-12	14 (4/28)	58
		K-*ras*-146	3 (1/28)	
C57BL/6J mouse[c]	Gamma radiation	K-*ras*	13 (2/15)	56
		N-*ras*	33 (5/15)	
AKR/J mouse[c]	MNU	K-*ras*-12	22 (8/13)	147
B6C3F1 mouse	1,3-butadiene	K-*ras*	18 (2/11)	59
C57BL/Lia mouse	ENU	K-*ras*-12	33 (4/12)	57
		K-*ras*-61	17 (2/12)	

Note: ENU, *N*-ethyl-*N*-nitrosourea and MNU, *N*-methyl-*N*-nitrosourea.

[a] Mutated codon is indicated when known.
[b] Tumors with *ras* mutated genes/tumors analyzed.
[c] Results refer to thymic lymphomas.
[d] Tumors induced by a single dose of MNU.

lymphomas up to 90% in some mouse strains.[53,54] It is noteworthy that in this system two completely different carcinogenic agents (alkylating chemical agent and ionizing radiation) produce an essentially identical disease.[18,32,55,56] This permits us to compare the spectrum of *ras* activation by different agents. The results shown in Table 3 indicate that K-*ras* seems to be the more prevalently activated *ras* family member for both MNU and gamma rays, although there is some strain variation. For example, in C57BL6, N-*ras* mutation is more frequent than in RF/J mice.[56] K-*ras* mutations are also detected in mouse lymphomas induced by *N*-ethyl-*N*-nitrosourea (ENU).[57] With respect to the activating mutation, both MNU and gamma rays induce tumors most frequently containing a G:C to A:T transition in the second base of the K-*ras* codon 12. On the contrary, neutron-induced tumors did not show that particular mutation and instead, a random pattern of *ras* mutations was detected.[16] This suggests a different mechanism of *ras* activation for the two types of ionizing radiation. Comparison of thymic lymphomas induced by a single injection of MNU with those induced by multiple injections indicate that, although the frequency of tumors containing activated K-*ras* is lower in the single-injection protocol, the spectrum of mutations remains essentially the same.[58]

E. LIVER TUMORS

Induction of mouse liver tumors is broadly used for carcinogenesis assays.[15] *ras* activation has been detected in hepatocarcinomas and adenomas induced by a broad variety of agents, as well as in spontaneous tumors of mice, as shown in Table 4. In most models, the carcinogenic agent was administrated orally, with one exception where the tumors were induced by inhalation of 1,3-butadiene.[59] As a model system, liver carcinogenesis takes advantage of the existence of several mouse strains that show a high incidence (10 to 20%) of spontaneous tumors, such as the C3H, CBA, and CF1 strains and the male B6C3F1 mice (C57BL × C3H).[15,60-63] These mouse strains are also very sensitive to liver carcinogenesis. Therefore, the incidence of *ras* activation can be compared in spontaneously vs. chemically induced liver tumors. In contrast, other mouse strains (e.g., C57BL, BALB/c, or B6C) as well as rats are relatively insensitive to hepatocarcinogenesis. H-*ras* activation incidence is lower in tumors induced in these "resistant" strains.[61,62,64] As in skin tumors, H-*ras* is the most prevalently activated oncogene in rodent liver tumors, although the inducing agents used in both types of model systems are different (see Tables 1 and 4). Despite the high incidence of H-*ras* activation in liver, the level of H-*ras* expression in this tissue is similar to other *ras* genes.[65] Conversely to the results obtained in mice, K-*ras*-mutated genes are found in rat tumors induced by aflatoxin B_1.[66] Also, K-*ras*-activated oncogenes are detected at high frequencies in spontaneous fish liver tumors, both in flounder[67] and tomcod.[68]

As in skin tumors, there is no relationship between the frequency of *ras* activation and the type of liver tumor. When analyzed,[15,35] the percentage of adenomas or precancerous lesions containing activated H-*ras* was similar to that of hepatocarcinomas.

F. LUNG TUMORS AND OTHER INDUCED TUMORS

In some instances, lung tumors were induced by intraperitoneal injections of the carcinogen (MNU, ethylcarbamate)[69-71] or, if the carcinogen is volatile such as 1,3-butadiene or tetranitromethane, by inhalation. These latter compounds are produced in industrial processes and the experimental tumors were induced under conditions resembling human exposure to volatile carcinogens.[59,72] As shown in Table 5, K-*ras* is the only member of the *ras* family found activated in lung tumors thus far. K-*ras* activation has also been detected in 30% of human lung carcinomas.[73] This is in contrast to skin, liver, and mammary tumors, where H-*ras* was the most prevalent activated *ras* gene. These differences will be discussed below.

The *ras*-mutated genes have been detected in other chemically induced animal tumors. These tumors include rat renal tumors,[52,74] rat thyroid tumors,[75] rat esophageal,[76] rat colonic,[77] and hamster pancreatic tumors.[78] In all these cases K-*ras* is the mutated oncogene prevalently found (Table 6). It is noteworthy that K-*ras* is also the mutated *ras* gene detected in a high percentage of human colonic and pancreatic carcinomas.[30,31,79] However, H-*ras* is the oncogene found in

TABLE 4
ras Activation in Chemically Induced Liver Tumors

Animal and type of tumor	Carcinogen	Activated ras genes[a]	Frequency[b] (%)	Ref.
Mouse tumors[c]	AFTB$_1$	H-ras-61	37 (3/8)	60
Mouse adenomas	Spontaneous	H-ras-61	25 (1/4)	60
Mouse carcinomas	HO-AAF	H-ras-61	100 (7/7)	148
Mouse carcinomas	Vinyl-C	H-ras-61	100 (7/7)	148
Mouse carcinomas	HO-DHE	H-ras-61	91 (10/11)	148
Mouse adenomas	Furan	H-ras-61	21 (4/19)	15
		H-ras-117	16 (3/19)	
		K-ras	10 (2/19)	
Mouse carcinomas	Furan	H-ras-61	10 (1/10)	15
		H-ras-117	10 (1/10)	
Mouse adenomas	Furfural	H-ras-61	33 (1/3)	15
		H-ras-13	33 (1/3)	
Mouse carcinomas	Furfural	H-ras-61	38 (5/13)	15
		H-ras-13	8 (1/13)	
		H-ras-117	8 (1/13)	
		K-ras	8 (1/13)	
Mouse adenomas	Spontaneous	H-ras-61	17 (3/10)	15
Mouse carcinomas	Spontaneous	H-ras-61	70 (12/17)	15
Mouse preneoplastic lesions	DEN	H-ras-61	21 (12/57)	35
Mouse tumors[c]	DEN	H-ras-61	33 (5/15)	35
Mouse adenomas	DEN	H-ras-61	36 (8/22)	64
Mouse carcinomas	DEN	H-ras-61	54 (6/11)	64
Mouse adenomas	Benzidine	H-ras-61	53 (8/15)	63
Mouse carcinomas	Benzidine	H-ras-61	66 (4/6)	63
Mouse adenomas	Spontaneous	H-ras-61	83 (15/18)	63
Mouse carcinomas	Spontaneous	H-ras-61	71 (10/14)	63
Mouse carcinomas	DMBA	H-ras-61	66 (21/32)	83
Mouse carcinomas	1,3-butadiene	H-ras	33 (4/12)	59
		K-ras	25 (3/12)	
Mouse tumors[d]	DEN	H-ras-61	36 (20/55)	61
	Spontaneous	H-ras-61	60 (9/15)	
Mouse preneoplastic lesions	DEN	H-ras-61	13 (9/69)	61
	ENU	H-ras-61	14 (4/29)	
Mouse tumors[c]	DEN	H-ras-61	13 (4/31)	62
Rat carcinomas	AFTB$_1$	K-ras-12	60 (3/5)	66
Flounder carcinomas	Spontaneous	K-ras-12	54 (7/13)	67
Tomcod carcinomas	Spontaneous	K-ras	40 (2/5)	68

Note: AFTB$_1$, aflatoxin B$_1$; DEN, diethylnitrosamine; DMBA, 7,12-dimethyl-benz[a]anthracene; HO-AAF, N-hydroxy-2-acetylaminofluorene; HO-DHE, 1'-hydroxy-2',3'-dehydroestragole; and Vinyl-C, vinyl-carbamate.

[a] Mutated codon is indicated when known.
[b] Tumors with ras mutated codons/tumors analyzed.
[c] Adenomas and carcinomas.
[d] Results from tumors induced in two mice strains.

TABLE 5
ras Activation in Chemically Induced Lung Tumors

Animal and type of tumor	Carcinogen	Activated ras gene[a]	Frequency[b] (%)	Ref.
Mouse adenomas	MNU	K-*ras*-12	100 (15/15)	71
Mouse adenomas	BAP	K-*ras*-12	81 (13/16)	71
Mouse tumors[c]	Ethylcarbamate	K-*ras*-61	81 (9/11)	71
		K-*ras*-12	9 (1/11)	
Mouse tumors[c]	Spontaneous	K-*ras*-12	54 (6/11)	71
		K-*ras*-61	27 (3/11)	
Mouse adenocarcinomas	NNK	K-*ras*-12	73 (8/11)	69
		K-*ras*-61	18 (2/11)	
Mouse tumors[c]	NDMA	K-*ras*-12	70 (7/10)	69
		K-*ras*-61	30 (3/10)	
Mouse adenocarcinomas	1,3-butadiene	K-*ras*-13	67 (6/9)	59
Mouse tumors[d]	DMBA	K-*ras*-61	40 (4/10)	83
Mouse tumors[d]	Spontaneous	K-*ras*-61	80 (4/5)	83
Mouse tumors[d]	NDMA	K-*ras*-12	100 (11/11)	70
	NNK	K-*ras*-12	100 (11/11)	70
	Spontaneous	K-*ras*-12	40 (3/7)	70
Mouse tumors[c]	TNM	K-*ras*-12	100 (11/11)	72
Rat tumors[c]	TNM	K-*ras*-12	100 (9/9)	72

Note: BAP, Benzo[a]pyrene; DMBA, 7,12-dimethylbenz[a]anthracene; MNU, *N*-methyl-*N*-nitrosourea; NDMA, *N*-nitrosodimethylamine; NNK, 4-(*N*-methyl-*N*-nitrosamino)-1-(3-pyridyl)-1-butanone; and TNM, tetranitromethane.

[a] Mutated codon is indicated when known.
[b] Tumors with *ras* mutated genes/tumors analyzed.
[c] Adenomas and adenocarcinomas.
[d] Not histologically identified.

induced rat bladder carcinomas,[20] again in concordance with the human situation.[80]

III. *ras* ACTIVATION AND TISSUE SPECIFICITY

One of the most striking conclusions that can be drawn from the above-described experiments is the close relationship that seems to exist between the tissue where the tumor develops and the member of the *ras* gene family that appears activated. The most relevant data can be summarized as follows:

1.   In tumors of epithelial origin, H-*ras* is the prevalently activated gene. For example, from the data of Tables 1-3 the frequency of activated H-*ras* with respect to K- or N-*ras*, is 20:1 for skin carcinomas, papillomas, and keratoacanthomas, 8:1 for liver tumors, and 20:1 for breast tumors. In this latter case, the relationship does not include mouse tumors obtained after

TABLE 6
ras Activation in Other Chemically Induced Tumors

Animal and type of tumor	Carcinogen	Activated ras genes[a]	Frequency[b] (%)	Ref.
Rat kidney tumors	MMMN	K-ras-12	40 (10/25)	70
		N-ras-12	4 (1/25)	
Rat kidney tumors	MNU[c]	K-ras-12	100 (5/5)	52
Rat thyroid tumors	NMU + ATA	H-ras	87 (13/15)	75
Rat thyroid tumors	Gamma radiation	K-ras	53 (8/15)	75
Rat esophageal papillomas	MBNA	H-ras-12	100 (18/18)	76
Rat bladder tumors	BBN	H-ras-12	19 (4/21)	20
		H-ras-61	33 (7/21)	
Rat colon carcinomas	DMH	K-ras-12	45 (20/44)	77
		K-ras-13	18 (8/44)	
		K-ras-59	2 (1/44)	
Hamster pancreas carcinomas	NBOP	K-ras-12	100 (10/10)	78

Note: ATA, aminotriazole; BBN, N-butyl-N-(4-hydroxybutyl)nitrosamine; DMH, 1,2-di-methylhydrazine; MBNA, methylbenzylnitrosamine; MNU, N-methyl-N-nitroso-urea; MMMN, methyl(methoximethyl)nitrosamine; and NBOP, N-nitrosobis(2-ox-opropyl)amine.

[a] Mutated codon is indicated when known.
[b] Tumors with ras mutated genes/tumors analyzed.
[c] Transplacental exposure to MNU.

in vitro exposure of cells to MNU, all of them containing K-*ras* oncogene.[81] Unlike the findings in rats, MNU does not induce mammary tumors in mice, so it is unknown whether *in vivo* exposure to MNU will induce activated H-*ras*-containing tumors.

2. Activated H-*ras* has not been found in tumors from hematopoietic origin (Table 3). In one study, H-*ras* mutation was not detected in more than 150 MNU-induced mouse lymphomas analyzed (Table 3).[32,56,58,143] This situation resembles the human scenario, where N-*ras* activation is frequent in acute myeloid leukemias and myelodysplastic syndromes, but H-*ras* mutation is rarely detected.[27]

3. Activated K-*ras* is the *ras* gene found activated in lung tumors. As shown in Table 5, lung tumors induced by seven different carcinogens as well as spontaneous tumors contained activated K-*ras*. These results again compare well with the human situation, where K-*ras* is the activated *ras* gene found.[73] However, as noted above, this correlation between the results in animal models and human cancer is not universal. For example, activated K-*ras* genes are found in 90% of carcinomas of human pancreas[31] and in a high percentage of hamster pancreatic tumors.[78] However, no K-*ras* activation is detected in chemically induced rat pancreatic tumors.[82]

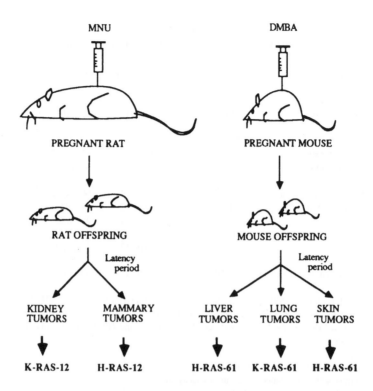

FIGURE 1. Preferential activation of *ras* genes depending on the target tissue. Tumors were induced by transplacental exposure to MNU in rats[52] or DMBA in mice.[83] The particular mutated *ras* gene is related to the tissue rather than to the carcinogenic agent used.

Model systems based on transplacental exposure to carcinogens offer conclusive evidence of the tissue-specific preferential activation of *ras* gene family members. It has been found that, using a single carcinogen transplacentally, different activated *ras* genes are detected in the different types of tumors developed. Rats exposed *in utero* to MNU developed mammary carcinomas and kidney mesenchymal tumors, among others. Activated H-*ras* (mutated in codon 12) was detected in all breast tumors analyzed (3/3) while activated K-*ras* was the oncogene found in the renal tumors (5/5). In all cases the activating mutation was a G:C → A:T transition in the 12th codon of either H- or K-*ras* (Figure 1).[52] In a similar study carried out in mice, transplacental exposure to DMBA induced lung, liver, and skin tumors (in this case after promotion with TPA on the newborn mice skin). The activated *ras* genes found were K-*ras*, H-*ras*, and N-*ras*, respectively. Furthermore, in mice that developed both lung tumors and hepatomas activated K-*ras* was found in the lung tumors and activated H-*ras* in the hepatomas.[83] The activating mutation was an A:T → T:A transversion in all activated *ras* genes, typical for DMBA-induced tumors, suggesting that both lung and liver tumors resulted from the action of the carcinogen rather than having a spontaneous origin.

The reason for this preferential activation of *ras* genes is not clear at present. All three members of the *ras* gene family are ubiquitously expressed in the rodent and human tissues tested. However, there are important differences in the relative expression levels among the three genes.[65,84]

It has been proposed that the particular *ras* gene activated in each type of tumor is related to the relative level of expression of that *ras* gene in the target tissue. For example, H-*ras* expression is high in skin while it is lower in hemopoietic tissues such as bone marrow, thymus, and spleen.[65] These differences would be in agreement with the data obtained in various carcinogenesis model systems, where H-*ras* activation is prevalent in skin tumors (Table 1) and never detected in lymphomas (Table 3). However, the introduction of mutated H-*ras* into bone marrow cells via retroviral gene transfer results in thymic lymphomas. These lymphomas appear at high frequencies and with a short latency.[85] Also, in mice there are no major differences among N- and K-*ras* expression at the mRNA level in lung and neither are there differences among the three *ras* genes in liver.[65] There is a clear preferential activation of H-*ras* in liver (Table 4) and K-*ras* in lung tumors (Table 5). Finally, no *ras* activation has been detected in MNU-induced neuroectodermal tumors,[52] although *ras* genes, especially H-*ras*, are highly expressed in brain.[65,86] Therefore, preferential activation may well be related to differences in the functional importance of *ras* genes in the different organs.

Compared to the data obtained on mutational *ras* activation, very limited information is available on the relative expression level of *ras* in carcinogen target tissues. On the other hand, other factors, apart from *ras* expression, may play a role in the preferential activation of *ras* genes depending on the tissue. These include the putative differential signal transduction activities of *ras* proteins and the tissue-specific differential expression of tumor suppressor genes. Last but not least, there are important species-dependent factors. For example, in a comparative study H-*ras* activation was not detected in chemically induced rat liver tumors, while this activation is prevalent in liver tumors from a closely related species such as mouse.[61,64] More studies with animal model systems are required to clarify the important issue of preferential *ras* gene activation.

IV. CARCINOGEN-SPECIFIC MUTATION OF *ras* GENES

Results from many model systems lead to the conclusion that while the member of the *ras* gene family found activated depends mostly on the target tissue, the particular mutation depends on the carcinogenic agent used (reviewed by Sukumar[19]). MNU (a methylating agent) and DMBA (a polyciclic hydrocarbon that acts as an intercalating agent) are two chemicals with known different mutational mechanisms. Their activities on *ras* mutation have been compared in the same model systems: mouse skin,[46] rat breast,[34,49] and mouse lung tumors.[71,83] From comparing *ras* activation data in tumors induced by both agents (Table 7) we can extract the following conclusions:

TABLE 7
ras **Activation and Carcinogen Specificity**

Agent[a]	Gene	Mutation	Animal and type of tumor	Frequency[b]	Ref.
MNU	H-ras	$G^{35} \to A$	Mouse skin tumors	5/12	46
MNU	K-ras	$G^{35} \to A$	Mouse thymic lymphomas	22/23	32
		$G^{35} \to A$		4/28	58
		$G^{35} \to A$		8/37	147
MNU	K-ras	$B^{35} \to A$	Mouse mammary tumors	19/21	81
MNU	H-ras	$G^{35} \to A$	Rat mammary tumors	61/71	34
		$G^{35} \to A$		3/3	52
		$G^{35} \to A$		15/28	51
	K-ras	$G^{35} \to A$		5/28	51
MNU	K-ras	$G^{35} \to A$	Rat kidney tumors	5/5	52
DMBA	H-ras	$A^{182} \to T$	Mouse skin papillomas	45/48	46
		$A^{182} \to G$		1/48	46
		$A^{183} \to T$		2/48	46
DMBA	H-ras	$A^{182} \to T$	Mouse skin carcinomas	10/13	46
DMBA	H-ras	$A^{182} \to T$	Rabbit skin keratoacanthomas	18/22	42
DMBA	H-ras	$G^{35} \to A$	Rabbit skin carcinomas	1/10	42
		$A^{182} \to T$		4/10	
DMBA	H-ras	$A^{182} \to T$	Mouse mammary tumors	4/4	39
		$A^{182} \to T$		12/18	146
		$A^{182} \to T$		6/12	46
		$A^{183} \to G$		5/12	36
DMBA	H-ras	$A^{181} \to N$	Rat mammary tumors	5/25[b]	34
		$A^{182} \to N$			

[a] MNU and DMBA as in Table 3.
[b] Tumors with ras-mutated allele/tumors analyzed.

1. Tumors induced by MNU carry a ras gene mutated on the 12th codon, either H-ras (Tables 1 and 2), K-ras (Tables 3, 5, and 6) or N-ras (Table 3). This mutation is a $G^{35} \to A$ transition in almost every case. One exception is thymic lymphoma containing a codon 61-mutated N-ras.[32,33]
2. Tumors induced by DMBA carry a ras gene mutated on the 61st codon, either H-ras (Table 1, 2, and 4) or K-ras (Table 5). The mutations are $A^{182} \to T$ and $A^{183} \to T$ transversions in the second and third bases of codon 61.

What is the molecular basis for these consistent mutagenesis spectra? In the case of the MNU, there is compelling evidence that the $G \to A$ mutation may be caused by direct interaction of the carcinogen with the target DNA, as originally postulated by M. Barbacid's group based on findings with rat breast tumors.[34,49] MNU is a short-lived methylating agent that preferentially produces O^6-methylguanine, which results in $G \to A$ transitions after DNA replication.[87,88] This is the mutation found in the 12th codon of ras genes in breast tumors induced by a single injection of MNU, while the use of DMBA results in *ras* mutation in codon 61.[34] A direct attack of MNU to guanine 35 of *ras* genes is further supported by transfection experiments with H-*ras* genes containing O^6-methylguanine in position 35, resulting in transformed cells carrying the G^{35} → A mutation in H-*ras*.[89]

Although model systems using MNU or DMBA have produced most of the data on *ras* mutation specificity, there are other examples consistent with a direct relationship between a carcinogen and a particular point mutation. Mouse lung tumors induced by benzo[a]pyrene contain K-*ras* mutated on codon 12 while lung tumors induced by ethylcarbamate contain K-*ras* mutated on codon 61 (Table 5).[71] The same *ras* mutation is detected in lung tumors induced by 4-(*N*-methyl-*N*-nitrosamine)-1-(3-pyridyl)-1-butanone (NNK) and N-nitrosodimethyl-amine (NDMA), two nitrosamines that appear to activate the K-*ras* gene through a direct mechanism.[70] A high percentage of lymphomas, liver tumors, and lung tumors induced by 1,3-butadiene contain K-*ras* genes mutated on codon 13.[59] The reason for this mutation specificity is still unknown.

For other carcinogens, no such consistent results are found. There are many model systems where a single carcinogenic agent induced tumors carrying *ras* genes with different mutations. This is the case, among others, for rat bladder carcinomas induced by *N*-butyl-*N*-(4-hydroxybutyl)nitrosamine (BBN),[20] skin tumors induced by 3-methylcholanthrene,[48] and hepatomas induced by furan, furfural, or aflatoxin B$_1$ (Tables 1, 4, and 6).[15,60,66] In these tumors *ras* mutations might not result from a direct interaction between the DNA and the carcinogen. To understand these effects more information on the carcinogen metabolism and DNA repair systems in higher eukaryotes will be needed.

As discussed above, the particular *ras* gene that becomes activated seems to depend largely on the target tissue rather than on the carcinogenic agent. However, the latter seems to be the case in some model systems. For example, rat thyroid tumors induced by MNU contain activated H-*ras* at high frequencies (87%), while tumors induced by gamma radiation (^{131}I) contain activated K-*ras*.[75]

The finding of specific point mutations in *ras* genes in spontaneous tumors also suggests that factors other than the particular carcinogen used play a role in mutation specificity. K-*ras* genes mutated on codons 12 and 61 have been found in spontaneous lung tumors.[71,83] H-*ras* mutated on codon 61 has been detected in spontaneous mouse liver and lung tumors[63,71,83,90] as well as tumors induced by nongenotoxic tumor promoters such as chloroform, ciprofibrate, or phenobarbital for liver tumors or TPA and mezerein for skin tumors.[21,60,63] "Spontaneous tumor" means tumor induced either by carcinogenic agents that remain unknown or by errors of DNA polymerases and DNA repair enzymes. In both cases an array of different mutations should be expected. However, the results mentioned above suggest that, at least for H-*ras*, not all the mutations are equivalent in conferring a selective growth advantage. Detailed *in vitro* experiments comparing the growth effects of different *ras* mutations are needed to clarify this point.

In most model systems, only one *ras* gene is found activated in a particular tumor type for each carcinogen used. The major exceptions thus far are mouse liver tumors induced by furan and 1,3-butadiene (Table 4)[15,59] and mouse thymic lymphomas induced by MNU and gamma radiation. In this model it was found that a majority of MNU-induced lymphomas carried activated K-*ras*, while activated N-*ras* genes are more infrequent.[32,56] Conversely to other MNU-induced tumors, a significant fraction of the activated N-*ras* genes found in thymic lymphomas carried the activating mutation on codon 61 (Table 3). As a model system, mouse thymic lymphomas are interesting in that two completely different agents (MNU and gamma radiation) induce lymphomas carrying activated K-*ras* genes with the same point mutation (G → A transition on the second base of codon 12).[32] As discussed above, MNU may interact directly with the DNA but gamma radiation is thought to act through the production of free radicals that, in turn, would ionize cellular structures like DNA bases or sugar phosphate bonds. This mechanism is less likely to produce mutation specificity by itself, although it could result in mutation hot spots if repair is inefficient in particular DNA positions. This could also help to explain the discrepancy between gamma rays and neutrons on the *ras* mutation spectra (Table 3). Neutrons are known to produce extensive DNA damage which is difficult to repair and may result in a more random mutation pattern.

V. IS *ras* MUTATION AN EARLY OR LATE EVENT IN TUMORIGENESIS?

A large body of evidence indicates that tumorigenesis is a multistep process. The number of "steps" is unknown, although one report, based on statistical studies, determined that this number must be at least five for human and mouse tumors.[91] Each of these steps involves molecular changes whose nature is largely unknown. Despite considerable experimental efforts, it is not clear at present whether *ras* mutations occur early or late during carcinogenesis. Furthermore, there is evidence supporting both hypotheses depending on the model system analyzed. We will briefly review some of this evidence.

A. EVIDENCE IN FAVOR OF *ras* GENE MUTATION AS A LATE EVENT

Three pieces of evidence indicate that *ras* activation may occur late in tumorigenesis. The first one is *ras* activation detected in late but not early passages of tumor cells grown *in vitro*. For example, this has been shown for H-*ras* in mouse keratinocytes,[92] K-*ras* in mouse lymphoma cells,[93] and N-*ras* in human teratocarcinoma cells[94] and myeloid leukemia cells.[95] Also, N-*ras* appeared activated in metastasis but not in the primary human melanoma.[96]

Some data obtained with tumors also indicate a late activation of *ras*. For example, transgenic mice carrying the SV40-T antigen transgen under the control of the albumin promoter readily develop liver tumors,[97,98] but the endogenous

H-*ras* mutated on codon 61 has been detected in 40% of these tumors.[97] Similarly, mutated *ras* genes have been found in lymphomas from PIM-1 transgenic mice.[57] Along the same line is the existence of clonal tumors where only a minority of the cells carry *ras* mutations (e.g., chemically induced mouse thymic lymphomas[32]). Findings in some human tumors also suggest that *ras* mutation occurs during tumor progression. For example, K-*ras* mutations are more prevalent in big colorectal adenomas (>1 cm in diameter) than in the smaller ones[99] and H-*ras* activation is more frequent in uterine cervix carcinomas in advanced stages than in early tumor stages.[100]

Additional evidence for late activation of *ras* genes is the finding of concomitant *ras* mutations in the same tumor. This has been shown in a relatively low number of cases, but concomitant *ras* mutations have not been looked for in most studies. Concomitant K- and N-*ras*-mutated genes have been found in two thymic lymphomas induced in mouse by MNU. The clonality in both tumors has been demonstrated by T-cell receptor gene rearrangement analysis.[32] Also, double mutations in codons 12 and 61 of H-*ras* have been detected in a chemically induced bladder tumor.[20] Concurrent mutations in two *ras* genes[101,102] or in two codons of one *ras* gene[30,103,104] have also been reported in human tumors. The most logical explanation is that one *ras* mutation occurred late in tumor development.

B. EVIDENCE IN FAVOR OF *ras* ACTIVATION AS AN EARLY EVENT

The finding of *ras* activation in premalignant tumors is considered indicative that the mutation occurred early in tumorigenesis. This is the case in a significant fraction of human colon villous adenomas,[30,78] keratoacanthomas,[37] and myelodysplastic syndromes.[105,106] More conclusive data have emerged from work carried out with animal model systems. In skin carcinogenesis a high incidence of H-*ras*-activated genes has been demonstrated not only in squamous cell carcinomas (a frank malignant tumor) but also in papillomas,[22,46] which are considered premalignant tumors. As papillomas may progress to squamous cell carcinoma,[107] this result argues in favor of the hypothesis that *ras* activation is an early event in tumorigenesis. These models make use of an initiation-promotion protocol where single doses of a mutagenic initiating agent like DMBA, MNU, 3-methylcholanthrene, or *N*-MNNG is followed by repeated doses of TPA as a tumor promoter. When infection of mouse epidermal cells with Harvey sarcoma virus (which carries an activated H-*ras* gene) was substituted for an initiating agent in this carcinogenesis protocol this resulted in development of papillomas, some of them progressing to carcinomas.[45] This strongly suggests that the development of these tumors depends on *ras* mutation rather than in other genetic alterations caused by the mutagenic agents used as initiators. Keratoacanthomas are a type of benign skin tumor that undergo spontaneous regression. Rabbit keratoacanthomas induced by DMBA show a high frequency of H-*ras* activation (82%)[42,43] and these results compare well with those obtained with human keratoacanthomas, as seen below. As shown in Table 1, when DMBA or MNNG are used as initiating agents the incidence of H-*ras* activation in mouse papillomas is higher than in squamous cell carcinomas.[46] In rabbits, this frequency of H-*ras* activation in keratoacanthomas (82%) is also higher than in carcinomas (50%).[42] The close relationship with the human situation is noteworthy, where H-*ras* activated genes were found in 30% of keratoacanthomas but only in 13% of carcinomas.[37] The low incidence of H-*ras* activated genes in carcinomas with respect to papillomas or keratoacanthomas has led to the proposal that in this system activation is not needed to maintain the tumorigenic phenotype and may be involved in epidermal cell differentiation.[37,42,46]

Additional evidence in favor of an early *ras* mutational event is the finding of mutated *ras* in preneoplastic liver tissue.[35,61] The incidence of *ras* activation in liver adenomas (a benign tumor) was similar than in malignant carcinomas (Table 4).[15,60,63,64] K-*ras* activation is also found at high frequencies in benign lung adenomas induced by benzo[a]pyrene, ethylcarbamate and MNU.[71]

One of the model systems where early *ras* mutation has been more firmly established is in rat mammary carcinogenesis.[19] Female rats neonatally injected with MNU develop mammary tumors 3 to 4 months later and H-*ras* activation can be detected in mammary gland tissue only two weeks after MNU exposure.[38] Consistent with the former data are the results obtained in rat kidney and breast tumors induced by transplacental exposure to MNU. These tumors, that developed months after the carcinogenic insult, contained the mutation G → A in the codon 12, characteristic of MNU-induced tumors.[52]

Taken together, the former data suggest that *ras* genes are activated early at least in some tumors. This situation is consistent with the multistep nature of tumorigenesis, as discussed below. Even in those models where *ras* activation is prevalent, tumors may be initiated by events other than *ras* mutation. For example, the number of rat mammary tumors containing H-*ras*-mutated oncogene does not increase with increasing carcinogen (MNU) doses, while the total number of induced tumors is higher.[108]

VI. *ras* ACTIVATION IS NOT SUFFICIENT FOR CARCINOGENESIS

A. LESSONS FROM CARCINOGENESIS MODEL SYSTEMS

In the previous section evidence consistent with an early activation of *ras* genes in particular animal models was reviewed. In those models where the mutation of the *ras* gene may result from the direct action of the carcinogen, the question arises as to whether *ras* mutation is sufficient for tumorigenesis. A negative answer to this question was originally suggested by *in vitro* experiments. It has long been known that activated *ras* oncogenes are unable to transform most normal diploid cells upon transfection, unless very high expression levels of the oncogene are achieved.[13,109] However, they do transform already immortalized cells (e.g., NIH 3T3) or primary cells when cotransfected with other cooperating oncogenes (reviewed by Hunter[115] and Ruley[116]).[110-114] This idea is consistent with the finding that activated H-*ras* does not confer tumorigenic phenotype to human fibroblasts[117-120] and the report of a human colon adenoma cell line harboring a mutant K-*ras* gene which is nontumorigenic.[121]

In some models with an early *ras* activation, the cells harbor mutated *ras* genes for a long time before proliferation becomes deregulated. This deregulation may be caused by other genetic alterations largely unknown at present and other epigenetic changes that trigger proliferation in cells already carrying mutated *ras* genes. One of them can be the effect of tumor promoters used in skin carci-nogenesis or the estrogen in the mammary tumor model system. Elegant exper-iments have shown that the tumor onset depends on estrogen secretion at rat puberty, despite the previous presence of H-*ras* mutated gene in the mammary glands for several months.[19,51] Also, *ras* activation during the early stages of carcinogenesis does not make the tumor growth hormone-independent.[50] More evidence indicating that *ras* mutation may not be sufficient for tumorigenesis is the finding of H-*ras* mutated genes in a high percentage (82%) of DMBA-induced rabbit keratoacanthomas, a type of skin tumor that undergoes complete regression.[42,43] Furthermore, it has been shown that the mutated *ras* gene is expressed not only in growing keratoacanthomas but also in regressing tumors.[42] More studies on the expression of *ras*-mutated alleles in other model systems will be required.

B. LESSONS FROM TRANSGENIC ANIMALS

Transgenic mice offer the possibility of studying the effect of a single on-cogene in physiological conditions. The T-antigen oncogene from SV40 virus was one of the first transgenes used.[122] Since then, many transgenic mice lines have been created harboring a set of oncogenes including *ras, fos, myc, neu,* bovine papilloma virus E5 and E6 genes, and polyoma mT gene (reviewed in References 123 to 125). By using tissue-specific promoters, one can direct the expression to a particular tissue of the transgenic animal. Table 8 gathers data referring to transgenic mice lines carrying 13 different combinations of the promoter-*ras* gene. In most of them the *ras* gene used is the human H-*ras* mutated on codon 12 (derived from the EJ cell line). N-*ras* transgenic mice have been developed[126] but no K-*ras*. The major conclusions of these experiments can be summarized as follows.

1. Activated *ras* Genes Are Not Sufficient to Induce a Frank Malignant Tumor

If *ras* activation was sufficient for tumorigenesis, one should expect the appearance of tumors with a short latency period in the targeted tissue of all the transgenic animals. One general conclusion deduced from experiments with transgenic mice is that this is not the case for the oncogenes thus far tested. The *neu* oncogene could be an exception to this statement, although contradictory reports have appeared.[127,128] In *ras*-transgenic mice the tumor incidence is not

TABLE 8
Tumors Induced by *ras* Transgenes

Transgene(s)	Promoter	Type of induced tumor or lesion[a]	Incidence and latency[b]	Ref.
H-*ras*-12 (human)[c]	Elastase	Pancreas hyperplasia and dysplasia	High/0–8 weeks	131
H-*ras*-12 (human)	WAP[d]	Mammary tumors and salivary tumors	Low/9–10 months	129
v-H-*ras*	MMT, LTR[e]	Mammary adenocarcinomas	High/6–12 months	132
		Harderian gland hyperplasia, salivary gland carcinomas	Medium	
v-H-*ras* + c-*myc*[c] (murine)	MMTV, LTR	Mammary adenocarcinomas	High/1–3 months	132
	MMTV, LTR	Lymphomas, salivary gland hyperplasia, harderian gland hyperplasia	Medium/1–3 months	
v-H-*ras*	MMTV, LTR	Harderian gland hyperplasia, mammary adenocarcinoma, and myeloid cell hyperplasia, lung and salivary gland adenocarcinomas	High/4–10 months Low	133
H-*ras*-12 (human)	Albumin	Liver hyperplasia	High/≈1 week	98
H-*ras*-12 (human) + c-*myc*[c] (murine)	Albumin	Lung adenomas and liver dysplasia	High/1–5 months	98
	Albumin	Lung adenomas, liver tumors	High/1–4 months	98
H-*ras*-12 (human) + SV40 antigen[f]	Albumin	Liver hyperplasia	High/1 week	98
H-*ras*-12 (human)	Albumin Keratin K10	Hyperkeratosis in skin and forestomach, papillomas	High/10 weeks	134
H-*ras*-12 (human)	IgG-SV40[g]	Lung adenomas	High/2–6 months	130
H-*ras* (human)	H-*ras*	Angiosarcomas, lung carcinomas	High/18 months	138
H-*ras*-12 (human)	Insulin	Beta-cell degeneration and diabetes	High in males/5 months	136

H-ras-12 (human)	MoMuLV, LTR[e]	Benign skin tumors	Medium/≈4 weeks	135
		Spindle cell endoplasms	Low	
v-H-ras + v-myc[h]	MoMuLV, LTR	Brain tumors, benign skin tumors and spindle cell neoplasms	Medium/≈3 weeks	140
N-ras	MMTV, LTR[e]	Harderian gland hyperplasia, male infertility	Medium/3–5 months	126

[a] Only the most prevalent lesions are included.

[b] High — more than 50% of tumor incidence; medium — 10–20% of tumor incidence. Data refers either to founder mice or their descendant. Latency times are given in weeks or months.

[c] H-ras mutated on codon 12 h.

[d] Whey acidic protein.

[e] Mouse mammary tumor virus (MMTV) or Moloney murine leukemia virus (MoMuLV) long terminal repeats.

[f] F_1 generation from transgenic founder mice.

[g] Immunoglobulin heavy chain enhanced and SV40-T gene promoter.

[h] Both genes under the LTR promotor control.

very high or the tumors appear in all mice but after long latency periods, despite continuous *ras* expression in the targeted tissue. For example, mice carrying the H-*ras* oncogene downstream of the whey acidic protein (WAP) promoter develop mammary or salivary gland tumors after 9 to 10 months.[128] The latency period for lung tumors developed in mice carrying the H-*ras* oncogene downstream of the SV40-T-Ag promoter can be one year.[130] Other tissue-specific promoters used with *ras* transgenes are those from elastase,[131] MMTV,[126,132,133] albumin,[98] and keratin.[134] Altogether, these experiments indicate that activated H-*ras* induces proliferative hyperplasia. However, further secondary events are necessary for progression to a true malignant tumor. Moreover, the benign lesions induced in some *ras*-transgenic mice when the oncogene expression occurs in the epidermis show a high degree of differentiation.[134,135] As pointed out above, *ras* activation is prevalent in chemically induced keratoacanthomas, which undergo an extensive differentiation process.[42] It is noteworthy that transgenic mice carrying the H-*ras* oncogene under the control of the rat insulin II promoter suffer pancreatic β-cell degeneration and subsequent diabetes. Strikingly, this alteration is only detected in male transgenic mice.[136] At least in one reported case, transgenic mice expressed a *ras* mutated gene without any abnormality in the target tissue (brain).[137]

2. There Exists a Tissue Specificity for the Effects of *ras* Transgenes Not Dependent on the Promoter

When *ras* transgenes are expressed in more than one tissue, tumors or hyperplasias appear only or preferentially in a subset of tissues. Some representative examples are shown in Figure 2, including transgenic mice where the *ras* protooncogene is under the control of its own promoter.[138] It is noteworthy that transgenic mice carrying *ras* or *myc* transgenes under the control of the same regulatory region (immunoglobulin enhancer plus SV40-T gene promoter) develop different tumors (lung and lymphoid tumors for *ras* or *myc* transgenes, respectively; Figure 2).[130] This difference must depend on the oncogene, but its molecular basis is not yet understood. Tumors may occur in unexpected tissues, as lung adenomas develop in mice carrying the *ras* oncogenes under the albumin promoter control.[98] Mutated H-*ras* transgene under the MMTV promoter induces harderian gland, salivary gland, and mammary gland tumors.[132,133] The same tumor spectra is induced by the N-*ras* gene under the MMTV promoter.[128] Interestingly, it has been reported that angiosarcomas induced in transgenic mice for the H-*ras* protooncogene contained somatically acquired H-*ras* mutations.[138] On the contrary, tumors appearing in transgenic mice for the N-*ras* protooncogene do not contain *ras* mutations, despite the high expression of the transgen. Therefore, N-*ras* protooncogene overexpression is able to contribute to the tumorigenesis process.[138] Since similar data for other *ras* genes are not available we do not know if this is a specific effect of N-*ras*.

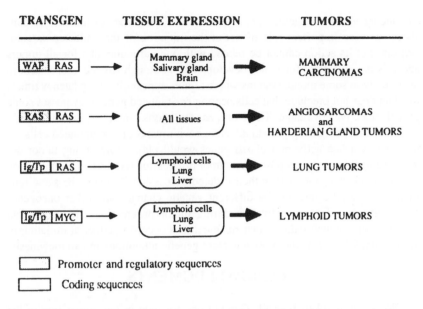

TRANSGEN   TISSUE EXPRESSION   TUMORS

WAP | RAS ⟶ Mammary gland / Salivary gland / Brain ⟹ MAMMARY CARCINOMAS

RAS | RAS ⟶ All tissues ⟹ ANGIOSARCOMAS and HARDERIAN GLAND TUMORS

Ig/Tp | RAS ⟶ Lymphoid cells / Lung / Liver ⟹ LUNG TUMORS

Ig/Tp | MYC ⟶ Lymphoid cells / Lung / Liver ⟹ LYMPHOID TUMORS

☐ Promoter and regulatory sequences

☐ Coding sequences

FIGURE 2. Tissue-specific effects of tumor development in *ras*-transgenic mice. Some examples are shown where the *ras* transgen is under the control of the promoter of whey acidic protein (WAP),[129] its own promoter,[138] or a regulatory region including an immunoglobulin enhancer and the SV40-T antigen promoter.[130] Only certain types of tumors appear, although the transgen is expressed in a wide array of tissues.

3. Other Oncogenes Cooperate with *ras* in Transgenic Tumor Development

Transgenic mice lines have been constructed with combinations of two co-operating oncogenes such as *ras* and *myc* or *ras* and SV40-T-antigen trans-genes.[98,132,138] These mice show shorter latency periods than the mice harboring only the *ras* transgene (Table 8). This is the case for mouse mammary tumors in mice with *ras* and *myc*,[132] liver dysplasias in mice with *ras* and SV40-T-antigen,[98] and skin hyperplasia with *ras* and c-*myc*.[139] In this latter case the oncogene cooperation also results in a higher percentage of malignant tumors with respect to the animals carrying only the *ras* transgene. These results are in agreement with those demonstrating cooperation *in vitro* of *ras* and c-*myc* on-cogenes for transformation of rodent embryo fibroblasts[110,111] or for development of carcinomas in reconstituted organs.[140] Altogether the data constitutes evidence of implication of different oncogenes in multistep carcinogenesis through com-plementary mechanisms.

VII. FINAL REMARKS

To study the involvement of *ras* genes in carcinogenesis, an array of animal model systems have been analyzed in the last few years. The tumors in these models are induced by the action either of carcinogenic agents or of activated *ras* oncogenes in transgenic mice. Results from both types of models are consistent with an important role of *ras* at the initiation of the multistep carcinogenesis, but its action cannot be related with one particular step for all tumors and it is not sufficient for tumor development. Although *ras* mutations seem to occur early in some model systems where tumors appear after long latency times, this fact does not preclude that cells bearing *ras*-mutated genes may posses some growth-selective advantage. Therefore, *ras* mutations can occur, in cells already initiated, thus, acquiring growth advantage with respect to nonmutated cells. In these cases a late activation of *ras* genes would play a role during tumor development. It has been proposed that tumor progression requires an accumulation of genetic changes rather than their ordered succession.[141] Thus, the same type of tumors may be initiated by different genetic changes and other oncogenes could substitute for *ras* activation during tumor progression. However, data shown in this review indicate that *ras* mutations occurring either at initiation or at later steps is one of the most prevalent genetic alterations in carcinogenesis.

ACKNOWLEDGMENTS

We are grateful to Juan M. García-Lobo for help in the preparation of the figures. Research in the authors laboratories is supported by Grants CICYT 91/SAL-0666 (Spain) and NIH CA36327, NIH CA50434, and ACS PDT393 (U.S.).

REFERENCES

1. **Lacal, J. C. and Tronick, S. R.,** The *ras* oncogene, in *The Oncogene Handbook,* Reddy, E. P., Skalka, A. M., and Curran, T., Eds., Elsevier Science, 1988, 257.
2. **Der, C. J., Finkel, T., and Cooper, G. M.,** Transforming genes of human bladder and lung carcinoma cell lines are homologous to the *ras* genes of Harvey and Kirsten sarcoma virus, *Cell,* 44, 167, 1982.
3. **Goldfarb, M. P., Shimuzu, K., Perucho, M., and Wigler, M.,** Isolation and preliminary characterization of a human transforming gene from T24 bladder carcinoma cells, *Nature,* 296, 404, 1982.
4. **Parada, L. F., Tabin, C. J., Shih, C. and Weinberg, R. A.,** Human EJ bladder carcinoma oncogene is homologue of Harvey sarcoma *ras* gene, *Nature,* 297, 474, 1982.
5. **Santos, E., Tronick, S. R., Aaronson, S. A., Pulciani, S., and Barbacid, M.,** T24 human bladder carcinoma oncogene is an activated form to the normal human homologue of BALB- and Harvey-*msv* transforming genes, *Nature* 298, 343, 1982.
6. **Bourne, H. R., Sanders, D. A., and McCormick, F.,** The GTPase superfamily: conserved structure and molecular mechanism, *Nature,* 349, 117, 1991.
7. **Hall, A.,** The cellular functions of small GTP-binding proteins, *Science,* 249, 635, 1990.
8. **Barbacid, M.,** *Ras* genes, *Annu. Rev. Biochem.,* 56, 779, 1987.
9. **Santos, E. and Nebreda, A. R.,** Structural and functional properties of *ras* proteins, *FASEB J.,* 3, 2151, 1989.

10. **Chang, E. H., Furth, M. E., Scolnick, E. M., and Lowy, D. R.,** Tumorigenic transformation of mammalian cells induced by a normal human gene homologous to the oncogene of Harvey murine sarcoma virus, *Nature,* 297, 479, 1982.

11. **Cichutek, K. and Duesberg, P. H.,** Recombinant BALB and Harvey sarcoma viruses with normal proto-*ras*-coding regions transform embryo cells in culture and cause tumors in mice, *J. Virol.,* 63, 1377, 1989.

12. **Pulciani, S., Santos, E., Long, L. K., Sorrentino, V., and Barbacid, M.,** *ras* gene amplification and malignant transformation, *Mol Cell. Biol.,* 5, 2836, 1985.

13. **Spandidos, D. A. and Wilkie, N. M.,** Malignant transformation of early passage rodent cells by a single mutated human oncogene, *Nature,* 310, 469, 1984.

14. **Velu, T. J., Vass, W. C., Lowy, D. R., and Tambourin, P. E.,** Harvey murine sarcoma virus: influences of coding and noncoding sequences on cell transformation in vitro and oncogenicity in vivo, *J. Virol.,* 63, 1384, 1989.

15. **Reynolds, S. H., Stowers, S. J., Patterson, R. M., Maronpot, R. R., Aaronson, S. A., and Anderson, M. W.,** Activated oncogenes in B6C3F1 mouse liver tumors: implications for risk assessment, *Science,* 237, 1309, 1987.

16. **Sloan, S. R., Newcomb, E. W., and Pellicer, A.,** Neutron radiation can activate K-*ras* via a point mutation in codon 146 and induces a different spectrum of *ras* mutations than does gamma radiation, *Mol. Cell. Biol.,* 10, 405, 1990.

17. **Balmain, A. and Brown, K.,** Oncogene activation in chemical carcinogenesis, *Adv. Cancer Res.,* 51, 147, 1988.

18. **Guerrero, I. and Pellicer, A.,** Mutational activation of oncogenes in animal model systems of carcinogenesis, *Mutat. Res.,* 185, 293, 1987.

19. **Sukumar, S.,** An experimental analysis of cancer: role of *ras* oncogenes in multistep carcinogenesis, *Cancer Cells,* 2, 199, 1990.

20. **Enomoto, T., Ward, J. M., and Perantoni, A. O.,** H-*ras* activation and *ras* p21 expression in bladder tumors induced in F344/NCr rats by *N*-butyl-*N*-(4-hydroxybutyl)nitrosamine, *Carcinogenesis,* 11, 2233, 1990.

21. **Pelling, J. C., Neades, R., and Strawhecker, J.,** Epidermal papillomas and carcinomas induced in uninitiated mouse skin by tumour promoters alone contain a point mutation in the 61st codon of the Ha-*ras* oncogene, *Carcinogenesis,* 9, 665, 1988.

22. **Quintanilla, M., Brown, K., Ramsden, N., and Balmain, A.,** Carcinogen-specific mutation and amplification of Ha-*ras* during mouse skin carcinogenesis, *Nature,* 322, 78, 1986.

23. **Field, J. K. and Spandidos, D. A.,** The role of *ras* and *myc* oncogenes in human solid tumours and their relevance in diagnosis and prognosis, *Anticancer Res.,* 10, 1, 1990.

24. **Bargmann, C. I., Hung, M.-C., and Weinberg, R.,** Multiple independent activations of the *neu* oncogene by a point mutation altering the transmembrane domain of p185, *Cell,* 45, 649, 1986.

25. **Shen-Ong, G. L., Keath, E. J., Piccoli, S. P., and Cole, M. D.,** Novel myc oncogene RNA from abortive immunoglobulin-gene recombination in mouse plasmacytomas, *Cell,* 31, 443, 1982.

26. **Sawey, M. J., Hood, A. T., Burns, F. J., and Garte, S. J.,** Activation of c-*myc* and c-K-*ras* oncogenes in primary rat tumors induced by ionizing radiation, *Mol. Cell Biol.,* 7, 932, 1987.

27. **Bos, J. L.,** The *ras* gene family and human carcinogenesis, *Mutat. Res.,* 195, 255, 1988.

28. **Blair, D. G., Cooper, C. S., Oskarson, M. K., Eader, L. A., and Vande Woude, G. F.,** New method for detecting cellular transforming genes, *Science,* 218, 1122, 1982.

29. **Fasano, O., Birnbaum, D., Edlund, K., Fogh, J., and Wigler, M.,** New human transforming genes detected by a tumorigenicity assay, *Mol. Cell. Biol.,* 4, 1695, 1984.

30. **Forrester, K. C., Almoguera, C., Itan, K., Grizzle, W. E., and Perucho, M.,** Detection of high incidence of K-*ras* oncogenes during human colon tumorigenesis, *Nature,* 327, 298, 1987.

31. **Almoguera, C., Shibata, D., Forrester, K., Martin, J., Arnheim, N., and Perucho, M.,** Most human carcinomas of the exocrine pancreas contain mutant c-K-*ras* genes, *Cell,* 53, 549, 1988.

32. **Diamond, L. E., Guerrero, I. and Pellicer, A.,** Concomitant K- and N-*ras* gene point mutation in clonal murine lymphoma, *Mol. Cell. Biol.,* 8, 2233, 1988.

33. **Guerrero, I., Villasante, A., Corces, V., and Pellicer, A.,** Loss of the normal N-*ras* allele in a mouse thymic lymphoma induced by a chemical carcinogen, *Proc. Natl. Acad. Sci. U.S.A.,* 82, 7810, 1985.

34. **Zarbl, H., Sukumar, S., Arthur, A. V., Martin-Zanca, D., and Barbacid, M.,** Direct mutagenesis of *Ha-ras*-1 oncogenes by *N*-nitroso-*N*-methylurea during initiation of mammary carcinogenesis in rats, *Nature,* 315, 382, 1985.

35. **Buchmann, A., Mahr, J., Baner-Hofman, R., and Schwarz, M.,** Mutations at codon 61 of the Ha-*ras* proto-oncogene in precancerous liver lesions of the B6C3F1 mouse, *Mol. Carcinogen.,* 2, 121, 1989.

36. **Kumar, R., Medina, D., and Sukumar, S.,** Activation of H-*ras* oncogenes in preneoplastic mouse mammary tissues, *Oncogene,* 5, 1271, 1990.

37. **Corominas, M., Kamino, H., León, J., and Pellicer, A.,** Oncogene activation in human benign tumors of the skin (keratoacanthomas): is HRAS involved in differentiation as well as proliferation?, *Proc. Natl. Acad. Sci. U.S.A.,* 86, 6372, 1989.

38. **Bizub, D., Wood, A. W., and Skalka, A. M.,** Mutagenesis of the Ha-*ras* oncogene in mouse skin tumors induced by polycyclic aromatic hidrocarbons, *Proc. Natl. Acad. Sci. U.S.A.,* 83, 6048, 1986.

39. **Dandekar, S., Sukumar, S., Zarbl, H., Young, L. J. T., and Cardiff, R.,** Specific activation of the cellular Harvey-*ras* oncogene in dimethylbenzanthracene-induced mouse mammary tumors, *Mol. Cell. Biol.,* 6, 4104, 1986.

40. **Kumar, R. and Barbacid, M.,** Oncogene detection at the single cell level, *Oncogene,* 3, 647, 1988.

41. **Drinkwater, N. R.,** Experimental models and biological mechanisms for tumor promotion, *Cancer Cells,* 2, 8, 1990.

42. **Corominas, M., León, J., Kamino, H., Cruz-Alvarez, M., Novick, S. C., and Pellicer, A.,** Oncogene involvement in tumor regression: H-*ras* activation in the rabbit keratoacanthoma model, *Oncogene,* 6, 645, 1991.

43. **León, J., Kamino, H., Steinberg, J. J., and Pellicer, A.,** H-*ras* activation in benign and self-regressing skin tumors (keratoacanthomas) in both humans and an animal model system, *Mol. Cell. Biol,* 8, 786, 1988.

44. **Balmain, A. and Pragnell, I. B.,** Mouse skin carcinoma, induced in vivo by chemical carcinogens have a transforming Harvey-*ras* oncogene, *Nature,* 303, 72, 1983.

45. **Brown, K., Quintanilla, M., Ramsden, M., Kerr, I. B., Young, S., and Balmain, A.,** v-*ras* genes from Harvey and BALB murine sarcoma viruses can act as initiators of two-stage mouse skin carcinogenesis, *Cell,* 46, 447, 1986.

46. **Brown, K., Buchmann, A. and Balmain, A.,** Carcinogen-induced mutations in the mouse c-Ha-*ras* gene provide evidence of multiple pathways for tumor progression, *Proc. Natl. Acad. Sci. U.S.A.,* 87, 538, 1990.

47. **Ramselaar, C. G. and van der Meer, J. B.,** Nonimmunological regression of dimethyl-benza(A)anthracene-induced experimental keratoacanthomas in the rabbit, *Dermatologica,* 158, 142, 1979.

48. **Gullino, P., Pettigrew, H. M., and Grantham, F. M.,** *N*-nitrosomethylurea as a mammary gland carcinogen in rats, *J. Natl. Cancer. Inst.,* 54, 404, 1975.

49. **Sukumar, S., Notario, V., Martin-Zanca, D., and Barbacid, M.,** Induction of mammary carcinomas in rats by nitrosomethylurea involves malignant activation of H-*ras*-1 locus by single point mutations, *Nature,* 306, 658, 1983.

50. **Sukumar, S., Carney, W. P., and Barbacid, M.,** Independent molecular pathways in initiation and loss of hormone responsiveness of breast carcinomas, *Science,* 240, 524, 1988.

51. **Kumar, R., Sukumar, S., and Barbacid, M.,** Activation of *ras* oncogenes preceding the onset of neoplasia, *Science,* 248, 1101, 1990.

52. **Sukumar, S. and Barbacid, M.,** Specific patterns of oncogene activation in transplacentally induced tumors, *Proc. Natl. Acad. Sci. U.S.A.,* 87, 718, 1990.
53. **Ball, J. K. and McCarter, J. A.,** Repeated demonstration of mouse leukemia virus after treatment with chemical carcinogens, *J. Natl. Cancer Inst.,* 46, 751, 1971.
54. **Kaplan, A. S.,** On the natural history of the murine leukemia, *Cancer Res.,* 27, 1325, 1967.
55. **Guerrero, I., Calzada, P., Mayer, A., and Pellicer, A.,** A molecular approach to leukemogenesis: mouse lymphomas contain an activated c-*ras* oncogene, *Proc. Natl. Acad. Sci. U.S.A.,* 81, 202, 1984.
56. **Newcomb, E. E., Steinberg, J. J., and Pellicer, A.,** *ras* oncogenes and phenotypic staging in *N*-methylnitrosurea- and γ-irradiation-induced thymic lymphomas in C57BL/6J mice, *Cancer Res.,* 48, 5514, 1988.
57. **Breuer, M., Wientjens, E., Verbeek, S., Slebos, R., and Berns, A.,** Carcinogen-induced lymphomagenesis in *pim*-1 transgenic mice: dose dependence and involvement of *myc* and *ras, Cancer Res.,* 51, 958, 1991.
58. **Corominas, M., Perucho, M., Newcomb, E. W., and Pellicer, A.,** Differential expression of the normal and mutated K-*ras* alleles in chemically induced thymic lymphomas, *Cancer Res.,* 51, 5129, 1991.
59. **Goodrow, T., Reynolds, S., Maronpot, R., and Anderson, M.,** Activation of K-*Ras* by codon 13 mutations in C57BL/6 X C3HF$_1$ mouse tumors induced by exposure to 1,3-butadiene, *Cancer Res.,* 50, 4818, 1990.
60. **Bauer-Hoffman, R., Buchmann, A., Wright, A. S., and Schwarz, M.,** Mutations in the Ha-ras proto-oncogen in spontaneous and chemically induced liver tumours of the CF1 mouse, *Carcinogenesis,* 11, 1875, 1990.
61. **Buchmann, A., Bauer-Hofmann, R., Mahr, J., Drinkwater, N. R., Luz, A., and Schwarz, M.,** Mutational activation of the c-Ha-*ras* gene in liver tumors of different rodent strains: correlation with susceptibility to hepatocarcinogenesis, *Proc. Natl. Acad. Sci. U.S.A.,* 88, 911, 1991.
62. **Dragani, T. A., Manenti, G., Colobo, B. M., Falvella, F. S., Gariboldi, M., Pierotti, M. A., and Della Porta, G.,** Incidence of mutations at codon 61 of the Ha-*ras* gene in liver tumors of mice genetically susceptible and resistant to hepatocarcinogenesis, *Oncogene,* 6, 333, 1991.
63. **Fox, T. R., Schumann, A. M., Watanabe, P. G., Yano, B. L., Maher, V. M., and McCormick, J. J.,** Mutational analysis of the H-*ras* oncogene in spontaneous C57BL/6 × C3H/He mouse liver tumors and tumors induced with genotoxic hepatocarcinomas, *Cancer Res.,* 50, 4014, 1990.
64. **Stowers, S. J., Wiseman, R. W., Ward, J. M., Miller, E. C., Miller, J. A., Anderson, M. W., and Eva, A.,** Detection of activated proto-oncogenes in *N*-nitrosodiethylamine-induced liver tumors: a comparison between B6C3F1 mice and Fisher 344 rats, *Carcinogenesis,* 9, 271, 1988.
65. **Leon, J., Guerrero, I., and Pellicer, A.,** Differential expression of the *ras* gene family in mice, *Mol. Cell. Biol.,* 7, 1535, 1987.
66. **McMahon, G., Davis, E. F., Huber, L. J., Kim, Y., and Wogan, G. N.,** Characterization of c-K-*ras* and N-*ras* oncogenes in aflatoxin B$_1$ induced rat liver tumors, *Proc. Natl. Acad. Sci. U.S.A.,* 87, 1104, 1990.
67. **McMahon, G., Huber, L. J., Moore, M. J., Stegeman, J. J., and Wogan, G. N.,** Mutations in c-Ki-*ras* oncogenes in diseased livers of winter flounder from Boston Harbor, *Proc. Natl. Acad. Sci. U.S.A.,* 87, 841, 1990.
68. **Wirgin, I., Currie, D., and Garte, S. J.,** Activation of the K-*ras* oncogene in liver tumors of Hudson River tomcod, *Carcinogenesis,* 10, 2311, 1989.
69. **Belinsky, S. A., Devereux, T. R., Maronpot, R. R., Stoner, G. D., and Anderson, M. W.,** Relationships between the formation of promutagenic adducts and the activation of the K-*ras* protooncogene in lung tumors from A/J mice treated with nitrosamines, *Cancer Res.,* 49, 5305, 1989.

70. **Devereux, T. R., Anderson, M. W., and Belinsky, S. A.,** Role of *ras* protooncogene activation in the formation of spontaneous and nitrosamine-induced lung tumors in the resistant C3H mouse, *Carcinogenesis,* 12, 299, 1991.

71. **You, M., Candrian, U., Maronpot, R. R., Stoner, G. D., and Anderson, M. W.,** Activation of the Ki-*ras* protooncogene in spontaneously occurring and chemically induced lung tumors of the strain A mouse, *Proc. Natl. Acad. Sci. U.S.A.,* 86, 3070, 1989.

72. **Stowers, S. J., Glover, P. L., Reynolds, S. H., Boone, L. R., Maronpot, R. R., and Anderson, M. W.,** Activation of the K-*ras* protooncogene in lung tumors from rats and mice chronically exposed to tetranitromethane, *Cancer Res.,* 47, 3212, 1987.

73. **Rodenhuis, S., Slebos, R. J., Boot, A. J., Evers, S. C., Mooi, W. J., Wagenaar, S. S., van Bodegom, P. C., and Bos, J. L.,** Incidence and possible clinical significance of K-*ras* oncogene activation in adenocarcinoma of the human lung, *Cancer Res.,* 48, 5738, 1988.

74. **Sukumar, S., Perantoni, A., Reed, C., Rice, J. M., and Wenk, M. L.,** Activated K-*ras* and N-*ras* oncogenes in primary renal mesenchymal tumors induced in F344 rats by methyl(methoxymethyl)nitrosamine, *Mol. Cell. Biol.,* 6, 2716, 1986.

75. **Lemoine, N. R., Mayall, E. S., Williams, E. D., Thurston, V., and Wynford-Thomas, D.,** Agent-specific *ras* oncogene activation in rat thyroid tumours, *Oncogene,* 3, 541, 1988.

76. **Wang, Y., You, M., Reynolds, S. H., Stoner, G. D., and Anderson, M. W.,** Mutational activation of the cellular Harvey *ras* oncogene in rat esophageal papillomas induced by methylbenzylnitrosamine, *Cancer Res.,* 50, 1591, 1990.

77. **Jacoby, R. F., Llor, X., Teng, B. B., Davidson, N. O., and Brasitus, T. A.,** Mutations in the K-*ras* oncogene induced by 1,2-dimethylhydrazine in preneoplastic and neoplastic rat colonic mucosa, *J. Clin. Invest.,* 87, 624, 1987.

78. **Fujii, H., Egami, H., Charney, W., Pour, P., and Pelling, J.,** Pancreatic ductal adenocarcinomas induced in Syrian hamster by *N*-nitrosobis(2-oxopropyl)amine contain a c-Ki-*ras* oncogene with a point mutated codon 12, *Mol. Carcinog.,* 3, 295, 1990.

79. **Bos, J. L., Fearon, E. R., Hamilton, S. R., de Vries, M. V., van Boom, J. H., van der Eb, A. J., and Volgestein, B.,** Prevalence of *ras* gene mutations in human colorectal cancers, *Nature,* 327, 293, 1987.

80. **Vis Vanathan, K. V., Pocock, R. D., and Summerhayes, I. C.,** Preferential and novel activation of H-*ras* in human bladder carcinomas, *Oncogene Res.,* 3, 77, 1988.

81. **Miyamoto, S., Sukumar, S., Guzman, R. C., Osborn, R. C., and Nandi, S.,** Transforming c-Ki-*ras* mutation is a preneoplastic event in mouse mammary carcinogenesis induced in vitro by *N*-methyl-*N*-nitrosourea, *Mol. Cell Biol.,* 10, 1593, 1990.

82. **Schaeffer, B. K., Zurlo, J., and Longnecker, D. S.,** Activation of c-Ki-*ras* not detectable in adenomas or adenocarcinomas in rat pancreas, *Mol. Carcinog.,* 3, 165, 1990.

83. **Loktionov, A., Hollstein, M., Martel, N., Galendo, D., Cabral, J. R., Tomatis, L., and Yamasaki, H.,** Tissue-specific activating mutations of Ha- and Ki-*ras* oncogenes in skin, lung and liver tumors induced in mice following transplacental exposure to DMBA, *Mol. Carcinog.,* 3, 134, 1990.

84. **Fiorucci, G. and Hall, A.,** All three human *ras* genes are expressed in a wide range of tissues, *Biochem. Biophys. Acta,* 950, 81, 1988.

85. **Dunbar, C. E., Crosier, P. S., and Nienhuis, A. W.,** Introduction of an activated RAS oncogene into murine bone marrow lymphoid progenitors via retroviral gene transfer results in thymic lymphomas, *Oncogene Res.,* 6, 39, 1991.

86. **Furth, M. E., Aldrich, T. H., and Cordon-Cardo, C.,** Expression of *ras* proto-oncogene proteins in normal human tissues, *Onocogene,* 1, 47, 1987.

87. **Burns, P. A., Gordon, A. J. E., and Glickman, B. W.,** Influence of neighbouring base sequence on *N*-methyl-*N'*-nitro-*N*-nitrosoguanidine mutagenesis in the *lacI* gene of *Escherichia coli, J. Mol. Biol.,* 194, 385, 1987.

88. **DuBridge, R. B., Tang, P., Hsia, H. C., Leong, P.-M., Miller, J. H., and Calos, M. P.,** Analysis of mutation in human cells by using an Epstein-Barr virus shuttle system, *Mol. Cell. Biol.,* 7, 379, 1987.

89. **Mitra, G., Pauly, G. T., Kumar, R., Pei, G. K., Hughes, S. H., Moschel, R. C., and Barbacid, M.**, Molecular analysis of O^6-substituted guanine-induced mutagenesis of *ras* oncogenes, *Proc. Natl. Acad. Sci. U.S.A.*, 86, 8650, 1989.

90. **Fox, T. R. and Watanabe, P. G.**, Detection of a cellular oncogene in spontaneous liver tumors of B6C3F1 mice, *Science*, 228, 596, 1985.

91. **Peto, R., Roe, F. J. C., Lee, P. N., Levy, L., and Clack, J.**, Cancer and aging in mice and men, *Br. J. Cancer*, 32, 411, 1975.

92. **Greenhalgh, D. A., Welty, D. J., Strickland, J. E., and Yuspa, S. H.**, Spontaneous Ha-*ras* gene activation in cultured primary murine keratinocytes: consequences of Ha-*ras* gene activation in malignant conversion and malignant progression, *Mol. Carcinog.*, 2, 199, 1989.

93. **Vousden, K. H. and Marshall, C. J.**, Three different activated *ras* genes in mouse tumours; evidence for oncogene activation during progression of a mouse lymphoma, *EMBO J.*, 3, 913, 1984.

94. **Tainsky, M. A., Cooper, C. S., Giovanella, B. C., Vande Woude, G. F.**, An activated *ras*[N] gene detected in late but not early passage of human PA1 teratocarcinoma cells, *Science*, 225, 643, 1984.

95. **Lübbert, M., Jonas, D., Miller, C. W., Hermann, F., Mertelsmann, R., McCormick, F., and Koeffler, H. P.**, Retrospective analysis of *ras* gene activation in myeloid leukemic cells, *Oncogene*, 5, 583, 1990.

96. **Albino, A. P., Le Strange, R., Oliff, A. I., Furth, M. E., and Old, L. J.**, Transforming *ras* genes from human melanoma: a manifestation of tumor heterogeneity, *Nature*, 308, 69, 1984.

97. **Lee, G. H., Li, H., Ohtake, K., Nomura, K., Hino, O., Furuta, Y., Aizawa, S., and Kitagawa, T.**, Detection of activated c-H-*ras* oncogene in hepatocellular carcinomas developing in transgenic mice harboring albumin promoter-regulated simian virus 40 gene, *Carcinogenesis*, 11, 1145, 1990.

98. **Sandgren, E. P., Quaife, C. J., Pinkert, C. A., Palmiter, R. D., and Brinster, R. L.**, Oncogene-induced liver neoplasia in transgenic mice, *Oncogene*, 4, 715, 1989.

99. **Vogelstein, B., Fearon, E. R., Hamilton, S. R., Kern, S. E., Preisinger, A. C., Leppert, M., Nakamura, Y., White, R., Smits, A. M. M., and Bos, J. L.**, Genetic alterations during colorectoral tumor development, *N. Engl. J. Med.*, 319, 525, 1988.

100. **Riou, G., Barrois, M., Sheng, Z. M., Duvillard, P., and Lhome, C.**, Somatic deletions and mutations in c-Ha-*ras* gene in human cervical cancers, *Oncogene*, 3, 329, 1988.

101. **Hiorns, L. R., Cotter, F. E., and Young, B. D.**, Co-incident N and K ras gene mutations in a case of AML, restricted to differing cell lineages, *Br. J. Haematol.*, 73, 165, 1989.

102. **Janssen, J. W. G., Lyons, J., Steenvoorden, A. C. M., Seliger, H., and Bartram, C. R.**, Concurrent mutations in two different *ras* genes in acute myelocytic leukemias, *Nucleic Acids Res.*, 15, 5669, 1987.

103. **Carter, G., Hughes, D. C., Clark, R. E., McCormick, F., Jacobs, A., Whittaker, J. A., and Padua, R. A.**, RAS mutations in patients following cytotoxic therapy for lymphoma, *Oncogene*, 5, 411, 1990.

104. **Radich, J. P., Kopecky, K. J., Willman, C. L., Weick, J., Head, D., Appelbaum, F., and Collins, S. J.**, N-*ras* mutations in adult de novo leukemia: prevalence and clinical significance, *Blood*, 76, 801, 1990.

105. **Ahuja, H. G., Foti, A., Bar-Eli, M., and Cline, M. J.**, The pattern of mutational involvement of RAS genes in hematological malignancies determined by DNA amplification and direct sequencing, *Blood*, 75, 1684, 1990.

106. **Yunis, J. J., Boot, A. J. M., Mayer, M. G., and Bos, J. L.**, Mechanisms of *ras* mutation in myelodysplastic syndrome, *Oncogene*, 4, 609, 1989.

107. **Hennings, H., Shores, R., Mitchell, P., Spangler, E. F., and Yuspa, S. H.**, Induction of papillomas with a high probability of conversion to malignancy, *Carcinogenesis*, 6, 1607, 1985.

108. **Zhang, R., Haag, J. D., and Gould, M. N.**, Reduction in the frequency of activated *ras* oncogenes in rat mammary carcinomas with increasing *N*-methyl-*N*-nitrosourea doses or increasing prolactin levels, *Cancer Res.*, 50, 4286, 1990.
109. **Pozzatti, R., Muschel, R., Williams, J., Padmanabhan, R., Howard, B., Liotta, L., and Khoury, G.**, Primary rat embryo cells transformed by one or two oncogenes show different metastatic potentials, *Science*, 232, 223, 1986.
110. **Land, H., Parada, L. F., and Weinberg, R. A.**, Tumorigenic conversion of primary embryo fibroblasts requires at least two cooperating oncogenes, *Nature*, 304, 596, 1983.
111. **Land, H., Chen, A. C., Morganstern, J. P., Parada, L. F., and Weinberg, R. A.**, Behavior of *myc* and *ras* oncogenes in transformation of rat embryo fibroblasts, *Mol. Cell Biol.*, 6, 1917, 1986.
112. **Parada, L. F., Land, H., Weinberg, R. A., Wolf, D., and Rotter, V.**, Cooperation between genes encoding p53 tumour antigen and *ras* in cellular transformation, *Nature*, 312, 649, 1984.
113. **Ruley, H. E.**, Adenovirus early region 1A enable viral and cellular transforming genes to transform primary cells in culture, *Nature*, 304, 602, 1983.
114. **Schwartz, R. C., Stanton, L. W., Riley, S. C., Marcu, K. B., and Witte, O. N.**, Synergism of v-*myc* and v-Ha-*ras* in the in vitro neoplastic progression of murine lymphoid cells, *Mol. Cell Biol.*, 6, 3221, 1986.
115. **Hunter, T.**, Cooperation between oncogenes, *Cell*, 64, 249, 1991.
116. **Ruley, H. E.**, Transforming collaborations between *ras* and nuclear oncogenes, *Cancer Cells*, 2, 258, 1990.
117. **Geiser, A. G., Der, C. J., Marshall, C. J., and Standbridge, E. S.**, Suppression of tumorigenicity with continued expression of the c-Ha-*ras* oncogene in EJ bladder carcinoma-human fibroblast hybrid cells, *Proc. Natl. Acad. Sci. U.S.A.*, 85, 5209, 1986.
118. **Harris, H.**, The analysis of malignancy in cell fusion: the position in 1988, *Cancer Res.*, 48, 3302, 1988.
119. **Marshall, C. J.**, Tumor supressor genes, *Cell*, 64, 313, 1991.
120. **Sager, R., Tamaka, K., Lau, C. C., Ebina, Y., and Anisowicz, A.**, Resistance of human cells to tumorigenesis induced by cloned ransforming genes, *Proc. Natl. Acad. Sci. U.S.A.*, 80, 7601, 1983.
121. **Farr, C. J., Marshall, C. J., Easty, D. J., Wright, N. A., Powell, S. C., and Paraskeva, C.**, A study of *ras* gene mutations in colonic adenomas from familial polyposis coli patients, *Oncogene*, 3, 673, 1988.
122. **Brinster, R. L., Chen, H. Y., Messing, A., Van Dyke, T., Levine, A. J., and Palmiter, R. D.**, Transgenic mice harbouring SV40 T-antigen genes develop characteristic brain tumors, *Cell*, 37, 367, 1984.
123. **Compere, S. J., Baldacci, P., and Jaenisch, R.**, Oncogenes in transgenic mice, *Biochim. Biophys. Acta*, 948, 129, 1988.
124. **Hanahan, D.**, Dissecting multistep tumorigenesis in transgenic mice, *Annu. Rev. Genet.*, 22, 479, 1988.
125. **Hanahan, D.**, Transgenic mice as probes into complex systems, *Science*, 246, 1265, 1989.
126. **Mangues, R., Seidman, I., Pellicer, A., and Gordon, J. W.**, Tumorigenesis and male sterility in transgenic mice expressing a MMTV/N-*ras* oncogene, *Oncogene*, 5, 1491, 1990.
127. **Bouchard, L., Lamarre, Tremblay, P. J., and Jolicoeur, P.**, Stochastic appearance of mammary tumors in transgenic mice carrying the MMTV/c-*neu* oncogene, *Cell*, 57, 931, 1989.
128. **Muller, W. J., Sinn, E., Pattengale, P. K., Wallace, R., and Leder, P.**, Single-step induction of mammary adenocarcinoma in transgenic mice bearing the activated c-*neu* oncogene, *Cell*, 54, 105, 1988.
129. **Andres, A.-A., Schönenberger, C.-A., Groner, B., Hennighausen, L., Le Meur, M., and Gerlinger, P.**, Ha-*ras* oncogene expression directed by a milk protein gene promoter: tissue specificity, hormonal regulation, and tumor induction in transgenic mice, *Proc. Natl. Acad. Sci. U.S.A.*, 84, 1299, 1987.

130. **Suda, Y., Aizawa, S., Hirai, S., Inoue, T., Furuta, Y., Suzuki, M., Hirohashi, S., and Ikawa, Y.,** Driven by the same Ig enhancer and SV40 T promoter *ras* induced lung adenomatous tumors, *myc* induced pre-B cell lymphomas and SV40 large T gene a variety of tumors in transgenic mice, *EMBO J.,* 6, 4055, 1987.

131. **Quaife, C. R., Pinkert, C. A., Ornitz, D. M., Palmiter, R. D., and Brinster, R. L.,** Pancreatic neoplasia induced by *ras* expression in acinar cells of transgenic mice, *Cell,* 48, 1023, 1987.

132. **Sinn, E., Muller, W., Pattengale, P., Tepler, I., Wallace, R., and Leder, P.,** Coexpression of MMTV/v-Ha-ras and MMTV/c-myc genes in transgenic mice: synergistic action of oncogenes in vivo, *Cell,* 49, 465, 1987.

133. **Tremblay, P. J. Pothier, F., Hoang, T., Tremblay, G., Brownstein, S., Liszaner, A., and Jolicoeur, P.,** Transgenic mice carrying the mouse mammary tumor virus *ras* fusion gene: distinct effects in various tissues, *Mol. Cell. Biol.,* 9, 854, 1989.

134. **Bailleul, B., Surani, M. A., White, S., Barton, S. C., Brown, K., Blessing, M., Jorcano, J., and Balmain, A.,** Skin hyperkeratosis and papilloma formation in transgenic mice expressing a *ras* oncogene from a suprabasal keratin promoter, *Cell,* 62, 697, 1990.

135. **Compere, S. J., Baldacci, P. A., Sharpe, A. H., and Jaenisch, R.,** Retroviral transduction of the human c-Ha-*ras*-1 oncogene into midgestation mouse embryos promotes rapid epithelial hyperplasia, *Mol. Cell. Biol.,* 9, 6, 1989.

136. **Efrat, S., Fleischer, N., and Hanahan, D.,** Diabetes induced in male transgenic mice by expression of human H-*ras* oncoprotein in pancreatic β cells, *Mol. Cell. Biol.,* 10, 1779, 1990.

137. **Botteri, F. M., van der Putten, H., Wong, D. E., Sauvage, C. A., and Evans, R. M.,** Unexpected thymic hyperplasia in transgenic mice harboring neuronal promoter fused with simian virus 40 large T antigen, *Mol. Cell. Biol.,* 7, 3178, 1987.

138. **Saitoh, A., Kimura, M., Takahashi, R., Yokoyama, M., Nomura, T., Izawa, M., Sekiya, T., Nishimura, S., and Katsuki, M.,** Most tumors in transgenic mice with human c-Ha-*ras* gene contained somatically activated transgenes, *Oncogene,* 5, 1195, 1990.

138a. **Mangues, R., Seidman, I., Gordon, J. W., and Pellicer, A.,** Overexpression of the N-*ras* protooncogene, not somatic mutational activation, associated with malignant tumors in transgenic mice, *Oncogene,* 7, 2073, 1992.

139. **Compere, S. J., Baldacci, P., Sharpe, A. H., Thompson, T., Land, H., and Jaenisch, R.,** The *ras* and *myc* oncogenes cooperate in tumor induction in many tissues when introduced into midgestation mouse embryos by retroviral vectors, *Proc. Natl. Acad. Sci. U.S.A.,* 86, 2224, 1989.

140. **Thompson, T. C., Southgate, J., Kitchener, G., and Land, H.,** Multistage carcinogenesis induced by *ras* and *myc* oncogenes in a reconstituted organ, *Cell,* 56, 917, 1989.

141. **Fearon, E. R. and Vogelstein, B. A.,** A genetic model for colorectal tumorigenesis, *Cell,* 61, 797, 1990.

142. **Bailleul, B., Brown, K., Ramsden, M., Akhurst, R. J., Fee, F., and Balmain, A.,** Chemical induction of oncogene mutations and growth factor activity in mouse skin carcinogenesis, *Env. Health. Persp.,* 81, 23, 1989.

143. **Anderson, L. M., Enomoto, T., Perantoni, A. O., Riggs, C. W., Kovatch, R. M., Reed, C. D., Giner-Sorolla, A., and Rice, J. M.,** *N*-nitrosocimetidine as an initiator of murine skin tumors with associated H-*ras* oncogene activation, *Carcinogenesis,* 10, 2009, 1989.

144. **Hochwalt, A. E., Solomon, J. J., and Garte, S. J.,** Mechanism of H-*ras* oncogene activation in mouse squamous carcinoma induced by an alkylating agent, *Cancer Res.,* 48, 556, 1987.

145. **Reynolds, S. H., Patterson, R. M., Mennear, J. M., Maronpot, R. R., and Anderson, M. W.,** *ras* gene activation in rat tumors induced by benzidine congeners and derived dyes, *Cancer Res.,* 50, 266, 1990.

146. **Cardiff, R. D., Gumerlock, P. H., Soong, M. M., Dandekar, S., Barry, P. A., Young, L. J. T., and Meyers, F. J.,** c-H-*ras*-1 expression in 7,12-dimethyl benzanthracene-induced Balb/c mouse mammary hyperplasias and their tumors, *Oncogene,* 3, 205, 1988.

147. **Warren, W., Clark, J. P., Gardner, E., Harris, G., Cooper, C. S., and Lawley, P. D.,** Chemical induction of thymomas in AKR mice: interaction of chemical carcinogens and endogenous nurine leukemia viruses. Comparison of N-methyl-N-nitrosurea and methyl-methanesulphonate, *Mol. Carcinog.,* 3, 126, 1990.
148. **Wiseman, R. W., Stowers, S. J., Miller, E. C., Anderson, M. W., and Miller, J. A.,** Activating mutations of the c-Ha-*ras* protooncogene in chemically induced hepatomas of the male B6C3F1 mouse, *Proc. Natl. Acad. Sci. U.S.A.,* 83, 5825, 1986.

Chapter 2

THREE-DIMENSIONAL STRUCTURE OF p21 AND ITS IMPLICATIONS

Fred Wittinghofer

TABLE OF CONTENTS

I. INTRODUCTION

New guanine nucleotide-binding (GNB) proteins keep on being discovered continuously and the number of individual members of this family of proteins may be 50 to 100 for a higher organism. In order to understand their function and mechanism of action, it is important to get a detailed description of the three-dimensional structure of these proteins down to the atomic level. This has been achieved for p21, the product of the *ras* protooncogene. It has enabled us to understand the basic features of this molecule whose general function is as a molecular switch. Although the three-dimensional structure of no other GNB protein except bacterial elongation factor Tu (EF-Tu) has been determined, albeit at lower resolution, the results of the structure determination have implications for other GNB proteins as well.

II. p21 AS A MEMBER OF THE GNB PROTEIN FAMILY

GNB proteins have been shown (for some) or are supposed (for the others) to be involved in many cellular processes such as signal transduction, protein transport and secretion, and polypeptide chain elongation (for recent reviews, see References 1–4). They possess several characteristics which classify them as members of this protein family.

They have four or five conserved sequence elements, one of which is a single amino acid. These are GXXXXGKS/T, T, DXXG, N/TKXD (where X is any amino acid), which appear in that order within the sequence.[1-4] Small GNB proteins have, in addition, an EXSAK/L element located after NKXD in the sequence.

They are all believed to act as molecular switches. In the "ON" or active state they are complexed to guanosine triphosphate (GTP) and in the "OFF" or inactive state to guanosine diphosphate (GDP). In the GTP-bound conformation they interact with an effector molecule, and the lifetime of this interaction is regulated either by the effector-mediated or by the intrinsic GTPase activity of the protein. Thus, the operational definition of the effector molecule is that it interacts only with the active form of the GNB protein. However, as has been pointed out recently,[1-2] there is *a priori* no reason to believe that the switch does not work in the opposite direction with the GDP-bound form as the active species.

GNB proteins have several biochemical properties in common. They bind the guanine nucleotides GDP and GTP each with high affinity and high specificity. With very few exceptions (i.e., the bacterial elongation factor G), the affinities are higher than 10^8 M^{-1} ranging up to 10^{12} M^{-1} for the p21-GTP interaction (5°C, 10 mM Mg^{2+}).[5] GMP is bound only very weakly,[5] stressing the importance of the β-phosphate group for substrate recognition, which had already been noticed by Rösch et al.[6] on the basis of its unusual phosphorous NMR resonance. Due to these high affinities, the GNB proteins usually contain bound nucleotide when isolated in their native state.[8-11] Since this aspect has often been neglected, the affinities between protein and nucleotide, that have been determined, frequently deviate by factors up to 10000-fold from the true values.[7]

The nucleotides are bound with very high specificity such that only a few modifications of the nucleotide are tolerated without a significant reduction of the corresponding affinity. Even the purine nucleotides ADP/ATP bind very weakly with affinities being several orders of magnitude lower. For tight binding of the nucleotide and for the GTPase reaction to occur, GNB proteins need, in addition, a divalent ion which *in vivo* is most likely Mg^{2+}.[3,8-15]

The X-ray structure determination of p21[16-20] and EF-Tu[21,22] has established that the three-dimensional structure of the G domain, which is the domain responsible for guanine nucleotide binding and hydrolysis, is identical for all the GNB proteins, as is further demonstrated below.

p21 proteins, the products of the N-, K-, or H-*ras* oncogenes, are believed to be involved in a growth-promoting signal transduction process.[23] Their affinities for GDP/GTP are in the order of 10^{11} to 10^{12} M^{-1}.[5,15] They possess low intrinsic GTPase activities which is 0.028 min^{-1} in the case p21 H-*ras*, the product of the H-*ras* gene.[24] In the triphosphate conformation they interact with the GTPase activating protein (GAP).[25,26] The GTPase in the presence of GAP is 20 s^{-1}, which means that the GTPase of p21 is accelerated approximately 10^5-fold.[27]

Mutant p21 proteins, which have been identified in human tumors, are effectively locked in the active GTP conformation because GAP does not stimulate their GTPase rate and are thus unable to recycle them quickly enough to the inactive GDP-bound form.[25,28,29] In order to understand how the oncogenic mutations interfere with hydrolysis, it is crucial to know the structures of both wild-type and mutant proteins in their active GTP-bound conformations.

III. THE THREE-DIMENSIONAL STRUCTURE OF p21

A. TRUNCATION AND CRYSTALLIZATION

More than one year of frustration was spent by us in trying to crystallize native p21. Crystallization is often not successful because mobile elements on the surface of the protein prevent it from being packed into crystals. The alignment of primary sequences of p21 proteins shows that the homology between the p21 proteins breaks down around His166.[23] Therefore, we reasoned that the C terminally located amino acids, the function of which is to attach the protein to the membrane,[23] constitute a flexible element, which is structurally independent of the G domain proper. Consequently, the last 23 amino acids were removed by mutating Lys167 to a stop codon. Both our truncated protein p21(1-166) = p21′, as well as a similar truncated protein p21(1-171) prepared by Jancarik et al.,[30] were finally crystallized and the crystals turned out to be suitable for X-ray analysis.[31]

In a detailed series of experiments, the biochemical properties of the truncated p21 protein p21′ were demonstrated to be basically unaltered as compared to

TABLE 1
List of Three-Dimensional Structures of p21
Protein-Nucleotide Complexes which have
been Solved by X-ray Analysis

Protein	Diphosphate	Triphosphate	Ref.
p21$_{wt}$	2.2 Å (78)	1.35 Å GppNHp	18
	2.8 Å (20)	2.2 Å GppCH$_2$p	19
		1.4 Å GppCH$_2$p	62
		3.0 Å cagedGTP	20
		2.8 Å GTP	20
p21(G12V)	2.2 Å 78	2.8 Å GTP	20
	2.8 Å (20)		
p21(G12R)		2.2 Å GppNHp	60
p21(G12P)		1.5 Å GppNHp	62
p21(Q61L)		2.0 Å GppNHp	60
p21(Q61H)		2.4 Å GppNHp	60
p21(D38E)		2.3 Å GppNHp	60

wild-type protein p21. These studies were performed on the interaction with guanine nucleotides,[24] with GAP,[27] and with the catalytic domain of the exchange factor SDC25 (A. Parmeggiani et al., manuscript in preparation). ^{31}P- and ^{1}H-NMR experiments also showed that the three-dimensional structure is unaltered on removing the C terminus. However, we find one property of p21 altered as a consequence of the truncation; the p21 · [^{3}H]GDP complex of truncated p21 is not quantitatively retained on nitrocellulose filters. This might be related to the fact that truncated p21 denatures more easily in solution and cannot be renatured easily after it has been blotted onto a nitrocellulose filter (H. Rensland and F. J. Klienz, unpublished observations). We assume that through the truncation of p21 normally buried hydrophobic residues become exposed on the surface with the result that the protein is less stable in solution. One candidate for such a residue has been noticed in the ^{1}H-NMR spectrum of truncated p21.[24] In general, however, it can be said that the structure of the G domain of p21 as determined by X-ray crystallography reflects very closely the structure of the G domain in intact p21 and that the extra tail of 23 amino acids does not markedly influence its properties.

B. CRYSTAL STRUCTURES DETERMINED

The structures of several different wild-type and mutant p21 complexes have been determined by S. H. Kim and co-workers as well as in our laboratory. A list of the published structures is shown in Table 1 together with the resolutions obtained in the crystallographic analyses. For cellular and one oncogenic mutant p21 both the triphosphate and diphosphate structures have been determined. From these the mechanism of the conformational change could be deduced. For the triphosphate complex of p21 the noncleavable analogs GppNHp and Gpp(CH$_2$)p

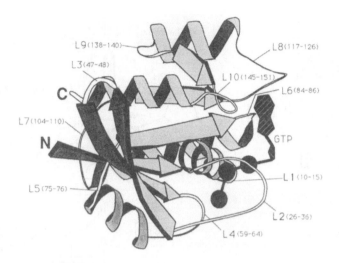

FIGURE 1. Scheme of secondary structure. Secondary structure elements and hydrogen-bonding pattern of the main chain for the present model were determined by the program DSSP.[80] The loops are labeled. (Modification of a drawing provided to us by D. Lowy.)

were used, because GTP and even GTPγS are hydrolyzed by p21 during crystallization. A complex of p21 with a noncleavable GTP analog having a photolabile protecting group on γ-phosphate, which was named cagedGTP, has also been crystallized.[32] This latter crystal type was used to determine the three-dimensional structure of the real p21-GTP complex after photolytic cleavage of the protecting group. Additionally, it also allowed us to directly follow the GTP hydrolysis reaction within the crystal and observe the conformational transition from the active to the inactive form of p21.[20]

C. OVERALL STRUCTURE

Figure 1 shows the overall topology of the polypeptide chain, indicating the secondary structure elements and their connections via loops. p21 has a central β-sheet consisting of six strands and five helices, two of which (α2 and α3) are below and three above the plane defined by the β-sheet. There are three loops which are important for the function of the protein (Plate 1B*): loop L1 contains Gly12, the residue which is the most frequently mutated residue in human tumors; loop L2 contains the residues believed to be involved in the interaction with the effector such as Asp33; loop L4 contains Gln61, the other residue which is highly important for the oncogenic activation of p21. In the three-dimensional structure, these three loops comprise the active site of the molecule centered around the γ-phosphate. Thus, the lower right corner of the molecule as depicted in Figure 1 is the business end of the molecule, where the important interactions and the GTPase reaction is seen to occur.[17]

---

* Plate 1 will appear after page 46.

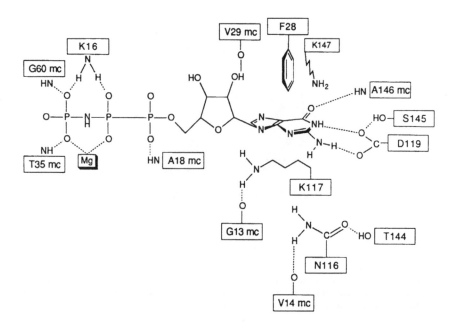

FIGURE 2. Scheme showing the interactions between GppNHp and p21 which are believed to significantly contribute to the binding interaction. (From Whittinghofer and Pai, *TIBS*, 16, 382, 1991. With permission.)

Most of the residues are well-defined in the crystal structure except residues 61 to 65 in loop L4, which face the solvent water and occupy several conformational positions. It is most likely, although no evidence has yet been presented for it, that this part of the structure is also highly mobile in solution. Conversely it must be considered that other regions of the molecule, which are immobile in the crystal due to intermolecular interactions, are mobile in solution.

In the high resolution structure we identified a great number of water molecules, which are more or less tightly bound by hydrogen bonds to the protein or other water molecules. Three of those are particularly tightly bound, as evidenced by their low temperature factors (which means low mobility). These water molecules are important for catalytic function and will be discussed in detail below.

D. THE NUCLEOTIDE-BINDING SITE

We have recently shown that guanine nucleotides with two or three phosphates bind to p21 with an affinity in the order of 10^{11} to $10^{12} M^{-1}$ in the presence of Mg^{2+} at 4°C, whereas the affinity to GMP is six orders of magnitude lower.[5] The high affinity for GDP and GTP is reflected by the large number of polar and nonpolar interactions between the protein and the nucleotide GppNHp. Plate 1A shows the nucleotide binding pocket in the atomic model of p21 and Figure 2 schematically shows the interactions between the nucleotide GppNHp and p21, which are believed to contribute significantly to the binding strength.

The guanine base of the nucleotide is anchored through interactions with the conserved sequence motifs NKXD (residues 116 to 119) and EXSAK/L (residue 143 to 147) within the polypeptide chain as shown in Figure 2. The first motif is identical in the great majority of GNB proteins and[1-4,33] the second only in small GNB proteins. Large GNB proteins may have a similar motif which performs a similar structural function, but has as yet been unrecognized due to the lack of structural information for these proteins. The guanine base is situated in a hydrophobic pocket (Plate 1C). The tight binding is due to the aromatic-aromatic interaction with Phe28 on one side of the base and the hydrophobic contact with the alkyl part of the side chain of Lys117 on the other side. Phe28 is held in place by a similar interaction with Lys147, which stretches out across the plane of the phenyl ring. The carboxylate group of Asp119 is hydrogen bonded to the exocyclic amino group, to the endocylcic nitrogen N1, and to the hydroxyl of Ser145. The keto group at position 6 of the guanine base makes a hydrogen bond to the main chain NH of Ala146. We have two examples here in which the conservation of lysines (Lys117, Lys147) is not a consequence of an interaction of the ϵ-amino group, but rather due to a hydrophobic interaction of the alkyl side chain of this residue.

In p21, the side chain carbamoyl group of Asn116, which based on the three-dimensional structure of EF-Tu·GDP has been proposed to be involved in the binding of the O^6-oxygen,[79] forms strong hydrogen bonds to the side chain of Thr144 and to the main chain oxygen atom of Val14. In conclusion, the main function of Asn116 is to tie together the three elements which are involved in nucleotide binding: the phosphate binding loop [10]GAGGVGKS, the [116]NKCD, and the [143]ETSAK motifs. A similar function is performed by the ϵ-amino group of Lys117, which links the phosphate binding loop and the NKCD motif by binding to the main chain carbonyl group of Gly13.

The ribose ring adopts the 2′-endo conformation. The angle of the N-gly-cosidic bond is $-112°$ which puts the base in an anticonformation relative to the ribose.[34] The 2′-hydroxyl group is hydrogen bonded to the main chain car-bonyl of Val29. Both the 2′- and 3′-hydroxyl groups of the ribose are exposed to the solvent. Thus, it is not surprising that the ribose can be labeled with fluorescent reporter groups such as the N-methylanthraniloyl = mant group and that mant-labeled nucleotides such as mant-GDP bind to p21 with similar affinity as unmodified GDP and are useful derivatives for the study of biochemical interactions.[5,35]

Each of the eight phosphate oxygens of GppNHp, except the pro-R oxygen of the α-phosphate and one γ-phosphate oxygen, is involved in either hydrogen bonds or metal ion coordination. Besides the strong interactions indicated in Figure 2, there are many more weak interactions. It is worth mentioning that all of the main chain nitrogens of residues 13 to 18 point toward the phosphate groups, thereby creating a positively polarized, local electrostatic field. The fact that only the pro-R oxygen and one γ-phosphate oxygen are devoid of electrostatic interactions nicely explains the finding that for p21 only GTPγS and the R_p analog of GTPαS are good analogs of GTP, but not the (S_p) isomer of GTPαS and both diastereomers of GTPβS.

There exists a hydrogen bond between the NH of Gly13 and the atom bridging the β- and γ-phosphate, an NH in the case of GppNHp. One would expect this hydrogen bond to be stronger with a bridging oxygen atom as in GTP, so that it seems likely that this interaction is responsible for the 30-fold lower affinity of GppNHp and for the 100-fold lower affinity of Gpp(CH$_2$)p as compared to GTP.[20,31] It could also be responsible for the unusual ^{15}N chemical shift of this residue as determined by NMR.[36]

E. SPECIFICITY OF NUCLEOTIDE BINDING

It has been shown by numerous investigators on many different GNB proteins that all of them interact very specifically with guanine nucleotides. p21 does not tolerate many substitutions on the guanine base, and it does not even look at adenine nucleotides. This is in contrast to many other phosphoryl transfer proteins like myosin ATPase, in which the phosphate-binding region is more important than the base-binding region. The affinity of ATP for p21 has been estimated to be 10^4 M^{-1}, about six orders of magnitude smaller than for GTP (R. S. Goody, unpublished observation). Considering the high specificity for the latter purine base, it is surprising that GMP is only weakly bound.[5] This seems to indicate that the presence of a second phosphate of the nucleotide induces a structural rearrangement, which results in the high specificity of the protein for GDP/GTP. The drastic increase of the thermal stability of p21 when going from the p21-GMP state to the p21-GDP state supports this notion.[5]

On the basis of the three-dimensional structure analysis it is not entirely clear why the GNB proteins are so specific for guanine nucleotides (having at least two phosphates) and do not bind adenine nucleotides efficiently. Thus, when the 2-amino group on GDP is removed (to obtain IDP) the affinity drops by a factor of only 30 to 50 which can be explained by the missing interaction between Asp119 and N2. Similarly, when Asp119 is mutated to Ala, the affinity between p21 and GDP drops only by a factor of 20, whereas the affinity to IDP is not altered.[37] Thus, we can assume that the contribution of the 2-amino group on guanine is not responsible for the six orders of magnitude drop in affinity when exchanging GTP with ATP. The only difference left between ATP and GTP is then the 6-amino group of ATP, which is a keto group on GTP. We have shown that h^{6}GDP, which has no keto group on guanine, binds with a 100-fold lower affinity than GDP as one would expect for the loss of a strong hydrogen bond.[38] Thus, it seems that the inability of ADP to bind to the nucleotide-binding site of GNB proteins stems from a steric or another incompatibility of the 6-amino group with the nucleotide-binding site of p21.

F. THE Mg^{2+}-BINDING SITE

It is generally believed that phosphotransferases require at least one divalent cation complexed directly to phosphoryl group oxygens for catalytic activity. One can postulate various possible catalytic functions of the Mg^{2+} ion, such as shielding of the negative charge of the attacked γ-phosphate, increasing the acid

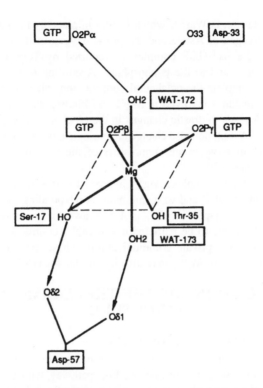

FIGURE 3. Schematic drawing of the Mg^{2+}-binding site showing the ligands of the first coordination sphere of the metal ion and some of the interactions of these ligands. (From Pai et al., *EMBO J.*, 9, 2351, 1990. With permission.)

strength of the leaving group (β-phosphate) or activation of the nucleophile.[39] Mg^{2+} could also be involved in the stabilization of the transition state of the reaction. The precise role of the metal ion for the GNB proteins has not been clarified and may, indeed, be different for different members of the protein family. Nevertheless, the GTPase activity of GNB proteins has been shown to be absolutely dependent on the presence of certain divalent ions.

In the three-dimensional structure of the p21-GppNHp complex Mg^{2+} is coordinated to one oxygen of each the β- and γ-phosphates, as well as to the side chain hydroxyl groups of Ser17 and Thr35, both of which are highly conserved in all GNB proteins (Figure 3, Plate 1A). In contrast to a recent report by Milburn et al.,[19] we find that amino acid Asp57, which is strictly conserved as part of the DXXG motif in GNB proteins and has been implicated before in metal ion binding, adopts a completely rigid conformation. The high resolution structure of p21 in the triphosphate conformation shows that Asp57 is not located in the first coordination sphere of Mg^{2+}. Instead, it is hydrogen bonded to a water molecule, Wat173, which in turn is liganded to the magnesium ion as shown in Figure 3. Asp57 further binds to the side chain of Ser17. The sixth ligand of the metal ion is Wat72, which is held in place by the interaction with the main chain oxygen of Asp33 and the pro-R oxygen of the α-phosphate. In the structure of the p21-GDP complex as reported by Tong et al.,[78] Mg^{2+} coordinates to Ser17 and to the β-phosphate. According to their results, four water molecules complete the octahedral coordination sphere of the metal ion. The mutation of residues involved in Mg^{2+} ion binding such as T35A, S17A, S17N, and D57A leads to drastic changes in the biochemical properties of p21. The S17N and S17A mutation also change the biological behavior of p21, because the resulting mutant proteins are suppressors of the transforming capacity of oncogene p21 mutants.[27,40]

In the EF-Tu-GDP complex, Mg^{2+} appears to be coordinated to the β-phosphate oxygen of GDP and to a threonine corresponding to Ser17. Asp80 which corresponds to Asp57 in p21, is 4.5 Å away from the metal ion and analogous to p21, seems to be participating in Mg^{2+} coordination via a water ligand also (J. Nyborg, private communication). These data suggest that the details of complexation of Mg^{2+} may also be similar for GNB proteins.

IV. IMPLICATIONS OF THE THREE-DIMENSIONAL STRUCTURE

A. MECHANISM OF GTP HYDROLYSIS

Feuerstein et al.[41] have shown that the mechanism of the GTPase reaction of p21 follows an associative in-line reaction pathway with inversion of the configuration at the γ-phosphate as shown schematically in Figure 4. By using a photolabile GTP-precursor nucleotide, caged GTP, Schlichting et al.[20,32] have found that p21 is competent to hydrolyze GTP within the crystal lattice. The rate of hydrolysis is basically unaltered as compared to hydrolysis in solution.[24] Thus, it seems reasonable to assume that there is a water molecule in the crystal structure of the p21-GppNHp complex, which would be in a position to attack the γ-phosphate. Since direct in-line displacements at phosphorus are known to proceed via a pentacoordinate transition state in which the nucleophile and the leaving group occupy the apical positions of the trigonal bipyramid (Figure 4; see also Knowles[39]), the attacking water molecule must be located at the opposite side of the leaving group, which is the β-phosphate. We could, indeed, identify such a water molecule (Wat175) as the possible attacking nucleophile. As expected, it is bound directly opposite to the β-γ bond.

In the electron density map of the GppNHp complex, we observed the highest temperature factors for amino acids 61 to 65 in loop L4, which indicates that they are highly mobile or disordered. In fact, we can identify alternative positions for the amino acids Gln61 (see Plate 1B), Glu62, Glu63, and Tyr64. In one of these conformations of Gln61, its side chain carbonyl group is close enough to Wat175 to make a hydrogen bond. Figure 5 shows a schematic drawing of this situation where Gln61, assisted by the carbonyl group of Thr35, could activate Wat175. In the same conformation, the amino group is proximal to the carboxylate

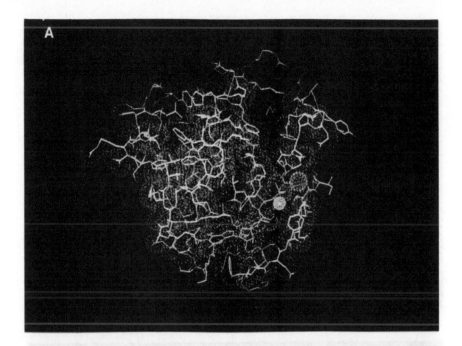

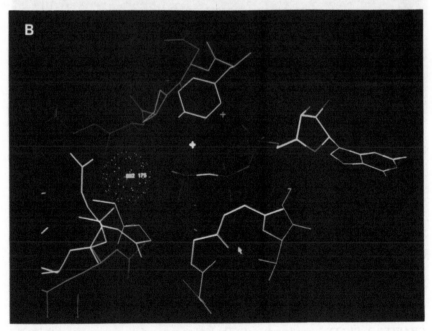

CHAPTER 2. PLATE 1. Various views of the three-dimensional structure. (A) Van der Waals representation of the protein containing bound GppNHp. The nucleotide is located in a cleft near the protein surface. The blue dot near γ-phosphate is the catalytic water molecule and the yellow dot the Mg^{2+} ion. (B) The active center of p21 shows the GppNHp molecule surrounded by the loops L1 (magenta), L2 (orange), and L4 with two alternative conformations of Gln61, normal position (purple), and as a general base catalyst (green) for water 175, which is indicated by the Van der Waals radius.

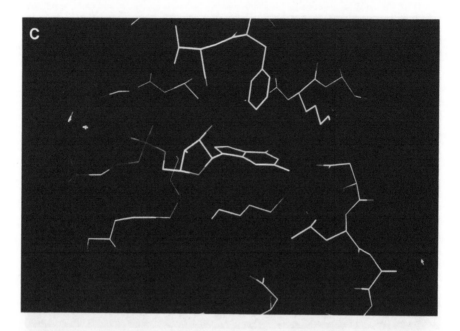

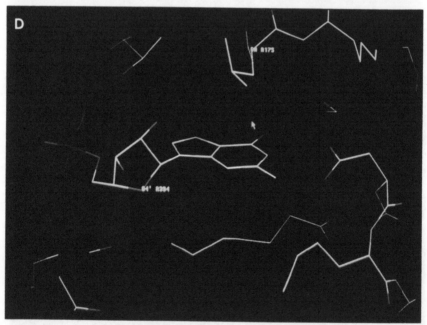

CHAPTER 2. PLATE 1 (continued). (C) The binding region of the guanine base with Phe28, Asp119, and Lys117 with (D) the same region in the three-dimensional structure of EF-Tu·GDP with Leu175 in place of Phe28. (The coordinates for EF-Tu·GDP were kindly provided to us by the Aarhus group.)

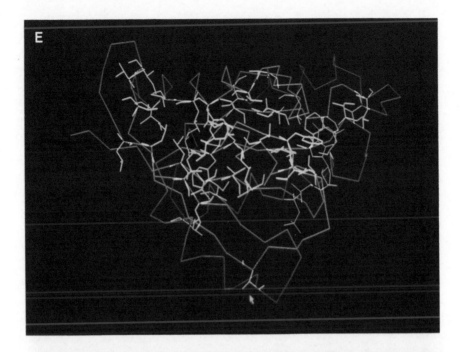

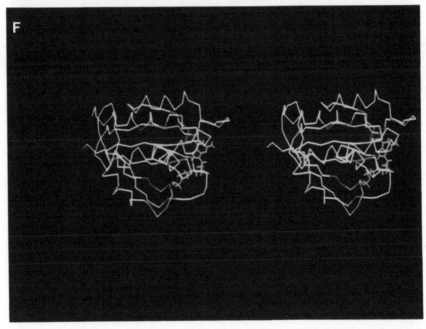

CHAPTER 2. PLATE 1 (continued). (E) C_α carbon (purple) plot of p21 with GppNHp (blue) showing the unequal distribution of hydrophobic residues (green). (F) Color stereo view of the p21 ribbonstructure, with colors indicating the variability of residues in small GNB proteins. Red = conserved (as listed in the sequence comparison of Table 1 of Valencia et al., *Biochemistry*, 30, 4637, 1991); yellow = intermediate; blue = most variable. (The photo is a kind gift of A. Valencia and C. Sander, EMBL.)

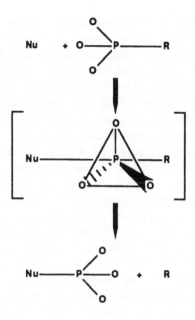

FIGURE 4. Scheme for the in-line associative mechanism for phosphoryl transfer with an S_N2-type mechanism. In such a mechanism the terminal phosphate undergoes an inversion of configuration, which has been proved for p21 by Feuerstein et al.[41]

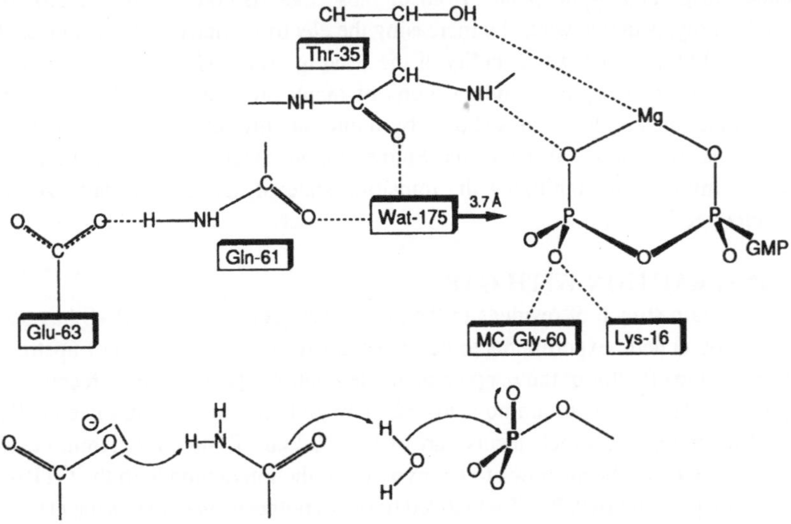

FIGURE 5. Schematic drawing of the active conformation, which is supposed to be competent for GTP hydrolysis, and the nucleophilic water molecules. Also shown is a possible mechanism of proton transfer in the hydrolysis reaction. (Adapted from Pai et al., *EMBO J.*, 9, 2351, 1990.)

side chain of Glu63, which in turn could assist in establishing the proper activating orientation for Gln61 and/or increase its proton withdrawing potential. The finding that a Glu63→Lys mutation partially activates the transforming potential of p21 seems to implicate this residue, at least indirectly, in GTP hydrolysis.[42] We have also found recently that a Glu63→His mutation severely affects p21-GAP interaction (P. Gideon et al., unpublished observation).

In addition to the structural arguments presented here, there are several other reasons to believe that Gln61 is involved in GTP hydrolysis. It is highly conserved in the small GNB proteins, except for the proteins encoded by the *rap* genes,[43] also called Krev or *smg* p21.[44,45] In the *rap*-p21 proteins Thr61 is exchanged for Gln61 and their GTPase activity is reduced as compared to *ras*-p21. Mutation of Thr61 to glutamine causes an increase of the GTPase rate up to the value found for *ras*-p21. The mutant protein is also partially responsive to GAP while the wild-type *rap*-p21 protein is not.[46] Substitution of Gln61 in *ras*-p21 by other amino acids *in vitro* reduces the GTPase rate constant, and the mutant proteins can no longer be activated by GAP.[47,29] Gln61 is conserved in the great majority of the GNB proteins. In EF-Tu, the corresponding amino acid is His84, and the *in vitro* GTPase rate has been found to be smaller than the GTPase of cellular p21, but similar to the one of p21(Q61H).[48,17] Mutation of His84 to glycine in the nucleotide-binding domain of EF-Tu further reduces the GTPase activity by a factor of more than 10,[49] which could indicate that histidine in EF-Tu has a function similar to glutamine in most other GNB proteins.

Figure 5 shows that Mg^{2+} and the ϵ-amino group of Lys16 are both coordinated to β- and γ-phosphate. Both of these ligands could therefore promote GTP hydrolysis in two ways: by increasing the electrophilicity of the γ-phosphate group and by increasing the acidity of the leaving group, GDP. A lysine residue is found in the phosphate-binding loops of many proteins which bind adenine or guanine nucleotides. It has shown by numerous investigators that this lysine is essential for binding or catalysis. Studies on adenylate kinase indicate that it may be involved in stabilizing the transition state of the reaction catalyzed by this enzyme.[50]

B. INTERACTION WITH GAP

A certain stretch of residues in the primary sequence of yeast RAS has been defined by genetic experiments to be important for the transduction of upstream signals to the effector of the *ras* pathway, adenylate cyclase.[51] The corresponding region in p21 consists of amino acids 32 to 40 and has consequently been called the effector region, which forms part of loop L2 and β-strand 2. Amino acids 32 to 40 have also been shown to be involved in the interaction with the GTPase-activating protein GAP,[26,52-54] which led to the hypothesis that GAP is the effector of p21 in mammalian cells. Residues Asp33, Thr35, Ile36, Asp38, and Tyr40 have been shown by mutational analysis to be involved in the GAP-stimulated GTPase activity. As mentioned above, Thr35 is involved in metal ion and γ-phosphate binding. This could mean that the effects observed on mutating this residue are only indirectly due to its involvement in p21-GAP interaction. The hydrophobic side chain of Ile36 is exposed to the aqueous solvent in the p21-GppNHp complex, which is usually unfavorable for protein stability and could mean that it is recognized by GAP or another effector protein. Plate 1E shows that it is the only hydrophobic residue in that part of the protein which contains the active site. Interestingly, it can indeed be mutated to Leu, but not to Ala, which has a much smaller side chain.[26] The I36L mutation is also temperature sensitive for cellular transformation.[82]

On the basis of structural arguments it would appear that L4 is also involved in the interaction with GAP because it is in close neighborhood to loop L2, and both comprise the active site. It is reasonable to assume that an interaction between p21 and such a large protein (115-kDa molecular mass) involves more than just loop L2. It has, indeed, been shown that the nature of the amino acid in position 61 in loop L4 is critical for tight binding, and that p21(Q61L) is the mutant with highest affinity for GAP described thus far.[29] This suggests that aside from loop L2, residues in loop L4 are also involved in the interaction. Recently, we have found that Glu63 is another residue in L4 critical for p21-GAP interaction (P. Gideon et al., manuscript submitted). Certain residues of loop L1 also form part of the active site and mutations of these residues, such as Gly12, affect p21-GAP interaction as well.

How can GAP stimulate the GTP hydrolysis of p21? At present two alternatives are being discussed for the mechanism by which GAP stimulates the GTPase of p21 by a factor of 10^5.[27] In one it is proposed that the mechanism of the GTPase reaction is the same in the presence or absence of GAP. The function of GAP would then be to catalyze and/or stabilize the structural rearrangement that renders p21 competent for hydrolysis. This structural rearrangement might be very slow in the absence of GAP. According to the other hypothesis, GAP supplies one or more residues which directly participate in the phosphoryl transfer reaction. In support of the latter idea it is argued that the α-subunits of heterotrimeric G proteins, which do not have a GAP-like molecule in their reaction pathway, hydrolyze GTP much faster than the small GNB proteins.[1,2] The latter proteins all seem to need a specific GAP-like molecule for a fast reaction. GTPase-activating proteins have already been found, except for *ras*-p21, for the *rho*,[55] *rap*,[56,57] and *ypt*[58] gene products.

It has been postulated recently by Neal et al.[59] that the rate-limiting step of the GTPase reaction of N-*ras* p21 is a conformational change of the p21·GTP complex preceding GTP hydrolysis.

$$p21 + GTP \rightarrow p21 \cdot GTP \rightarrow p21 \cdot GTP^* \rightarrow p21 \cdot GDP \cdot P_i \rightarrow p21 \cdot GDP$$

This change was observed using fluorescent GTP analogs, mant-GTP or mant-GppNHp, which have been used to measure association and dissociation kinetics.[5,35] Since H-*ras* p21 appears to show the same conformational change (H. Rensland et al., manuscript submitted) and since the crystallization of the p21-GppNHp complex takes several days, the protein in the crystal is assumed to have undergone this conformational change. Since GppNHp cannot be hydrolyzed by p21, the crystal structure of the p21-GppNHp complex is supposed to represent an equilibrium mixture of conformations, one of which is competent for hydrolysis.

If there is, indeed, a conformational change preceding GTP hydrolysis and if this reaction step is rate-limiting as proposed by the authors,[59] it would seem probable that GAP catalyzes this step to speed up the overall reaction. However, recent studies with mant-nucleotides from our laboratory (H. Rensland et al., manuscript submitted) and other studies with Trp mutants of p21 (P. Chardin, in press) have demonstrated that the spectroscopic signal observed in the non-stimulated and GAP-stimulated GTPase reaction is due to a step that is either the cleavage itself or a subsequent step such as product release. Therefore, if a conformational change exists, it is not rate-liming and cannot be observed by those methods.

Based on the structure of the p21-GppNHp complex, it cannot be excluded that GAP would supply catalytic residues to the active site. L2 and L4 are exposed to the solvent and this region is, therefore, easily accessible for another protein. One of the oxygens on the γ-phosphate is hydrogen-bonded to Tyr40 of a neighboring molecule in our structure of the p21-GppNHp complex, which means that around the γ-phosphate there is space enough for such a bulky residue. Therefore, it is conceivable that in solution a similarly contacting residue would be supplied by GAP.

It is interesting to note that in the structure of the p21-Gpp(CH$_2$)p complex of Milburn et al.,[19] which has a different crystallographic space group, a γ-phosphate oxygen of Gpp(CH$_2$)p is contacted by Tyr32 from the same molecule to which it is bound. Obviously, the active site of p21 offers enough space to accept at least one large side chain from another protein like GAP. Possible roles for a side chain of GAP in the active site could be to neutralize the charge on the terminal phosphate, to stabilize the transition state, and/or to protonate the leaving phosphate group.

C. MUTANT p21 PROTEINS

The three-dimensional structures of the p21 mutant proteins G12V, G12R, G12P, Q61H, and D38E complexed to GTP analogs have been determined.[60] The GDP-bound structure has also been determined for p21(G12V),[78] and indirectly by Schlichting et al.[20] All these mutants are oncogenic and have been found in human tumors except for D38E. p21 (D38E) is an effector mutant which is no longer activated by GAP but yet is not transforming. It was shown that the overall structure of all these proteins is very similar to that of wild-type p21. The rms differences between equivalent C$_\alpha$-positions, leaving out the mobile region of loop L4, are 0.41 Å for G12V and 0.24 Å for the other mutant proteins. Significant effects of the mutations on the three-dimensional structure are confined to the active site around the γ-phosphate. In the case of the Gln61 mutants, in which case the structure is most highly conserved, it seems that it is the absence of the catalytically important glutamine and not any change in structure, which is responsible for the reduced GTPase. In the G12R mutation, the long side chain of arginine and the nucleophilic water and glutamine 61 are pushed out of the way. In addition, the hydrogen bond between the γ-phosphate and the guanidinium group of arginine stabilizes this rearrangement of the structure. In G12V the bifurcated aliphatic side chain of valine is disturbing the catalytic configuration of Gly60 and Gln61 which is shown in Figure 5. Gly60 and Gln61 change their orientation in such a way that the GTPase reaction is inhibited. Since in all these mutants the positioning of the catalytic groups and the nucleophile of the GTPase reaction are affected, it is not surprising that GAP cannot stimulate the GTPase reaction of these proteins. This could support the notion that the mechanism of GTP hydrolysis is the same for the GAP-stimulated and the nonstimulated cleavage reaction.

For G12P it has been found that this is the only mutation of Gly12 which is not transforming.[61] The unstimulated GTPase of G12P has been found to be increased threefold as compared to wild-type p21 but is not stimulated by GAP.[54] In the three-dimensional structure of the p21(G12P)-GppNHp complex we find that the side chain of Pro12 does not interfere with the nucleophilic water or with the side chain of Gln61 contacting it.[62] It seems plausible, however, that the relatively bulky side chain inhibits the contact with GAP and would thus explain why GAP could not stimulate the GTP hydrolysis of p21(G12P).

D. THE CONFORMATIONAL CHANGE

The most important function of GNB proteins is to switch from the GTP to the GDP-bound conformation. This conformational change makes the protein unable to react with the effector molecule and, therefore, terminates the interaction. In the case of p21, the affinity towards GAP drops by a factor of 100 when the protein undergoes the structural change upon GTP hydrolysis.[29] From studies on G proteins and EF-Tu it has also been known for a long time that the conformational change does not have a drastic effect on the overall structure, but is rather confined to certain areas of the protein as evidenced by different sensitivities to protease digestion.

Since the structures of the p21-diphosphate and -triphosphate complexes (see Table 1) have been determined, the elements of the structural change were elucidated. The reports by Milburn et al.[19] and Schlichting et al.[20] do not completely agree on the details which might be due to the different approaches used by the two groups. It appears, however, that the structural rearrangement of p21 is confined to the two loops L2 and L4 and, according to Milburn et al.[19] also to helix $\alpha 2$, which is seen to change its relative orientation. With the structure of the triphosphate complex known in such great detail, it becomes clear how the conformational change is triggered (Figure 6). Loops L2 and L4 are connected to the γ-phosphate via Thr35 and Gly60, respectively. Obviously, this interaction is interrupted when the terminal phosphate is released during GTP hydrolysis.

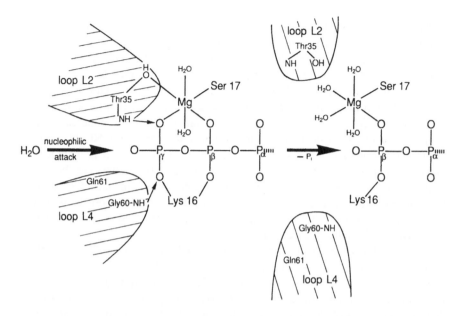

FIGURE 6. Conformational change triggered by the hydrolysis of γ-phosphate. The removal of γ-phosphate releases loop L2 and loop L4 which are bound to te terminal phosphate via Thr35 and Gly60, both of which are totally conserved in all GNB proteins. After hydrolysis the loops can adopt a new conformation. (From Whittinghofer and Pai, *TIBS*, 16, 382, 1991. With permission.)

As a result the loops can reorient and establish another conformation, which is more stable in the p21-GDP complex. Thus, the conformational change turns out to be a very straightforward reaction immediately related to the presence of the third phosphate. Since the elements involved in the interaction with γ-phosphate have been conserved in all GNB proteins, it is very likely that this reaction, which is central to the function of the GNB proteins, is triggered in the same way in all of them.

E. INTERACTION WITH EXCHANGE FACTORS

So far as it has been determined correctly, GNB proteins seem to have very high affinities toward nucleotides in the order of 10^{10} to 10^{12} M^{-1} (4°C, in the presence of Mg^{2+}) and correspondingly have very low dissociation rate constants for the nucleotide. For example, for p21-GDP one finds a half-life for the complex in the order of hours at 37°C. Therefore p21, like other GNB proteins, is dependent on nucleotide exchange factors in order to accelerate reloading of the protein with GTP after hydrolysis has taken place. From mutational analysis it has become clear that mutations affecting residues involved in the Mg^{2+}-binding site, such as S17N or S17A, or mutations around the guanine base-binding region, such as F28L, D119N, and N116I, drastically accelerate the nucleotide dissociation rate. According to these results, one would expect that an exchange factor would modify the interaction between nucleotide and protein at similar sites in order to increase the dissociation rate.

Exchange factors for p21 and for the yeast RAS gene products have been described both genetically and biochemically.[63-67] In the most thorough investigation on the interaction between a cytoplasmic exchange factor and mutants of p21 published so far, it was demonstrated that mutations on p21, such as Q61K, Q61P, D38A, and D119E, decrease the stimulation of guanine nucleotide release by the factor.[65] In a similar study dealing with p21 and the catalytic domain of SDC25 it was found that mutations in loop L2 and L4 reduce the efficiency of the p21-exchange factor interaction (A. Parmeggiani et al., manuscript submitted). It thus seems that the exchange factor is in contact with similar residues as GAP and might function by modifying the metal ion-binding site.

F. POSSIBLE OTHER INTERACTIONS OF p21, *let-60*

The debate is still ongoing whether GAP is the actual downstream target molecule of p21 or, as the properties of the yeast IRA gene products suggest, simply represents the negative regulator of p21, which keeps the concentration of active p21-GTP low. Recently the *let-60* gene of *C. elegans* has been shown to be the homolog of the mammalian *ras* gene product.[68,69] *let-60* is an essential gene and plays a role in the development of the vulva by specifying the fates of vulval precursor cells. This is a further indication that the *ras* gene product is mediating specific developmental processes.

Several mutations of the *let-60* gene were described[69] and Figure 7 shows where they occur in the three-dimensional structure of the mammalian p21. Dominant negative mutations, which lead to a multivulva phenotype, have Gly13 mutated to Glu. Although no mutation of Gly13 has yet been analyzed structurally, it is expected that it should lead to a disruption of the active site structure because Gly13 has unusual conformational main chain angles that are not allowed for other amino acids.[17] Biochemically, the mutant protein should be analogous to the oncogenic Gly12 and Gly13 mutant proteins which have a decreased GTPase activity which is not further stimulated by GAP. The recessive loss of function mutations at amino acids 37, 66, 75, and 84 are in positions that could potentially be recognized by either a GAP protein or a nucleotide exchange factor. However, mutations of residues 131, 132, and 136 on one side of helix α4 are far away from the active site and point out into the solvent. It seems that these residues specify a novel target region of p21 and similar small GNB proteins that is recognized by an unknown kind of interacting molecule. Such a molecule might be situated in the membrane, because based on the distribution of hydrophobic side chains (Plate 1E) it seems possible that the upper part of the molecule as shown in Figure 7 might interact with the membrane and the lower part may point toward the cytoplasm. An effector molecule acting on this part of the p21 structure does not seem to be plausible, because it is not modulated by the conformational change, which according to the results cited above, affects only loops L2 and L4 and helix α2.

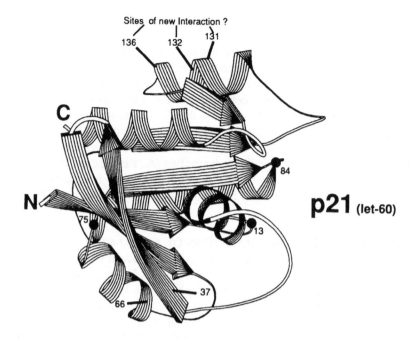

FIGURE 7. Mutations of the *let-60* gene of *C. elegans*, the *ras* gene homolog,[68,69] built into the model of *ras*[H]-p21.

V. A GENERAL G DOMAIN STRUCTURE

It was realized very early by Halliday[33] that there are conserved elements in the primary sequence of *ras*-p21 and EF-Tu which led her to suggest that these proteins might be structurally related. In the following two groups proposed structural models of p21, whose topology is identical to the structure of EF-Tu.[21,70] We now know that these model buildings were correct in that the chain tracing and the order of secondary structure elements is the same in both proteins. It is difficult to align the two sequences outside of the conserved elements and, therefore, it is difficult to give a value for the overall similarity of the primary sequence. If the alignment is based on homologous secondary structure elements, it has to be concluded that the similarity between the primary structures is less than 17%. As an example of structural similarity, Plate 1D shows the guanine-binding region of EF-Tu where Leu175 is situated above the aromatic base, instead of Phe28 in p21 (Plate 1C). Interestingly, it has been shown that the F28L mutation in p21 decreases the binding affinity of GDP such that it is similar to the EF-Tu-GDP affinity (J. Reinstein et al., in press).

Since p21 and EF-Tu with only limited similarity in the primary sequence, have very similar structures in their G domain we can predict that all the GNB proteins with the conserved sequence elements mentioned above have the same overall tertiary structure as the *ras*-p21 protein. The small GNB proteins, which have primary sequence homologies to p21 in the order of 30 to 80% also consist of only the G domain. The α-subunits of heterotrimeric G proteins, which have a molecular mass of between 45 and 50 kDa, elongation factor Tu (and its eukaryotic homolog EF1) with a molecular mass around 45 kDa, and the elongation factor G (or the eukaryotic homolog EF2) with a molecular mass around 80 kDa must have one or more additional domains which are needed for the particular function of this protein.

Sequence alignments of $G_α$ proteins have been proposed by a number of different authors. They all have variable insertions or deletions at different positions of the primary sequence in order to get a maximum homology. With the knowledge of the three-dimensional structure of p21 and the assumption that the overall structures of the G domains should be similar, we can modify these predictions to take into account structural considerations. Thus, it is now clear that there ought to be a Thr residue in all GNB proteins which is analogous to Thr35 of p21 and is of prime importance for the function. This residue, as shown above, performs at least two essential interactions — it coordinates to Mg^{2+} and to γ-phosphate, but only in the active triphosphate complex — and is thus one of the most important elements of the molecular switch. Considering that all GNB proteins have a corresponding Thr residue, we can propose a structural model for $G_α$ proteins based on the structure of p21 which has additional amino acids at the N terminus, an extra domain between helix α1 and β2, and additional amino acids at the C terminus. The length of the extra domain can be determined by the postulate that a threonine has to be situated at the beginning of β2. This threonine is most likely Thr204 in $G_αs$ as was proposed before.[1,2] The resulting alignment of $G_αs$ with the structure of p21 is shown schematically in Figure 8. It may turn out that all the extra sequences of $G_αs$ located at the N terminus, at the C terminus, and between α1 and β2, which are located on the same side of the molecule, together comprise one single extra domain. Such a domain in the general location of the effector region of p21 would thus represent a complete effector domain by itself interacting with a target molecule. It has been shown that a conserved arginine is important for the function of $G_α$-subunits. Incidentally, if Thr204 corresponds to Thr35 of p21, the conserved Arg201 of $G_αs$ would correspond to Tyr32 of p21 and would be positioned in the active site of the molecule. Model building on p21 shows that if Tyr32 is replaced by Arg, it can contact the γ-phosphate without any rearrangement in the structure (F. Wittinghofer, unpublished observations).

EF-Tu is a large GNB protein with 393 amino acids. The three-dimensional structure determination of the *Escherichia coli* trypsinized EF-Tu has shown that it contains two extra domains in addition to the G domain. These are both mainly comprised of β-sheets.[71] The sequence elements corresponding to the effector region of p21, amino acids 45 to 58, were removed by trypsin digestion from EF-Tu to facilitate crystallization.[72] This missing region in the structure was indicated by a dashed line as shown in Figure 9A.[71] However, on the basis of

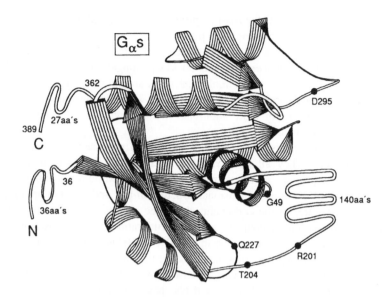

FIGURE 8. Schematic model of the GNB domain of G_αs based on considerations outlined in the text.

the structure comparison between p21 and EF-Tu, it seems reasonable to postulate that these amino acids form a loop connecting helix $\alpha1$ with β-strand $\beta2$ and that this loop is rather located as indicated in Figure 9B. In fact, biochemical cross-linking experiments have proven that this loop must be situated in an analogous position to p21, namely close to the nucleotide.[73] Therefore, on the basis of the homology and of these cross-linking experiments, the structure as shown in Figure 9B would be favored.

The fact that the overall structure of the G domain of GNB proteins is similar does not mean, however, that the details of the structure are similar too. Several studies dealing with the specificity of binding of thiophosphate analogs of GTP and GDP have appeared.[74] They showed that all of the proteins investigated have a low affinity for GTPβS, which is not surprising on the basis of the three-dimensional structure results. There are characteristic differences in the affinities for the R_p and S_p isomers of GTPβS. This has, in fact, been used to selectively activate calcium signaling in mast cells without affecting another G protein, which mediates degranulation.[75] Similar experiments have shown that the active site is somewhat different between transducin on one side and p21 and EF-Tu on the other.[14] It has also been reported that G_α proteins can be induced by GDP together with AlF_4^- or BeF_3 to adopt a GTP-like conformation.[76,77] It is thought that AlF_4^- and BeF_3 on those complexes occupy the terminal phosphate position. Similar experiments with EF-Tu and p21 show that these proteins behave differently.[79] (F. Wittinghofer et al., unpublished observations).

Since (to my knowledge) no other GNB protein has yet been crystallized such that it would be suitable for X-ray analysis, we do not know the structural

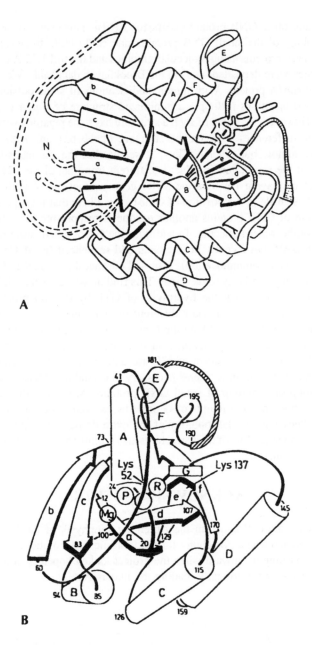

FIGURE 9. Location of the effector region in the three-dimensional structure of EF-Tu. This region has been removed from the protein by trypsin digestion to facilitate crystallization. (A) As taken from a paper from the Aarhus group[81] and (B) from a model which takes into account cross-linking studies between ribose oxidized GDP and the protein.[73] On the basis of the latter data and the homology to p21 the effector loop is most likely situated as shown in part B. (Part A from Wooley, P. and Clark, B. F. C., *Biotechnology*, 7, 913, 1989. Part B from Peter, M. E., Wittmann-Liebold, B., and Sprinzl, M., *Biochemistry*, 27, 9132, 1988.)

details of any other GNB protein except for EF-Tu. However, considering the high homology of the small GNB proteins one could build three-dimensional models of these proteins using molecular graphics and the full 1.35 Å coordinates of p21, which were deposited by us in the Brookhaven data bank. We have done this for the *rap1A* protein. With minor modifications of some outside residues and with two insertions in loops, a minimized structure of *rap1A* could easily be built.[46] However, the value of such a structure has to be regarded with caution, because the differences between the structure of *ras*-p21 and *rap1A*-p21 are presumably minor, but crucial. Thus, it seems that the mechanism of GTP hydrolysis of the *rap1A* gene product is different from that of *ras*-p21. All *rap*-related proteins have a Thr, and the RSR1 gene product (the yeast homolog of *rap*) even has an Ile in the position of Gln61. It is likely that for representatives of this group of GNB proteins another residue is responsible for activation of the nucleophilic water molecule. In addition, the GTPase-activating protein for *rap1A (rap*-GAP) has recently been cloned and sequenced (F. McCormick and P. Polakis, private communication). It turned out that the sequence of *rap*-GAP does not have any similarity to *ras*-GAP. This finding was unexpected, but seems to confirm the notion that the hydrolysis of GTP by the *rap* protein and its activation by *rap*-GAP occur via a different mechanism.

Building a model of *rap1A* on the basis of the *ras*-p21 structure would not be expected to reveal these differences in structure or mechanism. For a detailed understanding of the mechanism of GTPase and other questions related to the mechanism of action it, thus, seems necessary to crystallize at least one member of each group or subgroup of small GNB proteins. Modelling the large GNB proteins, such as the G_α-subunits of G proteins, is even more speculative since nothing is known about the topology of the extra domain(s) of these proteins. A complete understanding of their structures also requires at least one independent structure determination.

ACKNOWLEDGMENTS

I thank all my colleagues in the Abteilung Biophysik of the Max-Planck-Institut für Medizinische Forschung for the fantastic collaboration. I thank Ken Holmes for pushing us onward when the project looked hopeless. Many thanks to Lisa Wiesmüller for critically reading the manuscript.

REFERENCES

1. **Bourne, H. R., Sanders, D. A., and McCormick, F.,** The GTPase superfamily: a conserved switch for diverse cell functions, *Nature,* 348, 125, 1990.
2. **Bourne, H. R., Sanders, D. A., and McCormick, F.,** The GTPase superfamily: conserved structure and molecular mechanism, *Nature,* 349, 117, 1991.
3. **Gilman, A. G.,** G proteins: transducers of receptor-generated signals, *Annu. Rev. Biochem.,* 56, 615, 1987.
4. **Hall, A.,** Ras and ras-related guanine nucleotide binding proteins, in *G-proteins as Mediators of Cellular Signalling Processes,* Houslay, M. D. and Milligan, G. Eds., John Wiley & Sons, 1990, 173.
5. **John, J., Sohmen, R., Feuerstein, J., Linke, R., Wittinghofer, A., and Goody, R. S.,** Kinetics of interaction of nucleotides with nucleotide-free H-ras p21, *Biochemistry,* 29, 6058, 1990.
6. **Rösch, P., Wittinghofer, A., Tucker, J., Sczakiel, G., Leberman, R., and Schlichting, I.,** ^{31}P-NMR spectra of the Ha-ras p21·nucleotide complexes, *Biochem. Biophys. Res. Commun.,* 135, 549, 1985.
7. **Goody, R. S., Frech, M., and Wittinghofer, A.,** Affinity of guanine nucleotide binding proteins for their ligands: facts and artifacts, *TIBS,* 16, 327, 1991.
8. **Miller, D. L. and Weissbach, H.,** Studies on the purification of Factor Tu from *E. coli., Arch. Biochem. Biophys.,* 141, 26, 1970.
9. **Ferguson, K. M., Higashijima, T., Smigel, M. D., and Gilman, A. G.,** The influence of bound GDP on the kinetics of guanine nucleotide binding to proteins, *J. Biol. Chem.,* 261, 7393, 1986.
10. **Feuerstein, J., Goody, R. S., and Wittinghofer, A.,** Preparation and characterization of nucleotide-free and metal ion-free p21 "apoprotein", *J. Biol. Chem.,* 262, 8455, 1987.
11. **Poe, M., Scolnick, E. M., and Stein, R. B.,** Viral Harvey *ras* p21 expressed in *Escherichia coli* purified as a binary one-to-one complex with GDP, *J. Biol. Chem.,* 260, 3906, 1985.
12. **Hall, A. and Self, A. J.,** The effect of Mg^{2+} on the guanine nucleotide exchange rate of p21$^{\text{N-ras}}$, *J. Biol. Chem.,* 261, 10, 10963, 1986.
13. **Ivell, R., Sander, G., and Parmeggiani, A.,** Modulation by monovalent and divalent cations of the Guanosine-5'-triphosphatase activity dependent on elongation factor Tu, *Biochemistry,* 20, 6852, 1981.
14. **Tucker, J., Sczakiel, G., Feuerstein, J., John, J., Goody, R. S., and Wittinghofer, A.,** Expression of p21 proteins in *Escherichia coli* and stereochemistry of the nucleotide-binding site, *EMBO J.,* 5, 1351, 1986.
15. **Feuerstein, J., Kalbitzer, H. R., Goody, R. S., and Wittinghofer, A.,** Characterization of the metal ion-GDP complex at the active sites of transforming and nontransforming p21 proteins by observation of the ^{17}O-Mn superhyperfine coupling and by kinetic methods, *Eur. J. Biochem.,* 162, 49, 1987.
16. **Tong, L., Milburn, M. V., deVos, A. M., and Kim, S.-H.,** Structure of *ras* protein, *Science,* 245, 244, 1989.
17. **Pai, E. F., Kabsch, W., Krengel, U., Holmes, K. C., John, J., and Wittinghofer, A.,** Structure of the guanine-nucleotide-binding domain of the Ha-*ras* oncogene product p21 in the triphosphate conformation, *Nature,* 341, 209, 1989.
18. **Pai, E. F., Krengel, U., Petsko, G. A., Goody, R. S., Kabsch, W., and Wittinghofer, A.,** Refined crystal structure of the triphosphate conformation of H-ras p21 at 1.35 Å resolution: implications for the mechanism of GTP hydrolysis, *EMBO J.,* 9, 2351, 1990.
19. **Milburn, M. V., Tong, L., DeVos, A. M., Brünger, A., Yamizumi, Z., Nishimura, S., and Kim, S.-H.,** Molecular switch for signal transduction: structural differences between active and inactive forms of protooncogenic *ras* proteins, *Science,* 247, 939, 1990.

20. **Schlichting, I., Almo, S. C., Rapp, G., Wilson, K., Petratos, K., Lentfer, A., Wittinghofer, A., Kabsch, W., Pai, E. F., Petsko, G. A., and Goody, R. S.,** Time-resolved X-ray crystallographic study of the conformational change in Ha-ras p21 protein on GTP hydrolysis, *Nature,* 345, 309, 1990b.

21. **Jurnak, F.,** Structure of the GDP domain of EF-Tu and location of amino acids homologous to ras oncogene proteins, *Science,* 230, 32, 1985.

22. **LaCour, T. F. M., Nyborg, J., Thirup, S., and Clark, B. C. F.,** Structural details of the binding of guanosine diphosphate to elongation factor Tu from *E. coli* as studied by X-ray crystallography, *EMBO J.,* 4, 2385, 1985.

23. **Barbacid, M.,** *ras* genes, *Annu. Rev. Biochem.,* 56, 779, 1987.

24. **John, J., Schlichting, I., Schiltz, E., Rösch, P., and Wittinghofer, A.,** C-terminal truncation of p21^H preserves crucial kinetic and structural properties, *J. Biol. Chem.,* 264, 13086, 1989.

25. **Trahey, M. and McCormick, F.,** A cytoplasmic protein stimulates normal N-ras p21 GTPase, but does not affect oncogenic mutants, *Science,* 238, 542, 1987.

26. **McCormick, F.,** *ras* GTPase activating protein: signal transmitter and signal terminator, *Cell,* 56, 5, 1989.

27. **John, J.,** Structure-function relationships on the ras oncogene product p21, Ph.D. thesis, University of Heidelberg, 1990.

28. **Gibbs, J. B., Schaber, M. D., Allard, W. J., Sigal, I. S., and Scolnick, E. M.,** Purification of ras GTPase activating protein from bovine brain *Proc. Natl. Acad. Sci. U.S.A.,* 85, 5026, 1988.

29. **Vogel, U., Dixon, R. A. F., Schaber, M. D., Diehl, R. E., Marshall, M. S., Scolnick, E. M., Sigal, I., and Gibbs, J. B.,** Cloning of bovine GAP and its interaction with oncogenic ras p21, *Nature,* 335, 90, 1988.

30. **Jancarik, J., deVos, A., Kim, S.-H., Miura, K., Ohtsuka, E., Noguchi, S., and Nishimura, S.,** Crystallization of human c-H-ras oncogene products, *J. Mol. Biol.,* 200, 205, 1988.

31. **Scherer, A., John, J., Linke, R., Goody, R. S., Wittinghofer, A., Pai, E. F., and Holmes, K. C.,** Crystallization and preliminary X-ray analysis of the human c-H-ras oncogene product p21 complexed with GTP analogues, *J. Mol. Biol.,* 206, 257, 1989.

32. **Schlichting, I., Rapp, G., John, J., Wittinghofer, A., Pai, E. F., and Goody, R. S.,** Biochemical and crystallographic characterization of a complex of c-Ha-ras p21 and caged GTP with flash photolysis, *Proc. Natl. Acad. Sci. U.S.A.,* 86, 7687, 1989.

33. **Halliday, K. R.,** Regional homology in GTP-binding proto-oncogene products and elongation factors, *J. Cyclic Nucleotide Protein Phosphoryl. Res.,* 9, 435, 1983.

34. **Saenger, W.,** *Principles of Nucleic Acid Structure,* Springer Verlag, New York, 1984.

35. **John, J., Frech, M., Feuerstein, J., Goody, R. S., and Wittinghofer, A.,** in *Guanine-Nucleotide-Binding Proteins,* Bosch, L., Kraal, B., and Parmeggiani, A., Eds., Plenum Publishing, New York, 1989, 209.

36. **Redfield, A. G. and Papastavros, M. Z.,** NMR study of the phosphoryl binding loop in purine nucleotide proteins: evidence for a strong hydrogen bonding in human N-*Ras* p21, *Biochemistry,* 29, 3509, 1990.

37. **Sigal, I. S., Gibbs, J. B., D'Alonzo, J. S., Temeles, G. L., Wolanski, B. S., Socher, S. H., and Scolnick, E. M.,** Mutant *ras*-encoded proteins with altered nucleotide binding exert dominant biological effects, *Proc. Natl. Acad. Sci. U.S.A.,* 83, 952, 1986.

38. **Linke, R.,** Interaction of nucleotides with p21-ras oncogene product p21 as a contribution to the structure determination of the G domain, Diploma thesis, University of Heidelberg, 1988.

39. **Knowles, J.,** Enzyme-catalyzed phosphoryl transfer reactions, *Annu. Rev. Biochem.,* 49, 877, 1980.

40a. **Feig, L. A. and Cooper, G. M.,** Inhibition of NIH 3T3 cell proliferation by a mutant ras protein with preferential affinity for GDP, *Mol. Cell. Biol.,* 8, 3235, 1988.

40b. **Cai, H., Szeberenyi, J., and Cooper, G. M.**, Effect of a dominant inhibitory Ha-ras mutation on mitogenic transduction in NIH 3T3 cells, *Mol. Cell. Biol.*, 10, 5314, 1990.

41. **Feuerstein, J., Goody, R. S., and Webb, M. R.**, The mechanism of guanosine nucleotide hydrolysis by p21 c-Ha- *ras*, *J. Biol. Chem.*, 264, 6188, 1989.

42. **Fasano, O., Aldrich, T., Tamanoi, F., Taparowsky, E., Furth, M., and Wigler, M.**, Analysis of the transforming potential of the human H-ras gene by random mutagenesis, *Proc. Natl. Acad. Sci. U.S.A.*, 81, 4008, 1984.

43. **Pizon, V., Chardin, P., Lerosey, I., Olofsson, B., and Tavitian, A.**, Human cDNAs rap1 and rap2 homologous to the Drosophila gene Dras3 encode proteins closely related to ras in the "effector region", *Oncogene*, 3, 201, 1988.

44. **Kitayama, H., Sugiomoto, Y., Matsuzaki, T., Ikawa, Y., and Noda, M.**, A ras-related gene with transformation suppressor activity, *Cell*, 56, 77, 1989.

45. **Kawata, M., Matsui, Y., Kondo, J., Hishida, T., Teranishi, Y., and Takai, T.**, A novel small molecular weight GTP-binding protein with the same putative effector domain as the ras proteins in bovine brain membranes, *J. Biol. Chem.*, 263, 18,965, 1988.

46. **Frech, M., John, M., Pizon, P., Chardin, P., Tavitian, A., Clark, R., McCormick, F., and Wittinghofer, A.**, Inhibition of GTPase activating protein stimulation of ras-p21 GTPase by the Krev-1 gene product, *Science*, 249, 169, 1990.

47. **Der, C. J., Finkel, T., and Cooper, G. M.**, Biological and biochemical properties of human ras[H] genes mutated at codon 61, *Cell*, 44, 167, 1986.

48. **Jacquet, E. and Parmeggiani, A.**, Substitution of Val 20 by Gly in elongation factor Tu, *Eur. J. Biochem.*, 185, 341, 1989.

49. **Cool, R. H. and Parmeggiani, A.**, Substitution of histidine-84 and the GTPase mechanism of elongation factor Tu, *Biochemistry*, 30, 362, 1991.

50. **Reinstein, J., Schlichting, I., and Wittinghofer, A.**, Structurally and catalytically important residues in the phosphate binding loop of adenylate kinase of *Escherichia coli*, *Biochemistry*, 29, 7451, 1990.

51. **Sigal, I. S., Gibbs, J. B., D'Alonzo, J. S., and Scolnick, E. M.**, Identification of effector residues and a neutralizing epitope of Ha-ras encoded p21, *Proc. Natl. Acad. Sci. U.S.A.*, 83, 4725, 1986.

52. **Adari, H., Lowy, D. R., Willumsen, B. M., Der, C. J., and McCormick, F.**, Guanosine triphosphatase activating protein (GAP) interacts with the p21 *ras* effector binding domain, *Science*, 240, 518, 1988.

53. **Calés, C., Hancock, J. F., Marshall, C. J., and Hall, A.**, The cytoplasmic protein GAP is implicated as the target for regulation by the ras gene product, *Nature*, 332, 548, 1988.

54. **Gibbs, J. B., Schaber, M. D., Allard, W. J., Sigal, I. S., and Scolnick, E. M.**, Purification of ras GTPase activating protein from bovine brain, *Proc. Natl. Acad. Sci. U.S.A.*, 85, 5026, 1988.

55. **Garret, M. D., Self, A. J., van Oers, C., and Hall, A.**, Identification of distinct cytoplasmic targets for ras/R-ras and rho regulatory proteins, *J. Biol. Chem.*, 264, 10, 1989.

56. **Polakis, P. G., Rubinfeld, B., Evans, T., and McCormick, F.**, Purification of a plasma membrane associated GTPase-activating protein specific for rap1/Krev-1 from HL60 cells, *Proc. Natl. Acad. Sci. U.S.A.*, 88, 239, 1991.

57. **Ueda, T., Kikuchi, A., Ohga, N., Yammamoto, J., and Takai, Y.**, GTPase activating proteins for the smg-p21 GTP-binding protein having the same effector domain as the ras proteins in human platelets, *Biochem. Biophys. Res. Commun.*, 159, 1411, 1989.

58. **Becker, J., Tan, T. J., Trepte, H. H., and Gallwitz, D.**, Mutational analysis of the putative effector domain of the GTP binding Ypt1 protein in yeast suggest specific regulation by a novel GAP activity, *EMBO J.*, 10, 785, 1991.

59. **Neal, S. E., Eccleston, J. F., and Webb, M. R.**, Hydrolysis of GTP by p21[NRAS], the *NRAS* protooncogene product, is accompanied by a conformational change in the wild type protein: use of a single fluorescent probe at the catalytic site, *Proc. Natl. Acad. Sci. U.S.A.*, 87, 3652, 1990.

60. **Krengel, U., Schlichting, I., Scherer, A., Schumann, R., Frech, M., John, J., Kabsch, W., Pai, E. F., and Wittinghofer, A.,** Three-dimensional structures of H-ras p21 mutants: molecular basis for their inability to function as signal switch molecules, *Cell*, 62, 539, 1990.

61. **Seeburg, P. H., Colby, W. W., Capon, D. J., Goedel, D. V., and Levinson, A. D.,** Biological properties of human c-Ha-*ras* 1 genes mutated at codon 12, *Nature*, 312, 71, 1984.

62. **Krengel, U.,** Structure and GTP hydrolysis mechanism of the C-terminal truncated human cancer protein p21[H-ras], Ph.D. thesis, University of Heidelberg, 1991.

63. **Crechet, J.-B., Poullet, P., Mistou, M.-Y., Parmeggiani, A., Camonis, J., Boy-Marcotte, E., Damak, F., and Jacquet, M.,** Enhancement of the GDP-GTP exchange of RAS proteins by the carboxyl-terminal domain of SCD25, *Science*, 248, 866, 1990.

64. **Wolfman, A. and Macara, I.,** A cytosolic protein catalyzes the release of GDP from p21, *Science*, 248, 67, 1990.

65. **Downward, J., Riehl, R., Wu, L., and Weinberg, R. A.,** Identification of a nucleotide exchange-promoting activity for p21[ras,] *Proc. Natl. Acad. Sci. U.S.*, 87, 5998, 1990.

66. **Broek, D., Toda, T., Machael, T., Levin, L., Birchmeier, C., Zoller, M., Powers, S., and Wigler, M.,** The saccharomyces cerevisiae CDC25 gene product regulates the RAS/adenylate cyclase pathway, *Cell*, 48, 789, 1987.

67. **Huang, Y. K., Kung, H.-F., and Kamata, T.,** Purification of a factor capable of stimulating the guanine nucleotide exchange reaction of ras proteins and its effects on ras-related small molecular mass G proteins, *Proc. Natl. Acad. Sci. U.S.A.*, 87, 8008, 1990.

68. **Han, M. and Sternberg, P. W.,** let-60, a gene that specifies cell fates during *C. elgeans* vulval induction, encodes a ras protein, *Cell*, 63, 921, 1990.

69. **Beitel, G. J., Clark, S. G., and Horwitz, H. R.,** *Caenorhabditis elegans* ras gene let-60 acts as a switch in the pathway of vulval induction, *Nature*, 348, 503, 1990.

70. **McCormick, F., Clark, B. F. C., LaCour, T. F. M., Kjelgaard, M., Norskov-Lauritsen, L., and Nyborg, J.,** A model for the tertiary structure of p21, the product of the *ras* oncogene, *Science*, 230, 78, 1985.

71. **Clark, B. F. C., Kjelgaard, M., LaCour, T. F. M., Thirup, S., and Nyborg, J.,** Structural Changes of the functional sites of *E. coli* elongation factor Tu, *Biochem. Biophys. Acta*, 1050, 203, 1990.

72. **Gast, W. H., Leberman, R., Schulz, G. E., and Wittinghofer, A.,** Crystals of partially trypsin-digested elongation factor Tu, *J. Mol. Biol.*, 106, 943, 1976.

73. **Peter, M. E., Wittmann-Liebold, B., and Sprinzl, M.,** Affinity labelling of the GDP/GTP binding site in *thermus thermophilus* elongation factor Tu, *Biochemistry*, 27, 9132, 1988.

74. **Eckstein, F.,** Nucleoside phosphorothioates, *Annu. Rev. Biochem.*, 54, 367, 1985.

75. **Von Zur Mühlen, F., Eckstein, F., and Penner, R.,** Guanosine 5'-[β-thio]triphosphate selectively activates calcium signalling in mast cells, *Proc. Natl. Acad. Sci. U.S.A.*, 88, 926, 1991.

76. **Bigay, J., Deterre, P., Pfister, C., and Chabre, M.,** Fluoride complexes of aluminium or beryllium act on G-proteins as reversibly bound analogues of the γ-phosphate of GTP, *EMBO J.*, 6, 2907, 1987.

77. **Chabre, M.,** Aluminofluoride and beryllofluoride complexes: new phosphate analogs in enzymology, *TIBS*, 15, 6, 1990.

78. **Tong, L., DeVos, A. M., Milburn, M. V., and Kim, S.-H.,** Crystal structures at 2.2 Å resolution of the catalytic domains of normal and an oncogenic mutant complexed with GDP, *J. Mol. Biol.*, 217, 503, 1991.

79. **Kraal, B., DeGraaf, M., Mesters, J. R., VanHoff, P. J. M., Jacquet, E., and Parmeggiani, A.,** Fluoroaluminates do not affect the guanine-nucleotide binding centre of the peptide chain elongation factor EF-Tu, *Eur. J. Biochem.*, 192, 305, 1990.

80. **Kabsch, W. and Sander, C.,** Dictionary of protein secondary structure: pattern recognition of hydrogen-bonded and geometrical features, *Biopolymers*, 22, 2577, 1983.

81. **Wooley, P. and Clark, B. F. C.,** Homologies in the structure of G-binding proteins: an analysis based on elongation factor EF-Tu, *Biotechnology,* 7, 913, 1989.
82. **DeClue, J. E., Stone, J. C., Blanchard, R. A., Papageorge, A. G., Martin, P., Zhang, K., and Lowy, D. R.,** A ras effector domain mutant which is temperature sensitive for cellular transformation: interaction with GTPase-activating protein and NF-1, *Mol. Cell. Biol.,* 11, 3132, 1991.
83. **Valencia, A., Chardin, P., Wittinghofer, A., and Sander, C.,** The ras protein family: evolutionary tree and role of conserved amino acids, *Biochemistry,* 30, 4637, 1991.

Chapter 3

POSTTRANSLATIONAL PROCESSING OF *ras* AND *ras*-RELATED PROTEINS

John F. Hancock and Christopher J. Marshall

TABLE OF CONTENTS

I. INTRODUCTION

The knowledge that *ras* proteins require posttranslational modification both for plasma membrane association and for biological activity has long been established.[1] It is only over the last couple of years, however, that the true complexity of these modifications has been elucidated. Following the initial reports detailing the exact nature of *ras* posttranslational processing,[2-4] work in this field has expanded dramatically. In this chapter we will review and discuss the implications of recent data on the posttranslational processing of *ras* and *ras*-related proteins which have a CAAX C terminus. The posttranslational processing and membrane targeting of the *ras*-related *rab* proteins, which generally lack the CAAX motif, will be reviewed in Chapters 18 and 19.

II. CAAX MOTIFS

A CAAX motif (C, Cys; A, Aliphatic amino acid; and X, any amino acid) is found at the C terminus of a large number of cellular proteins including p21^{ras}, *ras*-related proteins (Figure 1), fungal mating factors, nuclear lamins, and certain α and γ-subunits of heterotrimeric G proteins. It is evident from Figure 1 that the acronym is somewhat misleading since nonaliphatic amino acids (e.g., Pro) can substitute for one of the aliphatic amino acids.

Certain CAAX motifs undergo a triplet of closely coupled posttranslational modifications. For the *Saccharomyces cerevisiae* a-type mating factor[5] and the transducin γ-subunit,[6,7] structural determination has shown that following processing: cysteine is the C terminal residue, C_{15} farnesyl is attached as a thioether to the sulfur, and the α-carboxyl group of the cysteine is methylesterified. Similarly, all *ras* proteins undergo farnesylation on the Cys residue of the CAAX motif,[3,4] proteolysis to remove the -AAX amino acids,[2,8] and methylesterification.[2,9] Although the site of methylation has not been formally identified on the *ras* proteins, it is assumed, by analogy with mating factor a, to be on the α-carboxyl group of the C terminal cysteine. Following this triplet of posttranslational modifications (Step I processing), p21^{N-ras}, p21^{H-ras}, and p21$^{K-ras(A)}$ are additionally palmitoylated on cysteine residues contained within the hypervariable domain (Step II processing). Acylation is a late event in posttranslational processing since Cys → Ser mutations at Cys181 in p21^{N-ras} and Cys181 and Cys184 in p21^{H-ras} which remove the acylation sites, block palmitoylation without effecting CAAX processing.[3] Conversely, palmitoylation is absolutely dependent on a normally processed CAAX motif; mutations which block CAAX processing (e.g., Cys186 → Ser, or premature termination of the protein at Cys186) also block palmitoylation.[3] There are no cysteine residues present in the hypervariable domain of p21$^{K-ras(B)}$ and, therefore, this *ras* protein is not palmitoylated.[3]

The α-subunits of G_i, G_o, and transducin terminate with the sequence CGLF but these proteins are not prenylated.[10] Furthermore, making a glycine substitution in the A_1 position of the CAAX motif of the transducin γ-subunit, a protein

PALMITOYLATED

H–ras	P D E S G P G C M S C K C V L S
N–ras	S D D G T Q G C M G L P C V V M
K–rasA	D D K T P G C V K I K K C V I M
RAS1	S A N A R K E Y S G G C C I I C
RAS2	T S E A S K S G S G G C C I I S
R–ras	P S A P R K K G G G C P C V L L
rap2	A Q P D K D D P C C S A C N I Q
rhoB	K R Y G S Q N G C I N C C K V L

POTENTIALLY PALMITOYLATED

spras	N S K T E D E V S T K C C V I C
RHO2	S L L M K K E P G A N C C I I L
TC21	P T R K E K D K K G C H C V I F
TC10	V K K R I G S R C I N C C L I T

POLYBASIC

K–rasB	E G K K K K K K S K T K C V I M
DdrasG	P E K G K K K R P L K A C T L L
Ddras	Q S S G K A Q K K K K Q C T I L
let60	R H D N N K P Q K K K K C Q I M
Dras1	G R K M N K P N R R F K C K M L
Xeras	N G K K K K S S K R K C V V L
rap1A	T P V E K K K P K K K S C L L L
rap1B	T P V P G K A R K K S S C Q L L
RSR1	S Q Q K K K K K N A S T C T I L
Dras2	I E Q D Y K K K G K R K C C L M
Dras3	R P R R N R R S R K V P C V L L
rhoA	A L Q A R R G K K K S G C L V L
rhoC	G L Q V R K N K R R R G C P I L
RHO1	A K K N T T E K K K K K C V L L
rac1	L C P P P V K K R K R K C L L L
rac2	L C P Q P T R Q Q K R A C S L L
CDC42	A L E P P V I K K S K K C T I L
G25K	A L E P P E T Q P K R K C C I F
ralA	K R K S L A K R I R E R C C I L
ralB	K S S K N K K S F K E R C C I L

FIGURE 1. The prenylated cysteine of the CAAX motif is shown in bold. Palmitoylated, potentially palmitoylated cysteines, and polybasic domains are underlined. RAS1, RAS2, RHO1, and RHO2 are the *ras* and *rho* proteins, respectively, of *Saccharomyces cerevisiae: spras* is the *ras* protein of *Schizosaccharomyces pombe;* DdrasG and Ddras are *Dictyostelium discodieum ras* proteins;[22] and Dras1, Dras2, and Dras3 are *Drosophila melanogaster* homologs of *ras* and *rap*, respectively. TC10 and TC21 were isolated by screening for *ras*-related genes;[89] CDC42 is an *S. cerevisiae* protein involved in bud site formation while RSR1 is involved in bud site selection,[90] G25K is the human homolog of CDC42,[91,92] *let-60* is a *Caenorhabditis elegans ras* protein,[93] and Xe-*ras* is a *Xenopus ras* gene.[94] Although *rac* 1 and *rac* 2 both have basic amino acids and cysteine residues upstream of the CAAX box, the cysteines are in a position which in H-*ras* would not lead to palmitoylation,[11] they have, therefore, been classified as polybasic.

which is normally farnesylated, blocks its prenylation.[10] A similar mutation in p21[H-ras] to change the wild-type CVLS motif to CGLS also abolishes farnesylation of the mutant protein,[11] suggesting that CGAX motifs are not substrates for isoprenyl-protein transferases.

Modification of cellular proteins with C_{20} geranylgeranyl is some 10 times more common than farnesylation.[12-14] A number of geranylgeranylated CAAX-containing proteins have now been identified including: the γ-subunits of brain heterotrimeric G proteins [15,16] and the *ras*-related proteins, p21*rap1B*,[17] p21*rap1A*,[18] p21*rho*,[100] and G25K.[19,20] Some of these proteins have also been shown to be methylesterified; the site being identified as the C terminal cysteine in the case of the G protein γ-subunits,[20] and presumed to be the same residue for the others. A common feature of the CAAX motifs of these proteins is the presence of a leucine residue in the X position, suggesting that Leu as the C terminal amino acid directs geranylgeranylation. This hypothesis has been validated by the observation that a Ser → Leu mutation in the CVLS motif of p21[H-ras] results in the mutant protein being geranylgeranylated.[21]

In summary then, there appear to be three types of CAAX motif: (1) motifs where amino acid A_1 = Gly which are not prenylated; (2) motifs where amino acid X = Leu which are geranylgeranylated, proteolyzed to remove the -AAL amino acids and methylesterified on the C terminal cysteine; and (3) motifs where amino acid $A_1 \neq$ Gly and X $\neq$ Leu which are farnesylated, -AAX proteolyzed and methylesterified. In general, *ras* proteins have a CAAX motif for farnesylation, but it should be noted that the sequences of two *Dictyostelium discoideum ras* genes, *Drosophila melanogaster Dras1*, and a *Xenopus ras* gene that have been isolated predict Leu as the C terminal amino acid.[22,94]

Most *rab* proteins terminate with a CXC or XCC C terminal amino acid sequence and these proteins also undergo geranylgeranylation.[23-25] Interestingly, however, whereas the CXC motif is methylated following geranylgeranylation the XCC motif is not.[25] This observation may, in part, account for the significant proportion of geranylgeranyl-cysteine that can be recovered from cells as the free carboxylic acid.[14]

The processing of CAAX-containing proteins and the *rab* proteins, therefore, calls for: a farnesyl-protein transferase, up to three different geranylgeranyl-protein transferases, an -AAX protease, and a prenyl-cysteine carboxyl-methyltransferase. These enzymes are of interest, not least because of the potential therapeutic exploitation of inhibitors as anti-*ras* agents in the treatment of malignancy.

III. PRENYL-PROTEIN TRANSFERASES

Farnesylation of p21*ras* is catalyzed by an enzyme present in the soluble fraction of all eukaryotic cells investigated.[26] The enzyme, farnesyl-protein transferase (FTase), farnesylates the cysteine residue of CAAX tetrapeptides using farnesylpyrophosphate (F-PP) as a prenyl donor. Reiss et al.[27] have purified

FTase activity from rat brain cytosol; the enzyme is a heterodimer, comprising an α and β_1 chain of 49 kDa and 46 kDa, respectively.[28,29] Others have purified *ras* FTase from bovine brain cytosol[29] and from porcine kidney cytosol.[26] The enzyme is heat labile and requires divalent cations, either Mg^{2+} or Mn^{2+}.[26,27,29]

In addition to recombinant *ras* proteins, FTase will farnesylate isolated CAAX peptides, indicating that the CAAX tetrapeptide sequence is the actual substrate for the transferase.[27,29,30] CAAX peptides at micromolar concentrations also function as competitive inhibitors of farnesylation. Peptides with the sequence SAAX are not substrates for the transferase and do not inhibit its activity.[27,29] Nonradioactive F-PP and geranylgeranyl-pyrophosphate (GG-PP) also competitively inhibit the incorporation of radiolabeled F-PP into *ras* protein at concentrations of 0.7 to 1.4 μM, whereas farnesol does not.[29] These data can now be rationalized in the light of experiments which show that the α-subunit is common to the geranylgeranyl transferase I enzyme (GGTaseI) which geranylgeranylates the CAA (X = L) tetrapeptide motif.[21] Moreover, as expected, the function of the α-subunit is to bind the prenyl pyrophosphate group, whereas the β_1-subunit binds the CAAX peptide.[28,31] Therefore, while the α-subunit can bind either farnesyl or geranygeranyl pyrophosphate, only productive transfer of a C_{15} chain can be effected to a CAAX peptide bound to the β_1-subunit.

Sequence analysis shows that the α and β_1 genes, which encode the respective subunits of the mammalian farnesyl transferase, are highly homologous to the two unlinked yeast genes *RAM2* and *DPR1/RAM1* respectively. The ram1/dpr1 mutation in *S. cerevisiae* blocks normal processing of *RAS* proteins and, thus, suppresses the heat shock phenotype associated with *RAS2* Val19 expression. In addition, ram1/dpr1 mutants are sterile due to an inability to process mating factor a[32,33] and cannot process the *STE18* gene product, a G protein γ-subunit.[34] Cell extracts from ram1/dpr1 mutants are significantly deficient in FTase activity, which can be fully reconstituted with *Escherichia coli* expressed recombinant *RAM1/DPR1* protein. However, recombinant *RAM1/DPR1* protein has no transferase activity in the absence of added yeast extract, indicating that it comprises only one component of the enzyme.[35] Similarly, ram2 mutant cells are deficient in *ras* FTase activity,[36] suggesting that the protein product of the *RAM2* gene may also be a component of the enzyme. The prediction that these genes, therefore, encode the *S. cerevisiae* FTase has recently been formally demonstrated, *RAM1/DPR1* and *RAM2*, when coexpressed in *E. coli*, generated FTase activity, whereas expression of either gene in isolation in *E. coli* does not result in any FTase activity.[37] As in the mammalian system, *RAM2* also encodes one subunit of the *S. cerevisiae* GGTaseI enzyme with *CDC43/CAL1* encoding the homolog of the β_2-subunit, which binds the CAA(X = L) tetrapeptide.[38]

Mammalian GGTaseI has recently been purified from bovine brain[30,39] and partially characterized. As discussed, the enzyme shares a common α-subunit with FTase, but has a different β-subunit designated β_2.[21,28,31] GGTaseI geranylgeranylates *ras* proteins if an X $\rightarrow$ L mutation is made in the CAAX motif and will also geranylgeranylate tetrapeptides if the X amino acid is Leu.[30] Certain tetrapeptides (when X = F, V, I, or M) are substrates for both FTase and GGTaseI, implying that they must have a productive interaction with both of the β-subunits.[30]

Recently, other geranylgeranyl transferases have been purified from bovine brain.[30,40] One, designated GGTaseII, geranylgeranylates recombinant YPT1 protein[30] and the other activity geranylgeranylates p25rab3A.[40] It is unclear whether these are the same or different enzymes since YPT1 has a XCC C terminus and p25rab3A has a CXC C terminus. Furthermore, p25rab3A is geranylgeranylated on both C terminal cysteine residues,[41] implying that this enzyme may carry out a double prenyl transfer. It is also clear from peptide inhibition studies[30] and from an analysis of *rab* proteins with C terminal deletions,[42] that the recognition unit for this GGTase involves much more of the protein than the C terminal tetrapeptide sequence. Finally, the recent observation that *BET2* is required for *SEC4/YPT* function in *S. cerevisiae* and is related to *DPR1/RAM1* suggests that this gene may encode one subunit of GGTaseII.[43]

IV. THE ras-AAX PROTEASE AND METHYLTRANSFERASE

Several lines of evidence indicate that proteolysis, to remove the -AAX amino acids, is the second event in the processing of the CAAX motif. A mutant p21^{H-ras} protein prematurely terminated at Cys186, which lacks the AAX amino acids, does not undergo prenylation *in vivo*.[3] Moreover, release of [^{3}H]-labeled amino acid 189W from a p21^{H-ras} 189S → W mutant protein is blocked if the cells are treated with mevinolin.[44] In reticulocyte lysates, p21^{ras} proteins and C terminal *ras* peptides undergo farnesylation, but not proteolysis.[35,45] Taken together, these data suggest that the substrate for the -AAX protease is a farnesylated CAAX motif and hence, proteolysis must follow prenylation. The processing of p21$^{K-ras(B)}$ in rabbit reticulocyte lysates is arrested following farnesylation and requires intracellular membranes (canine pancreatic microsomes) for both proteolysis and methylation to occur,[45] showing that while the FTase is cytosolic, the -AAX protease is membrane-associated. Nothing else is currently known about this enzyme and there are no candidate *STE* genes in *S. cerevisiae* which might be expected to encode it.

A methyltransferase has been identified which carries out α-carboxyl methylation of *ras* proteins and mating factor a. The activity has been found in the microsomal fractions of various rat tissues, including liver and brain,[46] and in membrane preparations from *S. cerevisiae*.[47] The methyltransferase has been characterized *in vitro* using synthetic peptides as substrates and *S*-adenosyl-L-methionine as the methyl donor. Hexapeptides mimicking the AAX proteolyzed C terminal sequence of *Dras1* were efficiently methylated when prenylated on the C terminal cysteine.[46] Peptides prenylated with farnesyl or geranylgeranyl were efficiently methylated, but the V_{max} for the C_{20} modified peptide was reduced to 28% of that for the C_{15}-modified peptide. Peptides prenylated with

C_{10} geranyl, or alkylated with nonprenyl n-alkyl chains with lengths of between 8 and 15 carbons, were very poor methyl acceptors.[46] These data show that both the prenoid structure itself and the chain length are important for enzyme recognition. Also, N-acetyl farnesyl cysteine is efficiently methylated *in vitro*, indicating that peptide sequences upstream of the prenylated cysteine are not required for substrate recognition by the methyltransferase.[48] Rando and coworkers[49] have taken these data further by showing that the acetoamide group is also not required as part of the recognition sequence and, therefore, farnesylthiopropionic acid is a good substrate for the transferase. Methylation of farnesylated peptides is readily inhibited by S-adenosylhomocysteine ($K_i = 9.2 \ \mu M$), farnesylated peptides competitively inhibit methylation of cGMP phosphodiesterase (a CAAX protein) in membranes derived from bovine retinal rod outer segments,[46] and farnesylthioacetic acid is a potent inhibitor of methylation *in vitro* ($K_i = 1 \mu M$).[49]

Genetic studies in *S. cerevisiae* have demonstrated that methyltransferase activity is dependent on the *STE14* gene product, for example, ste14 null cells are unable to methylate *RAS1*, *RAS2*, and mating factor a.[47] Furthermore, membranes prepared from ste14 mutant cells have no detectable methyltransferase activity unless transfected with a plasmid carrying the *STE14* gene. That the *STE14* gene product actually encodes the methyltransferase has recently been confirmed by the expression of *STE14* as a fusion protein in *E. coli* and the demonstration that extracts of these cells gain novel methyltransferase activity.[48] The *STE14* gene encodes a protein of 239 amino acids, which, appropriately, appears to contain multiple membrane-spanning domains.[50]

V. FUNCTIONAL SIGNIFICANCE OF CAAX MODIFICATIONS

The net effect of the triplet of posttranslational modifications is to significantly increase the hydrophobicity of the processed protein.[2,3] A common feature of CAAX-modified proteins is that they are associated with, or pass through, cellular membranes. Taken together, these observations suggest that the function of the hydrophobic posttranslational modifications is to facilitate association with cell membranes. However, in isolation a CAAX motif cannot target a protein to a specific membrane since, for example, p21^{ras} and lamins A and B are associated with different membranes. Additional targeting signals must, therefore, be involved (see below). However, in general terms how might the CAAX modifications contribute to membrane association and are all three CAAX modifications essential for membrane binding?

These questions have been addressed using a rabbit reticulocyte lysate to translate and process p21$^{K\text{-}ras(B)}$.[45] We have shown that the increase in hydrophobicity following Step I processing, assayed by the acquisition of detergent binding properties,[2,3] is fully accounted for by farnesylation alone. That is, -AAX proteolysis and methylation do not increase the proportion of p21$^{K\text{-}ras(B)}$

partitioning into the detergent phase of Triton® X-114 when the *ras* protein is translated and processed *in vitro*.[45] Farnesylated, non-AAXed, nonmethylated p21$^{K\text{-}ras(B)}$ binds inefficiently to cell membranes *in vitro*. If AAX proteolysis is allowed to occur, by translating with microsomal membranes in the presence of a methyltransferase inhibitor, there is a twofold increment in the extent of binding to cell membranes. If the protein is then methylated, there is a further twofold increment in membrane binding.[45] These data imply that efficient membrane binding of p21$^{K\text{-}ras(B)}$ requires all three CAAX modifications. The association of farnesylated proteins with cell membranes may involve the farnesyl chain binding to a specific receptor protein or simply interdigitating into the lipid bilayer. In either case the presence of the -AAX amino acids may sterically interfere with the farnesyl interaction, and the presence of a negative charge on nonmethylated protein may cause electrostatic repulsion from similarly charged phospholipid headgroups.

Similar questions can be addressed by studying the processing of *RAS* proteins in ste14 null cells, although interpreting the results is not straightforward. Hrycyna et al.[48] show that ste14 null cells have significantly reduced amounts of *RAS2* bound to membrane; 20% of the levels of *RAS2* associated with membranes in wild-type cells. However, in pulse chase experiments it is clear that the ste14 mutation also significantly affects processing events prior to methylation. During processing in wild-type cells p41 pro-*RAS2* rapidly chases into a faster migrating p40 form. In ste14 cells this occurs but much more slowly. The mobility shift must represent farnesylation and/or -AAX proteolysis so one or both of these processes is partially inhibited in the absence of the methyltransferase. The authors conclude that methylation is not required for membrane binding of *RAS2*. This is apparently inconsistent with our observations on the *in vitro* membrane-binding of p21$^{K\text{-}ras(B)}$. However, the study does not address what proportion of farnesylated *RAS2* is membrane-associated (and so these data are not directly comparable) nor whether any of the p40 *RAS* produced in the ste14 cells is palmitoylated. Palmitoylation increases the avidity of membrane binding of Step I-processed p21$^{H\text{-}ras}$.[3,51] However, p21$^{K\text{-}ras(B)}$ is not palmitoylated so it is possible that blocking methylation of p21$^{K\text{-}ras(B)}$ results in greater abrogation of membrane-binding than is the case for palmitoylated *ras* proteins.

The effect of ste14 on -AAX proteolysis and/or farnesylation is interesting. We have shown that the -AAX protease is membrane-associated,[45] therefore, a complex between the protease and the methyltransferase is a possibility. Lack of the *STE14* protein might then compromise the activity of the -AAX protease if, for example, the latter is functioning as a docking protein for prenylated CAAX proteins which are -AAXed and passed on to the methyltransferase. This model would predict that farnesylation is proceeding normally in the ste14 cells, a matter which has not yet been addressed. If farnesylation is impaired then the FTase must either be included in the complex, which is difficult to reconcile with its cellular location, or be subject to some sort of feedback regulation by partially processed *RAS* proteins. Alternatively, one component of the processing pathway may require methylation for full activity. It is intriguing to note, therefore, that the yeast HMGCoA reductase enzyme (but not the mammalian homolog) terminates in a CAAX motif,[52] raising the possibility that the ste14 cells may suffer from a deficiency of mevalonic acid.

VI. SUBCELLULAR LOCALIZATION: REQUIREMENT FOR CAAX AND A SECOND SIGNAL

Most of the CAAX-containing *ras*-related proteins have been shown to be associated with cell membranes on the basis that they are found in the particulate fraction (P100) after subcellular fractionation. However, an exception to this rule is p21^{rhoA} which is mainly cytosolic,[53] although a small fraction can be seen localized to the plasma membrane by immunofloresence.[54] Why p21^{rhoA} should be mainly cytosolic is not at all clear but, as will be discussed later, may reflect association with other proteins such as *rho*-GDI[55] which override its own membrane localization signals.

Although the cellular localization of small guanine nucleotide proteins is generally regarded as being central to their functions,[56] the precise localization to intracellular membranes is only known for a few of them. It has been known for some time that the *ras* proteins are localized on the inner surface of the plasma membrane,[57] *rap1* proteins are localized to a Golgi compartment, while the *rho* proteins have both plasma membrane and intracellular locations;[96] a proportion of p21*rhoB* is found in early endosomes/prelysosomes.[58]

Early studies showed that the CAAX motif was essential for membrane localization and transformation.[1] While all of the posttranslational modifications at the CAAX motif contribute to membrane-binding,[45] they are insufficient on their own to achieve targeting of p21^{ras} to the plasma membrane.[51] Placing a CAAX motif at the C terminus of the cytosolic protein, such as protein A, results in the CAAX processing steps but not plasma membrane localization.[3] The additional signal required for correct targeting of H-, N-, and K(A)*ras* proteins is palmitoylation.[3,51] However, K-*ras*(B) is not palmitoylated and the second signal is provided by a polybasic domain.[51] Interestingly these second signals, unlike the CAAX motif, are not required for transformation by oncogenically activated p21^{ras},[3,51] suggesting that plasma membrane localization per se may not be essential for transformation although the posttranslational modifications at the CAAX motif are absolutely required. It is striking that the second signals that combine with the CAAX motif to achieve plasma membrane localization are found within the C terminal hypervariable domain that distinguishes one *ras* protein from another.[59] This argues that one function of the hypervariable domain is to provide different signals for membrane localization.

Experiments with protein A fusions show that the CAAX motif, plus either an H-*ras* palmitoylation sequence (aa181 to aa184) or a K-*ras*(B) polybasic domain (aa175 to aa180), are sufficient to achieve plasma membrane localization.[60] These results demonstrate that no other regions of p21^{ras} are required for plasma membrane localization. Given the requirement of *ras* proteins for a second signal, in addition to the CAAX motif, for correct membrane localization it is likely that *ras*-related proteins also require a second signal. Inspection of the C-terminal sequences of *ras*-related proteins from many different species shows that they can be readily divided into those that have polybasic domains and those that have cysteine residues as potential palmitoylation sites (Figure 1), although to date only R-*ras*, *rap2*, and *rhoB* have been shown to be palmitoylated.[61,62,95] It is noteworthy that within each subfamily of small guanine nucleotide-binding proteins, there are members with polybasic domains and members with potential palmitoylation sites. This suggests that there is a functional requirement to have similar proteins which utilize different second signals for membrane localization, perhaps for association with different molecules in the membrane.

One of the outstanding questions concerning the association of small guanine nucleotide proteins with membranes is whether membrane receptors are involved to achieve targeting to specific intracellular membranes. Clearly, the simplest way to imagine that p21^{ras} is targeted to the plasma membrane is to invoke a plasma membrane receptor protein as has been suggested for the myristoylated pp60src.[63] However, we have not been able to find evidence for such a receptor for p21$^{K-ras(B)}$ since unlike pp60src, binding of posttranslationally modified protein to membranes is not inhibited by boiling or protease treatment of the membranes.[64] Similar results have been found by Isomura et al.[55] for the binding of p21^{rhoB} to membranes. Thus, it remains an unresolved question of how specificity of association is achieved.

VII. PALMITOYLATION OF p21^{ras} AND *ras*-RELATED PROTEINS

Site-directed mutagenesis and peptide mapping data have shown that the sites of palmitoylation in p21^{H-ras} are Cys181 and Cys184, while in p21^{N-ras} Cys181 is the site of palmitoylation.[3] Studies by Buss and colleagues have shown that p21$^{v-K-ras(A)}$ is palmitoylated[65] and presumably p21$^{K-ras(A)}$ is palmitoylated at Cys 180. The significance of the double palmitoylation of p21^{H-ras} is not clear since mutant proteins that have only one intact site localize to the plasma membrane and are biologically active.[3,51] However, it is worth noting that the sites of palmitoylation in each of the three mammalian *ras* proteins are slightly different relative to the CAAX motif, which may conceivably lead to differences in association with membrane components or other molecules. In *S. cerevisiae* RAS1 and RAS2, the sites of palmitoylation are immediately adjacent to the farnesylated cysteine of the CAAX box leading to an extremely hydrophobic C terminus. It has been reported that the *ras* proteins of *D. discoideum* are palmitoylated,[66] however, the sequences of the two *D. discoideum ras* genes that have been published[22] contain no potential palmitoylation sites close to the C terminus (see Figure 1), arguing that there must be an additional *ras* gene in this organism. Studies on p21rhoB show it to be palmitoylated at Cys189 and

Cys192. The predicted sites of palmitoylation of *ras*-related proteins vary within the C termini (Figure 1), again suggesting some functional significance to the site of palmitoylation.

Palmitoylation of p21^{H-ras} leads to a very strong attachment of the protein to the plasma membrane, such that the protein cannot be removed by high salt washing; breakage of the thioester bond with hydroxylamine is required to remove the protein from the membrane.[3,51] Early studies by Magee et al.[67] showed that there is a rapid turnover ($t_{1/2}$ = 20 to 40 min) of the palmitoyl moiety on p21^{N-ras}, presumably as the result of an esterase activity. This contrasts with the very stable nature of farnesylation ($t_{1/2}$ = 24 h).[3] Since p21^{N-ras} lacking the palmitoyl moiety is only weakly attached to membranes, this rapid turnover suggests that p21^{N-ras} and p21^{H-ras} might undergo cycles of plasma membrane attachment and detachment, however, no evidence has been reported for this to date. In spite of the importance of palmitoylation to *ras* protein function, virtually nothing is known about the enzymes involved. However, the position of the acceptor cysteine may be quite critical since site-directed mutagenesis experiments have shown that moving the cysteine residue for palmitoylation in p21^{H-ras} from aa181 to aa179, two residues N-terminal, results in a loss of palmitoylation.[11]

VIII. POLYBASIC DOMAINS OF p21$^{K-ras(B)}$ AND *ras*-RELATED PROTEINS

The polybasic domain of p21$^{K-ras(B)}$ contains six consecutive lysine residues (aa175-180; see Figure 1) and like the sites for palmitoylation in p21^{H-ras} and p21^{N-ras}, is found close to the CAAX motif. Some degree of plasma membrane localization is maintained when four of the six lysines of p21$^{K-ras(B)}$ are replaced with uncharged glutamine, however, replacing five or six lysines leads to a cytosolic localization.[51] The six lysines in the polybasic domain can be replaced with six arginines without compromising plasma membrane localization or biological activity as measured by the transforming activity of oncogenic p21$^{K-ras(B)}$.[60] This result suggests that the function of the polybasic domain is to give a net positive charge rather than to provide a specific amino acid sequence. Although the polybasic domain of p21$^{K-ras(B)}$ functions like palmitoylation of p21^{H-ras} and p21^{N-ras} to provide a second signal for plasma membrane localization, its consequences may be qualitatively different since p21$^{K-ras(B)}$ is readily removed from the plasma membrane by low concentration salt washes.[51] Interestingly modifying the CAAX motif of p21$^{K-ras(B)}$ so that it is geranylgeranylated makes it much more resistant to salt washing,[60] suggesting that this type of membrane association is determined by both the type of prenyl group and polybasic domain.

Polybasic domains are also found in *rap1a*, *rap1b*, *rho*A, *rho*C, etc., in a similar position to that in p21$^{K-ras(B)}$ (see Figure 1). The role of the polybasic domains in subcellular localization in *ras*-related proteins has not been investigated. However, it is striking that *rap1* proteins are localized to the Golgi rather than the plasma membrane.[58] It is not clear what is the basis for differences in subcellular localization between *ras* and *rap1* proteins. It is worth noting that the polybasic domains of K-*ras*(B) and *rap1A* and *rap1B* are not identical, and that *rap1a* is prenylated by geranylgeranyl[17] while K-*ras*(B) is farnesylated. However, geranylgeranylated p21$^{K-ras(B)}$ is still localized to the plasma membrane,[60] suggesting that the Golgi localization of the *rap1* proteins is not a consequence of geranylgeranylation, but that other domains in the *rap1* protein must be involved. It is possible that these Golgi localizing domains override the apparent plasma membrane localization signals at the C terminus of the *rap* proteins.

IX. SOME POLYBASIC C TERMINI CONTAIN PHOSPHORYLATION SITES

Lapetina and associates have shown that in platelets activation of protein kinase A leads to the phosphorylation of a large fraction of the membrane p21^{rap1B}.[68] Most interestingly the phosphorylated *rap* protein becomes cytosolic,[68] which raises the important issue that the membrane association of small guanine nucleotide-binding proteins may be under physiological control. Subsequent experiments have shown that p21^{rap1B} is also a substrate for a neuronal calmodulin-dependent kinase CaM kinase Gr.[69] The site of phosphorylation in p21^{rap1B} is Ser179 within its polybasic domain[70,71] and comparison of other polybasic domains shows that *rap1A*, K-*ras*(B), and *rho*A also contain potential phosphorylation sites. It has been known for some time that p21$^{K-ras(B)}$ becomes phosphorylated following activation of protein kinase C with phorbol esters.[72] This phosphorylation site has now been mapped to Ser181 in the polybasic domain.[64] However, unlike the phosphorylation of p21^{rap1B}, only a small proportion of p21$^{K-ras(B)}$ is phosphorylated.[72] The observation that phosphorylation of p21^{rap1B} leads to it becoming cytosolic suggests that other small guanine nucleotide-binding proteins with polybasic domains may cycle between membrane and cytosolic forms in a process regulated by phosphorylation. Such a cycle would be analogous to the potential cycling of palmitoylated *ras* proteins resulting from the turnover of the palmitoyl groups.[67]

X. INTERACTIONS WITH GUANINE NUCLEOTIDE EXCHANGE PROTEINS REQUIRES POSTTRANSLATIONAL MODIFICATIONS

Recent results from Takai and colleagues have provided new insights into how both the activity of small guanine nucleotide-binding proteins and their association with membranes might be regulated.[71] They have purified and characterized two classes of exchange factors for small guanine nucleotide-binding proteins. The first of these are GDP dissociation inhibitors (GDI) which inhibit the release of GDP from proteins. Such activities would keep a small guanine nucleotide-binding protein in the inactive GDP form. GDIs have been described for p21^{rab3A}[73] and for p21^{rho},[74-76] the latter being active on both p21^{rhoA} and p21^{rhoB}.[77]

The second class are guanine nucleotide dissociation stimulators (GDS) which stimulate the release of GTP and GDP. Since there is at least 10 times more GTP than GDP in the cell, the action of a GDS will be to convert an inactive protein in the GDP state to the active GTP state. A *rho*-GDS[78] and a *rap1*-GDS have been described.[79,80] Significantly, only small guanine nucleotide-binding proteins that have undergone postranslational modifications will interact with their GDI or GDS,[77,81,82] unlike the situation with GTPase-activating proteins whose effects do not require postranslational modifications. The *rap1* GDS is also active on p21$^{K-ras(B)}$ and p21^{rhoA},[82] both of which have polybasic domains at their C termini but not p21^{H-ras} which is palmitoylated. Thus, there may be some selectivity in the interactions of different members of each subfamily of small guanine nucleotide-binding proteins and different exchange factors.

The interaction of *rho*-GDI with p21^{rhoB}-GDP causes p21^{rhoB} to be released from membranes.[55] Similar observations have been made for *rab*3A-GDI.[83] These results suggest that when p21^{rho} or p25^{rab3a} are in the GDP form, presumably as the result of the action of their GTPase-activating proteins, a GDI binds and causes the protein to dissociate from membranes. Since a GDS has been described for p21^{rho}, it is likely that when p21^{rho}-GDP is dissociated from membranes it interacts with *rho*-GDS leading to the exchange of GDP for GTP and the protein may then reassociate with membranes. Whether such cycles could operate for p21^{rap} or p21^{ras} remains to be determined since as yet no GDI activity for these proteins has been discovered. However, the interaction of *rap1*-GDS with p21^{rap1} leads to a dissociation from membranes, arguing that it may be the interaction with a GDS rather than a GDI that regulates the association of p21^{rap} with membranes.[84] Strikingly, the binding of p21^{rap1B} to GDS and nucleotide exchange is stimulated by the phosphorylation of p21^{rap1B}.[71] A combination of phosphorylation and GDS binding may account for the release of phosphorylated p21^{rap1B} into the cytosol[68] and thereby, a stimulation of nucleotide exchange. These results, therefore, suggest that guanine nucleotide exchange and membrane binding could be intimately associated and for some *ras*-related proteins regulated through phosphorylation in the polybasic domain.

XI. DISCUSSION

We are now acquiring an increasingly detailed picture of the postranslational modifications that occur at the C termini of small guanine nucleotide-binding proteins. It is also becoming clearer how these modifications contribute to the association of the proteins with membranes by interacting with signals contained within the C terminal hypervariable domains. The idea that targeting to specific membranes was a consequence of the combination of prenylation and a second signal from within the hypervariable domain was first proposed for p21^{ras},[51] but also seems to be the case for different *rab* proteins.[85] However, a major unresolved issue is how specificity of interaction is achieved, for example, why p21^{ras} is targeted to the inner surface of the plasma membrane while p21^{rap1} is targeted to the Golgi. Specific protein receptors have not yet been described, indeed, experiments with both p21^{ras} and p21^{rho} indicate that a simple situation of protein receptors conferring specificity may not be involved because binding is still observed to membranes that have been boiled or treated with proteases.[55,64] An alternative possibility based on the resistance of binding to protease treatment or boiling of membranes is that lipid components provide the specificity. While there are clearly differences in lipid composition between membranes, there is no picture at present of how differences in lipid composition could affect association of small guanine nucleotide-binding proteins with membranes.

The localization of a particular small guanine nucleotide-binding protein to a specific intracellular membrane may serve two, not necessarily exclusive, functions. On one hand it may reflect that the protein is involved in a signal transduction event which occurs at that site, presumably by the guanine nucleotide-binding protein facilitating interactions between consecutive elements in the signaling pathway. Alternatively, a small guanine nucleotide-binding protein may be responsible for directing vesicular trafficking from one intracellular compartment to another.

While the majority of small guanine nucleotide-binding proteins are prenylated, it is not clear why some have C_{15} farnesyl, while others have C_{20} geranylgeranyl. In general, *ras* proteins are farnesylated while other small guanine nucleotide-binding proteins are geranylgeranylated. However, as more *ras* genes are discovered increasing numbers of *ras* proteins are predicted to be geranylgeranylated, for example, *Drosophila* Dras1, *Xenopus ras,* and two *Dictyostelium ras* genes, Dd *ras*D and Dd *ras*G (see Figure 1) all have X = leucine predicting a C_{20} modification. The functional consequences of C_{15} vs. C_{20} modification are not known, the intracellular localization and transforming capacity of oncogenic p21$^{K-ras(B)}$ with a geranylgeranylated C terminus is identical to the farnesylated protein.[60] In addition, the geranylgeranylated protein is more tightly bound to membranes presumably as a result of the longer alkyl side chain. However, normal p21^{H-ras} with a geranylgeranylated C-terminus inhibits cell proliferation, suggesting that it may block exchange factor or effector function.[97]

Two sets of molecules are known to interact with small guanine nucleotide-binding proteins: GTPase activators (GAPs) and guanine nucleotide exchange proteins (GEPs). There may also be a third class of interacting molecules: effectors. All studies thus far show no requirement for posttranslational modifications for the binding of GAPs or the acceleration of GTPase activity. However, it is clear that the GEPs for *rho* and *rap* proteins only interact with posttranslationally modified proteins. This is in spite of the fact that the GEPs themselves are not membrane-associated proteins. In the case of the GEPs, the need for posttranslational modification of the small guanine nucleotide-binding proteins may be a consequence of a requirement for protein-protein interactions. A precedent for such a situation is provided by the observation that the α-subunit of transducin (Tiα) has a much higher affinity for the farnesylated β-γ complex (Tiβγ) than unmodified Tiβγ.[7,86] This difference is seen in the absence of membranes so does not reflect membrane-binding of Tiβγ, but presumably a hydrophobic protein-protein interaction mediated by the farnesyl group. Similar considerations may explain why cytosolic mutants of oncogenically activated p21$^{H\text{-}ras}$ and p21$^{K\text{-}ras(B)}$ that retain the CAAX motif, but have lost the palmitoylation sites or the polybasic domain, still transform cells.[3,51] For generation of a downstream signal, the farnesylated C terminus may be required for interaction with a cytosolic effector molecule rather than association with the plasma membrane.

The observation that the posttranslational modifications and biological activity of p21ras could be blocked by inhibitors of HMG CoA reductase such as mevinolin[3,87] has raised the prospect that inhibition of posttranslational modifications of p21ras might have therapeutic potential. While inhibition of the mevalonate pathway is unlikely to have much selective capability, inhibitors of the farnesyl transferase could be more promising as they will only affect the relatively small number of proteins that are thought to be farnesylated. Analogs of farnesyl pyrophosphate or CAAX tetrapeptides are possible inhibitors, but as discussed by Gibbs,[88] problems of achieving tissue selectivity and cell entry must be overcome for effective therapy. Inhibitors of the AAX protease or the methyltransferase are also possible, but the design of such agents will have to take into account the fact that the substrates of the enzymes are farnesylated proteins.

ACKNOWLEDGMENTS

Work in the authors' laboratories is supported by grants from the Cancer Research Campaign and the Medical Research Council.

REFERENCES

1. **Willumsen, B. M., Norris, K., Papageorge, A. G., Hubber, N. L., and Lowy, D. R.,** Harvey murine sarcoma p21 *ras* protein: biological significance of the cysteine nearest the carboxy-terminus, *EMBO J.*, 3, 2581, 1984.
2. **Gutierrez, L., Magee, A. I., Marshall, C. J., and Hancock, J. F.,** Post-translation processing of p21^{ras} is two-step and involves carboxyl-methylation and carboxy-terminal proteolysis, *EMBO J.*, 8, 1093, 1989.
3. **Hancock, J. F., Magee, A. I., Childs, J. E., and Marshall, C. J.,** *All* ras proteins are polyisoprenylated but only some are palmitoylated, *Cell*, 57, 1167, 1989.
4. **Casey, P. J., Solski, P. A., Der, C. J., and Buss, J.,** p21^{ras} is modified by the isoprenoid farnesol, *Proc. Natl. Acad. Sci. U.S.A.*, 86, 1167, 1989.
5. **Anderegg, R. J., Betz, R., Carr, S. A., Crabb, J. W., and Duntze, W.,** Structure of *S. cerevisiae* mating hormone a-factor, *J. Biol. Chem.*, 263, 18236, 1988.
6. **Lai, R. K., Perez-Sala, D., Canada, F. J., and Rando, R. R.,** The gamma subunit of transducin is farnesylated, *Proc. Natl. Acad. Sci. U.S.A.*, 87, 7673, 1990.

7. **Fukada, Y., Takao, T., Ohuguro, H., Yoshizawa, T., Akino, T., and Shimonishi, Y.,** Farnesylated γ-subunit of photoreceptor G-protein indispensible for GTP binding, *Nature,* 346, 658, 1990.

8. **Fujiyama, A. and Tamanoi, F.,** RAS2 protein of *Saccharomyces cerevisiae* undergoes removal of methionine at N terminus and removal of three amino acids at C terminus, *J. Biol. Chem.,* 265, 3362, 1990.

9. **Clarke, S., Vogel, J. P., Deschenes, R. J., and Stock, J.,** Posttranslational modification of the Ha-ras oncogene protein: evidence for a third class of protein carboxyl methyltransferase, *Proc. Natl. Acad. Sci. U.S.A.,* 85, 4643, 1988.

10. **Maltese, W. A. and Robishaw, J. D.,** Isoprenylation of C-terminal cysteine in a G-protein gamma subunit, *J. Biol. Chem.,* 265, 18071, 1990.

11. **Magee, A. I. and Newman, C.,** personal communication.

12. **Rilling, H. C., Breunger, E., Epstein, W. W., and Crain, P. F.,** Prenylated proteins: the structure of the isoprenoid group, *Science,* 247, 318, 1990.

13. **Farnsworth, C. C., Gelb, M. H., and Glomset, J. A.,** Identification of geranylgeranyl modified proteins in HeLa cells, *Science,* 247, 320, 1990.

14. **Epstein, W. W., Lever, D. C., and Rilling, H. C.,** Prenylated proteins: synthesis of geranylgeranylcysteine and identification of this thioether as a component of proteins in CHO cells, *Proc. Natl. Acad. Sci. U.S.A.,* 87, 7352, 1990.

15. **Yamane, H. K., Farnsworth, C. C., Xie, H., Howald, W., Fung, B. K. K., Clarke, S., Gelb, M., and Glomset, J. A.,** Brain G protein γ-subunits contain an all-trans-geranylgeranyl-cysteine methyl ester, *Proc. Natl. Acad. Sci. U.S.A.,* 87, 5868, 1990.

16. **Mumby, S. M., Casey, P. J., Gilman, A. G., Gutowski, S., and Sternweis, P. C.,** G protein γ subunits contain a 20-carbon isoprenoid, *Proc. Natl. Acad. Sci. U.S.A.,* 87, 5873, 1990.

17. **Kawata, M., Farnsworth, C. C., Yoshida, Y., Gelb, M. H., Glomset, J. A., and Takai, Y.,** Posttranslationally processed structure of the human platelet protein smg p21B: evidence for geranylgeranylation and carboxy methylation of the C-terminal cysteine, *Proc. Natl. Acad. Sci. U.S.A.,* 87, 8960, 1990.

18. **Bus, J. E., Quillam, L. A., Kato, K., Casey, J., Solski, P. A., Wong, G., Clark, R., McCormick, F., Bokoch, G. M., and Der, C. J.,** The COOH terminal domain of the rap1a (Krev-1) protein is isoprenylated and supports transformation by an H-ras:rap1a chimeric protein, *Mol. Cell. Biol.,* 11, 1523, 1991.

19. **Maltese, W. A. and Sheridan, K. M.,** Isoprenoid modification of G25K (Gp), a low molecular mass GTP-binding protein distinct from p21ᵣᵃˢ, *J. Biol. Chem.,* 265, 17883, 1990.

20. **Yamane, H. K., Farnsworth, C. C., Xie, H., Evans, T., Howald, W., Gelb, M. H., Glomset, J. A., Clarke, S., and Fung, B. K. K.,** Membrane-binding domain of the small G protein G25K contains an S-(all-trans-geranylgeranyl) cysteine methyl ester at its carboxyl terminus, *Proc. Natl. Acad. Sci. U.S.A.,* 88, 286, 1991.

21. **Seabra, M. C., Reiss, Y., Casey, P. J., Brown, M. S., and Goldstein, J. L.,** Protein farnesyl transferase and geranylgeranyl transferase share a common α subunit, *Cell,* 65, 429, 1991.

22. **Robbins, S. M., Williams, J. G., Jermyn, K. A., Spiegelman, G. G., and Weeks, G.,** Growing and developing *Dictyostelium* cells express different ras genes, *Proc. Natl. Acad. Sci. U.S.A.,* 86, 938, 1989.

23. **Khoshari-Far, R., Lutz, R. J., Cox, A. D., Conroy, L., Bourne, J. R., Sinensky, M., Balch, W. E., Buss, J. E., and Der, C. J.,** Isoprenoid modification of rab proteins terminating in CC or CXC motifs, *Proc. Natl. Acad. Sci. U.S.A.,* 88, 6262, 1991.

24. **Kinsella, B. T. and Maltese, W. A.,** rab GTP-binding proteins implicated in vesicular transport are isoprenylated *in vitro* at cysteines within a novel carboxyterminal motif, *J. Biol. Chem.,* 266, 8540, 1991.

25. **Fawell, E. H., Hancock, J. F., Giannakourous, T., Newman, C. M. H., Armstrong, J., and Magee, A. I.,** Posttranslational processing of *Schizosaccharomyces pombe* YPT proteins, *J. Biol. Chem.,* 267, 11329, 1992.

26. Manne, V., Roberts, D., Tobin, A., O'Rourke, E., De Virgilio, M., Meyers, C., Ahmed, N., Kurz, B., Resh, M., Kung, H. F., and Barbacid, M., Identification and preliminary characterization of protein-cysteine farnesyltransferase, *Proc. Natl. Acad. Sci. U.S.A.*, 87, 7541, 1990.

27. Reiss, Y., Goldstein, J. L., Seabra, M. C., Casey, P. J., and Brown, M. S., Inhibition of purified p21^{ras} farnesyl:protein transferase by Cys-AAX tetrapeptides, *Cell*, 62, 81, 1990.

28. Reiss, Y., Seabra, M. C., Armstrong, S. A., Slaughter, C. A., Goldstein, J. L., and Brown, M. S., Non-identical subunits of p21^{ras} farnesyltransferase, *J. Biol. Chem.*, 266, 10672, 1991.

29. Schaber, M. D., O'Hara, M. B., Garsky, V. M., Mosser, S. C., Bergstrom, J. D., Moores, S. L., Marshall, M. S., Friedman, P. A., Dixon, R. A., and Gibbs, J. B., Polyisoprenylation of *ras in vitro* by a farnesyl-protein transferase, *J. Biol. Chem..*, 265, 14701, 1990.

30. Moores, S. L., Schaber, M. D., Mosser, S. D., Rands, E., O'Hara, M. B., Garski, V. M., Marshall, M. S., Pompliano, D. L., and Gibbs, J. B., Sequence dependence of protein prenylation, *J. Biol. Chem.*, 266, 14603, 1991.

31. Chen, W. J., Andres, D. A., Goldstein, J. L., Russel, D. W., and Brown, M. S., cDNA cloning and expression of the peptide binding β-subunit of rat p21^{ras} farnesyl transferase, the counterpart of the yeast DPR1/RAM, *Cell*, 66, 327, 1991.

32. Powers, S., Michaelis, S., Broek, D., Santa-Anna, A. S., Field, J., Herzkowitz, I., and Wigler, M., RAM a gene of yeast required for a functional modification of RAS proteins and for production of mating pheromone a factor, *Cell*, 47, 413, 1986.

33. Fujiyama, A., Matsumoto, K., and Tamanoi, F., A novel yeast mutant defective in the processing of *ras* proteins: assessment of the effect of the mutation on processing steps, *EMBO J.*, 6, 223, 1987.

34. Finegold, A. A., Schafer, W. R., Rine, J., Whiteway, M., and Tamanoi, F., Common modifications of trimeric G proteins and *ras* protein: involvement of polyisoprenylation, *Science*, 249, 165, 1990.

35. Schafer, W. R., Trueblood, C. E., Yang, C. C., Mayer, M. P., Rosenberg, S., Poulter, C. D., Kim, S. H., and Rine, J., Enzymatic coupling of cholesterol intermediates to a mating pheromone precursor and to the ras protein, *Science*, 249, 1133, 1990.

36. Goodman, L. E., Judd, S. R., Farnsworth, C. C., Powers, S., Gelb, M. H., Glomset, J. A., and Tamanoi, F., Mutants of *S. cerevisiae* defective in the farnesylation of *ras* proteins, *Proc. Natl. Acad. Sci. U.S.A.*, 87, 9665, 1990.

37. He, B., Chen, P., Chen, S.-Y., Vancura, K. L., Michelis, S., and Powers, S., RAM2, an essential gene of yeast, and RAM1 encode the two polypeptide components of the farnesyltransferase that prenylates a-factor and *ras* proteins, *Proc. Natl. Acad. Sci. U.S.A.*, 88, 11373, 1991.

38. Finegold, A. A., Johnson, D. I., Farnsworth, C. C., Gelb, M. H., Judd, S. R., Glomset, J. A., and Tamanoi, F., Protein geranylgeranyltransferase of *S. cerevisiae* is specific for cys-xaa-xaa-leu motif proteins and requires the CDC43 gene product but not the DPR1 product, *Proc. Natl. Acad. Sci. U.S.A.*, 88, 4448, 1991.

39. Casey, P. J. Thissen, J. A., and Moonmaw, J. F., Enzymatic modification of proteins with a geranylgeranyl isoprenoid, *Proc. Natl. Acad. Sci. U.S.A.*, 88, 8631, 1991.

40. Horiuchi, H., Kawata, M., Katayama, M., Yoshida, Y., Musha, T., Ando, S., and Takai, Y., A novel prenyltransferase for a small GTP binding protein having a C-terminal Cys-Ala-Cys structure, *J. Biol. Chem.*, 266, 16981, 1991.

41. Farnsworth, C. C., Kawata, M., Yoshida, Y., Takai, Y., Gelb, M. H., and Glomset, J. A., *Proc. Natl. Acad. Sci. U.S.A.*, 88, 6196, 1991.

42. Mathias, P., Chavrier, P., Nigg, E. A., and Zerial, M., Isoprenylation of *rab* proteins on structurally distinct cysteine motifs, *J. Cell Sci.*, 102, 857, 1992.

43. Rossi, G., Jiang, Y., Newman, A. P., and Ferro-Novick, S., Dependence of Ypt1 and Sec4 membrane attachment on Bet2, *Nature*, 351, 158, 1991.

44. Magee, A. I., personal communication.

45. **Hancock, J. F., Cadwallader, K. C., and Marshall, C. J.,** Methylation and proteolysis are essential for efficient membrane binding of prenylated p21$^{K-ras(B)}$, *Embo J.,* 10, 641, 1991.

46. **Stephenson, R. C. and Clarke, S.,** Identification of a C-terminal carboxylmethyl transferase in rat liver membranes utilizing a synthetic farnesyl cysteine containing peptide substrate, *J. Biol. Chem.,* 265, 16248, 1990.

47. **Hrycyna, C. A. and Clarke, S.,** Farnesyl cysteine C-terminal methyltransferase activity is dependent upon the STE14 gene product in *Saccharomyces cerevisiae, Mol. Cell. Biol.,* 10, 5071, 1990.

48. **Hrycyna, C. A., Sapperstein, S. K., Clarke, S., and Michaelis, S.,** The *Saccharomyces cerevisiae* STE14 gene encodes a methyltransferase that mediates C-terminal methylation of a-factor and *ras* proteins, *EMBO J.,* 10, 1699, 1991.

49. **Tan, E. W., Perez-Sala, D., Canada, F. J., and Rando, R. J.,** Identifying the recognition unit for G protein methylation, *J. Biol. Chem.,* 266, 10719, 1991.

50. **Sapperstein, S. K.,** CSH Abs. 1989.

51. **Hancock, J. F., Paterson, H., and Marshall, C. J.,** A polybasic domain or palmitoylation is required in addition to the CAAX motif to localize p21^{ras} to the plasma membrane, *Cell,* 63, 133, 1990.

52. **Basson, M. E., Thorness, M., Finer-Moore, J., Stroud, R. M., and Rine, J.,** Structural and functional conservation between yeast and human 3-hydroxy-3-methylglutaryl coenzyme A reductase, the rate limiting enzyme of sterol biosynthesis, *Mol. Cell. Biol.,* 8, 3797, 1988.

53. **Kawahara, Y., Kawata, M., Sunako, M., Araki, S., Koide, M., Ouda, T., Fukuzaki, H., and Takai, Y.,** Identification of a major GTP binding protein in bovine aortic smooth muscle cytosol as the rhoA gene product, *Biochem. Biophys. Res. Commun.,* 170, 673, 1990.

54. **Adamson, P. and Hall, A.,** personal communication.

55. **Isomura, M., Kikuchi, A., Ohga, N., and Takai, Y.,** Regulation of binding of rhoB p20 to membranes by its specific regulatory protein, GDP dissociation inhibitor, *Oncogene,* 6, 119, 1991.

56. **Hall, A.,** The cellular functions of small GTP binding proteins, *Science,* 249, 635, 1990.

57. **Willingham, M. C., Pastan, I., Shih, T. Y., and Scolnick, E. M.,** Localization of the src gene product of the Harvey strain of MSV to plasma membrane, *Cell,* 19, 1005, 1980.

58. **Beranger, F., Goud, B., Tavitian, A., and de Guzburg, J.,** Association of the Ras-antagonistic Rap1/Krev-1 proteins with the Golgi complex, *Proc. Natl. Acad. Sci. U.S.A.,* 88, 1606, 1991.

59. **Barbacid, M.,** ras genes, *Annu. Rev. Biochem.,* 56, 779, 1987.

60. **Hancock, J. F., Cadwallader, K. C., Paterson, H., and Marshall, C. J.,** A CAAX or a CAAL motif and a second signal are sufficient for plasma membrane targeting of *ras* proteins, *EMBO J.,* 10, 4033, 1991.

61. **Lowe, D. G., Capon, D. J., Delwart, E., Sakaguchi, A. Y., Naylor, S. L., and Goeddel, D. G.,** Structure of the human and mouse R-ras genes, novel genes closely related to ras proto-oncogenes, *Cell,* 48, 137, 1987.

62. **Beranger, F., Tavitian, A., and de Gunzberg, J.,** Post-translational processing and sub-cellular localization of the Ras-related Rap2 protein, *Oncogene,* 6, 1835, 1991.

63. **Resh, M. D. and Ling, H. P.,** Identification of a 32K plasma membrane protein that binds to the myristylated amino-terminal sequence of p60v-src, *Nature,* 346, 84, 1990.

64. **Hancock, J. F. and Marshall, C. J.,** unpublished data.

65. **Buss, J. E. and Sefton, B. M.,** Direct identification of palmitic acid as the lipid attached to p21 ras, *Mol. Cell. Biol.,* 6, 1116, 1986.

66. **Weeks, G., Lima, A., and Pawson, T. R.,** *ras* encoded protein in *Dictyosteleum discoideum* is acylated and membrane associated, *Mol. Microbiol.,* 1, 347, 1987.

67. **Magee, A. I., Gutierrez, L., McKay, I., Marshall, C. J., and Hall, A.,** Dynamic fatty acylation of p21^{N-ras}, *EMBO J.,* 6, 3353, 1987.

68. **Lapetina, E. G., Lacal, J. C., Reep, B. R., and Molina y Vedia, L.**, A ras-related protein is phosphorylated and translocated by agonists that increase cAMP levels in human platelets, *Proc. Natl. Acad. Sci. U.S.A.*, 86, 3131, 1989.

69. **Sahyoun, N., McDonald, O. B., Farrell, F., and Lapetina, E. G.**, Phosphorylation of a Ras-related GTP-binding protein, Rap-1b, by a neuronal Ca^{2+}/calmodulin-dependent protein kinase, CaM kinase Gr, *Proc. Natl. Acad. Sci. U.S.A.*, 88, 2643, 1991.

70. **Siess, W., Winegar, D. A., and Lapetina, E. G.**, Rap1-B is phosphorylated by protein kinase A in intact human platelets, *Biochem. Biophys. Res. Commun.*, 170, 944, 1990.

71. **Hata, Y., Kaibuchi, K., Kawamura, S., Hiroyoshi, M., Shirataki, H., and Takai, Y.**, Enhancement of the actions of smg p21 GDP/GTP exchange protein by the protein kinase A-catalyzed phosphorylation of smg p21, *J. Biol. Chem.*, 266, 6571, 1991.

72. **Ballester, R. M., Furth, M. E., and Rosen, O. M.**, Phorbol Ester and protein kinase C-mediated phosphorylation of the cellular Kirsten ras gene product, *J. Biol. Chem.*, 262, 2688, 1987.

73. **Matsui, Y., Kikuchi, A., Araki, S., Hata, Y., Kondo, J., Teranishi, Y., and Takai, Y.**, Molecular cloning and characterization of a novel type of regulatory protein, *Mol. Cell. Biol.*, 10, 4116, 1990.

74. **Ohga, N., Kikuchi, A., Ueda, T., Yamamoto, J., and Takai, Y.**, Rabbit intestine contains a protein that inhibits dissociation of GDP form and the subsequent binding of GTP to rhoB p20, a *ras* like GTP binding protein, *Biochem. Biophys. Res. Commun.*, 163, 1523, 1989.

75. **Ueda, T., Kikuchi, A., Ohga, N., Yamamoto, J., and Takai, Y.**, Purification and characterization from bovine brain cytosol of a novel regulatory protein inhibiting the dissociation of GDP from and the subsequent binding of GTP to rhoB p20, a *ras* p21-like GTP binding protein, *J. Biol. Chem.*, 265, 9373, 1990.

76. **Fukumoto, Y., Kaibuchi, K., Hori, Y., Fujioka, H., Araki, S., Ueda, T., Kikuchi, A., and Takai, Y.**, Molecular cloning and characterization of a novel type of regulatory protein (GDI) for the rho proteins, ras p21-like small GTP-binding proteins, *Oncogene*, 5, 1321, 1990.

77. **Hori, Y., Kikuchi, A., Isomura, M., Katayama, M., Miura, Y., Fujioka, H., Kaibuchi, K., and Takai, Y.**, Post-translational modifications of the C-terminal region of the rho protein are important for its interaction with membranes and the stimulatory and inhibitory GDP/GTP exchange proteins, *Oncogene*, 6, 515, 1991.

78. **Isomura, M., Kaibuchi, K., Yamamoto, T., Kawamura, S., Katayama, M., and Takai, Y.**, Partial purification and characterization of GDP dissociation stimulator (GDS) for the rho proteins from bovine brain cytosol, *Biochem. Biophys. Res. Commun.*, 169, 652, 1990.

79. **Yamamoto, T., Kaibuchi, K., Mizuno, T., Hiroyoshi, M., Shirataki, H., and Takai, Y.**, Purification and characterization from bovine brain cytosol of proteins that regulate the GDP/GTP exchange reaction of smg p21s, ras p21-like GTP-binding proteins, *J. Biol. Chem.*, 265, 16626, 1990.

80. **Kaibuchi, K., Mizuno, T., Fujioka, H., Yamamoto, T., Kishi, K., Fukumoto, Y., Hori, Y., and Takai, Y.**, Molecular cloning of the cDNA for stimulatory GDP/GTP exchange protein for smg p21s (ras p21-like small GTP-binding proteins) and characterization of stimulatory GDP/GTP exchange protein, *Mol. Cell. Biol.*, 11, 2873, 1991.

81. **Araki, S., Kaibuchi, K., Sasaki, T., Hata, Y., and Takai, Y.**, Role of the C-terminal region of smg p25A in its interaction with membranes and the GDP/GTP exchange protein, *Mol. Cell. Biol.*, 11, 1438, 1991.

82. **Mizuno, T., Kaibuchi, K., Yamamoto, T., Kawamura, M., Sakoda, T., Fujioka, H., Matsuura, Y., and Takai, Y.**, A stimulatory GDP/GTP exchange protein for smgp21 is active on the post-translationally processed from of c-Ki ras p21 and rhoA p21, *Proc. Natl. Acad. Sci. U.S.A.*, 88, 6442, 1991.

83. **Araki, S., Kikuchi, A., Hata, Y., Isomura, M., and Takai, Y.**, Regulation of reversible binding of smg p25A, a ras p21-like GTP-binding protein, to synaptic plasma membranes and vesicles by its specific regulatory protein, GDP dissociation inhibitor, *J. Biol. Chem.*, 265, 13007, 1990.

84. **Kawamura, S., Kaibuchi, K., Hiroyoshi, M., Hata, Y., and Takai, Y.,** Stoichiometric interaction of smg p21 with its GDP/GTP exchange protein and its novel action to regulate the translocation of smg p21 between membrane and cytoplasm, *Biochem. Biophys. Res. Commun.,* 174, 1095, 1991.

85. **Chavrier, P., Gorvel, J.-P., Steizer, E., Simons, K., Gruenberg, J., and Zerial, M.,** Hypervariable C-terminal domain of rab proteins acts as a targeting signal, *Nature,* 353, 769, 1991.

86. **Ohguro, H., Fukada, Y., Yoshizawa, T., Saito, T., and Akino, T.,** A specific beta gamma-subunit of transducin stimulates ADP-ribosylation of the alpha-subunit by pertussis toxin, *Biochem. Biophys. Res. Commun.,* 167, 1235, 1990.

87. **Schafer, W. R., Kim, R., Sterene, R., Thorner, J., Kim, S. H., and Rine, J.,** Genetic and pharmacological suppression of oncogenic mutations in *ras* genes of yeast and humans, *Science,* 245, 379, 1989.

88. **Gibbs, J. B.,** Ras C-terminal processing enzymes — new drug targets? *Cell,* 65, 1, 1991.

89. **Drivas, G. T., Shih, A., Coutavas, E., Rush, M. G.,and D'Eustachio, P.,** Characterization of four novel ras-like genes expressed in a human teratocarcinoma cell line, *Mol. Cell. Biol.,* 10, 1793, 1990.

90. **Drubin, D. G.,** Development of cell polarity in budding yeast, *Cell,* 65, 1093, 1991.

91. **Shinjo, K., Koland, J. G., Hart, M. J., Narasimhan, V., Johnson, D. I., Evans, T., and Cerione, R. A.,** Molecular cloning of the gene for the human placental GTP-binding protein Gp (G25K): identification of this GTP-binding protein as the human homolog of the yeast cell-division-cycle protein CDC42, *Proc. Natl. Acad. Sci. U.S.A.,* 87, 9853, 1990.

92. **Munemitsu, S., Innis, M. A., Clark, R., McCormick, F., Ullrich, A., and Polakis, P.,** Molecular cloning and expression of a G25K cDNA, the human homolog of the yeast cell cycle gene CDC42, *Mol. Cell. Biol.,* 10, 5977, 1990.

93. **Han, M. and Sternberg, P. W.,** let-60, a gene that specifies cell fates during *C. elegans* vulval induction, encodes a *ras* protein, *Cell,* 63, 921, 1990.

94. **Andeol, Y., Gusse, M., and Mechali, M.,** Characterisation and expression of a *Xenopus ras* during oogenesis and development, *Dev. Biol.,* 139, 24, 1990.

95. **Adamson, P., Paterson, H. F., and Hall, A.,** Intracellular localization of p21[rho] proteins, *J. Cell Biol.,* 119, 617, 1992.

96. **Adamson, P., Marshall, C. J., Hall, A., and Tilbrook, P. A.,** Post-translational modifications of p21[rho] proteins, *J. Biol. Chem.,* 267, 20033, 1992.

97. **Cox, A. D., Hisaka, M. M., Buss, J. E., and Der, C. J.,** Specific isoprenoid modification is required for function of normal, but not oncogenic, Ras protein, *Mol. Cell. Biol.,* 12, 2606, 1992.

Chapter 4

BIOLOGICAL ACTIVITIES OF *ras* GENES

Steven Sorscher, Axel Schönthal, Arthur S. Alberts, Judy L. Meinkoth,
and James R. Feramisco

TABLE OF CONTENTS

I. INTRODUCTION

ras oncogenes came into the spotlight of cancer research in 1982 when several independent groups discovered that the transforming genes of certain human tumors were activated counterparts of the normal cellular homologs of viral *ras* oncogenes.[1-8] Subsequent analyses revealed that *ras* gene mutations can be found in a variety of tumor types. Its presence and expression are well documented in many of the most common and devastating human malignancies including carcinomas of the pancreas, lung, colon, kidney, skin, thyroid, bladder, and testicle, as well as in hematologic malignancies including acute myelogenous leukemia[9-12] (for a review see Reference 13).

Several lines of evidence suggest that activation of one of the *ras* family genes (H-*ras*, K-*ras*, N-*ras*) may be an early event in tumorigenesis, including its expression in premalignant tissues.[14-18] Activation of *ras* may result from direct interaction of a carcinogen with the *ras* sequence,[19] and particular carcinogens may cause tumorigenesis only by inducing highly specific *ras* mutations. For example, urethan, a carcinogen, induces K-*ras* mutations in adenomas and adenocarcinomas in the mouse A/T strain. In this case a leucine substitution at codon 61 correlated with the more benign disease, while an arginine substitution correlated with the more malignant tumors or cell lines.[20] In general, however, requirements for *ras*-induced transformation may be more complex. For example, *ras* mutations induced with nitromethylurea (NMU) in neonatal rats primed the animals for the development of tumors only later in life as their ovaries produced estrogen, or if estrogen was given to ovariectomized animals. In contrast, tumorigenesis was blocked if their ovaries were removed early in life without estrogen replacement.[14]

ras expression is becoming an important clinical consideration as well. *ras* activation may correlate with resistance to natural killer cell activity,[21] radiation resistance,[22,23] metastatic potential,[22,24,25] the effectiveness of cytotoxic therapy against malignancy,[26] and with prognosis in various malignancies.[27-31]

Thus, mutationally activated members of the *ras* gene family are likely to play an important role in cellular growth control as well as in the initiation and progression of malignancy. Extensive research activity has been dedicated to analyzing the normal functions of *ras* protooncogene products and how these functions are changed after conversion of the genes into transforming oncogenes. It is now known that *ras* proteins play key roles in a variety of cellular activities, including cell growth, differentiation, secretion, and protein trafficking. The intriguing observations that oncogenic *ras* causes transformation in many cellular settings, and terminal differentiation in others, demonstrates that the biological effects of *ras* are truly pleiotrophic. In this chapter we discuss some of the apparent requirements for transformation or for differentiation induced by *ras*. Furthermore, we discuss some of the signal transduction pathways affected by *ras*, and describe our studies of potential sequences which may serve as nuclear targets for *ras* action.

II. TRANSFORMATION

Several years ago it became clear that human tumors include genes which are able to transform nonmalignant cells. Groups have demonstrated that many of these are *ras* oncogenes.[9,32-34] Introduction of the *ras* oncoprotein into cells by microinjection or overexpression of cellular *ras* protein by transfection of expression plasmids results in cellular transformation and increased DNA synthesis.[35-37] In contrast, introduction of antisense *ras* constructs results in growth inhibition of a human lung cancer cell line with a homozygous spontaneous K-*ras* mutation.[38] Similarly, microinjection of an anti-*ras* antibody resulted in transient reversion of a transformed cell line to a nontransformed phenotype with a decreased growth rate.[39,40]

Since cellular transformation is likely to require changes in gene expression, one approach has been to study the expression of proteins potentially induced by *ras*, that might be related to the transformed phenotype. Indeed, there are now numerous examples of proteins produced in response to *ras* expression which are probably involved in generating the transformed or metastatic phenotype. One of these proteins is stromelysin, a secreted metalloprotease which degrades components of the basement membrane and stroma.[41-44] Evidence suggests that stromelysin may be involved in tumor invasion and metastasis (for review see Reference 45). Stromelysin mRNA, for example, is found more predominantly in carcinogen-induced malignant tissue than in benign tissue,[45,47] and more predominantly in metastatic tissue as compared with primary tumors.[48] H-*ras* expression is able to induce stromelysin expression.[49,50] Experiments using recombinant tissue inhibitor of metalloproteinases (TIMP) to reduce metastatic lung colonization by B12 murine melanoma cells imply a role for metalloproteases in the metastatic phenotype.[51]

Type IV collagenase has been proposed to allow basement membrane penetration by malignant cells through the degradation of type IV collagen (for review see Reference 52). Immunocytochemical studies of colon, gastric, and breast cancer demonstrated that progressive disease correlated with an increased proportion of neoplastic cells positive for type IV collagenase.[53] Unlike spontaneously transformed NIH 3T3 cells, N-*ras* and H-*ras* transformed NIH 3T3 cells show significant collagen IV degrading activity.[54] Similarly, collagenase IV activity is induced in H-*ras* transformed human bronchoepithelial cells.[55] Furthermore, Garbisa has shown that collagenase IV secretion in H-*ras* transformed early passage rat embryo fibroblasts correlated with metastatic potential when these cells were introduced into nude mice.[56]

One of the best studied examples of a protein apparently important to the metastatic phenotype and regulated by *ras* expression is cathepsin L. Expression of procathepsin, the secreted precursor, correlates with metastatic potential.[57] Cathepsin L, also known as major excreted protein (MEP),[58-62] is secreted from transformed but not nontransformed NIH Swiss 3T3 cells.[63] It degrades extracellular matrix components including fibronectin, collagen, laminin,[64] and elastin,[65] as well as α1 proteinase inhibitors,[66] thereby allowing other proteases to become activated. In fact, Yagel has demonstrated that specific inhibition of cathepsin L produced by murine cancer cells blocks amniotic basement membrane penetration.[67] Both K-*ras* and H-*ras* can induce transformation and procathepsin L expression,[63,68] whereas revertants of K-*ras* transformed cells lack procathepsin L expression.[63] Finally, while T24 H-*ras* transformed C3H 10T1/2 cell lines form tumors *in vivo* following injection into mice, only the subset of cells producing procathepsin L also form lung metastasis. Furthermore, survival of these foci in the lung correlates with cathepsin L expression.[57]

Other examples of proteins potentially important to the malignant phenotype and regulated by *ras* expression include transforming growth factor α(TGFα),[69-75] transforming growth factor β (TGFβ),[76-79] epidermal growth factor receptor (and its relation to TGFα expression),[71,72,75,80,81] hyaluronon (HA) and hyaluronon-binding protein (HABP),[82] cytokines,[83] and plasminogen activators such as urokinase-type plasminogen activator (UPA).[84,85] Finally, some extracellular matrix proteins may show decreased expression in *ras*-transformed cells.[86-90]

How can *ras* proteins control cellular growth and regulate expression of proteins associated with tumorigenic transformation? Extensive analyses have shown that the *ras* proteins are GTP-binding proteins with GTPase activity. Their enzymatic activity is controlled by specific types of regulatory proteins, the GTPase-activating proteins (GAP) as well as by posttranslational modification events. This functional similarity to G proteins (enzymes that serve to transmit signals generated by ligand interactions with specific cell surface receptors into changes in intracellular functions) suggested a role for *ras* proteins in intracellular signal transduction pathways. Indeed, it has been shown that overexpression of *ras* protein stimulates the transcription of several genes whose protein products are thought to play crucial roles in the control of cell growth (for reviews see Chapters 5, 6, 10, 21, and 22 and References 91 and 92).

The most intensively studied of these genes is the c-*fos* protooncogene, one member of a large family of immediate early genes. C-*fos* is induced in different cell types by mitogens, differentiation agents, and other stimuli. *fos* plays a vital role in cell growth and differentiation (for review, see Reference 93). Overexpression or mutation of c-*fos* has been implicated in cellular transformation and tumorigenesis as well.[94] One target in the c-*fos* promoter that is responsive to *ras* activity is the serum response element (SRE), a short sequence that is also required for serum induction of the c-*fos* gene.[95] Other elements of the c-*fos* promoter, such as the cAMP response element (CRE), do not appear to be regulated by *ras*. The c-*fos* gene product plays a crucial role in cellular growth regulation, since inhibition of c-*fos* expression has been shown to block cellular proliferation.[96] The *fos* protein is part of a transcription factor, AP-1, that binds to and activates genes via a defined promoter element, the AP-1 site or TRE. The TRE (12-*O*-tetradecanoyl-phorbol-13 acetate [TPA] responsive element) was originally identified as the sequence necessary and sufficient for the activation of the collagenase gene by the tumor promoter TPA.[97,98] By using antisense *fos* expression vectors to block c-*fos* expression, it has been shown that overexpressed *ras* (or TPA treatment) is no longer able to stimulate collagenase expression.[99] These experiments established the *ras* protein as one component of regulatory signal transduction pathways, and suggested that mutated *ras* oncoproteins might continuously stimulate proliferation by sustained activation of this pathway. Since TRE sequences have been found in other genes as well (notably, for example, in the stromelysin gene), it is conceivable that activated *ras* might contribute to cell transformation by stimulating indirectly (via *fos*) the expression of a set of genes involved in cellular growth and proliferation (see Chapters 1, 5, 6, and 21 for details).

Regulation of the activity of the normal *ras* protein and its associated GAP protein is, despite intensive research efforts, not yet fully understood (see Chapters 21 and 22 for details). It does seem clear that transformation by *ras* can depend on many factors. The observation that in different cellular settings *ras* can cause phenotypic changes which vary from transformation to differentiation, suggests that the cell type is crucial in determining the biological consequences of *ras*.

A striking example of the importance of the cellular setting in determining *ras* effects is seen in the so-called nonterminally differentiated cells. These cells are said to display anticancer activity and "nonterminally differentiated" refers to previously transformed cells which revert to a benign state and become resistant to retransformation by chemical and physical carcinogens, but maintain proliferation. Native 3T3 mesenchymal stem cells in this group are resistant to transformation by the human bladder carcinoma-derived EJ *ras* oncogene, although the oncogene is integrated and expressed from the host genome.[100]

In other settings, activated *ras* can induce partial transformation. Transformation has been noted to be partial in thyroid epithelial cells. In work by Lemoine and others, *ras* expression using retroviral vectors to achieve Ha-*ras* gene transfer into primary thyroid epithelial cells prevented senescence for up to 20 rounds of cell division. Also, changes in morphology, growth factor, and anchorage dependence were seen in these cells. This system will allow further study of *ras*-induced transformation.[101]

III. DIFFERENTIATION

ras proteins are also involved in cell differentiation processes. Many examples for this function now exist including neuronal,[102-104] endocrine,[105] and lymphoid[106] cells and tissue. As is the case for transformation induced by *ras*, transcriptional activation of specific genes including nuclear protooncogenes may relate to its role in differentiation.

Probably the most extensively studied example of *ras*-induced differentiation is the rat pheochromocytoma PC12 cells, adult chromaffin-like cells that differentiate into neuron-like cells when exposed to nerve growth factor (NGF).[107] Differentiated properties of these cells include changes in neurotransmitter synthesis, neurite growth, and electrical excitability.[107] Using either microinjection of purified *ras* oncoproteins or infection with sarcoma viruses carrying the *ras* oncogene, similar patterns of differentiation are seen.[102,104,108] Furthermore, microinjection of a *ras*-specific antibody blocks differentiation of PC12 cells exposed to NGF.[109] In addition, dominant inhibitory mutant *ras* blocks NGF induced expression of c-*fos* and differentiation of PC12 cells, though lower levels of expression of the dominant inhibitory mutant *ras* could block PC12 differentiation without inhibiting *fos* induction.[110] Further evidence implicating c-*fos* expression in PC12-induced differentiation includes the repeated observation that activated *ras* and NGF are powerful stimulators of c-*fos* expression.[107,111-116] In fact, one of the first genes expressed after exposure to NGF is c-*fos*.[107,114,115] Also, *ras* oncoprotein induces a 20-fold increase in the expression of the c-*fos* gene in PC12 cells.[117] Using *ras* mutants, it has been shown that domains important for NIH 3T3 transformation overlapped with those needed for PC12 differentiation. In addition, all *ras* mutants causing PC12 differentiation activated c-*fos* transcription.[117] These observations are important since the role of *fos* in the differentiation of PC12 cells expressing *ras* is incompletely understood. Thomson has shown that a PC12 subline containing a stably integrated transforming mouse N-*ras* gene expressed c-*fos* mRNA normally in response to NGF. This expression is inhibited by activated N-*ras* expression, again suggesting interaction among these pathways.[118] All of these findings suggest that at least some portion of the signal transduction pathways used by *ras* and NGF is common.[117,119,120]

There is accumulating evidence that the activation of immediate early genes plays a key role in regulating expression, not only of later genes more directly tied to phenotypic features, but in fact to other immediate early genes as well. For example, fra-1, an immediate early gene product related to *fos*, and expressed at a later time point than *fos* in response to NGF-induced differentiation of PC12 cells, appears to not only block NGF-induced differentiation if constitutively expressed in PC12 cells, but also to downregulate transcriptional activation of c-*fos*.[121] This experimental system demonstrated how the regulation of immediate early genes may be controlled by other immediate early genes. The result of this interaction may relate to the phenotype seen.

IV. PROMOTER ELEMENTS

DeFeo-Jones and others have used an alkaline phosphatase reporter gene under the control of TRE promoter elements to study whether lovastatin might interfere with *ras*-induced transformation, probably by affecting posttranslational modifications of *ras*.[122] This work could naturally have far-reaching implications because of the known association of oncogenic ras to human malignancy.

Our laboratory has studied the role of activated *ras* proteins in signal transduction pathways which impinge upon particular promoter elements in order to ascertain which classes of genes might be potential targets of *ras* action. In an attempt to establish that specific promoter elements are targets of *ras*-mediated signaling pathways, we analyzed the transcriptional activity of defined promoter elements in response to microinjection of purified T24 *ras* protein into living cells.

TABLE 1
CRE, TRE, and SRE Cell Lines Exhibit
Regulated β-Galactosidase Expression

	CRE	TRE	SRE
8brcAMP	+ + + +	—	—
dbcAMP	+ + + +	ND	ND
Forskolin	+ + + +	ND	ND
IBMX	+ + + +	—	—
Serum (20% FCS)	—	+ + +	+ + + +
TPA	—	+ + +	+ +
PDGF	ND	+ + +	+ +
EGF	ND	+ + + +	+ +
TGFα	ND	+ + + +	ND
bFGF	ND	+	+ + + +
Insulin	—	+	+ + +
IGF-1	—	+	+ + +

Note: Cells were plated and starved for 24 h in DMEM + 0.05% FCS before stimulation. Test agents were added in DMEM + 0.05% FCS for 6 h. The cells were subsequently fixed in 3.7% formaldehyde for 5 min at 22°C and stained overnight at 37°C in 5 mM K$_3$Fe(CN)$_6$, 5 mM K$_4$ Fe (CN)$_6$-3H$_2$O, 2 mM MgCl$_2$, 1 mg/ml X-gal in PBS. ND = not determined.

To establish cell lines useful for the analysis of changes in gene expression in response to injected *ras* proteins, we isolated stably transfected cell lines containing integrated *Escherichia coli lacZ* genes under the control of a minimal promoter containing various enhancer elements.[123,123] The plasmid vector used contained the *E. coli lacZ* gene under the control of a small fragment of the RSV promoter (RSV/Z/NEO). This plasmid also contained a unique Xho I restriction site just upstream of the promoter into which synthetic oligonucleotides homologous to consensus cAMP, phorbol ester, and serum response elements (CREs, TREs, and SREs) were cloned. In addition, this plasmid contained the neomycin resistance gene to allow for G418 selection in mammalian cells. The consensus CRE element used is homologous to the CRE element found in the human vasoactive intestinal peptide gene promoter. The TRE utilized is derived from the human collagenase gene and the SRE from the mouse c-*fos* gene promoter. These plasmids were introduced into both rat and mouse fibroblasts by calcium phosphate coprecipitation and stable cell lines exhibiting regulated expression of β-galactosidase were isolated.

Cell lines containing CREs cloned upstream from the *lacZ* gene express β-galactosidase when intracellular cAMP levels are elevated, for example, following treatment with 8bromocyclicAMP (8brcAMP), dibutyryl cAMP (dbcAMP), forskolin, or isobutyl methylxanthine (IBMX); (Table 1). In contrast, serum and growth factors do not induce β-galactosidase expression in this cell line. Cell lines containing TREs exhibit different properties. Treatment of these cells with phorbol esters (TPA), EGF, TGFα, PDFG, or serum induces β-galactosidase expression, while cAMP analogs do not. Cell lines containing SREs express β-galactosidase strongly in response to serum, basic FGF (bFGF), PDGF, IGF-1, insulin, weakly in response to phorbol esters, and are unresponsive to elevations in cAMP. The specificity with which each cell line responds to agents demonstrated to act via distinct promoter elements suggested that these cells lines would provide a means by which to assess the role of putative signaling molecules in signaling pathways that result in changes in gene expression.

As described above, earlier experiments demonstrated that activated *ras* proteins induced gene expression from promoters containing TRE[117,125] and SRE[117,126] elements. More recently, a number of endogenous cellular promoters responsive to *ras* proteins have been identified.[127] These genes contain a consensus sequence element, a *ras*-responsive element (RRE) which is similar, but not identical, to a consensus TRE. In order to determine which sequence elements serve as primary targets for *ras*-mediated signaling pathways in living cells, we introduced activated T24 *ras* protein into cell lines containing CRE, TRE, and SRE cloned upstream from the *lacZ* gene and measured β-galactosidase activity in injected cells using a histochemical staining procedure.

Injection of T24 *ras* protein had no effect upon β-galactosidase expression in the CRE cell line (data not shown). However, in at least one example, transient transfection of both v-Ki-*ras* and c-Ki-*ras* expression vectors has been shown to activate CREs.[128] Whether these results reflect the analysis used (transient transfection vs. microinjection), the type of *ras* protein analyzed (Ki vs. Ha) or other flanking sequence elements remains unclear. Microinjection of T24 *ras* into the TRE cell line resulted in a strong stimulation of β-galactosidase expression (Figure 1). The *ras*-injected cells contained an abundant dark blue precipitate as well as morphological changes including membrane ruffling and increased vesicle formation typically induced following microinjection of *ras* protein. The same results were obtained following injection of activated N-*ras* protein (data not shown). In contrast to the TRE cell line, there were either small or no increases in β-galactosidase expression in response to introduction of T24 *ras* into BALB/c fibroblasts containing SREs upstream from the *lacZ* gene (Figure 1). This was in stark contrast to treatment with 20% fetal calf serum (FCS) which stimulated high levels of β-galactosidase expression in these cells. Since microinjection of the *ras* protein induces endogenous c-*fos* expression presumably through SREs,[117,125] we introduced the T24 *ras* protein into a second SRE-containing cell line, Rat 2 fibroblasts. Microinjection of T24 *ras* into the Rat 2 SRE cell line consistently failed to induce β-galactosidase expression. Although the BALB/c and rat 2 SRE cell lines are equally serum-responsive, they differ in the extent to which they respond to phorbol esters (Table 2).

These results suggest that one of the primary transcriptional targets of *ras*-mediated signaling pathways is the TRE or AP-1-binding site. These results are consistent with a large number of others in which TRE-containing promoters have been shown to be inducible by *ras* proteins. Moreover, the similarity of

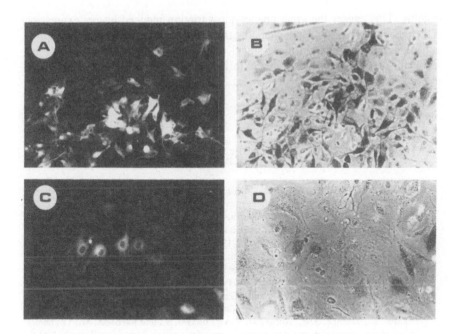

FIGURE 1. Microinjection of T24 *ras* into TRE and SRE β-galactosidase indicator cell lines. T24 *ras* stimulates the TRE cell line strongly while the SRE is induced slightly. (A) Immunofluorescent staining of marker antibody (Sheep IgG; 2 mg/ml) injected along with T24 *ras* (1 mg/ml) into a Rat 2 fibroblast cell line containing a stably transfected TRE-lac Z reporter gene; (B) phase-contrast photomicrograph of the same field following histochemical staining with the chromogenic substrate X-gal. T24 microinjected into the BALB/c 3T3 SRE β-galactosidase indicator cell line; (C) immunofluorescence of coinjected marker antibody; and (D) the same field stained with X-gal.

TABLE 2
Quantitation of Serum and TPA-Inducible
β-Galactosidase Expression in TRE- and
SRE-Containing Cell Lines

	(% blue cells)		
Stimulation	Rat 2 TRE	B/c SRE	Rat 2 SRE
0.05% FCS	10.3	11.9	3.5
20% FCS	33.8	47.4	46.9
200 ng/ml TPA	39.0	24.8	7.8

RREs to TREs provides further experimental support that the TRE is one of the primary targets of *ras*-mediated signal transduction.

In summary, the biological effects of *ras* are diverse and depend on the complex signal transduction pathways which have common features, but ultimately result in the activation of unique combinations of genes in different cell systems. This causes the great phenotypic variation induced by *ras* expression in different cell systems. Understanding the transcriptional control of the expression of important genes in these signal transduction pathways will enable investigators to plan rational therapies to influence the effects of *ras* on the initiation and progression of human malignancies.

REFERENCES

1. **Balmain, A. and Pragnell, I. B.**, Mouse skin carcinoma induced in vivo by chemical carcinogens have a transforming Harvey-ras oncogene, *Nature,* 303, 72, 1983.
2. **Sukumar, S., Notario, V., Martin-Zanca, D., and Barbacid, M.**, Induction of mammary carcinomas in rats by nitroso-methylurea involves malignant activation of H-ras-1 locus by single point mutations, *Nature,* 306, 658, 1983.
3. **Eva, A. and Aaronson, S. A.**, Frequent activation of c-kis as a transforming gene in fibrosarcomas induced by methylcholantrene, *Science,* 220, 955, 1983.
4. **Shimizu, K., Goldfarb, M., Suard, Y., Perucho, M., Li, Y., Kamata, T., Feramisco, J. R., Stavnezer, E., Fogh, J., and Wigler, M.**, Three human transforming genes are related to the viral ras oncogenes, *Proc. Natl. Acad. Sci. U.S.A.,* 80, 2112, 1983.
5. **Santos, E., Tronick, S. R., Aaronson, S. A., Pulciani, S., and Barbacid, M.**, T24 human bladder carcinoma oncogene is an activated form of the normal human homologue of BALB- and Harvey-MSV transforming genes, *Nature,* 298, 343, 1982.
6. **Der, C. J., Krontiris, T. G., and Cooper, G. M.**, Transforming genes of human bladder and lung carcinoma cell lines are homologous to the ras genes of Harvey and Kirsten sarcoma viruses, *Proc. Natl. Acad. Sci. U.S.A.,* 79, 3637, 1982.
7. **Parada, L. F., Tabin, C. J., Shih, C., and Weinberg, R. A.**, Human EJ bladder carcinoma oncogene is homologue of Harvey sarcoma virus ras gene, *Nature,* 297, 474, 1982.
8. **Guerrero, I., Villasante, A., D'Eustachio, P., and Pellicer, A.**, Isolation, characterization, and chromosome assignment of mouse N-ras gene from carcinogen-induced thymic lymphoma, *Science,* 225, 1041, 1984.
9. **Barbacid, M.**, ras genes, *Annu. Rev. Biochem.,* 56, 779, 1987.
10. **Bos, J. L., Fearon, E. R., Hamilton, S. R., Verlaan-deVries, M., Van Boom, J. H., Van der Eb, A. J., and Vogelstein, B.**, Prevalence of ras gene mutations in human colorectal cancers, *Nature,* 327, 293, 1987.
11. **Bos, J. L.**, The ras gene family and human carcinogenesis, *Mutat. Res.,* 195, 255, 1988.
12. **Shibata, D., Capella, G., and Perucho, M.**, Mutational activation of the c-K-ras gene in human pancreatic carcinomas, *Baillieres Clin. Gastroenterol.,* 4, 151, 1990.
13. **Kahn, S., Yamamoto, F., Almoguera, C., Winter, E., Forrester, K., Jordano, J., and Perucho, M.**, The c-K-ras gene and human cancer, *Anticancer Res.,* 7, 639, 1987.
14. **Kumar, R., Sukumar, S., and Barbacid, M.**, Activation of ras oncogenes preceding the onset of neoplasia, *Science,* 248, 1101, 1990.
15. **Sagae, S., Kudo, R., Kuzumaki, N., Hisada, T., Mugikura, Y., Nihei, T., Takeda, T., and Hashimoto, M.**, Ras oncogene expression and progression in intraepithelial neoplasia of the uterine cervix, *Cancer,* 66, 295, 1990.
16. **Minami, T., Himeno, S., Shinomura, Y., Kariya, Y., Tarui, S.., and Sakurai, M.**, Rearrangement, amplification and overexpression of c-Ha-Ras gene in premalignant lesion of Turcot's syndrome, *Jpn. J. Med.,* 29, 391, 1990.
17. **Meltzer, S. J., Mane, S. M., Wood, P. K., Resau, J. H., Newkirk, C., Terzakis, J. A., Korelitz, B. I., Weinsten, W. M., and Needleman, S. W.**, Activation of c-Ki-ras in human gastrointestinal dysplasias determined by direct sequencing of polymerase chain reaction products, *Cancer Res.,* 50, 3627, 1990.

18. **Saski, M., Sugio, K., and Sasazuki, T.,** K-ras activation in colorectal tumors from patients with familial polyposis coli, *Cancer,* 65, 2576, 1990.

19. **Sukumar, S.,** An experimental analysis of cancer: role of ras oncogenes in multistep carcinogenesis, *Cancer Cells,* 7, 199, 1990.

20. **Nuzum, E. O., Malkinson, A. M., and Beer, D. G.,** Specific Ki-ras codon 61 mutations may determine the development of urethan-induced mouse lung adenomas or adenocarcinomas, *Mol. Carcinogenesis,* 3, 287, 1990.

21. **Bagli, D. J., D'Emmilia, J. C., Summerhayes, I. C., Steele, G. D., and Barlozzari, T.,** c-Ha-ras-I oncogene-induced differentiation and natural killer cell resistance in a human colorectal carcinoma cell line, *Cancer Res.,* 50, 2518, 1990.

22. **McKenna, W. G., Weiss, M. C., Bakanauskas, V. J., Sandler, H., Kelsten, M. L., Biaglow, J., Tuttle, S. W., Endlich, B., Ling, C. C., and Muschel, R. J.,** The role of the H-ras oncogene in radiation resistance and metastasis, *Int. J. Radiat. Oncol. Biol. Phys.,* 18, 849, 1990.

23. **Sklar, M. D.,** The ras oncogenes increase the intrinsic resistance of NIH 3T3 cells to ionizing radiation, *Science,* 239, 645, 1988.

24. **Deng, G. R., Liu, X. H., and Wang, J. R.,** Correlation of mutations of oncogene C-Ha-ras at codon 12 with metastasis and survival of gastric cancer patients, *Oncogene Res.,* 6, 33, 1991.

25. **Kyprianou, N. and Isaacs, J. T.,** Relationship between metastatic ability and H-ras oncogene expression in rat mammary cancer cells transfected with the v-H-ras oncogene, *Cancer Res.,* 50, 1449, 1990.

26. **Carter, G., Hughes, D. C., Clark, R. E., McCormick, F., Jacobs, A., Whittaker, J. A., and Padua, R. A.,** RAS mutations in patients following cytotoxic therapy for lymphoma, *Oncogene,* 5, 411, 1990.

27. **Tanaka, T., Slamon, D. J., Shimada, H., Shimoda, H., Fujisawa, T., Ida, N., and Seeger, R. C.,** A significant association of Ha-ras p21 in neuroblastoma cells with patient prognosis. A retrospective study of 103 cases, *Cancer,* 68, 1296, 1991.

28. **Watson, D. M., Elton, R. A., Jack, W. J., Dixon, J. M., Chetty, U., and Miller, W. R.,** The H-ras oncogene product p21 and prognosis in human breast cancer, *Breast Cancer Res. Treat.,* 17, 161, 1991.

29. **Lubbert, M., Mirro, J. J., Miller, C. W., Kahan, J., Isaac, G., Kitchingman, G., Mertelsmann, R., Herrmann, F., McCormick, F., and Koeffler, H. P.,** N-ras gene point mutations in childhood acute lymphocytic leukemia correlate with a poor prognosis, *Blood,* 75, 1163, 1990.

30. **Field, J. K. and Spandidos, D. A.,** The role of ras and myc oncogenes in human solid tumours and their relevance in diagnosis and prognosis (review), *Anticancer Res.,* 10, 1, 1990.

31. **Sagae, S., Kuzumaki, N., Hisada, T., Mugikura, Y., Kudo, R., and Hashimoto, M.,** ras oncogene expression and prognosis of invasive squamous cell carcinomas of the uterine cervix, *Cancer,* 63, 1577, 1989.

32. **Shih, C., Padhy, L. C., Murray, M., and Weinberg, R. A.,** Transforming genes of carcinomas and neuroblastomas introduced into mouse fibroblasts, *Nature,* 290, 261, 1981.

33. **Perucho, M., Goldfarb, M., Himizu, K., Lama, C., Fogh, J., and Wigler, M.,** Human tumor-derived cell lines contain common and different transforming genes, *Cell,* 27, 467, 1981.

34. **Pulciani, S., Santos, E., Lauver, A. V., Long, L. K., and Robins, K. C.,** Oncogenes in human tumor cell lines: molecular cloning of a transforming gene from human bladder carcinoma cells, *Proc. Natl. Acad. Sci. U.S.A.,* 79, 2845, 1982.

35. **Kinsella, A. R., Fiszer-Maliszewska, L., Mitchell, E. L., Guo, Y. P., Fox, M., and Scott, D.,** Introduction of the activated N-ras oncogene into human fibroblasts by retroviral vector induces morphological transformation and tumorigenicity, *Carcinogenesis,* 11, 1803, 1990.

36. **Feramisco, J. R., Gross, M., Kamata, T., Rosenberg, M., and Sweet, R. W.,** Microinjection of the oncogene form of the human H-ras (T24) protein results in rapid proliferation of quiescent cells, *Cell,* 38, 109, 1984.

37. **Stacey, D. W. and Kung, H. F.,** Transformation of NIH-3T3 cells by microinjection of Ha-ras p21 protein, *Nature,* 310, 508, 1984.

38. **Mukhopadhyay, T., Tainsky, M., Cavender, A. C., and Roth, J. A.,** Specific inhibition of K-ras expression and tumorigenicity of lung cancer cells by antisense RNA, *Cancer Res.,* 51, 1744, 1991.

39. **Feramisco, J. R., Clark, Wong, G., Arnheim, N., Milley, R., and McCormick, F.,** Transient reversion of ras oncogene-induced cell transformation by antibodies specific for amino acid 12 of ras protein, *Nature,* 314, 639, 1985.

40. **Bar-Sagi, D., McCormick, F., and Feramisco, J. R.,** Inhibition of cell surface ruffling and fluid-phase pinocytosis by microinjection of anti-ras antibodies into living cells, *J. Cell. Physiol.,* 5, 69, 1987.

41. **Wilhelm, S. M., Collier, I. E., Kronberger, A., Eisen, A. Z., Marmer, B. L., Grant, G. A., Bauer, E. A., and Goldberg, G. I.,** Human skin fibroblast stromelysin: structure, glycosylation, substrate specificity, and differential expression in normal and tumorigenic cells, *Proc. Natl. Acad. Sci. U.S.A.,* 84, 6725, 1987.

42. **Chin, J. R., Murphy, G., and Werb, Z.,** Stromelysin, a connective tissue-degrading metalloendopeptidase secreted by stimulated rabbit synovial fibroblasts in parallel with collagenase, *J. Biol. Chem.,* 260, 12367, 1985.

43. **Nicholson, R., Murphy, G., and Breathnach, R.,** Human and rat malignant-tumor-associated mRNAs encode stromelysin-like metalloproteinases, *Biochemistry,* 28, 5195, 1989.

44. **Okada, Y., Nagese, H., and Harris, E. D.,** A metalloproteinase from human rheumatoid synovial fibroblasts that digests connective tissue matrix components, *J. Biol. Chem.,* 261, 14245, 1986.

45. **McDonnell, S. and Matrisian, L. M.,** Stomelysin in tumor progression and metastasis, *Cancer Metastasis Rev.,* 9, 305, 1990.

46. **Matrisian, L. M., Bowden, G. T., Krieg, P., Fürstenberger, G., Briand, J.-P., Leroy, P., and Breathnach, R.,** The mRNA coding for the secreted protease transin is expressed more abundantly in malignant than in benign tumors, *Proc. Natl. Acad. Sci. U.S.A.,* 83, 9413, 1986.

47. **Ostrowski, L. E., Finch, J., Krieg, P., Matrisian, L., Patskan, G., O'Connell, J. F., Phillips, J., Slaga, T. J., Breathnach, R., and Bowden, G. T.,** Expression pattern of a gene for a secreted metalloproteinase during late stages of tumor progression, *Mol. Carcinog.,* 1, 13, 1988.

48. **Muller, D., Quantin, B., Gesnel, M. C., Millon, C. R., Abecassis, J., and Breathnach, R.,** The collagenase gene family in humans consists of at least four members, *Biochem. J.,* 253, 187, 1988.

49. **Matrisian, L. M., Glaichenhaus, N., Gesnel, M.-C., and Breathnach, R.,** Epidermal growth factor and oncogenes induce transcription of the same cellular mRNA in rat fibroblasts, *EMBO J.,* 4, 1435, 1985.

50. **Sistonen, L., Holtta, E., Makela, T. P., Keski, O. J., and Alitalo, K.,** The cellular response to induction of the p21 c-Ha-ras oncoprotein includes stimulation of jun gene expression, *EMBO J.,* 8, 815, 1989.

51. **Schultz, R. M., Silberman, S., Persky, B., Bajkowski, A. S., and Carmichael, D. F.,** Inhibition by human recombinant tissue inhibitor of metalloproteinases of human amnion invasion and lung colonization by murine B16-F10 melanoma cells, *Cancer Res.,* 48, 5539, 1988.

52. **Stetler-Stevenson, W. G.,** Type IV collagenases in tumor invasion and metastasis, *Cancer Metastasis Rev.,* 9, 289, 1990.

53. **D'Errico, A., Garbisa, S., Liotta, L. A., Castronovo, V., Stetler-Stevenson, W. G., and Grigioni, W. F.,** Augmentation of type IV collagenase, laminin receptor, and Ki67 proliferation antigen associated with human colon, gastric, and breast carcinoma progression, *Mod. Pathol.,* 4, 239, 1991.

54. **Thorgeirsson, U. P., Turpeenniemi-Hujanen, T., William, J. E., Westin, E. H., Heilman, C. A., Talmadge, J. E., and Liotta, L. A.,** NIH/3T3 cells transfected with human tumor DNA containing activated ras oncogenes express the metastatic phenotype in nude mice, *Mol. Cell. Biol.,* 5, 259, 1985.

55. **Collier, I. E., Wilhelm, S. M., Eisen, A. Z., Marmer, B. L., Grant, G. A., Seltzer, J. L., Kronberger, A., He, C. S., Bauer, E. A., and Goldberg, G. I.,** H-ras oncogene-transformed human bronchial epithelial cells (TBE-1) secrete a single metalloprotease capable of degrading basement membrane collagen, *J. Biol. Chem.,* 263, 6579, 1988.

56. **Garbisa, S., Pozzatti, R., Muschel, R. J., Saffiotti, U., Ballin, M., Goldfarb, R. H., Khoury, G., and Liotta, L. A.,** Secretion of type IV collagenolytic protease and metastic phenotype: induction by transfection with c-Ha-ras but not c-Ha-ras plus Ad2-E1a, *Cancer Res.,* 47, 1523, 1987.

57. **Denhardt, D. T., Greenberg, A. H., Egan, S. E., Hamilton, R. T., and Wright, J. A.,** Cysteine proteinase cathepsin L expression correlates closely with the metastatic potential of H-ras-transformed murine fibroblasts, *Oncogene,* 2, 55, 1987.

58. **Gal, S. and Gottesman, M. M.,** The major excreted protein (MEP) of transformed mouse cells and cathepsin L have similar protease specificity, *Biochem. Biophys. Res. Commun.,* 139, 156, 1986.

59. **Gal, S. and Gottesman, M. M.,** Isolation and sequence of a cDNA for human pro(cathepsin L), *Biochem. J.,* 253, 303, 1988.

60. **Mason, R. W., Gal, S., and Gottesman, M. M.,** The identification of the major excreted protein (MEP) from a transformed mouse fibroblast cell line as a catalytically active precursor form of cathepsin L, *Biochem. J.,* 248, 449, 1987.

61. **Joseph, L., Lapid, S., and Sukhatme, V.,** The major ras induced protein in NIH3T3 cells is cathepsin L, *Nucleic Acids Res.,* 15, 3186, 1987.

62. **Joseph, L. J., Chang, L. C., Stamenkovich, D., and Sukhatme, V. P.,** Complete nucleotide and deduced amino acid sequences of human and murine preprocathepsin L. An abundant transcript induced by transformation of fibroblasts, *J. Clin. Invest.,* 81, 1621, 1988.

63. **Gottesman, M. M.,** Transformation-dependent secretion of a low molecular weight protein by murine fibroblasts, *Proc. Natl. Acad. Sci. U.S.A.,* 75, 2767, 1978.

64. **Kirschke, H., Kembhavi, A. A., Bohley, P., and Barrett, A. J.,** Action of rat liver cathepsin L on collagen and other substrates, *Biochem. J.,* 201, 367, 1982.

65. **Mason, R. W., Johnson, D. A., Barrett, A. J., and Chapman, H. A.,** Elastinolytic activity of human cathepsin L, *Biochem. J.,* 233, 925, 1986.

66. **Johnson, D. A., Barrett, A. J., and Mason, R. w.,** Cathepsin L inactivates α1-proteinase inhibitor by cleavage in the reactive site region, *J. Biol. Chem.,* 261, 14748, 1986.

67. **Yagel, S., Warner, A. H., Nellans, H. N., Lala, P. K., Waghorne, C., and Denhardt, D. T.,** Suppression by cathepsin L inhibitors of the invasion of amnion membranes by murine cancer cells, *Cancer Res.,* 49, 3553, 1989.

68. **Doherty, P. J., Hua, L., Liau, G., Gal, S., Graham, D. E., Sobel, M., and Gottesman, M. M.,** Malignant transformation and tumor promoter treatment increase levels of a transcript for a secreted glycoprotein, *Mol. Cell. Biol.,* 5, 466, 1985.

69. **Ciardiello, F., Kim, N., McGeady, M. L., Liscia, D. S., Saeki, T., Bianco, C., and Salomon, D. S.,** Expression of transforming growth factor alpha (TGFα) in breast cancer, *Ann. Oncol.,* 2, 169, 1991.

70. **Ciardiello, F., McGeady, M. L., Kim, N., Basolo, F., Hynes, N., Langton, B. C., Yokozaki, H., Saeki, T., Elliott, J. W., and Masui, H.,** Transforming growth factor-alpha expression is enhanced in human mammary epithelial cells transformed by an activated c-Ha-ras protooncogene but not by the c-neu protooncogene, and overexpression of the transforming growth factor-alpha complementary DNA leads to transformation, *Cell Growth Differ.,* 1, 407, 1990.

71. **Ciardiello, F., Hynes, N., Kim, N., Valvarius, E. M., Lippman, M. E., and Salomon, D. S.,** Transformation of mouse mammary epithelial cells with the Ha-ras but not with the neu oncogene results in a gene dosage-dependent increase in transforming growth factor alpha production, *FEBS Lett.,* 250, 474, 1989.

72. **Ciardiello, F., Kim, N., Hynes, N., Jaggi, R., Redmond, S., Liscia, D. S., Sanfilippo, B., Merlo, G., Callahan, R., and Kidwell, W. R.**, Induction of transforming growth factor alpha expression in mouse mammary epithelial cells after transformation with a point-mutated c-Ha-ras protooncogene, *Mol. Endocrinol.*, 2, 1202, 1988.

73. **Tortora, G., Ciardiello, F., Ally, S., Clair, T., Salomon, D. S., and Cho-Chung, Y. S.**, Site-selective 8-chloroadenosine 3′,5′-cyclic monophosphate inhibits transformation and transforming growth factor alpha production in Ki-ras-transformed rat fibroblasts, *FEBS Lett.*, 242, 363, 1989.

74. **Yamamoto, K. K., Gonzalez, G. A., Biggs, W. H., and Montminy, M. R.**, Phosphorylation-induced binding and transcriptional efficacy of nuclear factor CREB, *Nature*, 334, 494, 1988.

75. **Salomon, D. W., Perroteau, I., Kidwell, W. R., Tam, J., and Derynck, R.**, Loss of growth responsiveness to epidermal growth factor and enhanced production of alpha-transforming growth factors in ras-transformed mouse mammary epithelial cells, *J. Cell. Physiol.*, 130, 397, 1987.

76. **Geiser, A. G., Kim, S. J., Roberts, A. B., and Sporn, M. B.**, Characterization of the mouse transforming growth factor-beta 1 promoter and activation by the Ha-ras oncogene, *Mol. Cell. Biol.*, 11, 84, 1991.

77. **Moses, H. L., Branum, E. B., Proper, J. A., and Robinson, R. A.**, Transforming growth factor production by chemically transformed cells, *Cancer Res.*, 41, 2842, 1981.

78. **Roberts, A. B., Anzano, M. A., Lamb, L. C., Smith, J. M., and Sporn, M. B.**, New class of transforming growth factors potentiated by epidermal growth factor: isolation from nonneoplastic tissues, *Proc. Natl. Acad. Sci. U.S.A.*, 78, 5339, 1981.

79. **Leof, E.B., Proper, J. A., and Moses, H. L.**, Modulation of transforming growth factor type beta action by activated ras and c-myc, *Mol. Cell. Biol.*, 7, 2649, 1987.

80. **Theodorescu, D., Cornil, I., Sheehan, C., Man, M. S., and Kerbel, R. S.**, Ha-ras induction of the invasive phenotype results in up-regulation of epidermal growth factor receptors and altered responsiveness to epidermal growth factor in human papillary transitional cell carcinoma cells, *Cancer Res.*, 51, 4486, 1991.

81. **Ciardiello, F., Valverius, E. M., Colucci-D'Amato, G. L., Kim, N., Bassin, R. H., and Salomon, D. S.**, Differential growth factor expression in transformed mouse NIH-3T3 cells, *J. Cell. Biochem.*, 42, 45, 1990.

82. **Turley, E. A., Auston, L., Vandeligt, K., and Clary, C.**, Hyaluronon and cell-associated hyaluronon binding protein regulate the locomotion of ras-transformed cells, *J. Cell Biol.*, 112, 1041, 1991.

83. **Demetri, G. D., Ernst, T. J., Pratt, E. S., Zenzie, B. W., Rheinwald, J. G., and Griffin, J. D.**, Expression of ras oncogenes in cultured human cells alters the transcriptional and posttranscriptional regulation of cytokine genes, *J. Clin. Invest.*, 86, 12561, 1990.

84. **Unkeless, J., Dano, K., Kellerman, G. M., and Reich, E.**, Fibrinolysis associated with oncogenic transformation. Partial purification and characterization of the cell factor, a plasminogen activator, *J. Biol. Chem.*, 249, 4295, 1974.

85. **Testa, J. E., Medcalf, R. L., Cajot, J. F., Schleuning, W. D., and Sordat, B.**, Urokinase-type plasminogen activator biosynthesis is induced by the EJ-Ha-ras oncogene in Cl26 mouse colon carcinoma cells, *Int. J. Cancer*, 43, 816, 1989.

86. **Devouge, M. W., Mukherjee, B. B., and Pena, S. D. J.**, Kirsten murine sarcoma virus-coded p21^{ras} may act on multiple targets to effect pleotrophic changes in transformed cells, *Virology*, 121, 327, 1982.

87. **Liau, G., Yamada, Y., and De Crombrugghe, B.**, Coordinate regulation of the levels of type III and type I collagen RNA in most but not all mouse fibroblasts, *J. Biol. Chem.*, 260, 531, 1985.

88. **Sistonen, L., Keski-Oja, J., Ulmanen, I., Holtta, E., Wikgren, B. J., and Alitalo, K.**, Dose effects of transfected c-Ha-ras Val 12 oncogene in transformed cell clones, *Exp. Cell Res.*, 168, 518, 1987.

89. **Godwin, A. K. and Lieberman, M. W.**, Early and late responses to induction of rasT24 expression in Rat-1 cells, *Oncogene*, 5, 1231, 1990.

90. Gingras, M. C., Jarolim, L., Finch, J., Bowden, G. T., Wright, J. A., and Greenberg, A. H., Transient alterations in the expression of protease and extracellular matrix genes during metastatic lung colonization by H-ras-transformed 10T1/2 fibroblasts, *Cancer Res.*, 50, 4061, 1990.

91. Evans, T., Hart, M. J., and Cerione, R. A., The Ras superfamilies: regulatory proteins and post-translational modifications, *Curr. Opinion Cell Biol.*, 3, 185, 1991.

92. Chardin, P., Small GTP-binding proteins of the ras family: a conserved functional mechanism, *Cancer Cells*, 3, 117, 1991.

93. Verma, I. M., Ransone, L. J., Visvader, J., Sassone, C. P., and Lamph, W. W., fos-jun conspiracy: implications for the cell, *Ciba Found. Symp.*, 150, 128, 1990.

94. Schönthal, A., Nuclear protooncogene products: fine-tuned components of signal transduction pathway, *Cell. Signalling*, 2, 215, 1990.

95. Treisman, R., The SRE: a growth factor responsive transcriptional regulator, *Semin. Cancer Biol.*, 1, 47, 1990.

96. Holt, J. T., Gopal, T. V., Moulton, A. D., and Nienhuis, A. W., Inducible production of c-fos antisense RNA inhibits 3T3 cell proliferation, *Proc. Natl. Acad. Sci. U.S.A.*, 83, 4794, 1986.

97. Curran, T. and Franza, B. R., Jr., Fos and jun: the AP-1 connection, *Cell*, 55, 395, 1988.

98. Angel, P., Baumann, I., Stein, B., Delius, H., Rahmsdorf, H. J., and Herrlich, P., 12-*O*-tetradecanoyl-phorbol-13-acetate induction of the human collagenase gene is mediated by an inducible enhancer element located in the 5'-flanking region, *Mol. Cell. Biol.*, 7, 2256, 1987.

99. Schönthal, A., Herrlich, P., Rahmsdorf, H. J., and Ponta, H., Requirement for fos gene expression in the transcriptional activation of collagenase by other oncogenes and phorbol esters, *Cell*, 54, 325, 1988.

100. Tzen, C. Y., Filipak, M., and Scott, R. E., Metaplastic change in mesenchymal stem cells induced by activated ras oncogene, *Am. J. Pathol.*, 137, 1091, 1990.

101. Lemoine, N. R., Staddon, S., Bond, J., Wyllie, F. S., Shaw, J. J., and Wynford-Thomas, D., Partial transformation of human thyroid epithelial cells by mutant Ha-ras oncogene, *Oncogene*, 5, 1833, 1990.

102. Bar-Sagi, D. and Feramisco, J. R., Microinjection of the ras oncogene protein into PC12 cells induces morphological differentiation, *Cell*, 42, 841, 1985.

103. Guerrero, I., Wong, H., Petlicer, A., and Burstein, D. E., Activated N-ras gene induces neuronal differentiation of PC12 rat pheochromocytoma cells, *J. Cell. Physiol.*, 129, 71, 1986.

104. Noda, M., Ko, M., Ogura, A., Liu, D., Amano, T., Takano, T., ad Ikawa, Y., Sarcoma viruses carrying ras oncogenes induce differentiation-associated properties in a neuronal cell line, *Nature*, 318, 73, 1985.

105. Nakagawa, T., Mabry, M., de Bustros, A., Ihle, J. H., Nelkin, B. D., and Baylin, S. B. Introduction of v-Ha-ras oncogene induces differentiation of cultured human medullary thyroid carcinoma cells, *Proc. Natl. Acad. Sci. U.S.A.*, 84, 5923, 1987.

106. Seremetis, S., Inghirami, G., Ferrero, D., Newcomb, E. W., Knowles, D. M., Dotto, G. P., and Dalla-Favra, R., Transformation and plasmacytoid differentiation of EBV-infected human B lymphoblasts by ras oncogenes, *Science*, 243, 660, 1989.

107. Greenberg, M. E., Greene, L. A., and Ziff, E. B., Nerve growth factor and epidermal growth factor induce rapid transient changes in proto-oncogene transcription in PC12 cells, *J. Biol. Chem.*, 260, 14, 1010, 1985.

108. Alema, S., Casalbore, P., Agostini, E., and Tato, F., Differentiation of PC12 phaeochromocytma cells induced by v-srv oncogene, *Nature (London)*, 316, 557, 1985.

109. Hagag, N., Halegoua, S., and Viola, M., Inhibition of growth factor-induced differentiation of PC12 cells by microinjection of antibody to ras p21, *Nature*, 319, 680, 1986.

110. Szebrenyi, J., Cai, H., and Cooper, G. M., Effect of a dominant inhibitory Ha-ras mutation on neuronal differentiation of PC12 cells, *Mol. Cell. Biol.*, 10, 5324, 1990.

111. Greenberg, M. E. and Ziff, E. B., Stimulation of 3T3 cells induces transcription of the c-*fos* proto-oncogene, *Nature*, 311, 433, 1984.

112. **Bravo, R., Burckhardt, J., Curran, T., and Müller, R.,** Stimulation and inhibition of growth by EGF in different A431 cell clones is accompanied by the rapid induction of c-fos and c-myc proto-oncogenes, *EMBO J.,* 4, 1193, 1985.

113. **Tsuda, T., Hamamori, Y., Yamashita, T., Fukumoto, Y., and Takai, Y.,** Involvement of three intracellular messenger systems, protein kinase C, calcium ion and cyclic AMP, in the regulation of c-fos gene expression in Swiss 3T3 cells, *FEBS Lett.,* 208, 39, 1986.

114. **Curran, T. and Morgan, J. I.,** Superinduction of c-fos by nerve growth factor in the presence of peripherally active benzodiazepines, *Science,* 229, 1265, 1985.

115. **Kruijer, W., Schubert, D., and Verma, I. M.,** Induction of the proto-oncogene fos by nerve growth factors, *Proc. Natl. Acad. Sci. U.S.A.,* 82, 7330, 1985.

116. **Stacey, D. W., Watson, T., Kung, H. S., and Curran, T.,** Microinjection of transforming ras protein induces c-fos expression, *Mol. Cell. Biol.,* 7, 523, 1987.

117. **Sassone-Corsi, P., Der, C. J., and Verma, I. M.,** ras-induced neuronal differentiation of PC12 cells: possible involvement of fos and jun, *Mol. Cell. Biol.,* 9, 3174, 1989.

118. **Thomson, T. M, Green, S. H., Trotta, R. J., Burstein, D. E., and Pellicer, A.,** Oncogene N-ras mediates selective inhibition of c-fos induction by nerve growth factor and basic fibroblast growth factor in a PC12 cell line, *Mol. Cell. Biol.,* 10, 1556, 1990.

119. **Nishizuka, Y.,** Studies and perspectives of protein kinase C, *Science,* 233, 305, 1986.

120. **Nishizuka, Y.,** The molecular heterogeneity of protein kinase C and its implications for cellular regulation, *Nature,* 334, 661, 1988.

121. **Ito, E., Sweterlitsch, L. A., Tran, P. B., Rauscher, F. J., and Narayanan, R.,** Inhibition of PC-12 cell differentiation by the immediate early gene fra-1, *Oncogene,* 5, 1755, 1990.

122. **Defeo-Jones, D., McAvoy, E. M., Jones, R. E., Vuocolo, G. A., Haskell, K. M., Wegrzyn, R. J., and Oliff, A.,** Lovastatin selectively inhibits ras activation of the 12-*O*-tetradecanoylphorbol-13-acetate response element in mammalian cells, *Mol. Cell. Biol.,* 11, 2307, 1991.

123. **Meinkoth J., Alberts, A., and Feramisco, J.,** Construction of mammalian cell lines with indicator genes driven by regulated promoters, *Protooncogenes Cell Dev. (CIBA Found. Symp.),* 150, 47, 1990.

124. **Meinkoth, J. L., Montminy, M. R., Fink, J. S., and Feramisco, J. R.,** Induction of a cyclic AMP-responsive gene in living cells requires the nuclear factor CREB, *Mol. Cell. Biol.,* 11, 1759, 1991.

125. **Imler, J. L., Schatz, C., Wasylyk, C., Chatton, B., and Wasylyk, B.,** A Harvey-ras responsive transcription element is also responsive to a tumour-promoter and to serum, *Nature,* 332, 275, 1988.

126. **Gauthier-Rouviere, C., Fernandez, A., and Lamb, N. J. C.,** ras-induced c-fos expression and proliferation in living rat fibroblasts involves C-kinase activation and the serum response element pathway, *EMBO J.,* 9, 171, 1990.

127. **Owen, R. D., Bortner, D. M., and Ostrowski, M. C.,** ras oncogene activation of a VL30 transcriptional element is linked to transformation, *Mol. Cell. Biol.,* 10, 1, 1990.

128. **Galien, R., Mercier, G., Garcette, M., and Emanoil-Ravier, R.,** RAS oncogene activates the intracisternal A particle long terminal repeat promoter through a c-AMP response element, *Oncogene,* 6, 849, 1991.

Chapter 5

p21*ras* AND RECEPTOR SIGNALING

Boudewijn M. Th. Burgering and Johannes L. Bos

TABLE OF CONTENTS

I. p21*ras*

A. p21*ras* IS INVOLVED IN SIGNAL TRANSDUCTION

p21*ras* belongs to the family of GTPases, that cycle between an active GTP-bound state and an inactive GDP-bound state. This family includes, among others, the α-subunit of heterotrimeric G proteins, initiation and elongation factors of protein synthesis, and a large number of small GTPases. Analogous to the pathway in which the heterotrimeric G proteins functions, the "p21*ras* pathway" can be divided into an upstream and a downstream part, taking the GTP/GDP switch as a reference point. All events that are directed to activate p21*ras* (the dissociation of GDP or the inhibition of GTP hydrolysis) are considered to be upstream elements of the *ras* pathway and all cellular events that result from p21*ras* activation are considered to be downstream elements.

In several lower eukaryotes, upstream and downstream elements of the *ras* pathway have been identified by genetic approaches. In *Saccharomyces cerevisiae* the products of two genes, IRA1 and IRA2, stimulate the intrinsic GTPase of yeast RAS proteins,[1,2] while guanine nucleotide exchange is most likely stimulated by the CDC25 gene product.[3] All these proteins may serve solely to regulate RAS activity. Downstream signaling is through interaction with adenylate cyclase, although other proteins may be involved as well. In *Caenorhabditis elegans* genetic evidence has been obtained that *let-60*, a protein very similar to p21*ras*, is involved in directing the differentiation of the vulva, a process in which a tyrosine kinase receptor (EGF receptor-like) upstream of *let-60* and a GTPase-activating protein downstream of *let-60* have been implicated.[4-7]

In mammalian cells p21*ras* function was initially studied by microinjection of neutralizing antibodies, in particular, the rat monoclonal antibody Y13-259.[8] These studies have shown that in a variety of cell types p21*ras* is required for growth factor-induced ^{3}H-thymidine incorporation. These growth factors include serum, TPA, prostaglandin $F_{2\alpha}$ and phosphatidic acid.[9] It is also required for serum-induced c-*fos* expression in NIH 3T3 cells,[10] insulin-induced maturation of *Xenopus* oocytes,[11,12] and neural growth factor (NGF)-induced differentiation of PC12 cells.[13] Furthermore, oncogenic transformation by several tyrosine kinase oncogenes (v-*src*, v-*fes*) was transiently reverted after microinjection of Y13-259, indicating that p21*ras* is required for transformation by these oncogenes. Cellular transformation by cytoplasmic serine/threonine kinase oncogenes (v-*raf* and v-*mos*) appears not to require p21*ras*.[14]

The function of p21*ras* in growth factor signal transduction pathways has also been studied using so-called interfering mutants.[15-19] Two mutants have been described in detail. p21*ras*^{asn17} displays a reduced affinity for GTP, but normal affinity for GDP. This mutant may interfere in the interaction between p21*ras* and its exchange protein.[15,16] p21*ras*^{leu61ser186} has a high affinity for GAP (and maybe for other proteins), but has lost its farnesylation site necessary for membrane attachment and biological activity. This mutant may interfere in the interaction between p21*ras* and its effector.[17,18] Coexpression of these interfering mutants inhibits cellular responses induced by activated p21ras, such as the induction of *Xenopus* oocyte germinal vesicle breakdown,[17] yeast heat shock response,[19] and the induction of gene expression.[15] These interfering mutants have also been used to demonstrate the involvement of activated p21ras in the induction of c-*fos* expression after EGF and TPA treatment.[15]

A third approach to study the involvement of p21ras in signal transduction is to overexpress p21ras in cells and to investigate the effect of this overexpression on signal transduction. This approach has been explored first by Marshall and co-workers using NIH 3T3 cells overexpressing the N-*ras* gene. They found that elevated levels of normal p21N-*ras* augment the signaling of bombesin to phosphatidylinositol breakdown.[20] In this case the effect may not be direct, however.[21] Also, in Rat 1 cells overexpressing p21H-*ras* the effect of bradykinine on phosphatidylinositol breakdown is increased. This effect appears to be indirect as well and due to an increase in the number of bradykinine receptor numbers.[22,23] Using Rat 1 cells overexpressing p21-H*ras* it was found that the cells became responsive to insulin and IGF-I.[24] In this case, the induction of DNA synthesis and the induction of the early response genes c-*fos* and c-*myc* was measured. Furthermore, it was shown that the suppression of 8-bromo cAMP inhibition of DNA synthesis by mutant p21H-*ras* could be mimicked by overexpression of p21H-*ras* in the presence of insulin.[24]

The examples described above show that p21ras is necessary for proper signal transduction of several growth factors. However, this does not necessarily imply that p21ras is directly activated by these stimuli. Maybe a basal level of the GTP-form of p21ras is sufficient for most of these pathways. Alternatively, only a small increase in the level of GTP may be sufficient for activation. This possibility was fed by the fact that, until recently, none of these cellular stimuli were reported to activate p21ras as measured by an increase of GTP bound to it.

B. GROWTH FACTORS CAN ACTIVATE p21ras

Convincing evidence that receptor stimulation can indeed activate p21ras by increasing the ratio of GTP/GDP bound to it, was provided by experiments analyzing the activation of T lymphocytes. Stimulation of the T-cell receptor by phytohemagglutinin or specific monoclonal antibodies rapidly increases the percentage GTP-bound p21ras from 7 to 60%.[25] In this type of experiment, the cells were labeled with ^{32}P-orthophosphate *in vivo* and after the proper stimulation, p21ras was collected by immunoprecipitation. Labeled GDP/GTP was eluted from p21ras and separated. The activation of p21ras occurred very rapidly, was maximal within 2 min and sustained for at least 15 min. This strong induction was in sharp contrast to the very weak induction of GTP bound to p21ras after serum, PDGF and EGF stimulation of NIH 3T3 cells (7 to 15% GTP),[26] or Swiss 3T3 cells overexpressing normal p21H-*ras* (0.5 to 1% GTP).[27,28] This may imply that for the mitogenic stimulation of T lymphocytes a higher level of p21ras-GTP is necessary than for normal fibroblasts.

Insulin is a very weak mitogen for NIH 3T3 cells, presumably due to low numbers of receptors. However, in an NIH 3T3 cell line expressing elevated numbers of human insulin receptor, insulin is a mitogen as efficient as serum or other potent mitogens. In these cells it was found that p21ras was activated rapidly upon insulin stimulation (15 to 70% GTP), but not upon serum stimulation.[29] This activation was also observed in chinese hamster ovary (CHO) cells expressing elevated levels of the human insulin receptor.[112] These results together have provided the first direct proof that p21ras can become activated upon growth factor stimulation and, thus, is involved actively in growth factor signal transduction.

C. MECHANISM OF p21ras ACTIVATION BY GROWTH FACTOR STIMULI

The dissociation of GDP is considered to be the rate-limiting step in the exchange of GDP for GTP. In the heterotrimeric G proteins this dissociation is induced by the ligand-activated receptor followed b binding of GTP of which the concentration in the cell is 10-fold higher than GDP. Recent reports indicate the existence of proteins that stimulate the dissociation of GDP from p21ras, denoted as *ras* guanine nucleotide releasing factor (rasGRF),[30] *ras* guanine nucleotide exchange factor (rGEF),[31] and *ras* exchange promoting protein (REP).[32] The rGEF is reported to be localized in the plasma membrane and appears to have a broad specificity and GNRF and REP are apparently predominantly localized in the cytosol. At this moment the molecular characterization of these proteins is limited. Obviously, the next step will be to study the effect of activated growth factors receptors on rGEF/rasGRF activity. The possible existence of other proteins, similar to the guanine nucleotide dissociation inhibitor of *rho* (*rho*-GDI)[33] or *smg* p25-GDI[34] inhibiting the dissociation of GDP, suggests that counteracting regulatory elements may also be involved in growth factor regulation of GDP dissociation.

Inhibition of the hydrolysis of GTP will also result in p21ras activation. In this respect one has to consider the possibility of growth factor-induced inactivation of the GTPase activating protein (GAP) or the GAP-related neurofibromatosis type 1 (NF-1) protein. It has been demonstrated that lipids, the cellular concentration of which changes after stimulation of cells with growth factors, inhibit the interaction of GAP with p21ras *in vitro*, resulting in a decrease of GTP hydrolysis.[35,36] However, in the initial studies the concentration of lipid required for half maximal inhibition of p21ras-GAP interaction is rather high (up to 100 μg/ml). Although it can be forwarded that *in vivo* local concentrations of lipid may be within the same order, a more likely explanation is raised by recent results that indicate that these lipids *in vivo* probably act in concert with an as yet unidentified cellular proteins.[37] Furthermore, it has been reported that the generation of putative inhibitory lipid(s) is stimulated by growth factor (serum) treatment of quiescent fibroblasts.[38] Thus, growth factor treatment would stimulate the transient release of an inhibitory lipid, that in combination with a cellular protein would prevent p21*ras*-GAP interaction and, consequently, this would result in the activation of p21*ras*. This mechanism of p21*ras* activation is remarkably similar to the mechanism that has been proposed for the activation of p21*ras* in T lymphocytes.[25] In these cells, accumulation of GTP on p21*ras* appears to be mediated by inactivation of GAP. The proposed pathway is that stimulation of the T-cell receptor stimulates lipid breakdown resulting in the production of (diacylglycerol) DAG and the subsequent activation of protein kinase C (PKC). It appears that at least one other, unidentified, cellular protein is required to mediate GAP inhibition, since GAP is not modified directly by activated PKC. Activated PKC may affect lipid metabolism resulting in the activation of a protein which inhibits GAP.[37]

For the activation of p21*ras* by insulin in insulin receptor expressing NIH 3T3 cells the mechanism is still unclear. It is different from p21*ras* activation in T lymphocytes, since PKC does not seem to mediate the activation.[111] First, downmodulation of PKC by prolonged treatment with TPA does not affect insulin-induced p21*ras* activation and second, activation of PKC by a short incubation with TPA does not activate p21*ras* in these cells.

II. GTPase-ACTIVATING PROTEIN

A. INTERACTION OF TYROSINE KINASE RECEPTORS WITH GAP

An important role for GAP in PDGF signaling is suggested by the observation that following PDGF stimulation, GAP is rapidly phosphorylated on tyrosine[39] and associated with the activated PDGF receptor.[40,41] *In vitro* PDGF receptor autophosphorylation, but not GAP tyrosine phosphorylation, is necessary for association.[40,41] Furthermore, certain PDGF receptor mutants, defective in signal transduction, do not bind GAP.[41] Also, after EGF[42] and CSF-1 stimulation,[43] GAP becomes phosphorylated on tyrosine, but stable association with the respective receptor is not observed. Also, in cells transformed by tyrosine kinase oncogenes (v-*src*, v-*abl*, v-*fms*, v-*fps*) GAP is phosphorylated on tyrosine and complexes to two other tyrosine-containing phosphoproteins, p190 and p62. These proteins have been cloned recently.[113,114]

GAP was originally discovered as a cytosolic protein with an approximate molecular weight of 120 kDa.[44] The GAP coding sequence has been cloned and its predicted amino acid sequence revealed few clues as to its function.[45,46] It showed the presence of SH2 (src homology) and SH3 domains. SH2 domains are found in nonreceptor tyrosine kinases (*src, yes*, etc.), phopsholipase C-γ, phosphatidylinositol-3-kinase, and the *crk* oncogene.[47] The SH2 domains are probably involved in the interaction with phosphotyrosine-containing domains, either intra- or intermolecularly.[48,49] Deletion analysis revealed that the C terminal part (334 amino acids) of GAP is sufficient for the GTPase activating function and that the N terminal part, including the SH2 and 3 domains, probably fulfills a regulatory function.[50]

A variety of observations have indicated that the effector region of p21*ras* (amino acids 32 to 40) provides the binding site for GAP on p21*ras*.[51-56] Evidence that GAP may be the effector of p21ras, however, is limited. An interfering mutant (p21ras^{leu61ser186}, see above) that is thought to sequester GAP, inhibits cellular responses induced by activated p21ras, suggesting a role for GAP in downstream signaling.[17,19] Also, studies analyzing the activation of K$^+$ channels of atrial cells show that a complex between p21ras in its GTP form and GAP inhibits the coupling of activated muscarinic receptors to a G protein involved in the regulation of K$^+$ channel opening.[57] In addition to serving as an effector, GAP may downregulate the level of GTP bound to p21ras and as such, provides an obligate negative feedback to signal transmission by p21ras. Alternatively, GAP may fulfill an upstream function. In this case inhibition of GAP activity may be either permissive for the activation of p21ras by other signals or, the signal that inactivate GAP directly results in p21ras activation. This possibility is suggested by the inhibition of GAP activity after stimulation of the T-cell receptor and the subsequent activation of p21ras.[25] It is not excluded that GAP functions both upstream and downstream of p21ras.

It is clear that tyrosine phosphorylation of GAP and the association of GAP to the PDGF receptor and other cellular proteins may be involved in the various possible functions of GAP. For instance, tyrosine phosphorylation of GAP or its association to cellular proteins may inactivate GAP, resulting in the activation of p21ras. Alternatively, tyrosine phosphorylation and complex formation may activate the effector function of GAP.[39] Thus far, no clear evidence exists to support these possibilities.

III. p21ras AND INSULIN SIGNALING

A. p21ras INVOLVED IN INSULIN SIGNAL TRANSDUCTION

Several distinct observations have indicated a possible involvement of p21ras in insulin signal transduction. Initially it was suggested that, like EGF, insulin would stimulate phosphorylation of v-H-ras.[58,59] However, as for PKC-mediated phosphorylation,[60] this phosphorylation is unlikely to have any relevance to the kinetics of GTP/GDP cycling of normal p21ras. A strong indication for a possible function of p21ras in insulin signal tranduction is provided by the hormone-induced maturation of *Xenopus* oocytes.[11,12] It was shown that microinjection of purified p21ras proteins also induces oocyte maturation. Furthermore, oncogenic proteins are more potent in inducing maturation than normal p21ras proteins. Insulin-induced, but not progesterone-induced maturation of *Xenopus* oocytes is blocked by microinjection of the neutralizing Y13-259 antibody to p21ras. These observations clearly suggest that activation of p21ras induces oocyte maturation and that within this system insulin stimulation is likely to activate p21ras. It has been reported that both purified p21H-ras[59] and purified GAP[46] are substrates for the insulin receptor tyrosine kinase *in vitro*. However, in neither case has insulin-stimulated phosphorylation *in vivo* been demonstrated. Purified p21H-ras has also been reported to bind to the insulin receptor *in vitro*,[61] but p21ras needs to be denatured and renatured several times before this association occurs and, thus, does not appear to be very relevant. Finally, cellular responses of both insulin and p21ras are often very similar. For instance, in fibroblasts both insulin action and p21ras function are indicated to be necessary for G1/S passage at approximately the same time point.[8] Furthermore, both insulin and oncogenic p21ras induce accumulation of mRNA for the glucose transporter Glut-1.[62] As described above, we have shown recently that insulin can activate p21ras and, thus, it is reasonable to conclude that p21ras functions in the insulin signal transduction pathway. To put into perspective a role of p21ras in insulin signal transduction a brief overview of insulin signal transduction is given below.

B. THE INSULIN SIGNAL TRANSDUCTION PATHWAY
1. The Insulin Receptor

The cellular response to insulin is highly pleiotropic.[63] It includes stimulation of amino acid and hexose uptake, increases in lipid, glycogen, protein, RNA and DNA synthesis, and modulation of enzymatic activity of key-regulatory enzymes by phosphorylation and dephosphorylation. The effects of insulin are initiated by its interaction with specific, high affinity cell surface receptors.[64] The insulin receptor consists of disulfide linked α- and β-subunits that are cleaved products of a common precursor. During precursor biosynthesis the insulin receptor becomes extensively modified. The α-subunit is located entirely at the extracellular face of the plasma membrane and contains the insulin-binding site. The β-chain has a small extracellular part, a membrane spanning domain, and a cytoplasmic part. This subunit possesses insulin-stimulated tyrosine kinase activity.[65] At least five tyrosine phosphorylation sites have been identified in two regions of the β-subunit. Three of these sites are within a putative regulatory region (Tyr1146, 1150, and 1151) and two are located close to the C terminal end (Tyr1326 and 1322).[66] Phosphorylation of the three residues within the regulatory domain is required for full activity of the receptor tyrosine kinase. Tyrosine phosphorylation of the residues located within the C terminal region occurs, but does not appear to play a significant role.[67-71] In addition, the region around tyrosine 960 which is not autophosphorylated, seems to interact with cellular proteins that are essential for signal transduction.[72]

2. Insulin-Induced Protein Phosphorylation

Since the insulin receptor exhibits tyrosine kinase activity, most likely one of the earliest events is the phosphorylation on tyrosine of substrate molecules. However, only very few proteins have been found to become phosphorylated on tyrosine after insulin stimulation. The most abundant substrate is a 185 kDa protein found first in hepatoma cells.[73] Recently, Blackshear and co-workers[74] used the pharmacological compound phenylarsine oxide (PAO) and found, that upon insulin stimulation,[26] proteins become rapidly phosphorylated. At least 10 of these proteins were phosphorylated on tyrosine. These proteins did not become phosphorylated after serum induction or PDGF induction in the presence of PAO, indicating some specificity for insulin. In addition, PAO was found to inhibit

CD45, a tyrosine-specific phosphatase in T lymphocytes,[75] and it is likely that the observed induction of insulin-induced tyrosine phosphorylation is due to the inhibition of a tyrosine-specific phosphatase. After insulin-stimulation a variety of proteins become phosphorylated on serine and threonine as well, but the kinetics are slower.[74] A reasonable explanation for this delayed effect is that insulin first activates a serine/threonine kinase by tyrosine phosphorylation. For instance, MAP2 kinase has been reported to become phosphorylated on tyrosine after insulin stimulation[76] and, thus, may provide the link between insulin receptor tyrosine kinase activity and the serine/threonine phosphorylation cascade as suggested by Sturgill et al.[77] A large number of other serine/threonine kinases have been identified, which become activated after insulin stimulation. These include: HMG kinase; protease activated kinase II; casein kinase kinase II; glycogen synthase kinase 3; adipocyte Mn^{2+}-dependent kinase; raf-1 kinase; and S6-kinase (for a review see Reference 78). These proteins become phosphorylated mostly on serine and, presumably, play a role more downstream in the signal transduction pathway.

Recently, an important connection between the phosphorylation and dephosphorylation events induced by insulin has been established by the observation that an insulin-stimulated kinase activated protein phosphatase 1 by phosphorylating its regulatory subunit at a specific serine residue. This phosphatase controls glycogen metabolism by promoting the dephosphorylation of glycogen synthase and the β-subunit of phosphorylase kinase.[79]

3. Insulin-Induced Lipid Metabolism

Insulin causes the increased labeling of several phospholipids, including phosphoinositides and phosphatidic acid[41,80-84] and stimulates the production of DAG.[85-87] However, insulin has not been found to stimulate the hydrolysis of polyphosphoinositides, such as PIP2, and it does not induce Ca^{2+} mobilization through the generation of IP_3.[88] Insulin does activate a phospholipase C able to hydrolyze glycosyl phosphoinositide, resulting in the release of DAG and the polar head group called phosphooligosaccharide (POS) or phosphoinositol glycan.[89] For POS it has been reported that it can mimic a variety of effects attributed to insulin stimulation,[90,91] but it is still unclear whether POS, indeed, is a genuine second messenger in the insulin signal transduction pathway. Insulin has been found to increase DAG, both by increased *de novo* synthesis of phosphatidic acid, which may be directly converted to DAG, and by the hydrolysis of phospholipids including phosphatidylcholine and the aforementioned insulin-sensitive glycosyl phosphoinoside.[81,82,86] Insulin-induced activation of PKC, a likely consequence of DAG increase, however, is only poorly documented, and the role of PKC in insulin action remains a controversial issue.[92]

A link between the phosphorylation cascade and insulin-induced changes in lipid metabolism may be provided by the observation that insulin directly tyrosine phosphorylates and activates PI-3 kinase.[93,94] This kinase phosphorylates (poly)phosphatidylinositols on the 3-position of the inositol ring. The biological function of this molecule is unclear, however.

4. Insulin-Induced Gene Expression

The induction of gene expression by insulin largely depends on the responsiveness of the cell. For instance, insulin stimulation of 3T3-L1 adipocytes, which express high numbers of insulin receptors, results in expression of early response genes, e.g., c-*fos*.[95] Also in hepatoma cells, which are extremely sensitive to insulin as a growth factor, insulin can induce gene expression.[96,97] Fibroblasts, however, do not express sufficient insulin receptors to show this effect. By introducing high numbers of the (human) insulin receptor in fibroblasts, insulin becomes a potent mitogen and is able to induce the expression of early response genes.[29] A similar effect of insulin receptor numbers on the ability of insulin to induce gene expression is observed in CHO cells stably transfected with human insulin receptor constructs.[98]

In those cells that are responsive, insulin is known to increase as well as decrease the levels of certain cellular mRNAs after its addition to cells. Genes of which mRNA levels increase after insulin stimulation include c-*fos*, c-*myc*, p33, and other (immediate) early genes as well as late genes such as the genes for ornithine decarboxylase and a 70 kDa heat shock protein. A recent extensive survey revealed about 300 genes, among which 52 early response genes that were induced after insulin stimulation of hepatoma cells (H35).[99] Most of these genes were also induced by serum in NIH 3T3 cells. Genes of which the expression is decreased after insulin stimulation are relatively rare. They include phosphoenolpyruvate carboxykinase and IGF-1 binding protein.

The signal transduction pathway used by insulin to induce gene expression is largely unknown. Stumpo et al.[95,98] have investigated the possible involvement of PKC in the induction of c-*fos* by insulin in 3T3-L1 adipocytes and in insulin receptor overexpressing CHO cells. They concluded that PKC is not necessary for the induction of this gene by insulin.

C. THE ROLE OF p21*ras* IN INSULIN SIGNAL TRANSDUCTION

From the above discussion it is clear that insulin stimulation has a large variety of effects on the cell, some of which are dependent on the specificity of the cell. Several distinct pathways are involved in mediating this diversity of cellular responses. At present, p21*ras* activation following insulin administration has only been demonstrated in NIH 3T3 fibroblasts and in CHO cells, both overexpressing the human insulin receptor. In these cells it is not known yet whether p21*ras* activation mediates all insulin-induced pathways. Some indications suggest that the activation of PI-3 kinase by insulin is not mediated by p21*ras*.[29] Also, insulin-induced uptake of hexose and production of DAG may be independent of p21*ras*.[112] However, the induction of the expression of genes like c-*myc* c-*fos*, p33, and c-*jun* and the induction of mitogenicity are presumably mediated through p21*ras*.[29] The involvement of p21*ras* in a part of the insulin signal transduction pathway may also explain the observation that under certain experimental conditions insulin can synergize with p21*ras*-elicited signals,[100] maybe through the increase in glucose and/or metabolite uptake.

The role p21*ras* may play in insulin signaling, i.e., mediating the signal to induce early gene expression and mitogenicity, may be similar as the possible role p21*ras* plays in signal transduction of other growth factors.[15,38,101] Clearly, p21*ras* appears to be more sensitive to activation by insulin than by other strong mitogens, such as PDGF or EGF, when similar levels of the respective receptors are present,[29] but it is still unknown which level of p21*ras*-GTP is required to achieve stimulation of the downstream part of the *ras*-pathway. Overexpression of the insulin receptor is "pathological" for fibroblasts and in the "normal" situation insulin may have only a minor effect on p21*ras*, similar as observed for the other growth factors. Therefore, the strong mitogenicity of these growth factors is probably due to synergistic pathways, which are not induced by insulin. As observed for mutant p21*ras*, the strong activation of p21*ras* by insulin makes these synergistic pathways superfluous.

IV. CONCLUDING REMARKS

p21*ras* is involved in a variety of signal transduction pathways. All these pathways have in common that they transmit signals from cell surface receptors to the nucleus, regulating proliferation and/or differentiation of the cell. Another common feature appears to be that all the cell surface receptors that are somehow linked to p21*ras* belong to the tyrosine kinase receptors or to receptors that are directly linked to tyrosine kinases (e.g., T-cell receptor). Therefore, the elucidation of the interaction between p21*ras* and tyrosine kinases will be of paramount importance to understand the functioning of p21*ras* in signal transduction. Very recently, more progress has been made in our understanding of signal transduction that is mediated by p21*ras*. First, several proteins that may be involved in receptor tyrosine kinase-induced activation of p21*ras* have been identified. One of them, grb2, is a 23-kDa protein with an src-homology 2 (SH2) domain in between two SH3 domains. The grb2 protein is homologous to *C. elegans* SEM-5, and this latter protein has been shown genetically to be involved in signaling to p21*ras*. Furthermore, overexpression of grb2 stimulates (normal) H-*ras* in the induction of [³H]-thymidine incorporation.[102] shc is a protein of 46 to 65 kDa, which contains an SH2 and a gly-pro rich region homologous to collagen. This protein binds to and is a substrate of the activated EGF receptor.[103] Furthermore, shc associates to grb2 after stimulation of the EGF receptor.[104] Also, after PDGF and insulin stimulation, shc is phosphorylated on tyrosine and forms a complex with grb2.[105] Second, in fibroblasts, the activation of p21*ras* by receptor tyrosine kinases may be mediated, at least in part, by guanine nucleotide exchange.[106] Finally, by using dominant negative mutants of p21*ras*, in particular p21*ras*[asn17] it has been demonstrated that p21*ras* is involved in the activation of extracellular signal-regulated kinases (ERKs) and raf-1.[107-109] Since c-raf has been identified as an ERK kinase-kinase[110] the picture is emerging that receptor tyrosine kinases activate p21*ras* using several proteins, including shc, grb2, and GNRF. This results in the activation of ERK2, presumably through the activation of raf-1.

The variety of different possibilities to activate p21*ras* and the possible involvement of different effector molecules, indicate that p21*ras* might be involved in a complex interplay of different signal transduction pathways. Furthermore, it should be noted that there are four p21*ras* proteins (p21 H-*ras*, p21 K-*ras*-4A and -4B, and p21 N-ras) which may have different specificity. The last few years have been tremendously exciting in the p21*ras* field. With all the unsolved issues, there are many exciting years to come before we know the function of p21*ras* in signal transduction in all its aspects.

ACKNOWLEDGMENTS

We thank A. D. M. van Mansfeld, J. A. Maassen, G. J. Pronk, and R. Medema, for critically reading the manuscript. This project is supported by the Dutch Cancer Society (KWF).

REFERENCES

1. **Tanaka, K., Matsumoto, K., and Toh-E., A.,** IRA1, an inhibitory regulator of the RAS-cyclic AMP pathway in *Saccharomyces cerevisiae, Mol. Cell. Biol.,* 9, 757, 1989.
2. **Tanaka, K., Nakafuku, M., Satoh, T., Marshall, M. S., Gibbs, J. B., Matsumoto, K., Kaziro, Y., and Toh-E, A.,** *S. cerevisiae* genes *IRA1* and *IRA2* encode proteins that may be functionally equivalent to mammalian *ras* GTPase activating protein, *Cell,* 60, 803, 1990.
3. **Broek, D., Toda, T., Michaeli, T., Levin, L., Birchmeier, C., Zoller, M., Powers, S., and Wigler, M.,** The *S. cerevisiae* CDC25 gene product regulates the *RAS*/adenylate cyclase pathway, *Cell,* 48, 789, 1987.
4. **Aroian, R. V., Koga, M., Mendel, J. E., Ohshima, Y., and Sternberg, P. W.,** The let-23 gene necessary for *Caenorhabditis elegans* vulval induction encodes a tyrosine kinase of the EGF receptor subfamily, *Nature,* 348, 693, 1990.
5. **Beitel, G. J., Clark, S. G., and Horvitz, H. R.,** *Caenorhabditis elegans ras* gene let-60 acts as a switch in the pathway of vulval induction, *Nature,* 348, 503, 1990.
6. **Han, M. and Sternberg, P. W.,** *let-60,* a gene that specifies cell fates during *C. elegans* vulval induction, encodes a *ras* protein, *Cell,* 63, 921, 1990.
7. **Han, M., Aroian, R. V., and Sternberg, P. W.,** The *let-60* locus controls the switch between vulval and nonvulval cell fates in *Caenorhabditis elegans, Genetics,* 126, 899, 1990.
8. **Stacey, D. W., Tsai, M.-H., Yu, C.-L., and Smith, J. K.,** Critical role of cellular ras proteins in proliferative signal tranduction, *CSHSQB,* 53, 871, 1988.
9. **Yu, C.-L., Tsai, M.-H., and Stacey, D. W.,** Cellular *ras* activity and phospholipid metabolism, *Cell,* 52, 63, 1988.
10. **Stacey, D. W., Watson, T., Kung, H.-F., and Curran, T.,** Microinjection of transforming ras protein induces c-fos expression, *Mol. Cell. Biol.,* 7, 523, 1987.
11. **Deshpande, A. K. and Kung, H.-F.,** Insulin induction of *Xenopus laevis* oocyte maturation is inhibited by monoclonal antibody against p21ras proteins, *Mol. Cell. Biol.* 1, 1285, 1987.
12. **Korn, L. J., Siebel, C. W., McCormick, F., and Roth, R. A.,** Ras p21 as a potential mediator of insulin action in *Xenopus* oocytes, *Science,* 236, 840, 1987.
13. **Hagag, N., Halegoua, S., and Viola, M.,** Inhibition of growth factor-induced differentiation of PC12 cells by microinjection of antibody to *ras* p21, *Nature,* 319, 680, 1986.
14. **Smith, M. R., DeGudicibus, S. J., and Stacey, D. W.,** Requirement for c-ras proteins during viral oncogene transformation, *Nature,* 320, 540, 1986.

15. **Cai, H., Szeberényi, J., and Cooper, G. M.,** Effect of a dominant inhibitory Ha-*ras* mutation on mitotic signal transduction in NIH 3T3 cells, *Mol. Cell. Biol.,* 10, 5314, 1990.

16. **Szeberényi, J., Cai, H., and Cooper, G. M.,** Effect of a dominant inhibitory Ha-*ras* mutation on neuronal differentiation of PC12 cells, *Mol. Cell. Biol.,* 10, 5324, 1990.

17. **Gibbs, J. B., Schaber, M. D., Schofield, T. L., and Scolnick, E. M.,** *Xenopus* oocyte germinal-vesicle breakdown induced by [Val¹²] Ras is inhibited by a cytosol-localized Ras mutant, *Proc. Natl. Acad. Sci. U.S.A.,* 86, 6630, 1989.

18. **Powers, S., O'Neill, K., and Wigler, M.,** Dominant yeast and mammalian *RAS* mutants that interfere with the *CDC25*-dependent activation of wild-type *RAS* in *Saccharomyces cerevisiae, Mol. Cell. Biol.,* 9, 390, 1989.

19. **Michaeli, T., Field, J., Ballester, R., O'Neill, K. O., and Wigler, M.,** Mutants of H-*ras* that interfere with *RAS* effector function in *Saccharomyces cerevisiae, EMBO J.,* 8, 3039, 1989.

20. **Wakelam, M. J. O., Davies, S. A., Houslay, M. D., McKay, I., Marshall, C. J., and Hall, A.,** Normal p21$^{N\cdot ras}$ couples bombesin and other growth factor receptors to inositol phosphate production, *Nature,* 323, 173, 1986.

21. **Wakelam, M. J. O.,** Inhibition of the amplified bombesin-stimulated inositol phosphate response in N-ras transformed cells by high density culturing, *FEBS Lett.,* 228, 182, 1988.

22. **Downward, J., De Gunzburg, J., Riehl, R., and Weinberg, R. A.,** p21*ras*-induced responsiveness of phosphatidylinositol turnover to bradykinin is a receptor number effect, *Proc. Natl. Acad. Sci. U.S.A.,* 85, 5774, 1988.

23. **Alonso, T., Morgan, R. O., Marvizon, J. C., Zarbl, H. and Santos, E.,** Malignant transformation by *ras* and other oncogenes produces common alterations in inositol phospholipid signaling pathways, *Proc. Natl. Acad. Sci. U.S.A.,* 85, 4271, 1988.

24. **Burgering, B. M. T., Snijders, A. J., Maassen, J. A., Van der Eb, A. J., and Bos, J. L.,** Possible involvement of normal p21 H-*ras* in the insulin/insulinlike growth factor 1 signal transduction pathway, *Mol. Cell. Biol.,* 9, 4312, 1989.

25. **Downward, J., Graves, J. D., Warne, P. H., Rayter, S., and Cantrell, D. A.,** Stimulation of p21*ras* upon T-cell activation, *Nature,* 346, 719, 1990.

26. **Gibbs, J., Marshall, M. S., Scolnick, E. M., Dixon, R. A. F., and Vogel, U. S.,** Modulation of guanine nucleotides bound to ras in NIH/3T3 cells by oncogenes, growth factors, and the GTPase activating protein, *J. Biol. Chem.,* 265, 20437, 1990.

27. **Satoh, T., Endo, M., Nakafuku, M., Nakamura, S., and Kaziro, Y.,** Platelet-derived growth factor stimulates formation of active p21$^{ras\cdot}$ GTP complex in Swiss mouse 3T3 cells, *Proc. Natl. Acad. Sci. U.S.A.,* 87, 5993, 1990.

28. **Satoh, T., Endo, M., Nakafuku, M., Akiyama, T., Yamamoto, T., and Kaziro, Y.,** Accumulation of p21^{ras}-GTP in response to stimulation with epidermal growth factor and oncogene products with tyrosine kinase activity, *Proc. Natl. Acad. Sci. U.S.A.,* 87, 7926, 1990.

29. **Burgering, B. M. T., Medema, R. H., Maasen, J. A., van de Wetering, M. L., McCormick, F., van der Eb, A. J., and Bos, J. L.,** Insulin stimulation of gene expression mediated by p21ras activation, *EMBO J.,* 10, 1103, 1991.

30. **Wolfman, A. and Macara, I. G.,** A cytosolic protein catalyzes the release of GDP from p21^{ras}, *Science,* 248, 67, 1990.

31. **West, M., Kung, H.-F., and Kamata, T.,** A novel membrane factor stimulates guanine nucleotide exchange reaction of *ras* proteins, *FEBS Lett.,* 259, 245, 1990.

32. **Downward, J., Riehl, R., Wu, L., and Weinberg, R. A.,** Identification of a nucleotide exchange-promoting activity for p21^{ras}, *Proc. Natl. Acad. Sci. U.S.A.,* 87, 5998, 1990.

33. **Ohga, N., Kikuchi, A., Ueda, T., Yamamoto, J., and Takai, Y.,** Rabbit intestine contains a protein that inhibits the dissociation of GDP from and the subsequent binding of GTP to rhoP p20, a *ras* p21-like GTP-binding protein, *Biochem. Biophys. Res. Commun.,* 163, 1523, 1989.

34. **Araki, S., Kikuchi, A., Hata, Y., Isomura, M., and Takai, Y.**, Regulation of reversible binding of p25A, a *ras* p21-like GTP-binding protein, to synaptic plasma membranes and vesicles by its specific regulatory protein, GDP dissociation inhibitor, *J. Biol. Chem.*, 265, 13007, 1990.

35. **Tsai, M.-H., Yu, C. L., Wei, F. S., and Stacey, D. W.**, The effect of GTPase activating protein upon ras is inhibited by mitogenically responsive lipids, *Science*, 243, 522, 1989.

36. **Tsai, M.-H., Hall, A., and Stacey, D. W.**, Inhibition by phospholipids of the interaction between R-ras, rho, and their GTPase activating proteins, *Mol. Cell. Biol.*, 9, 5260, 1989.

37. **Tsai, M.-H., Yu, C.-L., and Stacey, D. W.**, A cytoplasmic protein inhibits the GTPase activity of H-ras in a phospholipid dependent manner, *Science*, 250, 982, 1990.

38. **Yu, C.-L., Tsai, M. H., and Stacey, D. W.**, Serum stimulation of NIH 3T3 cells induces the production of lipids able to inhibit GTPase-activating protein activity, *Mol. Cell. Biol.*, 10, 6683, 1990.

39. **Molloy, C. J., Bottaro, D. P., Fleming, T. P., Marshall, M. S., Gibbs, J. B., and Aaronson, S. A.**, PDGF induction of tyrosine phosphorylation of GTPase activating protein, *Nature*, 342, 711, 1989.

40. **Kaplan, D. R., Morrison, D. K., Wong, G., McCormick, F., and Williams, L. T.**, PDGF β-receptor stimulates tyrosine phosphorylation of GAP and association with a signalling complex, *Cell*, 61, 125, 1990.

41. **Kazlauskas, A., Ellis, C., Pawson, T., and Cooper, J. A.**, Binding of GAP to activated PDGF receptors, *Science*, 247, 1578, 1990.

42. **Ellis, C., Moran, M., McCormick, F., and Pawson, T.**, Phosphorylation of GAP and GAP-associated proteins by transforming and mitogenic kinases, *Nature*, 343, 377, 1990.

43. **Reedijk, M., Liu, X., and Pawson, T.**, Interactions of phosphatidyl kinase, GTPase activating protein (GAP) and GAP-associated proteins with the colony-stimulating factor 1 receptor, *Mol. Cell. Biol.*, 10, 5601, 1990.

44. **Trahey, M. and McCormick, F.**, A cytoplasmic protein stimulates normal N-*ras* p21 GTPase, but does not affect oncogenic mutants, *science*, 238, 542, 1987.

45. **Trahey, M., Wong, G., Halenbeck, R., Rubinfeld, B., Martin, G. A., Ladner, M., Long, C. M., Crosier, W. J., Watt, K., Koths, K., and McCormick, F.**, Molecular cloning of two types of GAP complementary DNA from human placenta, *Science*, 242, 1697, 1988.

46. **Vogel, U. S., Dixon, R. A., Schaber, M. D., Diehl, M. S., Marshall, M. S., Scolnick, E. M., Sigal, I. S., and Gibbs, J. B.**, Cloning of bovine GAP and its interaction with oncogenic *ras* p21 *Nature*, 335, 90, 1988.

47. **Cantley, L. C., Auger, K. R., Carpenter, C., Duckworth, B., Graziani, A., Kapeller, R., and Soltoff, S.**, Oncogenes and signal transduction, *Cell*, 64, 281, 1991.

48. **Anderson, D., Koch, A. C., Grey, L., Ellis, C., Moran, M. F., and Pawson, T.**, Binding of SH2 domains of phospholipase Cγ1, GAP, and src to activated growth factor receptor, *Science*, 250, 979, 1990.

49. **Pawson, T.**, Non-catalytic domains of cytoplasmic protein-tyrosine kinases: regulatory elements in signal transduction, *Oncogene*, 3, 491, 1988.

50. **Marshall, M. S., Hill, W. S., Ng, A. S., Vogel, U. S., Schaber, M. D., Scolnick, E. M., Dixon, R. A. F., Sigal, I. S., and Gibbs, J. B.**, A c-terminal domain of GAP is sufficient to stimulate *ras* p21 GTPase activity, *EMBO J.*, 8, 1105, 1989.

51. **Adari, H., Lowy, D. R., Willumsen, B. M., Der, C. J., and McCormick, F.**, Guanosine triphosphatase activating protein (GAP) interacts with the p21^*ras* effector binding domain, *Science*, 240, 518, 1988.

52. **Cales, C., Hancock, J. F., Marshall, C. J., and Hall, A.**, The cytoplasmic protein GAP is implicated as the target for regulation by the *ras* gene product, *Nature*, 332, 548, 1988.

53. **Rey, I., Soubigou, P. Debussche, L., David, C., Morgat, A., Bost, P. E., Mayaux, J. F., and Tocque, B.**, Antibodies to synthetic peptide from the residue 33 to 42 domain of c-Ha-*ras* p21 block reconstitution of the protein with different effectors, *Mol. Cell. Biol.*, 9, 3904, 1989.

54. **Garrett, M. D., Self, A. J., van Oers, C., and Hall, A.,** Identification of distinct cytoplasmic targets for ras/R-ras and rho regulatory proteins, *J. Biol. Chem.,* 264, 10, 1989.
55. **Frech, M., John, J., Pizon, V., Chardin, P., Tavitian, A., Clar, R., McCormick, F., and Wittinghofer, A.,** Inhibition of GTPase activating protein stimulation of ras-p21 GTPase by the K*rev*-1 gene product, *Science,* 29, 169, 1990.
56. **Hart, P. A. and Marshall, C. J.,** Amino acid 61 is a determinant of sensitivity of *rap* proteins to the ras GTPase activating protein, *Oncogene,* 5, 1099, 1990.
57. **Yatani, A., Okabe, K., Polakis, P., Halenbeck, R., McCormick, F., and Brown, A. M.,** Ras p21 and GAP inhibit coupling of muscarinic receptors to atrial K+ channel, *Cell,* 61, 769, 1990.
58. **Kamata, T. and Feramisco, J. R.,** Epidermal growth factor stimulates guanine nucleotide binding activity and phosphorylation of *ras* oncogene protein, *Nature,* 310, 147, 1984.
59. **Kamata, T., Kathuria, S., and Fujita-Yamaguchi, Y.,** Insulin stimulates the phosphorylation level of v-Ha-*ras* protein in membrane fraction, *Biochem. Biophys. Res. Commun.,* 144, 19, 1987.
60. **Ballester, R., Furth, M. E., and Rosen, O. M.,** Phorbol ester- and protein kinase C-mediated phosphorylation of the cellular kirsten *ras* gene product, *J. Biol. Chem.,* 262, 2688, 1987.
61. **O'Brien, R. M., Siddle, K., Houslay, M. D., and Hall, A.,** Interaction of the human insulin receptor with the ras oncogene product p21, *FEBS Lett.,* 217, 253, 1987.
62. **Flier, J. S., Mueckler, M. M., Usher, P., and Lodish, H. F.,** Elevated levels of glucose transport and transporter messenger RNA are induced by ras or src oncogenes, *Science,* 235, 1493, 1987.
63. **Rosen, O. M.,** After insulin binds, *Science,* 237, 1452, 1987.
64. **White, M. F. and Kahn, C. R.,** The insulin receptor and tyrosine phosphorylation, *Enzymes,* 18, 247, 1986.
65. **Chou, C. K., Dull, T. J., Russell, D. S., Cherzi, R., Lebwohl, D., Ullrich, A., and Rosen, O. M.,** Human insulin receptors mutated at the ATP-binding site lack protein tyrosine kinase activity and fail to mediate postreceptor events of insulin, *J. Biol. Chem.,* 262, 1842, 1987.
66. **White, M. F., Shoelson, S. E., Keutmann, H., and Kahn, C. R.,** A cascade of tyrosine autophosphorylation in the β-subunit activates the phosphotransferase of the insulin receptor, *J. Biol. Chem.,* 263, 2969, 1988.
67. **Tornqvist, H. E., Pierce, M. W., Frackelton, A. R., Nemenoff, R. A., and Avruch, J. A.,** Identification of insulin receptor tyrosine residues autophosphorylated in vitro, *J. Biol. Chem.,* 262, 10212, 1987.
68. **Tornqvist, H. E., Gunsalus, J. R., Nemenoff, R. A., Frackelton, A. R., Pierce, M. W., and Avruch, J.,** Identification of the insulin receptor residues undergoing insulin-stimulated phosphorylation in intact rat hepatoma cells, *J. Biol. Chem.,* 263, 350, 1988.
69. **Tornqvist, H. E. and Avruch, J.,** Relationship of site-specific β subunit tyrosine autophosphorylation to insulin activation of the insulin receptor (tyrosine) protein kinase activity, *J. Biol. Chem.,* 263, 4593, 1988.
70. **Rosen, O. M., Herrera, R., Olowe, Y., Petruzzelli, L. M., and Cobb, M. H.,** Phosphorylation activates the insulin receptor tyrosine protein kinase., *Proc. Natl. Acad. Sci. U.S.A.,* 80, 3237, 1983.
71. **Yu, K.-T. and Czech, M. P.,** Tyrosine phosphorylation of the insulin receptor β subunit activates the receptor associated tyrosine kinase activity, *J. Biol. Chem.,* 259, 5277, 1984.
72. **White, M. F., Livingston, J. N., Backer, J. M., Lauris, V., Dul, T. J., Ullrich, A., and Kahn, R. C.,** Mutation of the insulin receptor at tyrosine 960 inhibits signal transmission but does not affect its tyrosine kinase activity, *Cell,* 54, 641, 1988.
73. **White, M. F., Maron, R., and Kahn, M. R.,** Insulin rapidly stimulates tyrosine phosphorylation of a Mr-185,000 protein in intact cells, *Nature,* 318, 183, 1985.
74. **Levenson, R. M. and Blackshear, P. J.,** Insulin-stimulated protein tyrosine phosphorylation in intact cells evaluated by giant two dimensional gel electrophoresis, *J. Biol. Chem.,* 254, 19984, 1989.

75. **Garcia-Moralis, P., Minami, Y., Luong, E., Lausner, R. D., and Samuelson, L. E.,** Tyrosine phosphorylation in T-cells is regulated by phosphatase activity: studies with phenylarsine oxide, *Proc. Natl. Acad. Sci. U.S.A.,* 87, 9255, 1990.

76. **Ray, L. B. and Sturgill, T. W.,** Insulin-stimulated microtubule-associated protein kinase is phosphorylated on tyrosine and threonine *in vivo, Proc. Natl. Acad. Sci. U.S.A.,* 85, 3753, 1988.

77. **Sturgill, T. W., Ray, L. B., Erikson, E., and Maller, J. L.,** Insulin-stimulated MAP-2 kinase phosphorylates and activates ribosomal S6 kinase II, *Nature,* 334, 715, 1988.

78. **Czech, M. P., Klarlund, J. K., Yagaloff, K. A., Bradford, A. P., and Lewis, R. E.,** Insulin receptor signaling, *J. Biol. Chem.,* 263, 11017, 1988.

79. **Dent, P., Lavoinne, A., Nakielny, S., Caudwell, F. B., Watt, P., and Cohen, P.,** The molecular mechanism by which insulin stimulates glycogen synthesis in mammalian skeletal muscle, *Nature,* 348, 302, 1990.

80. **Farese, R. V., Barnes, D. E., Davis, J. S., Standaert, M. L., and Pollet, R. J.,** Effects of insulin and protein synthesis inhibitors on phospholipid metabolism, diacylglycerol levels and pyruvate dehydrogenase activity in BC3H-1 cultured myocytes, *J. Biol. Chem.,* 259, 7094, 1984.

81. **Farese, R. V., Davis, J. S., Barnes, D. E., Standaert, M. L., Babischkin, J. S., Hock, R., Rosic, N. K., and Pollet, R. J.,** The *de novo* phospholipid effect of insulin is associated with increases in diacylglycerol, but not inositol phosphates or cytosolic Ca^{2+}, *Biochem. J.,* 231, 269, 1985.

82. **Farese, R. V., Kuo, J. Y., Babishchkin, J. S., and Davis, J. S.,** Insulin provokes a transient activation of phospholipase C in the rat epididymal fat pad, *J. Biol. Chem.,* 261, 8589, 1986.

83. **Saltiel, A. R., Sherline, P., and Fox, J. A.,** Insulin-stimulated diacylglycerol production results from the hydrolysis of a novel phosphatidylinositol glycan, *J. Biol. Chem.,* 262, 1116, 1987.

84. **Standaert, M. L., Fares, R. V., Cooper, D. R., and Pollet, R. J.,** Insulin-induced glycerolipid mediators and the stimulation of glucose transport in BC3H-1 myocytes, *J. Biol. Chem.,* 263, 8696, 1988.

85. **Saltiel, A. R., Fox, J. A., Sherline, P., and Cuatrecasas, P.,** Insulin-stimulated hydrolysis of a novel glycolipid generates modulators of cAMP phosphodiesterase, *Science,* 967, 1986.

86. **Farese, R. V., Sumon Konda, T., Davis, J. S., Standaer, M. L., Pollet, R. J., and Cooper, D. R.,** Insulin rapidly increases diacylglycerol by activating de novo phosphatidic acid synthesis, *Science,* 236, 586, 1987.

87. **Fraese, R. V., Cooper, D. R., Konda, T. S., Nair, G., Standaert, M. L., Davis, J. S., and Pollet, R. J.,** Mechanisms whereby insulin increases diacylglycerol in BC3H-1 myocytes, *Biochem. J.,* 256, 175, 1988.

88. **Taylor, D., Uhing, R. J., Blackmore, P. F., Prpic, V., and Exton, J.,** Insulin and epidermal growth factor do not affect phosphoionositide metabolism in rat liver plasma membranes and hepatocytes, *J. Biol. Chem.,* 260, 2011, 1985.

89. **Saltiel, A. R. and Cautrecasas, P.,** In search of a second messenger for insulin, *Am. J. Physiol.,* 255, cl, 1988.

90. **Villalba, M., Kelly, K. L., and Mato, J. M.,** Inhibition of cyclic AMP-dependent protein kinase by the polar head group of an insulin-sensitive glycophospholipid, *Biochim. Biophys. Acta,* 968, 69, 1988.

91. **Alemany, S., Mato, J. M., and Strålfors, P.,** Phospho-dephospho-control by insulin is mimicked by a phospho-oligosaccharide in adipocytes, *Nature,* 330, 77, 1987.

92. **Ishizuka, T., Cooper, D. R., Hernandez, H., Buckley, D., Standaert, M., and Frese, R. V.,** Effects of insulin on diacylglycerol-protein kinase C signaling in rat diaphragm and soleus muscles and relationship to glucose transport, *Diabetes,* 39, 181, 1990.

93. **Endemann, G., Yonezawa, K., and Roth, R. A.,** Phosphatidylinositol kinase or an associated protein is a substrate for the insulin receptor tyrosine kinase, *J. Biol. Chem.,* 265, 396, 1990.

94. **Ruderman, N. B., Kapeller, R., White, M. F., and Cantley, L. W.,** Activation of phosphatidylinositol 3-kinase by insulin, *Proc. Natl. Acad. Sci. U.S.A.,* 87, 1411, 1990.

95. **Stumpo, D. J. and Blackshear, P. J.,** Insulin and growth factor effects on c-fos expression in normal and protein kinase C-deficient 3T3-L1 fibroblasts and adipocytes, *Proc. Natl. Acad. Sci. U.S.A.,* 83, 9453, 1986.

96. **Messina, J. L., Hamlin, J., and Larner, J.,** Effect of insulin alone on the accumulation of a specific mRNA in rat hepatoma cells, *J. Biol. Chem.,* 260, 16418, 1985.

97. **Ting, L.-P., Tu, C.-L., and Chour, C.-K.,** Insulin-induced expression of human heat-shock protein gene hsp70, *J. Biol. Chem.,* 264, 3404, 1989.

98. **Stumpo, D. J., Steward, T. N., Gilman, M. Z., and Blackshear, P. J.,** Identification of *c-fos* sequences involved in induction by insulin and phorbol esters, *J. Biol. Chem.,* 263, 1611, 1988.

99. **Mohn, K. L., Laz, T. M., Hsu, J.-C., Melby, A. E., Bravo, R., and Taub, R.,** The immediate-early growth response in regenerating liver and insulin-stimulated H-35 cells: comparison with serum-stimulated 3T3 cells and the identification of 41 novel immediate-early genes, *Mol. Cell. Biol.,* 11, 381, 1990.

100. **Morris, J. D. H., Price, B., Lloyd, A. C., Self, A. J., Marshall, C. J., and Hall, A.,** Scrape-loading of Swiss 3T3 cells with *ras* protein rapidly activates protein kinase C in the absence of phosphoinositide hydrolysis, *Oncogene,* 4, 27, 1989.

101. **Mulcahy, L. S., Smith, M. R., and Stacey, D. W.,** Requirements for *ras* protooncogene function during serum-stimulated growth of NIH 3T3 cells, *Nature,* 313, 241, 1985.

102. **Lowenstein, E. J., Daly, R. J., Batzer, A. G., Li, W., Margolis, B., Lammers, R., Ullrich, A., Skolnik, D., and Schlessinger, J.,** The SH2 and SH3 domain-containing protein grb2 links receptor tyrosine kinases to ras signaling, *Cell,* 70, 431, 1992.

103. **Pelicci, G., Lanfrancone, L., Grignani, F., McGlade, J., Cavallo, F., Forni, G., Nicoletti, I., Grignani, F., Pawson, T., and Pelicci, P. G.,** A novel transforming protein (SHC) with an SH2 domain is implicated in mitogenic signal transduction, *Cell,* 70, 93, 1992.

104. **Rozalis-Adcock, M., McGlade, J., Mbamalu, G., Pelicci, G., Daly, R., Li, W., Balzer, A., Thomas, S., Brugge, J., Pelicci, P. G., Schlessinger, J., and Pawson, T.,** Association of the shc and grb2/sem5 SH2-containing proteins is implicated in activation of the ras pathway by tyrosine kinases, *Nature,* 360, 689, 1992.

105. **Pronk, G. J., McGlade, J. M., Pelicci, G., Pawson, T., and Bos, J. L.,** Insulin-induced phosphorylation of the 46 and 52 kDa shc proteins, *J. Biol. Chem.,* in press, 1993.

106. **Medema, R. H., de Vries-Smits, A. M. M., van der Zon, G. C. M., Maassen, J. A., and Bos, J. L.,** Stimulation of a guanine nucleotide exchange factor mediates activation of p21ras, *Mol. Cell. Biol.,* in press, 1993.

107. **Thomas, S. M., DeMarco, M., D'Arcangelo, G., Halegoua, S., and Brugge, J. S.,** Ras is essential for nerve growth factor- and phorbol ester-induced tyrosine phosphorylation of MAP kinases, *Cell,* 48, 525, 1992.

108. **Wood, K. W., Sarnecki, C., Roberts, T. M., and Blenis, J.,** Ras mediates nerve growth factor receptor modulation of three signal-transducing protein kinases: MAP kinase, raf-1, and RSK, *Cell,* 68, 1041, 1992.

109. **de Vries-Smits, A. M. M., Burgering, B. M. T., Leevers, S. J., Marshall, C. J., and Bos, J. L.,** Involvement of p21ras in activation of extracellular signal-regulated kinase 2, *Nature,* 357, 602, 1992.

110. **Kariakis, J. M., App, H., Zhang, X., Banerjee, P., Brautigan, D. L., Rapp, U. R., and Avruch, J.,** Raf-1 activates MAP kinase-kinase, *Nature,* 388, 417, 1992.

111. **Medema, R. H., Burgering, B. M. T., and Bos, J. L.,** Insulin-induced p21ras activation does not require protein kinase C, but a protein sensitive to phenylarsine oxide, *J. Biol. Chem.,* 266, 21186, 1991.

112. **Osterop, A. P. R. M., Medema, R. H., Bos, J. L., van der Zon, G. C. M., Moller, D. E., Flier, J. S., Möller, W., and Maassen, J. A.,** Relation between the insulin receptor number in cells, autophosphorylation and insulin-stimulated ras.GTP formation, *J. Biol. Chem.,* 267, 14647, 1992.

113. **Settleman, J., Narashimhan, V., Foster, L. C., and Weinberg, R. A.,** Molecular cloning of cDNAs encoding the GAP-associated protein p190: implications for a signaling pathway from ras to the nucleus, *Cell,* 69, 539, 1992.

114. **Wong, G., Müller, O., Clark, R., Conroy, L., Moran, M. F., Polakis, P., and Mc-Cormick, F.,** Molecular cloning and nucleic acid binding properties of the GAP-associated tyrosine phosphoprotein p62, *Cell,* 69, 551, 1992.

Chapter 6

ras-p21 PROTEINS: SWITCH DEVICES FOR SIGNAL TRANSDUCTION

Antonio Cuadrado, Amancio Carnero, and Juan Carlos Lacal

TABLE OF CONTENTS

0-8493-5214-2/93/$0.00 + $.50
© 1993 by CRC Press, Inc.

I. INTRODUCTION

The *ras* gene family has been the focus of intensive research since 1982, when their transforming versions were first identified in human tumors.[1] In addition to their role in cancer, this family of genes has been implicated in the regulation of key steps of cell proliferation and differentiation, suggesting that *ras* proteins may function as pivotal elements in signal transduction mechanisms. Despite a vivid research of both the biological and the biochemical properties of the *ras* gene products, there is not yet a precise knowledge about their real functions and the nature of the alterations that follow *ras*-mediated cell transformation. However, there is abundant information that locates *ras* proteins as essential switches involved in regulating processes as diverse as cell growth, cell differentiation, and cell fate during development. In this chapter, we will discuss the most relevant contributions aimed at the deciphering of the function of *ras* proteins in the signaling pathways that regulate proliferation, differentiation, and development as well as their implications in neoplastic transformation.

II. *ras* PROTEINS: THE BASIS FOR ITS SIGNAL TRANSDUCTION IMPLICATION

All information generated so far indicates that *ras* proteins may be essential for the transmission of signals that control cell behavior from the external milieu. In the following sections, we will summarize the most important discoveries supporting this hypothesis.

A. *ras* GENES ARE UBIQUITOUS IN EUKARYOTES

The *ras* oncogenes were initially discovered as the transforming genes of the Harvey and Kirsten murine sarcoma viruses (Ha-MSV, Ki-MSV).[2-3] These retroviruses were generated during passage of murine leukemia viruses through laboratory rat strains. The molecular analysis of Harvey and Kirsten retroviral genomes indicated that these sarcoma viruses contained sequences which were not present in their parental murine leukemia viruses. Further studies indicated that these sequences were transduced into the viral genome from the host cellular genetic material. At present, five different sarcoma viruses carrying *ras* sequences have been identified. Their structural properties and biochemical similarities to endogenous mammalian *ras* genes have been reviewed recently elsewhere[4] and will not be discussed any further in this review.

Three *ras* genes have been identified as single genes in mammalian genomes and designated as H-*ras*-1, K-*ras*-2, and N-*ras*, respectively.[5-9] These genes are contained in four exons separated by introns of very different lengths. An exception is the K-*ras*-2 gene which contains two alternative fourth exons, generating two transcripts by alternative splicing. Two pseudogenes have also been described in rats and humans, one of which belongs to K-*ras* (K-*ras*-1) and the other one to H-*ras* (H-*ras*-2).[10,11] Several mouse and hamster subspecies possess additional *ras* pseudogenes, probably due to germ line amplification.[12] Each of the three functional *ras* genes have been cloned and sequenced in several mammalian species.[1,4]

Homologs to mammalian *ras* genes have been found broadly in eukaryotes from yeast to fungus, snails, and insects. Several *ras* genes have been found in lower eukaryotes such as *Saccharomyces cerevisiae* (RAS1 and RAS2),[13,14] *Schizosaccharomyces pombe* (SPRAS),[15] and *Dictyostelium discoideum* (Ddras and DdrasG).[16,17] In *Drosophila melanogaster*, three *ras* genes named Dras1, Dras2, and Dras3 have been identified.[18-20] Finally, at least one *ras* gene has been found in fungi *(Mucor racemosus)*,[21] goldfish,[22] *Caenorhabditis elegans*[23,24] and mollusks (*Aplysia,* Apl-*ras* gene).[25] All *ras* genes isolated thus far show a very high degree of homology in their codified products (see References 1 and 4 and Chapters 2, 10, and 20), indicating a possible conserved essential function for cell viability in all described organisms.

B. *ras*-p21 PROTEINS EXHIBIT HIGH SEQUENCE SIMILARITY

The three mammalian *ras* genes codify for closely related proteins of 188 or 189 amino acids, with a mass of 21 kDa.[26] All *ras* proteins are membrane-associated (see Chapter 3), being this location essential for their biological function.[27,28]

The first 80 amino terminal residues share more than 80% sequence homology. From residue 80 to 165 there is a lower, but still significant, degree of conservation among all *ras* proteins. However, the carboxyl region of the protein has diverged considerably, since there is almost no significant sequence similarity (revised in References 4 and 13). An important sequence is found in all *ras* proteins at their carboxyl end, where the information for their membrane association is located.[27,28]

Most *ras* proteins found in eukaryotes show a similar sequence conservation pattern although there are important differences as well. As a matter of fact, it has been proposed that the highly conserved aminoterminal region might be the effector domain for all *ras* proteins, and the highly variable carboxyl terminal domain might contain determinants of their respective physiological specificity.[29]

Thus, different stimuli could couple to the same biological effect through activation of different *ras* proteins. If this were the case, the specificity and possible role for different functions could be located in their variable region. This region ranges from 20 residues among the three mammalian proteins, to 120 as in the RAS1 protein from *S. cerevisiae* (revised in References 1 and 4).

The possibility that the *ras* function might be conserved in distantly evolved species has been studied by complementation experiments in heterologous systems.[20,30,31] However, this approach may have some pitfalls. For instance, it has been shown that *ras* proteins regulate adenylate cyclase function in *S. cerevisiae* (see Chapter 7). Mammalian *ras*-p21 proteins can complement a lethal mutation in the RAS1 locus of *S. cerevisiae*[30,31] despite the fact that *ras* mutants similar to those which constitutively activate yeast adenylate cyclase do not affect cAMP levels in mammalian cells.[32]

C. DOMAINS OF *ras*-p21 PROTEINS WITH FUNCTIONAL SIGNIFICANCE

Common biochemical properties such as binding of guanine nucleotides (GTP and GDP) and GTPase activity are exhibited in *ras*-p21 proteins. In addition, two viral *ras* proteins (Harvey and Kirsten p21 *ras* proteins) exhibit autophosphorylation at position Thr59.[33,34] The nature of this phosphorylation is related to the proximity of the side chain of the residue Thr59 to the catalytic GTPase domain.[35] It functions as a good phosphate acceptor that in the normal protein is transferred to a water molecule closed to the side chain of Ala59, the normal residue. Although the meaning of this phosphorylation is poorly understood, it has been shown to increase the stability of the protein.[36] It is by itself a weak activating mutation, and increases the transforming efficiency of the viral versions with mutations at position 12 (J. C. Lacal, unpublished).

Our knowledge about the biochemical properties of these small GTP-binding proteins relied for several years on genetic studies with point and deletion mutants, and the use of neutralizing antibodies to define epitopes with specific functions. Extensive *in vitro* mutagenesis allowed the definition of domains responsible for GTP binding and GTPase activities of *ras* proteins. Thus, it was soon observed that mutations around codons 12 and 61, the most frequent alterations in the transforming alleles, had little effect on the GTP-binding activity (revised in References 1 and 4). However, an antibody against Ser12 of v-K-*ras* blocked the ability of the protein to bind GTP,[37] suggesting its implication in the biochemical function of the protein. Furthermore, small deletions within residues 5 to 23, as well as deletion of the first 22 aminoterminal residues, completely impaired binding of GTP to the *ras*-p21 protein.[38]

Other regions of the protein involved in the formation of the GTP-binding pocket have been identified in a similar fashion. Substitution of Ala for Asp119 decreased the binding of both GTP and GDP by 50-fold and results in transforming activity due to a higher rate of replacement of GDP for GTP, more abundant under intracellular conditions.[39] Similar results were observed in a

Thr59 mutant, where an increased GTP interchange was observed.[40] Substitution of Lys to Asn16, drastically impaired both GTP-binding and transforming activity.[39] Deletion of residues 165 to 186 did not alter the GTP-binding properties of *ras*-p21. However, a longer deletion (residues 152 to 186) completely abolished this activity.[38] Although the three-dimensional structure of the protein might be altered in such a large deletion mutant, it is also possible that part of this region might be involved in GTP binding. In fact, a polyclonal antibody raised against amino acids 161 to 176 blocks GTP binding.[41] Thus, several mutations involving residues around positions 12, 59 to 61, 116 to 119, and 147, were reported to affect GDP/GTP binding and interchange (revised in References 1 and 4).

Activating point mutations at residues 12 reduce by 2- to 10-fold the *in vitro* GTPase activity of the wild-type protein.[42-44] All substitutions analyzed at position 61 of the *ras*-p21 protein also reduced the *in vitro* GTPase activity. However, when analyzed, there was not a good correlation between GTPase and transforming activities,[45,46] indicating that other factors are involved in the regulation of *ras*-p21 signaling *in vivo* (see Section III).

Another important mechanism of regulation of *ras*-p21 function relates to its GDP/GTP interchange (see Chapter 25). An Ala to Thr substitution at position 59 reduced by three-fold its affinity for both GDP and GTP.[40] Monoclonal antibody Y13-259, which recognizes a proximal region of the *ras*-p21 molecule (residues 63 to 73), blocks GTP/GDP interchange (off rate) without affecting its binding (on rate) properties.[47] Thus, genetic analysis made it possible to conclude that regions around codons 12 and 61 contribute to the formation of the GTP binding pocket.

The recent deciphering of the three-dimensional structure of the *ras*-p21 protein has provided a better understanding of the residues involved in its GTP-binding and GTPase function[48-53] (see also Chapter 2). However, we have to be cautious with the conclusions provided from crystallographic studies, since they have been performed with truncated versions of the *ras* proteins.[48-53] Although binding properties and GTPase rates of full length and truncated versions seem to be similar, there is still a possibility that the C terminal region could stabilize the L6, L8, and L10 loops (residues 84 to 86, 117 to 126, and 145 to 151, respectively) for direct interaction with the GTP molecule. In any case, these studies established that there are five important regions on the *ras*-p21 molecule required for GTP binding. These regions correspond to consensus regions found in all small GTPases, including the guanine base recognition sites (116 to 119 and 145 to 147), and the phosphate-binding domains (10 to 17, 32 to 35, and 57 to 62). For a more detailed explanation, see Chapter 2.

Besides the involvement of specific residues of the *ras*-p21 protein in GTP bindings and GTPase activities, other regions of the molecule are candidates to interact with regulatory factors. In this regard, it has been proposed that the carboxyl end of the *ras*-p21 molecule could be involved in the interaction of regulatory factors, as described for *smg* p21/*rap1A*.[54]

D. *ras*-p21 PROTEINS ARE MEMBRANE-ASSOCIATED

In order to be fully functional, *ras* proteins need to be attached to the plasma membrane[27] (see also Chapter 3). The anchorage to the membrane occurs after proteolysis of the three last residues of the molecule, a process that is accompanied by isoprenylation of the cysteine located at the carboxyl terminal region (Cys186) which is conserved in all *ras* proteins.[55] Cys186 is also conserved in all known members of the *ras* as well as *rho* branches of the superfamily of *ras*-related proteins. It is contained within the CAAX consensus sequence, where C is a cysteine, AA are two aliphatic residues, and X represents one last residue that contains the specificity for farnesylation or geranylgeranylation.[56] The only exception for this rule is the K-*ras* protein, which is not farnesylated. In addition, all *ras* proteins, including K-*ras*-p21, are also palmitoylated at a second Cys residue proximal to Cys186.[28]

Other membrane-associated oncoproteins are located at the plasma membrane by a similar, but distinct mechanism. This is the case for the p60[src] tyrosine-kinase, which is located at the plasma membrane by a myristoylation process through an amino terminal Gly residue, after proteolytic removal of the traditional Met at the amino end.[57] It has been reported that a membrane factor is responsible for the actual location of p60[src] to the membrane, probably by a protein-protein interaction.[58] The possibility of a polypeptide anchorage device for *ras* proteins has not been investigated yet. The fact that chimeric proteins containing the first 15 residues from p60[src] and the full length or truncated versions of *ras*-p21 proteins, devoid of their natural C terminal region are fully transforming,[59] suggests a close location for both p60[src] and *ras*-p21 proteins which may be mediated by similar protein factors.

E. G PROTEINS ARE INVOLVED IN SIGNAL TRANSDUCTION

Guanine nucleotide-binding proteins have been identified in several cell systems as regulatory molecules coupling cell surface receptors to a variety of cytoplasmic effector enzymes.[60,61] The most well-characterized systems where G proteins have been identified include the control of the adenylate cyclase by Gs (for stimulatory G protein) and G_i (for inhibitory G protein),[60,61] regulation of the cyclic GMP-specific phosphodiesterase activity by transducin in the retinal rod outer segment,[62] and activation of the β-isoenzyme of phosphatidylinositol (PI)-specific phospholipase C by G_q.[63] Other biochemical reactions where evidence has been provided to implicate unidentified G proteins include muscarinic cholinergic receptors, atrial potassium channels, neuronal potassium channels, protein translocation, phospholipase A2 activation, and PI-3-kinase.[60-64]

G_s, G_i, transducin, and G_q belong to a large family of proteins involved in receptor-induced signal transduction processes. They function as heterotrimeric complexes (α,β,γ) where only their α-subunits are able to bind guanine nucleotides with high efficiency.[60-64] These classical, heterotrimeric G proteins respond to activated receptors by liberating their α-subunits from the complex

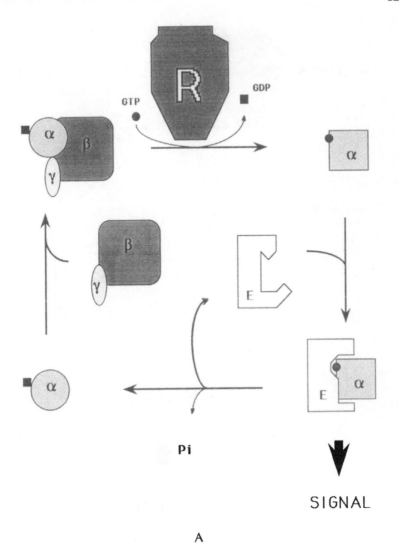

FIGURE 1. Schematic representation of the GTP/GDP cycle of heterotrimeric (A) and monomeric (B) GTPases. While replacement of GDP for GTP is accomplished by receptor activation of heterotrimeric GTPases, monomeric GTPases are activated by specific GNRPs. Monomeric GTPases also require the presence of specific GAPs for inactivation and probably transmission of their signals (see text for details).

at the same time that bound GDP is replaced by GTP (see Figure 1). The activated, GTP-bound α-subunits acquire an active state which allows them to interact with their effector molecules. All known G proteins return to the inactive state by virtue of their intrinsic GTPase activity which converts GTP to GDP. As indicated above, amino acid sequence comparison of *ras*-p21 with that of α-chains of

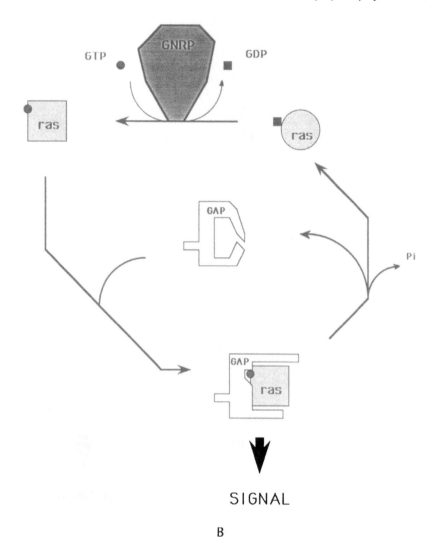

SIGNAL

B

FIGURE 1 (continued).

GTP-binding proteins, such as G_s and G_i, revealed a consistent pattern of homology, especially in regions of the molecule involved in their GTP-binding and GTPase properties.[65,66]

The α-subunits of the classical G proteins, contain not only the sites for GTP binding, but also for ADP-ribosylation by pertussis and cholera toxins[67,68] and for specific recognition of receptor and effector molecules.[64] The characterization of specific mutations in *ras*-p21 as well as in the α-subunits of heterotrimeric G proteins indicates that the two proteins use parallel molecular mechanisms and structural elements for GTP hydrolysis. Thus, replacement of Gly49 in $G_{i\alpha}$-subunits reduces its *in vitro* GTP hydrolysis rate.[69] Also, several mutations in the $G_{s\alpha}$-subunit (Gln227 to Arg, His, or Leu, or Arg201 to Cys or His) and $G_{i\alpha}$-subunit (arg179 to Cys or His) may contribute to the generation of human tumors by altering the regulation of adenylate cyclase.[70,71] These results resemble those found in activated *ras* genes.

It was originally thought that *ras*-p21 proteins could represent G proteins for some of the above systems. However, the evidence generated thus far does not seem to support this hypothesis. While the *ras* genes are ubiquitously expressed, some of these metabolic reactions are restricted to very specific cell types. Also, some of these reactions are related to housekeeping activities or to specialized functions that are probably not involved in the control of cell proliferation. Deregulation of a housekeeping function may result in the loss of cell proliferation control. This is the case for activating point mutations in two G proteins of the classical family of GTP-binding proteins (G_i and G_s) which have been reported to induce oncogenic transformation.[69-71] However, a more attractive hypothesis might suggest that small GTP-binding proteins regulate a subset of activities directly involved in the control of cell proliferation and differentiation.

F. FACTORS REGULATING *ras*-p21 FUNCTION

The hypothesis that *ras*-p21 proteins could be real GTPases with regulatory functions, was launched initially by comparison of their primary sequences to those of authentic GTPases involved in the regulation of well-defined systems such as adenylate cyclase. This hypothesis was mostly supported by the discovery of mutated versions of the *ras* protein with reduced *in vitro* GTPase activities.[51-53] However, some strong discrepancies were soon reported where no correlation was found between *in vitro* GTPase activities and the transforming efficiency of several mutant proteins (revised in References 1 and 4).

McCormick and collaborators described a factor, designated as GAP (GTPase-activating protein) that specifically accelerated the intrinsic GTPase activity of wild-type *ras* proteins, with no affect on transforming proteins[72] (see Section III). Recently, a new protein, neurofibromin, NF-1 the product of the gene, implicated in the development of neurofibromatosis, has also been shown to regulate intrinsic GTPase activity of *ras* proteins,[73] suggesting that *ras*-p21 may have several regulators of its GTPase activity *in vivo*. Finally, several nucleotide exchange factors have been described for *ras* and *ras*-related proteins implying a second way of regulating its function (see Chapter 25).

The regions responsible for interaction with these molecules are being investigated. Two regions of *ras*-p21 have been identified as responsible for its interaction with GAP. The so-called "effector domain" 32 to 40 region,[74-76] and the 58 to 68 region[77] are both important for GAP-activated GTPase and biological activities. Moreover, they are not relevant for p21 processing and membrane localization. Thus, these regions should be designated as putative "effector domain-1'' (32 to 40 residues) and ''effector domain-2'' (58 to 58 residues). As described in detail in Section III, it is important to note that the C terminal region of GAP seems to be sufficient for its enzymatic interaction with *ras*-p21,[78] suggesting that the N terminal part of GAP could be its regulatory domain.

Interestingly, the antibody Y13-259, which neutralizes *ras* activity, seems to interact with *ras* proteins at the same site as GAP,[77] suggesting that the antibody inhibits its biological activity by preventing the interaction with its effector molecule(s). An important observation is that the antibody Y13-259 does not affect GTP binding to p21 but somehow interferes with its exchange rate,[47] suggesting that the region recognized by Y13-259 could be also critical for GDP/GTP interchange. A region with chances to become an important domain for nucleotide exchange is the very C terminus end of the molecule, as proposed for *smg* p21/*rap1A*.[54] At lease three groups have described the characterization of interchange factors (see Chapter 25) specific for *ras*-p21 proteins. The implications of these findings are certainly an important challenge for the next years of research on *ras*.

It has been observed that GAP can associate with two additional proteins of 62 and 190 kDa.[79-81] The potential interaction of these proteins with *ras* has not been reported yet, but it is an important issue to investigate. Thus, it is conceivable that regulation of GAP function may be complex, involving both tyrosine as well as Ser/Thr phosphorylation.

III. *ras*-p21 PROTEINS ARE IMPLICATED IN THE TRANSDUCTION OF SIGNALS INITIATED BY PROTEIN TYROSINE KINASES

A large number of growth factors and morphogens have been identified whose function is performed through receptor tyrosine kinases.[82] The best known example is the induction of cell proliferation by soluble growth factors, such as platelet-derived growth factor (PDGF) or IGF-1.

Tyrosine kinase activity has also been identified in biological models for cell differentiation and development. That is the case of the high affinity receptor for nerve growth factor, which promotes differentiation of neuroblasts, and has been recently identified as the normal homolog of *trk*, a tyrosine kinase oncogene found in a human colon carcinoma.[83-84] Below are some of the evidences that imply that *ras* proteins might be important regulators of these processes involving tyrosine kinase receptors.

A. REGULATION OF PROLIFERATION THROUGH *ras* PROTEINS

In fibroblasts and epithelial cells, receptor activation by soluble growth factors such as PDGF or epidermal growth factor (EGF) results in cell proliferation.[82] Moreover, if the proliferative signal is sustained either as a result of autocrine stimulation (v-*sis*), receptor amplification (EGF receptor in some breast carcinomas), or receptor activation in the absence of ligand, (v-*fms,trk*) then cells acquire the transformed phenotype.[85]

Interestingly, a very strong correlation can be made between the effect of such enzymes on the target cells and that induced by the active variants of *ras* genes. Activated *ras* genes have been found in a large variety of sarcomas, including those of retroviral origin and human carcinomas.[1] Microinjection of activated *ras*-p21 in quiescent fibroblasts induced initiation of DNA synthesis and cell proliferation.[86,87] Moreover, microinjection of a monoclonal antibody raised against *ras*-p21, T13-259, neutralized the mitogenic activity of serum.[88] These results imply that there must be a connection between *ras* function and growth factors, most likely involving their surface receptors.

The location of *ras*-p21 in the receptor tyrosine kinase signaling pathway may bring some hope for those searching for the *ras* function. In practice, the cascade of known events that follow receptor activation is large and ill-defined. The *ras* proteins could interact with the receptor kinase itself, activating its intrinsic kinase activity or potentiating the ligand induced stimulation. This hypothesis is supported by the observation that *ras*-transformed cells display a higher content in tyrosine-phosphorylated substrates than normal fibroblasts, although the nature of this increase is still unknown.[89] A trivial explanation could be that a large number of *ras*-transformed cells exhibit an autocrine production of growth factors, specially EGF-like growth factors.[90,91] If this were the case, this phenomenon should be irrelevant to its transforming activity, since it has been demonstrated that *ras* is a potent oncogene even to cells that lack EGF receptors.[92]

Studies conducted *in vitro* have recently indicated that polylysine-like basic peptides, including the C terminus of K-*ras* which carries a basic domain between residues 172 to 182, were able to increase the affinity of the receptor kinase for substrates such as calmodulin.[93] A related gene such as H-*ras*, which does not contain this basic domain, failed to activate the insulin receptor.[93] An independent group has also described that the carboxyl terminal segment of c-Ki-*ras*-2 stimulated both the *in vitro* phosphorylation of calmodulin and autophosphorylation of the β-subunit of the insulin receptor, independently of the presence of insulin.[94] Whether this phenomenon takes place *in vivo* and whether this is the regulatory function of K-*ras* remains to be elucidated. Interestingly, the K-*ras* protein has a unique mechanism to be located at the membrane which distinguishes it from H- and N-*ras* proteins.[95] It is also the only one among *ras* proteins that has been shown to be phosphorylated by protein kinase C (PKC) *in vivo*,[96] although the H-*ras* p21 protein can also be phosphorylated by PKC *in vitro*.[97] As we will describe later, PKC is activated after generation of diacylglycerol (DAG), a process intimately linked to activation of some tyrosine kinase receptors.[82]

A link between tyrosine kinases and *ras* signaling pathways has been recently shown. The *ras*-specific GTPase-activating factor, 120-GAP, is rapidly phosphorylated in tyrosine residues following stimulation with a series of growth factors such as PDGF, EGF, or CSF-1.[98-100] Although the functional relevance of this modification is still unclear, phosphorylated GAP may exhibit a conformational change that modifies its interaction with *ras*-p21. As a matter of fact, stimulation of cells with PDGF, EGF, or CSF-1, growth factors that promote tyrosine phosphorylation of GAP, increases the amount of p21 molecules bound to GTP, suggesting a *ras*-activating process.[101,102] However, addition of insuline to cells overexpressing its receptor also induces a rapid interchange of GDP for GTP without an apparent phosphorylation of GAP, suggesting an alternative pathway for regulation of *ras*-p21 function.[103]

There may still be other mechanisms for GAP and neurofibromin regulation. Stacey and co-workers have found a regulatory role for some lipids and lipid derivatives to *in vitro* GAP activity.[104] Some of these lipids are generated as a function of ligand-dependent activation of surface receptors.[105,106] Thus, the *ras*-specific GTPase activator proteins, GAP or neurofibromin, could be the direct or indirect target of regulation by growth factors. However, there has been some controversy regarding this possibility, since at least some of these lipids may function as simple detergents, inhibiting GAP function by trapping the molecule into micelles.[107]

Xenopus laevis oocytes have proved to be extremely useful tools for the dissection of the signaling pathway of cell proliferation. This biological model allows for the investigation of events leading to progression along meiosis and egg maturation. Moreover, these events have been shown to closely correlate with those that govern the entry of mammalian cells from quiescence into the mitotic cycle. Activation of tyrosine kinases, such as the insulin receptor, induce germinal vesicle breakdown (GVBD).[108] Also, microinjection of oncogenic *ras*-p21 proteins promote GVBD in a similar fashion as with the activation of some receptor tyrosine kinases.[109]

The development of a neutralizing antibody for *ras* activity (Y13-259) has allowed the further location of *ras*-p21 proteins in the tyrosine kinase signaling pathways beyond the point of the strong correlations described above. Y13-259 interacts with the *ras*-p21 protein at one of the two regions considered part of the effector domain neutralizing their interaction with GAP, the putative effector molecule. Microinjection of the neutralizing antibody Y13-259 into quiescent fibroblasts rendered cells unable to proliferate in response to the mitogenic signals initiated by extracellular agonists that interact with receptor tyrosine kinases.[88] This is strong evidence for supporting the involvement of *ras* in regulating signal transduction mediated by tyrosine kinase receptors. In addition, several growth factors have been shown to activate *ras* proteins by increasing its association to GTP (see Chapters 5, 21, and 25.)

Increasing evidence has implicated some *ras*-related proteins with intracellular trafficking as discussed in Chapter 18. In addition, it has been suggested that the mitogenic signal involves internalization of mitotic complexes which might contain oligomeric forms of receptor tyrosine kinases together with putative kinase substrates such as phospholipase C-γ, GAP, or PI-3 kinase.[110] It is possible that *ras*-p21 might be implicated in the release of these complexes from the plasma membrane. Some light about how a GTP-binding protein might be involved in this process stems from the observation that phosphatidylinositol glycan-associated proteins are cleaved at the sugar backbone and released from the plasma membrane in response to insulin.[111] Insulin-induced effects are inhibited by pertussis toxin, suggesting a G protein in the process.[112]

B. SIGNAL TRANSDUCTION BY *ras*-PROTEINS DURING DIFFERENTIATION AND DEVELOPMENT

Tyrosine phosphorylation is an important event associated to cell differentiation and development in a variety of cell systems. Also shown to be important for some of these events are *ras* proteins. For instance, when *ras* oncogenes were introduced into the rat pheochromocytoma cell line PC12, these cells produced neurite outgrowth and differentiation to sympathetic neurons in a way similar to NGF.[113-115] These *ras* proteins have been also implicated in the differentiation program of other cell systems, such as lymphoid, endocrine, adipocite, and muscle cells.[116-119]

An important observation which provides further support to the hypothesis of *ras* being an important transmitter of growth factor receptors signaling involved in regulation of cell fate has come from the *ras*-homolog gene *let-60* from *C. elegans*. In a set of genetic experiments it has been shown that *let-60* is essential for the fate of those cells involved in the formation of the vulva.[23,24] This phenomenon requires both a functional *let-23* gene, which codifies for a tyrosine kinase equivalent to the EGF receptor, and the *let-60* gene. Dominant activating mutations in *let-60* similar to those found in human tumors bypass the requirement for *let-23*, producing a multivulva phenotype.

Recently, a functional relationship between *ras* proteins and the receptor tyrosine kinase pathway has been found also in the fly *Drosophila melanogaster*. RAS1 and *sevenless* (*sev*) are both required for the development of photoreceptor cells in the *Drosophila* eye.[120] A mutated RAS1-Val[12] suppresses the absence of *sev* function, which codifies for the *Drosophila* analog of the mammalian EGF receptor. Thus, there is a striking similarity to the *C. elegans* model system described above, suggesting a functional connection between *ras* proteins and the EGF-receptor signaling.

Another system where *ras* proteins may have a role in development is the slime mold *Dictyostelium discoideum*. Two genes related to mammalian *ras* have been identified. Dd *ras* and Dd *ras*G, which may have different roles during proliferation and differentiation (see Chapter 8).

Similarly to the experiments described earlier in serum-induced cell proliferation, microinjection of the Y13-259 monoclonal antibody into the pheochromocytoma cell line PC12, aborted the differentiating response to sympathetic neurons induced by NGF.[121] Recently, it has been reported that normal *trk*, a receptor-like tyrosine kinase protooncogene, may constitute the NGF receptor or one of its components.[83,84] Expression of *trk* into *X. laevis* induces NGF-dependent maturation,[122] a biological activity previously found when activated *ras*-p21 was microinjected into oocytes.[109] Insuline may be coupled to *ras* proteins in this system[123] since *ras* can be phosphorylated by insuline-receptor stimulation. All these results strongly support the hypothesis of a functional relationship between *ras* proteins and tyrosine kinase receptors signaling in multiple systems during differentiation and development.

IV. *ras* PROTEINS AND REGULATION OF PHOSPHOLIPID METABOLISM

A. PHOSPHATIDYLINOSITOL TURNOVER

One of the earliest biochemical events that occurs following mitogen stimulation of many cell types is the activation of phosphoinositide-specific phospholipase C (PLC).[82,85,124] The substrate of this enzyme, phosphatidylinositol 4,5-trisphosphate (PIP2), is hydrolyzed to yield two well-characterized second messengers: membrane-bound DAG and cytosolic inositol 1,4,5-trisphosphate (IP$_3$). Diacylglycerol has been shown to specifically activate some of the known PKC isozymes,[125,126] whereas IP$_3$ mediates release of free Ca^{2+} from nonmitochondrial intracellular stores.[124]

It was initially speculated that *ras*-p21s could be the G proteins for phospholipase C isoenzymes. Cantley and co-workers found an increased PI turnover in *ras*-transformed cells.[127] Moreover, transfection of Swiss 3T3 cells with a construct where the expression of the oncogenic N-*ras* protein was under the control of the glucocorticoid inducible promoter of the mouse mammary tumor virus (MMTV-LTR), allowed the investigation of the temporal relationship between *ras*-p21 appearance and PI breakdown.[128] Although these experiments originally suggested that *ras*-p21 might be coupling growth factor receptors to a PI-specific phospholipase C, it was shown later that it is an effect related to the number of receptors themselves.[129] Moreover, these results could not be reproduced either in other cell types or in other clones expressing the same *ras*-inducible system.[130]

The finding that some *ras*-transformed cells exhibit higher levels of DAG and IP$_3$ than normal cells was further extended to several cell systems.[131-133] However, this finding was not universal, since it was also reported that other cell lines expressing oncogenic *ras*-p21 proteins did not have altered intracellular IPs levels.[134-137] Some of these cell lines contained oncogenes unrelated to *ras* and, therefore, it is unlikely that their increased PI turnover was due to *ras*-p21.[138]

Finally, a G protein termed Gq has been shown to activate PI hydrolysis in rat brain extracts. Isolation and cloning of the Gq protein demonstrated that it is an α-subunit of the family of classic G proteins.[63] Therefore, the small GTP-binding proteins such as *ras*-p21 do not seem to be directly involved in coupling with phosphoinositide-specific PLC.

B. PHOSPHATIDYLCHOLINE METABOLISM

Several groups have reported that mitogenic stimulation of receptor tyrosine kinases results in a bimodal production of DAG.[139] The first wave occurs immediately after stimulation and is maintained for a few minutes. The second wave follows the first one and is much more sustained in time. These results provided the support to the hypothesis that another source, different from hydrolysis of PIP_2, is responsible for the mitogenic activity of growth factors. This finding is also in good agreement with the fact that, for mitogenic activity, growth factors are required for longer periods of time than that required to produce PI-derived metabolites.[140]

Several groups have reported that serum, PDGF, and phorbol esters activate the hydrolysis of phosphatidylcholine (PC), as detected by the production of DAG and phosphorylcholine (revised in References 141 and 142). The most abundant membrane phospholipid in mammalian tissues is PC.[143] Hydrolysis of PC generates DAG in a way similar to that of PIs, without release of inositolphosphates.

Phosphatidylcholine hydrolysis is much more sustained than PI hydrolysis, in agreement with the second wave of DAG observed in cells stimulated with growth factors and serum.[141,142] However, while PC-derived DAG has the potential to activate PKC in a way similar to PI-derived DAG, there is a recent report suggesting that PC-derived DAG is innocuous for PKC activation in IIC9 cells, a subclone of Chinese hamster embryo (CHO) fibroblasts.[144]

Although *ras* proteins do not seem to be directly involved in the metabolism of PI, several studies have implicated *ras* proteins in the metabolism of other phospholipids on the basis that *ras*-transformed cells display increased levels of DAG. Fleishman et al., observed differences in the ratio of DAG/PIP_2 vs. PI/PIP_2 between normal and transformed cells.[127] Wolfman and Macara found that despite normal levels of IPs, *ras*-transformed cells showed increased levels of DAG.[135] Other laboratories have obtained similar results in H-*ras*- or K-*ras*-transformed cells, with elevated levels of DAG without significant alterations in the levels of PI-specific metabolites.[136,137]

Lacal et al. observed increased levels of DAG in several *ras*-transformed cell lines, with no accompanying increase in the levels of IPs.[136] They postulated that DAG in *ras*-transformed cells could have been generated from the hydrolysis of other major membrane phospholipids such as PC. Indeed, they observed increased levels of phosphorylcholine and phosphorylethanolamine but not phosphorylserine in these cell lines, when compared to normal cells or to cells expressing transformation-deficient *ras* mutants. In contrast, v-*sis*-transformed cells exhibited normal levels of these compounds. Hall and collaborators also showed a rapid formation of DAG, phosphorylcholine, and choline without appearance of IPs, after scrape-loading oncogenic *ras* protein into NIH 3T3 cells.[137]

It has been proposed that the increase in DAG observed in *ras*-transformed cells might be the result of synthesis *de novo*.[145] In fact, neosynthesis of DAG is constitutively active in transformed cell lines and this pathway may be reversed by blocking the glycolytic pathway. However, the largest body of evidence to explain the altered levels of DAG of *ras*-transformed cells suggests that it proceeds from the metabolism of PC.

Microinjection of H-*ras*-p21 proteins into *Xenopus laevis* oocytes induced a rapid increase in 1,2-DAG levels, consistent with the ability to induce GVBD.[146] In these studies, microinjection of high doses of activated H-*ras*-p21 protein produced the release of large amounts of DAG and lower, but still significant amounts of IPs. However, minimal doses of activated *ras*-p21 were sufficient to induce a biphasic increase in the amounts of phosphorylcholine and CDP-choline, with no detectable increase on IPs.[147] The first burst of phosphorylcholine and CDP-choline preceded the increase in DAG levels. The second peak paralleled the appearance of DAG. These results suggested the interesting possibility that *ras*-p21 mediates the activation of enzymes involved in PC metabolism other than the PC-specific PLC (PC-PLC).

Additional evidence in favor of this hypothesis is being accumulated. For instance, Macara et al. have observed that *ras*-transformed NIH 3T3 cells exhibit higher choline kinase activity than exponentially growing NIH 3T3 control cells with no detectable alteration in PC-PLC.[148] Kent and collaborators also found that C3H 10T1/2 fibroblasts transformed by the H-*ras* oncogene display increased levels of choline kinase activity (1.9-fold) and decreased levels of CTP phosphorylcholine cytidyltransferase activity, therefore, contributing to increased intracellular levels of phosphorylcholine.[149]

Alternatively, it was originally postulated that PDGF could induce the activation of a PC-PLC, and that a direct effect of *ras* proteins on a PC-PLC could explain the generation of both DAG and phosphorylcholine.[136,150] This hypothesis has been later supported by the evidence that addition of a PC-specific PLC to fibroblasts is a potent mitogen[151] and that microinjection of a neutralizing antibody to this PLC blocks biological activity of both PC-specific PLC and *ras* proteins.[152] However, no characterization has been published of the utilized antibody and some cross-reactivity of this antibody to other enzymes could also explain the observed results. While it was originally reported that mitogenic activity by PC-PLC it is PKC-independent,[151] the same group has also reported that it may be a process involving activation and translocation of PKC.[153] Another strong contradiction with the *ras*-PLC hypothesis is that it has been shown that oncogenic *ras*-p21 requires functional PKC for mitogenic activity.[154,155] Preliminary evidence suggests that GAP cooperates with *ras*-p21 to activate PC-PLC.[156] These results await confirmation by other investigators. In fact, accumulating evidence indicates that there is at least an alternative and contrasting explanation, since *ras* proteins, as well as independent growth factors, induce phosphorylcholine and DAG in a PLD-dependent, rather than in a PLC-dependent, fashion (see Note Added in Proof at end of chapter).

C. *ras* PROTEINS AS MODULATORS OF PLA$_2$

Another potentially relevant connection between *ras*-p21 proteins and the phospholipid metabolism has been suggested that implies the activation of phospholipase A_2.[157,158] This enzyme releases lysophospholipids and free fatty acids (specifically arachidonic acid) which may be utilized as precursors for the synthesis of eicosanoids and prostaglandins.[159]

It has been reported that microinjection of both normal and transforming H-*ras*-p21 proteins induces membrane ruffling which correlates with a rapid increase in the levels of lysophosphatidylcholine (lyso-PC) and lysophosphatidylethanolamine (lyso-PE).[157] This increase is more sustained after microinjection of the oncogenic version. Moreover, PLA_2 colocalized with *ras* proteins in normal and *ras*-transformed cells.[158] The possible activation of the synthesis of eicosanoids, thromboxanes, and prostaglandins is especially interesting since recently it has been reported that these lipids could modulate the interaction between *ras*-p21 and GAP.[106]

V. PKC-DEPENDENT PATHWAYS IN *ras*-INDUCED SIGNAL TRANSDUCTION

Protein kinase C is a family of cytoplasmic Ser/Thr protein kinases that at present counts with eight members (revised in References 125 and 126). The most well-characterized isoenzymes, α, βI, βII, and γ exhibit strict requirements for phospholipids and calcium. DAG is an allosteric activator that decreases the calcium requirements for these PKC isoenzymes. Since one of the earliest events following mitogen stimulation is the activation of phospholipase C and subsequent production of DAG, it has been concluded that PKC is a component of the mitogenic signaling pathways.[124-126,160]

Although a clear connection has been established for *ras* in the pathway of tyrosine kinases, a marked difference exists between their respective signaling pathways in their requirement for PKC. The PKC pathway does not appear to be essential for induction of DNA synthesis in the case of growth factors that interact with receptor tyrosine kinases.[125,126,160] Zagari et al. have reported that the mitogenic responsiveness to PDGF remains unchanged in cells depleted of PKC.[160] In agreement with this observation, PKC downregulation does not block DNA synthesis induced by serum.[154]

Increasing evidence is being accumulated suggesting that receptor tyrosine kinases trigger a number of biochemical reactions which are redundant in their ability to induce cell proliferation. Such redundancy might allow a proliferative response despite a nonfunctional PKC pathway. On the contrary, *ras*-p21 proteins exhibit a strict requirement for the presence of PKC in order to induce mitogenesis. Lacal et al. found that cells depleted of PKC by chronic phorbol ester treatment were unable to incorporate [^{3}H]-thymidine following microinjection of *ras*-p21.[154] When normal PKC levels were restored by microinjection, these cells recovered the mitogenic responsiveness to p21, indicating an absolute functional requirement of PKC for mitogenic activity of *ras* proteins.

Similar results have been reported by Morris et al. who have shown that scrape-loading of H-*ras* p21 induced DNA synthesis only in cells with functional

PKC, and in the presence of insulin.[155] Therefore, these observations suggest that the induction of DNA synthesis by *ras* proteins is related to a subset of biochemical reactions which are rigorously dependent on PKC.

The involvement of PKC on DNA synthesis has been reported by many groups using different approaches. Phorbol esters, such as tetradecanoyl phorbol acetate, which are potent activators of PKC induced DNA synthesis in a PKC-dependent manner.[125,126] Cells overexpressing moderate levels of PKC exhibited a parallel increase in the amount of DNA synthesis induced by TPA.[161]

As discussed earlier, activation of *ras*-p21 induces a rapid increase in the levels of DAG. Therefore, these observations suggest that the most evident connection between *ras* and PKC could rely on the release of DAG. However, increasing evidence is being accumulated indicating that not all DAG species are able to activate PKC. Leach et al. have analyzed the production of DAG from PI or PC in IIC9 fibroblasts.[144] Treatment with thrombin at high concentrations (500 ng/ml) stimulated both PI and PC hydrolysis, whereas lower doses (100 pg/ml) stimulated only PC breakdown. Using these two conditions they have shown that, in intact cells, activation of PKC parallels the formation of DAGs from PI, but not PC hydrolysis.

Issandou and Rozengurt have reported that DAGs, unlike phorbol esters, do not induce homologous desensitization or downregulation of PKC in Swiss 3T3 cells.[162] The addition of 1-olyl-2-acetyl-glycerol (OAG), 1,2-dioctanoyl-glycerol (DiC8) or PDBu to Swiss 3T3 fibroblasts rapidly increases the PKC activity, but only prolonged incubation with PDBu completely downregulated PKC activity.

In PC12 cells, Lacal et al. have shown that induction of activated N-*ras* protein under the glucocorticoid-inducible promoter, protected PKC from degradation induced by chronic treatment with phorbol esters.[163] These results suggest a functional relationship between *ras* and PKC also in the PC12 cells system. A trivial explanation could be that the generated DAG induced after expression of the activated *ras*-p21 protein, could activate PKC without inducing its degradation as it would have been the case with phorbol esters. Indeed, this explanation would be in good agreement with the results obtained in fibroblasts by Issandon and Rozengurt,[162] but the real nature of this effect remains to be explained. Activation of PKC could also be mediated by some specific DAGs produced after *ras* induction. Finally, other alternative ways of *ras*/PKC interaction may also occur by mechanisms that remain to be elucidated.

An additional evidence for the connection between PKC and *ras* in the same signaling pathway stems from the characterization of flat revertants of K-*ras*-transformed cells. These revertants exhibited lower PKC activity than the parental transformed cells, indicating again a functional relationship between these two proteins.[164] Furthermore, overexpression of PKC-βI increases the efficiency of the H-*ras* oncogene to transform NIH 3T3 fibroblasts.[165,166]

VI. PKC-INDEPENDENT PATHWAYS IN *ras*-INDUCED SIGNAL TRANSDUCTION

Since receptor tyrosine kinases require functional *ras* but not PKC to transmit the mitogenic signal, it seems clear that *ras* proteins stimulate other pathways in addition to PKC. Depletion of cellular PKC by chronic treatment with phorbol esters has allowed further exploration of this possibility. Such PKC-depleted cells, although unable to initiate DNA synthesis or proliferate in response to activated *ras*,[154] exhibit membrane raffling, increased mobility, and a transformed appearance.[167] Therefore, the morphological changes associated with *ras* transformation appear to be PKC-independent. The dissection of *ras* requirements for DNA synthesis and for morphological transformation may provide a useful way to further separate those reactions involved in proliferation from those which, in addition, promote cell transformation.

The observation that receptor tyrosine kinases require normal *ras*-p21 for DNA synthesis[88] but not PKC,[160] together with the finding that mutated *ras*-p21 exhibits a strict requirement for PKC for induction of DNA replication,[154] involves a complex paradox. Moreover, it has been reported that microinjection of Y13-259 blocks the mitogenic activity of phorbol esters, suggesting that *ras*-p21 may be located downstream of PKC.[168] An explanation could come by assuming a mutual interdependence between *ras* and PKC (Figure 2). Thus, some element of the PKC-independent pathways of *ras* activity could be necessary, but not sufficient for induction of replication. In the case of *ras*-p21, this element would cooperate with PKC to induce DNA synthesis. In the case of receptor tyrosine kinases, such element would cooperate with *ras*-p21-independent pathways as well as with either PKC-dependent or PKC-independent routes. Indirect evidence for the necessity of this element in induction of DNA synthesis stems from the observation that phorbol esters are weak mitogens in fibroblasts, and therefore, other signals must be required in addition to PKC activation in order to achieve an optimal replication response. A good candidate for such a molecule is the *raf*-1 kinase as discussed next.

The *raf*-1 kinase is a cytoplasmic serine-threonine kinase which is translocated to the cell membrane and quickly activated following stimulation with mitogens which interact with tyrosine kinases.[169] Although PKC can activate *raf*-1 kinase by phosphorylation, cells treated chronically with TPA also exhibit mitogen activation of *raf*-1 kinase, indicating that this kinase can also be activated in a PKC-independent manner. Depletion of *raf*-1 kinase by the use of antisense RNAs or dominant recessive mutants drastically reduces the ability of NIH 3T3 cells to grow in serum.[170] Therefore, *raf*-1 kinase appears to be essential for cell proliferation.

Some evidence suggests a functional relationship between *ras*-p21 and *raf*-1 kinase indicating that both proteins are interconnected in the same signaling

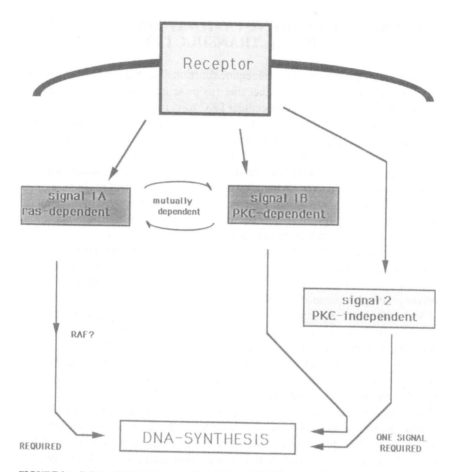

FIGURE 2. Relationship between *ras*-p21 proteins and PKC in the regulation of signal transduction in cell proliferation. DNA synthesis requires at least two signals. One of them is absolutely *ras*-dependent (signal 1A), since serum cannot induce cell proliferation in the absence of *ras*. The second one has at least two alternatives. The first alternative is PKC-dependent, since downregulation of PKC blocks *ras*-induced proliferation. Also, direct activation of PKC by phorbol esters requires *ras*. Thus, *ras*-dependent and PKC-dependent signals are mutually interdependent. A second alternative, signal 2, is PKC-independent since serum is active even in cells where PKC has been downregulated.

pathway. Microinjection experiments with antibody Y13-259 which blocks *ras* function prevented transformation by membrane bound oncogenes but had no effect on the transformation induced by v-*raf*, the oncogenic variant of c-*raf*-1.[88,171] In addition, the mouse cell line C127 is efficiently transformed by v-H-*ras*, v-K-*ras*, and v-*raf* oncogenes but not by tyrosine kinase oncogenes.[172] These results indicated that *raf*-1 is either downstream from *ras* or in an independent, parallel pathway.

The *ras* transformed cells exhibit a constitutive activation of *raf*-1 kinase even if cellular PKC has been removed.[173] The use of an inducible expression vector where the H-*ras* oncogene is transcribed in the presence of dexamethasone indicates that activation of *raf*-1 is concomitant with the appearance of significant levels of active *ras*-p21, suggesting that *raf*-1 should be located downstream of *ras* in a PKC-independent pathway.

Again, more conclusive evidence for this hypothesis stems from the use of antisense RNAs and dominant recessive mutants of c-*raf*-1 which partially reverted the transformed phenotype induced by K-*ras* in NIH 3T3 cells.[171] Finally, overexpression of normal c-H-*ras* or transfection with a plasmid-carrying oncogenic H-*ras* under the control of a weak promoter induced a much larger number of transformed foci when c-*raf*-1 was cotransfected in the same experiments.[173] These observations strongly support the idea that *raf*-1 is a key element in the PKC-independent events that lead *ras* activation to morphological transformation.

VII. *ras* ACTIVATION OF NUCLEAR PROTOONCOGENES

ras oncogenes cannot induce full transformation of primary human cells by themselves. They require the concurrence of one of several nuclear oncogenes (revised in Reference 174). Although the nature of this cooperation is still unknown, a clear connection between *ras*-p21 and nuclear oncogenes has been described. Microinjection of *ras*-p21 protein into NIH 3T3 cells induces a transient but significant expression of c-*fos*.[175] In PC12 cells, induction of c-*fos* and c-*jun* may also be affected by *ras*.[176] In fact, expression of a mutated N-*ras* protein under an inducible glucocorticoid receptor results in a blockage of the c-*fos* induction mediated by NGF.[177]

Induction of c-*fos* in fibroblasts seems to be mediated by a sequence similar, if not identical, to the TPA-responsive element (TRE)[178] (see also Chapter 4), and most likely implies PKC activation,[179] in agreement with the previously discussed evidence linking *ras*-p21 and PKC (Section V). Recently Binétruy et al. have shown that the expression of Ha-*ras* in several cell lines induces transcriptional activation of c-*jun* independently of c-*fos*.[180] This activation is related to *ras*-induced phosphorylation of two N terminal sites on the c-*jun* molecule, providing an excellent model to investigate cooperation between oncogenes.[180]

VIII. CONCLUDING REMARKS

The finding of activated *ras* genes in a variety of human and animal tumors stimulated investigations aimed at the discovery of their biochemical properties to explain biological function both in normal and transformed cells. Ever since, *ras* proteins have been the focus of intensive research. However, we may still be at the frontispiece of the real world governed by *ras* proteins. The future of the *ras* field depends on new breakthroughs to provide more light within the complex network that regulates the intricate and contradicting pathways involved in signal transduction. There may be more roads and highways than we suspect, interconnecting all cell departments to each other. Redundancy may very well be an obscure trap making generalization a dangerous and slipper path. Although we have managed to make some progress in locating the major pathways where *ras* function is most likely engaged, we still have a long journey.

NOTE ADDED IN PROOF

Since submission of this chapter, abundant information has been generated strengthening the concept that *ras* proteins lie within the signalling pathway triggered by growth factors associated to tyrosine kinase receptors. Three important issues have been addressed further: (1) upstream activation of *ras* tyrosine kinases; (2) interaction of *ras* proteins with the serine/threonine kinases cascade; and (3) regulation of phospholipid metabolism by growth factors and *ras* oncogenes.

One of the strongest evidences to connect tyrosine kinase to *ras* stems from genetic studies in *C. elegans* and *Drosophila*. In *C. elegans let*-60, the *ras*-gene homolog, is functionally related to the EGFR gene homolog, *let*-23. The linking gene has been identified and cloned as *sem*5 in *C. elegans*[181] and independently, as Grb2 from mammalian cells.[182] These genes codify for proteins of 23 to 25 kDa that contain two SH2 (*src* homology 2) and one SH3 (*src* homology 3) motifs. A new oncogene with a similar structure to that of *sem*5/GRB-2 has been isolated, designated as *shc*.[183] At least three products related to the *shc* gene have been identified with relative molecular weights of 46, 52, and 66 kDa that are phosphorylated at tyrosine in cells stimulated with a variety of growth factors and in cells transformed by v-*src* or v-*fps*.[183,184] The product of the Grb2/*sem*5 is not phosphorylated at tyrosine residues, but *sem*5/Grb2 and *shc* form a complex which seems to be responsible for growth factors-mediated activation of *ras*-p21.[185] The SH-2 regions of Grb2/*sem*5 interact with the phosphotyrosine residues of *shc*, and the SH-3 domain may be responsible for activation of a new member of the cascade, *sos* (son of sevenless), an exchange factor for *ras*-p21, recently identified first in *Drosophila* and later on in mammalian cells (T. Pawson, personal communication). In *Drosophila*, a functional connection has been found also for *ras* and *raf* in the sevenless signal transduction system.[186] Finally, a gene that codifies for a GAP-like protein for *Drosophila ras* proteins has been cloned,[187] which most likely has a signal terminator function. This provides an almost complete picture of the structural components involved in the activation of *ras* proteins by growth factors, but it would not be surprising if other components are found in the next few years, especially if searched for in other cell systems.

On the other hand, increasing evidence is accumulating suggesting that *ras* connects the signal initiated at the plasma membrane by tyrosine kinases to cytoplasmic switch kinases that transduce the mitogenic or differentiating signal to the cell nucleus. The development of a dominant negative mutant of p21-H-

ras where Asn substitutes Ser at position 17 has been instrumental in this field.[188] When NIH 3T3 cells were transfected with the Asn17 mutant under the control of the glucocorticoid inducible promoter MMTV-LTR, cells were not responsive to growth factors and *raf*-1 kinase was not activated, suggesting that *raf* lies downstream from *ras* in the mitogenic signalling cascade. Similar results were obtained in PC12 cells,[189-191] where treatment with NGF failed to induce neuronal differentiation in the presence of p21-*ras*-Asn17. Moreover, in both systems p21-Asn17 blocked activation of *raf*-1 kinase, although it is not clear where PKC should be placed in this cascade, and maybe each system has its own alternative hierarchical network.[192]

Recently, it has been demonstrated that one of the relevant substrates for *raf*-1 kinase are MAP kinases (also designated as extracellular signal-regulated kinases, ERKs).[193,194] Phosphorylation of *raf*-1 activates its intrinsic kinase activity and produces phosphorylation of a family of 40- to 45-kDa kinases known as MAP kinases/ERKs. They are activated by phosphorylation in both Thr and Tyr residues in response to a large variety of stimuli including growth factors.[195,196] Therefore, *ras* proteins might activate a cascade of kinases starting with *raf*-1 followed by activation of MAP/ERK kinases. This signal also activates ribosomal S6 kinase, leading to initiation of protein synthesis[192] and is transmitted to the nucleus. As a consequence, phosphorylation of *c-jun* and *c-myc* takes place leading to activation of transcription from specific responsive elements.[197] Finally, *ras* proteins are involved in the regulation of the activity of *raf*, PKC, and MAP/ERK kinases. PKC seems to be placed either downstream or upstream of *ras*, but always upstream of MAP/ERK kinases in the signalling pathway depending on the cell system studied.[188,190,192,198,199] In Xenopus oocytes, *ras*-induced activation of MAPK and S6 kinase requires a functional interaction between *ras* and GAP proteins.[200] Thus, elucidation of this cascade provides a clear connection between *ras* function and activation of transcription factors. However, a clear mechanism of activation of the serine/threonine kinases cascade by *ras* proteins has not yet emerged and further investigation is required.

Finally, additional evidence is being provided to link *ras* proteins to the regulation of phospholipid metabolism. It is becoming clear that growth factors interacting with tyrosine kinases activate a D-type phospholipase (PLD), as demonstrated by the generation of phosphatidic acid,[200] even in the presence of inhibitors of diacylglycerol kinase.[201] Moreover, the second wave of DAG, which has been related to a PC-PLD function, appears to be the result of PA hydrolysis by a PA phosphatase since propranolol, an inhibitor of this enzyme, blocks DAG formation after stimulation with growth factors or after transformation by oncogenes such as *src* or *ras*[201-203] (Cuadrado, A., Carnero, A., and Lacal, J. C., submitted).

The relevance of phospholipid metabolites other than DAG in cell signalling is becoming more evident after the observation that PLD is activated by growth factors and phorbol esters,[201-205] and the fact that PA and lyso-PA are potent mitogens.[206,207] We have recently observed that both in *ras*-transformed and in serum-stimulated NIH 3T3 cells, generation of phosphorylcholine and DAG depends on a PC-PLD activity followed by choline kinase and PA phosphatase (Cuadrado, A., Carnero, A., and Lacal, J. C., submitted). This finding is based on the observation that generation of phosphorylcholine is sensitive to a choline kinase inhibitor, hemicholinium, and that DAG generation is sensitive to propranolol. Moreover, *ras*-transformed NIH 3T3 cells exhibit a five-fold constitutive increase in PLD activity, which renders a constitutive increase in both choline and PA content (Cuadrado, A., Carnero, A., and Lacal, J. C., submitted). We have also found that phosphorylcholine may be one of the most important molecules generated as a consequence of growth stimulation in response to growth factors and oncogenes. Thus, it seems feasible that in the NIH 3T3 cells, growth factors activate PLD activity through *ras* with the generation of both DAG and phosphorylcholine, two important second messengers. Although at this point these results should be interpreted only as an attractive basis for speculation, the finding that phosphorylcholine may be a new second messenger essential for cell proliferation in fibroblasts opens a new avenue for understanding regulation of cell growth by *ras* proteins.

ACKNOWLEDGMENTS

This work was supported by grants from Dirección General de Investigación Científica y Técnica (DGICYT), Fundación Ramón Areces, and Laboratorios Serono. A. Cuadrado is an Investigador Contratado (CSIC); A. Carnero is a Fellow from the Spanish Department for Education and Science; J. C. Lacal is Investigador Científico (CSIC) and Honorary Professor of the Universidad Autónoma de Madrid.

REFERENCES

1. **Barbacid, M.,** *ras* genes, *Annu. Rev. Biochem.,* 56, 779, 1987.
2. **Harvey, J. J.,** An unidentified virus which causes the rapid production of tumors in mice, *Nature,* 204, 1104, 1964.
3. **Kirsten, W. H. and Mager, L. A.,** Morphologic responses to a murine erythroblastosis virus, *J. Natl. Cancer Inst.,* 39, 311, 1967.
4. **Lacal, J. C. and Tronick, S. R.,** The *ras* oncogene, in *The Oncogene Handbook,* Reddy, E. P., Skalka, A. M., and Curran, T., Eds., Elsevier Science Publishers, 1988, 255.
5. **Parada, L. F., Tabin, C. J., Shih, C., and Weinberg, R. A.,** Human EJ bladder carcinoma oncogene is homologue of Harvey sarcoma virus ras gene, *Nature,* 297, 474, 1982.
6. **Santos, E., Tronick, S. R., Aaronson, S. A., Pulciani, S., and Barbacid, M.,** T24 human bladder carcinoma oncogene is an activated form of the normal human homologue of BALB- and Harvey-MSV transforming genes, *Nature,* 298, 343, 1982.

7. **Der, C. J., Finkel, T., and Cooper, G. M.**, Transforming genes of human bladder and lung carcinoma cell lines are homologous to the ras genes of Harvey and Kirsten sarcoma viruses, *Cell*, 44, 167, 1982.

8. **Goldfarb, M., Shimizu, K., Perucho, M., and Wigler, M.**, Isolation and preliminary characterization of a human transforming gene from T24 bladder carcinoma cells, *Nature*, 296, 404, 1982.

9. **Taparowsky, E., Shimizu, K., Goldfarb, M., and Wigler, M.**, Structure and activation of the human N-ras gene, *Cell*, 34, 581, 1983.

10. **McGrath, J. P., Capon, D. J., Smith, D. H., Chen, E. Y., Seeburg, P. H., Goeddel, D. V., and Levinson, A. D.**, Structure and organization of the human Ki-*ras* proto-oncogene and a related processed pseudogene, *Nature*, 304, 501, 1983.

11. **Miyoshi, J., Kagimoto, M., Soeda, E.-I., and Sakaki, Y.**, The human c-Ha-ras-2 is a processed pseudogene inactivated by numerous base substitutions, *Nucleic Acids Res.*, 12, 933, 1984.

12. **Chattopadyay, S. K., Chang, E. H., Lauder, M. R., Ellis, R. W., Scolnick, E. M., Koller, R., and Dhar, R.**, Amplification and rearrangement of onc genes in mammalian species, *Nature*, 296, 361, 1982.

13. **DeFeo, D., Scolnick, E. M., Koller, R., and Dhar, R.**, Ras-related gene sequences identified and isolated from Saccharomyces cerevisiae, *Nature*, 306, 707, 1983.

14. **Powers, S., Kataoka, T., Fasano, O., Golfarb, M., Strathern, J., Broach, J., and Wigler, M.**, Genes in S. cerevisiae encoding proteins with domains homologous to the mammalian ras proteins, *Cell*, 36, 607, 1984.

15. **Fukui, Y. and Kaziro, Y.**, Molecular cloning and sequence analysis of a ras gene from Schizosaccharomyces pombe, *EMBO J.*, 4, 687, 1985.

16. **Reymond, C. D., Gomer, R. H., Mehdy, M. C., and Fiortel, R. A.**, Developmental regulation of a Dictyostelium gene encoding a protein homologous to mammalian ras protein, *Nature*, 309, 174, 1984.

17. **Robbins, S. M., Williams, J. G., Jermyn, K. A., Spiegelman, G. B., and Weeks, G.**, Growing and developing *Dictyostellium* cells express different *ras* genes, *Proc. Natl. Acad. Sci. U.S.A.*, 86, 938, 1989.

18. **Neuman-Silberberg, F. S., Schejter, E., Hofman, F. M., and Shilo, B. Z.**, The Drosophila ras oncogenes: structure and nucleotide sequence, *Cell*, 37, 1027, 1984.

19. **Mozer, B., Marlor, R., Parkhurst, S., and Corces, V.**, Characterization and developmental expression of a *Drosophila ras* oncogene, *Mol. Cell. Biol.*, 5, 885, 1985.

20. **Schejter, E. D. and Shilo, B. Z.**, Characterization of functional domains of p21 ras by use of chimeric genes, *EMBO J.*, 407, 412, 1985.

21. **Casale, W. L., McConnell, D. G., Wang, S.-Y., Lee, Y.-L., and Linz, J. E.**, Expression of a gene family in the dimorphic fungus *Mucor racemosus* which exhibits striking similarity to human *ras* genes, *Mol. Cell. Biol.*, 10, 6654, 1990.

22. **Nemoto, N., Kodama, H., Tozawa, A., Matsumoto, J., Masahito, P., and Ishikawa, T.**, Nucleotide sequence comparison of the predicted first exonic region of goldfish ras gene between normal and neoplastic tissues, *J. Cancer Res. Clin. Oncol.*, 113, 56, 1987.

23. **Han, H. and Sternberg, P. W.**, Let-60, a gene that specifies cell fates during C. Elegans vulval induction, encodes a ras protein,, *Cell*, 63, 921, 1990.

24. **Beitel, G. J., Clark, S. G., and Horvitz, H. R.**, Caenorhabditis elegans ras gene let-60 acts as a switch in the pathway of vulval induction, *Nature*, 348, 503, 1990.

25. **Swanson, M. E., Elste, A. M., Greenberg, S. M., Schwartz, J. H., Aldrich, T. H., and Furth, M. E.**, Abundant expression of ras proteins in Aplysia neurons, *J. Cell. Biol.*, 103, 485, 1986.

26. **Shih, T. Y., Weeks, M. O., Young, H. A., and Scolnick, E. M.**, Identification of sarcoma virus-coded phosphoprotein in nonproducer cells transformed by Kirstein or Harvey murine sarcoma virus, *Virology*, 96, 64, 1979.

27. **Willumen, B. M., Norris, K., Papageorge, A. G., Hubbert, N. L., and Lowy, D. R.,** Harvey murine sarcoma virus p21 ras protein: biological and biochemical significance of the cystine nearest to the carboxyterminus, *EMBO J.,* 3, 2581, 1984.
28. **Hancock, J. F., Magee, A. I., Childs, J. E., and Marshall, C. J.,** All ras proteins are polyisoprenylated but only some are palmitoylated, *Cell,* 57, 1167, 11, 1989.
29. **Marshall, C. J.,** Oncogenes and growth control, *Cell,* 49, 723, 725, 1987.
30. **Defeo-Jones, D., Tatchell, K., Robinson, L. C., Sigal, I. S., Vass, W. C., Lowy, D. R., and Scolnick, E. M.,** Mammalian and yeast ras gene products: biological function in their heterologous systems, *Science,* 228, 179, 1985.
31. **Kataoka, T., Powers, S., Cameron, S., Fasano, O., Goldfarb, M., Broach, J., and Wigler, M.,** Functional homology of mammalian and yeast ras genes, *Cell,* 40, 19, 1985.
32. **Beckner, S. K., Hattori, S., and Shih, T. Y.,** The ras oncogene product p21 is not a regulatory component of adenoylate cyclase, *Nature,* 317, 71, 1985.
33. **Shih, T. Y., Papageorge, A. G., Stowes, P. E., Week, M. O., and Scolnick, E. M.,** Guanine nucleotide-binding and autophosphorylating activities associated with the p21[RAS] protein of Harvey murine sarcoma virus, *Nature,* 287, 686, 1980.
34. **Young, H. A., Shih, T. Y., Scolnick, E. M., Rasheed, S., and Gardner, M. B.,** Different rat-derived transforming retroviruses code for an immunologically related intracellular phosphoprotein, *Proc. Natl. Acad. Sci. U.S.A.,* 76, 3523, 1979.
35. **Shih, T. Y., Stokes, P. E., Smythers, G. W., Dhar, R., and Oroszlan, S.,** Characterization of the phosphorylation sites and surrounding amino acid sequences of the p21 transforming proteins coded for by the Harvey and Kirsten strains of murine sarcoma viruses, *J. Biol. Chem.,* 257, 11767, 1982.
36. **Ulsh, L. S. and Shih, T. Y.,** Metabolic turnover of human c-ras[H] p21 protein of EJ bladder carcicoma and its normal cellular and viral homologs, *Mol. Cell. Biol.,* 4, 1647, 1984.
37. **Clark, R., Wong, G., Arnhein, N., Nitecki, D., and McCormick, F.,** Antibodies specific for amino acid 12 of the ras oncogene product inhibit GTP binding, *Proc. Natl. Acad. Sci. U.S.A.,* 82, 5280, 1985.
38. **Lacal, J. C., Anderson, P. S., and Aaronson, S. A.,** Deletion mutants of Harvey ras-21 protein reveal the absolute requirement of at least two distant regions for GTP-binding and transforming activities, *EMBO J.,* 5, 679, 1986.
39. **Sigal, I. S., Gibbs, J. B., D'Alozo, J. S., Temeles, G. T., Wolanski, B. S., Socher, S. H., and Scolnick, E. M.,** Mutant ras-encoded proteins with altered nucleotide binding exert dominant biological effects, *Proc. Natl. Acad. Sci. U.S.A.,* 83, 952, 1986.
40. **Lacal, J. C. and Aaronson, S. A.,** Activation of ras p21 transforming properties associated with alteration of the interchange rate of bound guanine nucleotide, *Mol. Cell. Biol.,* 6, 1002, 1986.
41. **Srivastava, S. K., Lacal, J. C., Reynolds, H. S., and Aaronson, S. A.,** Antibody of predetermined specificity to a carboxy-terminal region of H-ras gene oncogene products inhibits their guanine nucleotide binding functions, *Mol. Cell. Biol.,* 5, 3316, 1988.
42. **Swett, R. W., Yokoyama, S., Kamata, T., Feramisco, J. R., Rosenberg, M., and Gross, M.,** The product of ras is a GTPase and the T24 oncogenic mutant is deficient in this activity, *Nature,* 311, 273, 1984.
43. **McGrath, J. P., Capon, D. J., Goeddel, D. V., and Levinson, A. D.,** Comparative biochemical properties of normal and activated human ras p21 protein, *Nature,* 310, 644, 1984.
44. **Gibbs, J. B., Sigal, I. S., Poe, M., and Scolnick, E. M.,** Intrinsic GTPase activity distinguishes normal and oncogenic ras p21 molecules, *Proc. Natl. Acad. Sci. U.S.A.,* 81, 5704, 1984.
45. **Seeburg, P. H., Colby, W. W., Capon, D. J., Goeddel, D. V., and Levinson, A. D.,** Biological properties of human c-Ha-*ras* 1 genes mutated at codon 12, *Nature,* 312, 71, 1984.
46. **Der, C. J., Finkel, T., and Cooper, G. M.,** Biological and biochemical properties of human ras[H] genes mutated at codon 61, *Cell,* 44, 167, 1986.

47. **Lacal, J. C. and Aaronson, S. A.**, Monoclonal antibody Y13-259 recognizes an epitope of p21 ras molecule not involved in GTP binding activity of the protein, *Mol. Cell. Biol.*, 6, 1002, 1986.

48. **Tong, L., Milburn, M. V., de Vos, A. M., and Kim, S.-H.**, Structure of *ras* protein, *Science*, 245, 244, 1989.

49. **Pai, E. F., Kabsch, W., Krengel, U., Holmes, K. C., John, J., and Wittinghofer, A.**, Structure of the guanine-nucleotide-binding domain of the Ha-*ras* oncogene product p21 in the triphosphate conformation, *Nature*, 341, 209, 1989.

50. **Milburn, M. V., Tong, L., Devos, A. M., Brünger, A., Yamaizumi, Z., Nishimura, S., and Kim, S. H.**, Molecular switch for signal transduction: structural differences between active and inactive forms of protooncogenic ras proteins, *Science*, 247, 939, 1990.

51. **Tong, L. A., de Vos, A. M., Milburg, M. V., Jancarik, J., Noguchi, S., Nishimura, S., Miura, K., Ohtsuka, E., and Kim, S. H.**, Structural differences between a ras oncogene protein and the normal protein, *Nature*, 337, 90, 1989.

52. **Pai, E. F., Krengel, U., Petsko, G. A., Goody, R. S., Kabsch, W., and Wittinghofer, A.**, Refined crystal structure of the triphosphate conformation of H-*ras* p21 at 1.35 Å resolution: implications for the mechanism of GTP hydrolysis, *EMBO J.*, 9, 2351, 1990.

53. **Schlichting, I., Almo, S. C., Rapp, G., Wilson, K., Petratos, K., Lentfer, A., Wittinghofer, A., Kabsch, W., Pai, E. F., Petsko, G. A., and Goody, R. S.**, Time-resolved X-ray crystallographic study of the conformational change in Ha-*rasxl* p21 protein on GTP hydrolysis, *Nature*, 345, 309, 1990.

54. **Hiroyoshi, M., Kaibuchi, K., Kawamura, S., Hata, Y., and Takai, Y.**, Role of the C-terminal region of smg p21, a ras p21-like small GTP-binding protein, in membrane and smg p21 GDP/GTP exchange protein interactions, *J. Biol. Chem.*, 15, 2665, 1991.

55. **Willumen, B. M., Papageorge, A. G., Kung, H., Bekesi, E., Robins, T., Johnsen, M., Vass, W. C., and Lowy, D.**, Multifunctional analysis of ras catalytic domain, *Mol. Cell. Biol.*, 6, 2646, 1986.

56. **Seabra, M. C., Reiss, Y., Casey, P. J., Brown, M. S., and Goldstein, J. L.**, Protein farnesyltransferase and geranylgeranyltransferase share a common subunit, *Cell*, 429, 434, 1991.

57. **Cross, F. R., Garber, E. A., Pellman, D., and Hanafusa, H.**, A short sequence in the p60[src] N terminus is required for p60[src] myristoylation and membrane association for cell transformation, *Mol. Cell. Biol.*, 4, 1834, 1985.

58. **Resh, M. D. and Ling, H. P.**, Identification of a 32K plasma membrane protein that binds to the myristoylated amino-terminal sequence of p-60[src], *Nature*, 346, 84, 1990.

59. **Lacal, P. M., Pennington, C. Y., and Lacal, J. C.**, Transforming activity of ras proteins translocated to the plasma membrane by a myristoylation sequence from the src gene product, *Oncogene*, 2, 533, 1988.

60. **Casey, P. J. and Gilman, A. G.**, G protein involvement in receptor effector coupling, *J. Biol. Chem.*, 263, 2577, 1988.

61. **Gilman, A. G.**, G proteins: transducers of receptor-generated signals, *Annu. Rev. Biochem.*, 56, 615, 1987.

62. **Stryer, L.**, Cyclic GMP of vision, *Annu. Rev. Neurosci.*, 9, 87, 1986.

63. **Taylor, S. J., Chae, H. Z., Rhee, S. G., and Exton, J. H.**, Activation of the beta 1 isozyme of phospholipase C by alpha subunits of the Gq class of G proteins, *Nature*, 350, 516, 1991.

64. **Heiderman, W. and Bourne, H. R.**, Structure and function of G-protein α-chains in G-proteins, *Academic Press*, 17, 40, 1990.

65. **Bourne, H. R., Sanders, D. A., and McCormik, F.**, The GTPase superfamily: conserved structure and molecular mechanism, *Nature*, 349, 117, 1990.

66. **Bourne, H. R., Sanders, D. A., and McCormick, F.**, The GTPase superfamily: a regulated switch for diverse cell functions, *Nature*, 348, 125, 1990.

67. **Fishman, P. H.,** Mechanism of action of cholera toxin, in *ADP Ribosilating Toxins and G Proteins,* 1st ed., Moss, J. and Vaughan, M., Eds., American Society for Microbiology, Washington, D.C., 1990, chap. 8.

68. **Vi, M.,** Pertussis toxin as a valuable probe for G-protein involvement in signal transduction, in *ADP Ribolisating Toxins and G Proteins,* 1st ed., Moss, J. and Vaughan, M., American Society for Microbiology, Washington, D.C., 1990, chap. 4.

69. **Landis, C. A., Harsh, G., Lyions, J., Davis, R. L., McCormick, F., and Bourne, H. R.,** Clinical characteristics of acromegalic patients whose pituitary tumors contain mutant Gs protein, *J. Clin. Endocrinol. Metab.,* 71(6), 1416, 1990.

70. **Lyions, J., Landis, C. A., Harsh, G., Vallar, L., Grunewald, K., Feichtinger, H., Duh, Q. Y., Clark, O. H., Kawasaki, E., and Bourne, H. R.,** Two G protein oncogenes in human endocrine tumors, *Science,* 249(4969), 655, 1990.

71. **Suarez, H. G., Du-Villard, J. A., Caillou, B., Schlumberger, M., Parmentier, C., and Monier, R.,** *gsp* mutations in human thyroid tumors, *Oncogene,* 6, 4, 677, 1991.

72. **Trahey, M. and McCormick, F. A.,** Cytoplasmic protein stimulates normal N-ras p21 GTPase, but does not affect oncogenic mutants, *Science,* 238, 542, 1987.

73. **Xu, G., Lin, B., Tanaka, K., Dunn, D., Wood, D., Gesteland, R., White, R., Weiss, R., and Tamanoi, F.,** The catalytic domain of the neurofibromatosis type 1 gene product stimulates ras GTPase and complements IRA mutants of S. cerevisiae, *Cell,* 63, 835, 1990.

74. **Calés, C., Hancock, J. F., Marshall, C. J., and Hall, A.,** The cytoplasmic protein GAP is implicated as the target for regulation by the *ras* gene product, *Nature,* 332, 548, 1988.

75. **Adari, H., Lowy, D. R., Willumsen, B. M., Der, C. H. J., and McCormick, F.,** Guanosine triphosphatase activating protein (GAP) interacts with the p21 ras effector binding domain, *Science,* 240, 518, 1989.

76. **Farnsworth, C. L., Marshall, M. S., Gibbs, J. B., Stacey, D. W., and Feig, L. A.,** Preferential inhibition of the oncogenic form of *ras*-H by mutations in the GAP binding/ "effector" domain, *Cell,* 64, 625, 1991.

77. **Srivastasa, S. K., Di Donato, A., and Lacal, J. C.,** H-ras mutants lacking the epitope for the neutralizing monoclonal antibody Y13-259 show decreased biological activity and are deficient in GTPase-activating protein interaction, *Mol. Cell. Biol.,* 9, 1779, 1989.

78. **Marshall, M. S., Hill, W. S., Neg, A. S., Vogel, U. S., Schaber, M. D., Scolnick, E. M., Dixon, R. A., Sigal, I. S., and Gibbs, J. B.,** A C-terminal domain of GAP is sufficient to stimulate ras p21 GTPase activity, *EMBO J.,* 8, 1105, 1989.

79. **Ellis, C., Moran, M., McCormick, F., and Pawson, T.,** Phosphorylation of GAP and GAP-associated proteins by transforming and mitogenic tyrosine kinases, *Nature,* 343, 377, 1990.

80. **Bouton, A. H., Kanner, S. B., Vines, R. R., Wang, H.-C.R., Gibbs, J. B., and Parsons, J. T.,** Transformation by pp60src or stimulation of cells with epidermal growth factor induces the stable association of tyrosine-phosphorylated cellular proteins with GTPase-activating protein, *Mol. Cell. Biol.,* 11, 945, 1991.

81. **Moran, M. F., Polakis, P., McCormick, F., Pawson, T., and Ellis, C.,** Protein-tyrosine kinases regulate the phosphorylation, subcellular distribution, and activity of p21-*ras* GTPase-activating protein, *Mol. Cell. Biol.,* 11, 1804, 1991.

82. **Cantley, L. C., Augeri, K. R., Carpenter, C., Duckworth, B., Graciani, A., Kapeler, R., and Soltoff, S.,** Oncogenes and signal transduction, *Cell,* 64, 281, 1991.

83. **Kaplan, D. R., Hempstead, B. L., Martin-Zanca, D., Chao, M. V., and Parada, L. F.,** The trk protooncogene product: a signal transducing receptor for nerve growth factor, *Science,* 252, 554, 1991.

84. **Klein, R., Jing, S. Q., Nanduri, V., O'Rourke, E., and Barbacid, M.,** The trk proto-oncogene encodes a receptor for nerve growth factor, *Cell,* 65, 189, 1991.

85. **Goastin, A. S., Leot, E. B., Shipley, G. D., and Moses, H. L.,** Growth factors and cancer, *Cancer Res.,* 46, 1015, 1986.

86. **Stacey, D. W. and Kung, H. F.**, Transformation of NIH 3T3 cells by microinjection of H-ras p21 protein, *Nature*, 310, 508, 1984.
87. **Feramisco, J. R., Gross, M., Kamata, T., Rosenberg, M., and Sweet, R. W.**, Microinjection of the oncogenic form of the human H-ras (T24) protein results in rapid proliferation of quiescent cells, *Cell*, 38, 109, 1984.
88. **Mulcahy, L. S., Smith, M. R., and Stacey, D. W.**, Requirement for ras protooncogene functions during serum stimulated in NIH/3T3 cells, *Nature*, 313, 241, 1985.
89. **Cuadrado, A.**, Increased tyrosine phosphorylation in ras-transformed fibroblasts occurs prior to manifestation of the transformed phenotype, *Biochem. Biophys. Res. Commun.*, 170, 526, 1990.
90. **Salomon, D. S., Perroteau, I., Kidwell, W. R., Tam, J., and Derynck, R.**, Loss of growth responsiveness to epidermal growth factor and enhanced production of alpha-transforming growth factors in *ras*-transformed mouse mammary epithelial cells, *J. Cell. Physiol.*, 130, 397, 1987.
91. **Leof, E. B., Proper, J. A., and Moses, H. L.**, Modulation of transforming growth factor type beta action by activated *ras* and *myc*, *Mol. Cell. Biol.*, 7, 2649, 1987.
92. **MacKay, I. A., Malone, P., Marshall, C. J., and Hall, A.**, Malignant transformation of murine fibroblasts by a human c-Ha-ras-1 oncogene does not require a functional epidermal growth factor receptor, *Mol. Cell*, 3382, 3387, 1986.
93. **Fujita-Yamaguchi, Y., Kathuria, S., Xu, Q. Y., McDonald, J. M., Nakano, H., and Kamata, T.**, In vitro tyrosine phosphorylation studies on RAS proteins and calmodulin suggest that polylysine-like basic peptides or domains may be involved in interactions between insulin receptor kinase and its substrate, *Proc. Natl. Acad. Sci. U.S.A.*, 86, 7306, 1989.
94. **Sacks, D. B., Glenn, K. C., and McDonald, J. M.**, The carboxyl terminal segment of the c-Ki-ras 2 gene product mediates insulin-stimulated phosphorylation of calmodulin and stimulates insulin-independent autophosphorylation of the insulin receptor, *Biochem. Biophys. Res. Commun.*, 161, 399, 1989.
95. **Hancock, J. F., Paterson, H., and Marshall, C. J.**, A polybasic domain of palmitoylation is required in addition to the CAAX motif to localize p21 ras to the plasma membrane, *Cell*, 63, 133, 1990.
96. **Ballester, R., Farth, M. E., and Rosen, O. M.**, Phorbol ester and protein kinase C-mediated phosphorylation of cellular Kirstein ras gene product, *J. Biol. Chem.*, 262, 2688, 1987.
97. **Jeng, A. Y., Srivastava, S. K., Lacal, J. C., and Blumberg, P. M.**, Phosphorylation of ras oncogene product by protein kinase C, *Biochem. Biophys. Res. Commun.*, 145, 782, 1987.
98. **Molloy, C. J., Bottaro, D. P., Fleming, T. P., Marshall, M. S., Gibbs, J. B., and Aaronson, S. A.**, PDGF induction of tyrosine phosphorylation of GTPase activating protein, *Nature*, 342, 711, 1989.
99. **Ellis, C., Morran, M., McCormick, F., and Pawson, T.**, Phosphorylation of GAP and GAP associated proteins by transforming and mitogenic tyrosine kinases, *Nature*, 343, 377, 1990.
100. **Liu, X. Q. and Pawson, T.**, The epidermal growth factor receptor phosphorylates GTPase activating protein at Tyr-460, adjacent to the GAP SH2 domains, *Mol. Cell. Biol.*, 11, 2511, 1991.
101. **Satoh, T., Endo, M., Nakafuku, M., Nakamura, S., and Kaziro, Y.**, Platelet-derived growth factor stimulates formation of active p21-*ras*-GTP complex in Swiss mouse 3T3 cells, *Proc. Natl. Acad. Sci. U.S.A.*, 87, 5993, 1990.
102. **Satoh, T., Endo, M., Nakafuku, M., Akiyama, T., Yamamoto, T., and Kaziro, Y.**, Accumulation of p21-*ras*-GTP in response to stimulation with epidermal growth factor and oncogene products with tyrosine kinase activity, *Proc. Natl. Acad. Sci. U.S.A.*, 87, 7926, 1990.

103. **Burgering, B. M. T., Medema, R. H., Maassen, J. A., van der Wetering, M. L., McCormick, F., van der Eb, A. J., and Boss, J. L.,** Insulin stimulation of gene expression mediated by p21-*ras* activation, *EMBO J.,* 10, 1103, 1991.
104. **Tsai, H.-H., Hall, A., and Stacey, D. W.,** Inhibition by phospholipids of the interaction between R-*ras, rho* and their GTPase activating proteins, *Mol. Cell. Biol.,* 9, 5260, 1989.
105. **Tsai, M.-H., Yu, C. L., Wei, F. S., and Stacey, D. W.,** The effect of GTPase activating protein upon *ras* is inhibited by mitogenically responsive lipids, *Science,* 243, 522, 1989.
106. **Llan, J. W., McCormick, F., and Macara, I. G.,** Regulation of Ras-GAP and the neurofibromatosis-1 gene produced by eicosanoids, *Science,* 252, 276, 1991.
107. **Serth, J., Lautwein, A., Frech, M., Wittinghofer, A., and Pingoud, A.,** The inhibition of the GTPase activating protein-Ha-ras interaction by acidic lipids is due to physical association of C-terminal domain of the GTPase, *EMBO J.,* 10, 1325, 1991.
108. **Maller, J. C. and Koontz, J. W.,** A study of the induction of cell division in amphibian oocytes by insulin, *Dev. Biol.,* 85, 309, 316, 1991.
109. **Birchmeyer, C., Broek, D., and Wigler, M.,** Ras proteins can induce meiosis in Xenopus ooxytes, *Cell,* 43, 615, 1985.
110. **Ulrich, A. and Schlessinger, J.,** Signal transduction by receptors with tyrosine kinase activity, *Cell,* 61, 3895, 1990.
111. **Saltiel, A. R., Sherline, P., and Fox, J. A.,** Insulin-stimulated diacylglycerol production results from the hydrolysis of a novel phosphatidylinositol glycan, *J. Biol. Chem.,* 262, 1116, 1987.
112. **Luttrell, L., Kilgour, E., Larner, J., and Romero, G.,** A pertussis toxin-sensitive G-protein mediates some aspects of insulin action in BC3H-1 murine myocytes, *J. Biol. Chem.,* 265(28), 16,873, 1990.
113. **Noda, M., Ko, M., Ogura, A., Liu, D., Amano, T., Takano, T., and Ikawa, Y.,** Sarcoma viruses carrying *ras* oncogenes induce differentiation-associated properties in neuronal cell line, *Nature,* 318, 73, 1985.
114. **Bar-Sagi, D. and Feramisco, J. R.,** Microinjection of the ras oncogene protein into PC12 cells induces morphological differentiation, *Cell,* 42, 841, 1985.
115. **Guerrero, I., Wong, H., Pellicer, A., and Burstein, D. E.,** Activated N-*ras* gene induces neuronal differentiation of PC12 rat pheochromocytoma cells, *J. Cell. Physiol.,* 129, 71, 1986.
116. **Seremetis, S., Inghirami, G., Ferrero, D., Newcomb, E. W., Knowles, D. M., Dotto, G. P., and Dalla-Favera, R.,** Transformation and plasmacytoid differentiation of EBV-infected human B lymphoblasts by *ras* oncogenes, *Science,* 243, 660, 1989.
117. **Nakawaga, T., Mabry, M., de Bustros, A., Ihle, J. H., Nelkin, B. D., and Baylin, S. B.,** Introduction of v-Ha-*ras* oncogene induces differentiation of cultured human medullary thyroid carcinoma cells, *Proc. Natl. Acad. Sci. U.S.A.,* 84, 5923, 1987.
118. **Benito, M., Porras, A., Nebreda, A. R., and Santos, E.,** Differentiation of 3T3-L1 fibroblasts to adipocytes induced by transfection of ras oncogene, *Science,* 253, 565, 1991.
119. **Olson, E. N., Spizz, G., and Tainsky, M. A.,** The oncogenic forms of N-ras or H-ras prevent skeletal myoblast differentiation, *Mol. Cell,* 2104, 2111, 1987.
120. **Fortini, M. E., Simon, M. A., and Rubin, G. M.,** Signalling by the sevenless protein tyrosine kinase is mimicked by Ras 1 activation, *Nature,* 355, 559, 1992.
121. **Hagag, N., Halegova, S., and Viola, M.,** Inhibition of growth factor induced differentiation of PC12 cells by microinjection of antibody to ras p21, *Nature,* 319, 680, 1986.
122. **Nebreda, A. R., Martin-Zanca, D., Kaplan, D. R., Parada, L. F., and Santos, E.,** Induction of NGF of meiotic maturation of *Xenopus* oocytes expressing the *trk* proto-oncogene product, *Science,* 252, 558, 1991.
123. **Korn, L. J., Scribel, C. W., McCormick, F., and Roth, R. A.,** Ras p21 as a potential mediator of insulin action in Xenopus oocytes, *Science,* 236, 840, 1987.
124. **Berridge, M.,** Inositol trisphosphate and diacylglycerol: two interacting second messengers, *Annu. Rev. Biochem.,* 56, 159, 1987.

125. **Nishizuka, Y.,** The molecular heterogeneity of protein kinase C and its implications for cellular regulation, *Nature,* 334, 661, 1988.
126. **Kikkawa, V., Kishimoto, A., and Nishizuka, Y.,** The protein kinase C family: heterogeneicity and its implication, *Annu. Rev. Biochem.,* 58, 31, 1989.
127. **Fleischman, L. F., Chahwala, S. B., and Cantley, L.,** Ras transformed cells: altered levels of phosphatidylinositol,4,5, bisphosphate and catabolites, *Science,* 231, 407, 1986.
128. **Wakelam, M. J., Davies, S. A., Houslay, M. D., McKay, I., Marshall, C. J., and Hall, A.,** Normal p21N-ras couples bombesin and other growth factor receptors to inositol phosphate production, *Nature,* 323, 6084, 1986.
129. **Downward, J., De Gunzberg, J., Riehl, R., and Weinberg, R. A.,** p21ras-induced responsiveness of phosphatidylinositol turnover to bradykinin is a receptor number effect, *Proc. Natl. Acad. Sci. U.S.A.,* 85, 5774, 1988.
130. **Wakelam, M. J.,** Inhibition of the amplified bombesin-stimulated inositol phosphate response in N-ras transformed cells by high density culturing, *FEBS Lett.,* 228, 182, 1988.
131. **Preiss, J., Loomis, C. R., Bishop, W. R., Stein, R., Niedel, S. E., and Bell, R. M.,** Quantitative measurement of sn-1,2-diacylglycerols present in platelets, hepatocytes, and ras- and sis-transformed normal rat kidney cells, *J. Biol. Chem.,* 261, 8597, 1986.
132. **Wilkinson, W. D., Sandgren, E. P., Palmiter, R. D., Briniter, R. L., and Bell, R. D.,** Elevation of 1,2 diacylglycerol in ras-transformed neonatal liver and pancreas of transgenic mice, *Oncogene,* 4, 625, 1989.
133. **Hancock, J. F., Marshall, C. J., McKay, I. A., Garlizer, J., Honslay, M. D., Hall, A., and Wakelam, M. I. O.,** Mutant but not normal p21[RAS] elevates inositol phospholipid breakdown in two different cell systems, *Oncogene,* 3, 187, 1988.
134. **Seuwen, K., Lagarde, A., and Pouysségur, J.,** Deregulation of hamster fibroblast proliferation by mutated ras oncogenes is not mediated by constitutive activation of phosphoinositide-specific phospholipase C, *EMBO J.,* 161, 168, 1988.
135. **Wolfman, A. and Macara, I. G.,** Elevated levels of diacylglycerol and decreased phorbol ester sensitivity in ras transformed fibroblasts, *Nature,* 325, 359, 1987.
136. **Lacal, J. C., Moscat, J., and Aaronson, S. A.,** Novel source of 1,2 diacylglycerol elevated in cells transformed by Ha-Ras oncogene, *Nature,* 330, 269, 1987.
137. **Morris, D. H., Price, B. D., Lloyd, A., Marshall, C. J., and Hall, A.,** Scrape loading of Swiss 3T3 cells with ras protein rapidly activates protein kinase C in the absence of phosphoinositide hydrolysis, *Oncogene,* 4, 27, 1989.
138. **Alonso, T., Morgan, R. O., Marvizon, J. C., Zarbl, H., and Santos, E.,** Malignant transformation by *ras* and other oncogenes produces common alterations in inositol phospholipid signalling pathways, *Proc. Natl. Acad. Sci. U.S.A.,* 85, 4271, 1988.
139. **Bockino, S. B., Blackmore, P. I., Exton, J. H.,** Stimulation of 1,2 diacylglycerol accumulation in hepatocytes by vasopressin, epinephrine and angiotensin II, *J. Biol. Chem.,* 260, 14201, 1985.
140. **Pledger, W. J., Stiles, C. D., Antoniades, H. N., and Scher, C. D.,** An ordered sequence of events is required before BALB/c-3T3 cells become committed to DNA synthesis, *Proc. Natl. Acad. Sci. U.S.A.,* 75, 2839, 1987.
141. **Billah, M. M. and Anthes, J. C.,** The regulation and cellular functions of phosphatidyl choline hydrolysis, *Biochem. J.,* 296, 281, 1990.
142. **Exton, J. H.,** Signalling through phosphatidyl choline breakdown, *J. Biol. Chem.,* 265, 1, 1990.
143. **Pelech, S. L. and Vance, D. E.,** Regulation of phosphatidylcholine biosynthesis, *Biochem. Biophys. Res. Commun.,* 217, 251, 1984.
144. **Leach, R. L., Ruff, V. A., Wright, T. M., Pessin, M. S., and Raben, D. M.,** Dissociation of protein kinase C activation and sn-1,2-diacylglycerol formation. Comparison of phosphatidylinositol and phosphatidylcholine-derived diglycerides in alpha-thrombin-stimulated fibroblasts, *J. Biol. Chem.,* 266, 3215, 1991.

145. **Chiarugi, V., Bruni, P., Pasquali, F., Magnelli, L., Basi, G., Ruggiero, M., and Farnararo, M.,** Synthesis of diacylglycerol de novo is responsible for permanent activation and downregulation of protein kinase in transformed cells, *Biochem. Biophys. Res. Commun.,* 164, 816, 1989.

146. **Lacal, J. C., Dela Peña, P., Moscat, J., García Barreno, P., Anderson, P. S., and Aaronson, S. A.,** Rapid stimulation of diacylglycerol production in Xenopus oocytes by microinjection of H-ras p21, *Science,* 238, 833, 1987.

147. **Lacal, J. C.,** Diacylglycerol production in *Xenopus laevis* oocytes after microinjection of p21ʳᵃˢ proteins is a consequence of activation of phosphatidylcholine metabolism, *Mol. Cell. Biol.,* 10, 333, 1990.

148. **Macara, I. C.,** Elevated phosphocholine concentration in ras-transformed NIH 3T3 cells araises from increased choline kinase activity, not from phosphatidylcholine breakdown, *Mol. Cell. Biol.,* 9, 325, 1989.

149. **Teegarden, D., Taparowsky, E. J., and Kent, C.,** Altered phosphatidylcholine metabolism in C3H10T½ cells transfected with Harvey-ras oncogene, *J. Biol. Chem.,* 265, 6042, 1990.

150. **Besterman, J. M., Duronio, V., and Cuatrecasas, P.,** Rapid formation of diacylglycerol from phosphatidylcholine: a pathway for generation of a second messenger, *Proc. Natl. Acad. Sci. U.S.A.,* 6785, 6789, 1986.

151. **Larrodera, P., Cornet, M. E., Díaz-Meco, M. T., Lopez-Barahona, M., Díaz-Laviada, I., Guddal, P. H., Johasen, T., and Moscat, J.,** Phospholipase C-mediated hydrolysis of phosphatidylcholine is an important step in PDGF stimulated DNA synthesis, *Cell,* 61, 1113, 1990.

152. **García de Herreros, A., Domínguez, I., Díaz-Meco, M. T., Graziani, G., Cornet, M. E., Guddal, P. H., Johansen, T., and Moscat, J.,** Requirement of phospholipase C-catalyzed hydrolysis of phosphatidylcholine for maturation of *Xenopus laevis* oocytes in response to insuli and *ras* p21, *J. Biol. Chem.,* 266, 6825, 1991.

153. **Laviada, I. D., Larrodera, P., Díaz-Meco, M. T., Cornet, M. E., Guddal, P. H., Johansen, T., and Moscat, J.,** Evidence for a role of phosphatidylcholine-hydrolising phospholipase C in the regulation of kinase C by ras and src oncogenes, *EMJO J.,* 3907, 3912, 1990.

154. **Lacal, J. C., Fleming, T. P., Warren, B. S., Blumberg, P. M., and Aaronson, S. A.,** Involvement of functional protein kinase C in the mitogenic response to the H-ras oncogene product, *Mol. Cell. Biol.,* 7, 4146, 1987.

155. **Morris, J. D. H., Price, B., Lloyd, A. C., Self, A. J., Marshall, C. J., and Hall, A.,** Scrape-loading of Swiss 3T3 cells with *ras* protein rapidly activates protein kinase C in the absence of phosphoinositide hydrolysis, *Oncogene,* 4, 27, 1989.

156. **Dominguez, I., Marshall, M. S., Gibbs, J. B., García de Herreros, A., Cornet, M. E., Graziani, G., Díaz-Meco, M. T., Johansen, T., McCornick, F., and Moscat, J.,** Role of GTPase activating protein in mitogenic signalling through phosphatidylcholine hydrolysing phospholipase C, *EMBO J.,* 32, 15, 1991.

157. **Bar-Sagi, B. and Feramisco, J. R.,** Induction of membrane ruffling and fluid-phase pinocytosis in quiescent fibroblast by ras proteins, *Science,* 233, 1061, 1986.

158. **Bar-Sagi, D., Suhan, J. P., McCormick, F., and Feramisco, J. R.,** Localization of phospholipase A₂ in normal and ras-transformed cells, *J. Cell. Biol.,* 106, 1649, 1988.

159. **Samochocki, M. and Strosznajder, J.,** Regulation of arachidonic acid release by enzyme(s) of rat brain cortex, *Acta Biochem. Pol.,* 37, 93, 1990.

160. **Zagari, M., Stephens, M., Earp, H. S., and Herman, B.,** Relationship of cytosolic ion fluxes and protein kinase C activation to platelet-derived growth factor induced competence in BALB/c-3T3 cells, *J. Cell. Physiol.,* 139, 167, 1989.

161. **Cuadrado, A., Molloy, C. J., and Pech, M.,** Expression of protein kinase CI in NIH 3T3 cells increases its growth response to specific activators, *FEBS Lett.,* 260, 281, 1990.

162. **Issandou, M. and Rozengurt, E.,** Diacylglycerols, unlike phorbol esters, do not induce homologous desensitization or down-regulation of protein-kinase C in Swiss 3T3 cells, *Biochem. Biophys. Res. Commun.,* 163, 201, 1989.

163. **Lacal, J. C., Cuadrado, A., Jones, J. G., Trotta, R., Burstein, D. E., Thomson, T., and Pellicer, A.**, Regulation of protein kinase C activity in neuronal differentiation induced by N-ras oncogene in PC-12 cells, *Mol. Cell. Biol.*, 10, 333, 1990.

164. **Kamata, T., Sullivan, N. Y., and Wooten, M. W.**, Reduced protein kinase C activity in a ras-resistant cell line derived from Ki-MSV transformed cells, *Oncogene*, 1, 37, 1987.

165. **Persons, D. A., Wilkinson, W. A., Bell, R. M., and Finn, O. S.**, Altered regulation and tumorigeniety of NIH 3T3 fibroblasts transfected with protein kinase C-I cDNA, *Cell*, 52, 447, 1988.

166. **Housey, G. M., Johnson, M. D., Hsiao, W. L., O'Brian, C. A., Murphy, J. P., Kirschmeier, P., and Weinstein, I. B.**, Overproduction of protein kinase C causes disordered growth control in rat fibroblasts, *Cell*, 52, 343, 1988.

167. **Lloyd, A. C., Paterson, H. F., Morris, J. D., Hall, A., and Marshall, C. J.**, p21-ras-induced morphological transformation and increases in c-myc expression are independent of functional protein kinase C, *EMBO J.*, 8, 4, 1099, 1989.

168. **Yu, C. L., Tsai, M. E., and Stacy, D. W.**, Cellular ras activity and phospholipid metabolism, *Cell*, 52, 63, 1988.

169. **Rapp, U. R.**, Role of Raf-1 serine/threonine protein kinase in growth factor signal transduction, *Oncogene*, 495, 500, 1991.

170. **Kolch, W., Heidecker, G., Lloid, P., and Rapp, U. R.**, Raf-1 kinase is required for growth of induced NIH 3T3 cells, *Nature*, 349, 426, 1991.

171. **Smith, M. R., DeGudicibus, S. J., and Stacey, D. W.**, Requirement for c-ras proteins during viral oncogene transformation, *Nature*, 320, 540, 1986.

172. **Cuadrado, A., Talbot, N., and Barbacid, M.**, C127 cells resistant to transformation by tyrosine protein kinase oncogenes, *Cell Growth Differ.*, 1, 9, 1990.

173. **Cuadrado, A., Fleming, T. P., App, H., Rapp, U. R., and Aaronson, S. A.**, Functional link between ras and raf in oncogenic transformation, *Oncogene*, in press, 1993.

174. **Hunter, T.**, Cooperation between oncogenes, *Cell*, 249, 270, 1991.

175. **Stacey, D. W., Watson, T., Kung, H.-F., and Curran, T.**, Microinjection of transforming ras protein induces c-fos expression, *Mol. Cell*, 523, 527, 1987.

176. **Sassone-Corsi, P., Der, C. J., and Verma, I.**, *ras*-induced neuronal differentiation of PC12 cells: possible involvement of *fos* and *jun*, *Mol. Cell. Biol.*, 9, 3174, 1989.

177. **Thomson, T. M., Green, S. H., Trotta, R. J., Burstein, D. E., and Pellicer, A.**, Oncogene N-*ras* mediates selective inhibition of c-*fos* induction by nerve growth factor and basic fibroblasts growth factor in a PC12 cell line, *Mol. Cell. Biol.*, 10, 1556, 1990.

178. **Imler, J. L., Schatz, C., Wasylyk, C., Chatton, B., and Wasyluk, B.**, A Harvey-*ras* responsive transcription element is also responsive to a tumor-promoter and to serum, *Nature*, 332, 275, 1988.

179. **Gauthier-Rouviere, C., Fernandez, A., and Lamb, N. J. C.**, *ras*-induced c-*fos* expression and proliferation in living rat fibroblasts involves C-kinase activation and the serum response element pathway, *EMBO J.*, 9, 171, 1990.

180. **Binétruy, B., Smeal, T., and Karin, M.**, Ha-*ras* augments c-*jun* activity and stimulates phosphorylation of its activation domain, *Nature*, 351, 122, 1991.

181. **Clark, S. G., Stern, M. J., and Horvitz, H. R.**, *C. elegans* cell-signalling gene sem-5 encodes a protein with SH-2 and SH-3 domains, *Nature*, 356, 340, 1992.

182. **Lowenstein, E. J., Daly, R. J., Batzer, A. G., Li., W., Margolis, B., Lammers, R., Ulrich, A., Skolnick, E. Y., Bar-Sagi, D., and Schlessinger, Y.**, The SH2 and SH3 domain-containing protein Grb2 links receptor tyrosine kinases to *ras* signalling, *Cell*, 70, 431, 1992.

183. **Pelicci, G., Lanfrancone, L., Grignani, F., Pawson, T., and Pelicci, G.**, A novel transforming protein (SHC) with an SH2 domain is implicated in mitogenic signal transduction, *Cell*, 70, 93, 1992.

184. **McGlade, J., Cheng, A., Pelicci, G., Pelicci, P. G., and Pawson, T.**, Shc proteins are phosphorylated and regulated by the v-src and v-fps protein-tyrosine kinase, *Proc. Natl. Acad. Sci. U.S.A.*, 89, 8869, 1992.

185. **Rozakis-Adcock, M., McGlade, J., Bamalin, G., Pelicci, G., Daly, R., Li, W., Batzar, A., Thomas, S., Brugge, J., Pelicci, P. G., Schlessinger, J., and Pawson, T. P.,** Association of the Shc and Grb2/sem5 SH2-containing proteins is implicated in activation of the ras pathway by tyrosine kinases, *Nature,* 360, 189, 1992.

186. **Dickson, B., Sprenger, F., Morrison, D., and Hafen, E.,** Raf-1 functions down stream of RAS-1 in sevenless signal transduction pathway, *Nature,* 360, 600, 1992.

187. **Gaul, U., Marbon, G., and Rubin, G. M.,** A putative ras GTPase activating protein acts as a negative regulator of signalling by sevenless receptor tyrosine kinase, *Cell,* 68, 1007, 1992.

188. **Troppmair, J., Bruder, J. T., App, H., Cai, H., Liptak, L., Szebernyi, J., Cooper, G. M., and Rapp, U. R.,** Ras controls coupling of growth factor receptors and protein kinase C in the membrane to Raf-1 and B-raf protein serine kinases in the cytosol, *Oncogene,* 7, 1867, 1992.

189. **Wood, K. W., Sarnecki, C., Roberts, T. M., and Blenis, J.,** ras mediates nerve growth factor receptor modulation of three signal-transducing protein kinases: MAP kinase, Raf-1, and RSK, *Cell,* 68, 1041, 1992.

190. **Thomas, S. M., De Marco, M., D'Arcangelo, G., Halegona, S., and Brugge, S. J.,** Ras is essential for nerve-growth factor and phorbol ester induced tyrosine phosphorylation of MAP kinases, *Cell,* 68, 1031, 1992.

191. **Qiu, M. S. and Green, S. H.,** PC12 cell neuronal differentiation is associated with prolonged p21ras activity and consequent prolonged RK activity, *Neurone,* 9, 705, 1992.

192. **Leevers, S. L. and Marshall, C. J.,** MAP-kinase regulation — the oncogene connection, *Trends Cell Biol.,* 2, 283, 1992.

193. **Kyriakis, J. M., App, H., Zhang, X., Banerjee, P., Brautigan, D. L., Rapp, U. R., and Avruch, J.,** Raf-1 activates MAP kinase-kinase, *Nature,* 358, 417, 1992.

194. **Dent, P., Haser, W., Haystead, T. A. J., Vincent, L. A., Roberts, T. M., and Sturgill, T. W.,** Activation of Mitogen-Activated kinase kinase by v-raf in NIH 3T3 cells and in vitro, *Science,* 257, 1404, 1992.

195. **Pelech, S. L. and Sanghera, J. S.,** MAP kinases: charting the regulatory pathways, *Science,* 257, 1355, 1992.

196. **Ray, L. B. and Sturgill, T. W.,** Insulin-stimulated microtubule-associated protein kinase is phosphorylated on tyrosine and threonine in vivo, *Proc. Natl. Acad. Sci. U.S.A.,* 85, 3753, 1988.

197. **Alvarez, E., Northwood, I. C., Gonzalez, F. A., Latour, D. A., Eth, A., Abate, C., Curran, T., and Davis, R. J.,** Pro-Leu-Ser/Thr-Pro is a consensus primary sequence for substrate protein phosphorylation. Characterizaton of the phosphorylation of c-myc and c-jun proteins by an epidermal growth factor receptor threonine 669 kinase, *J. Biol. Chem.,* 266, 15277, 1991.

198. **Vries-Smits, A. M. M., Burgering, B. M. T., Leevers, S. J., Marshall, C. J., and Bos, J. L.,** Involvement of p21ras in activation of extracellular signal-regulated kinase 2, *Nature,* 357, 602, 1992.

199. **Leevers, S. J. and Marshall, C. J.,** Activation of extracellular signal-regulated kinase, ERK-2 by p21ras oncoprotein, *EMBO J.,* 11, 569, 1992.

200. **Pomerance, M., Schweighoffer, F., Tocque, B., and Pierre, M.,** Stimulation of mitogen-activated protein kinase by oncogenic ras p21 in Xenopus oocytes, *J. Biol. Chem.,* 267, 16155, 1992.

201. **Cook, S. J. and Wakelam, J. O.,** Epidermal growth factor increases sn-1,2-diacylglycerol levels and activates phospholipase D-catalysed phosphatidylcholine breakdown in Swiss 3T3 cells in the absence of inositol-lipid hydrolysis, *Biochem J.,* 285, 247, 1992.

202. **Fukami, K. and Takenawa, T.,** Phosphatidic acid accumulates in platelet-derived growth factor-stimulated Balb/c 3T3 cells is a potential mitogenic signal, *J. Biol. Chem.,* 267, 10988, 1991.

203. **Song, J., Pfeffer, M. L., and Foster, D. A.,** v-src increases diacylglycerol levels via a type D phospholipase mediated hydrolysis of phosphatidylcholine, *Mol. Cell. Biol.,* 1, 2018, 1991.
204. **Cook, S. J., Briscoe, C. P., and Wakelam, M. J. O.,** The regulation of PLD activity and its role in sn-1,2-DAG formation in bombesin and PMA stimulated Swiss 3T3 cells, *Biochem. J.,* 280, 431, 1992.
205. **Hii, C. S. T., Edwards, Y. S., and Murray, A. U.,** Phorbol ester-stimulated hydrolysis of PC and PE by PLD in HeLa cells, *J. Biol. Chem.,* 266, 20238, 1992.
206. **Moolenaar, W. H., Kruijer, W., Tilly, B. C., Verlaan, I., Bierman, A. J., and McCall, C. E.,** Growth factor-like action of phosphatidic acid, *Nature,* 323, 171, 1986.
207. **Van Coven, E. J., Groenink, A., Jalink, K., Eichholtz, T., and Moolenaar, W. H.,** Lysophosphatidate induced cell proliferation: identification and dissection of signalling pathways mediated by G proteins, *Cell,* 59, 45, 1989.

Chapter 7

THE RAS SYSTEM IN YEASTS

Michael H. Wigler

TABLE OF CONTENTS

0-8493-5214-2/93/$0.00 + $.50

I. INTRODUCTION

The *ras* oncogenes are highly conserved in evolution. Homologs of these genes have been found in many eukaryotes, including unicellular[18,23,32,75,78] and complex multicellular organisms.[26,40,70] It is likely that *ras* homologs will be found in all eukaryotic organisms. Certainly there are no other oncogenes presently known that are as highly conserved. The degree of identify between the mammalian oncogenic RAS proteins and the RAS proteins of the yeasts *Saccharomyces cerevisiae* (RAS1 and RAS2) and *Schizosaccharomyces pombe* (ras1) in the N terminal halves of the molecules is about 90%. To put this into perspective, the degree of identity between the cAMP-dependent protein kinase regulatory subunits from yeasts and mammals is in the range of 50%.[95] These subunits bind two molecules of cAMP apiece, dimerize, bind the catalytic subunit, release the catalytic subunit from inhibition upon binding cAMP, and retain the ability to interact with the catalytic subunit of the divergent organisms.[42,111] From this example it is reasonable to infer that the level of conservation of primary structure observed for RAS is associated with the conservation of many biochemical functions. This inference is correct.

RAS proteins are sufficiently conserved at the structural level that mammalian RAS can function in yeast,[19,48] and this makes the study of RAS in yeast of particular importance. Many of the interactions between RAS and other cellular components have also been conserved, including elements of the processing, regulatory, and effector systems. Only these latter two systems will be discussed in detail below. The processing of RAS, which is much the same in yeast and vertebrates, is reviewed elsewhere in this volume.

This review reflects my biases due to familiarity with the literature and due to opinions of significance. I have not tried to avoid this, and in particular I have not attempted to comprehensively reference all the relevant literature. Rather, I hope to communicate a coherent and comprehensible picture of a complex subject, and to leave the reader with a sense of what might come next. I also include a brief speculative section. I begin with a brief overview.

II. AN OVERVIEW OF RAS IN *Saccharomyces cerevisiae*

In the budding yeast, *S. cerevisiae*, there are two *ras* homologs, RAS1 and RAS2.[23,75] At lease one *ras* gene is required for viability.[47] Yeast lacking endogenous RAS are viable if they express mammalian H-*ras*, although they grow more slowly.[48] This slow growth is due almost entirely to a prolonged G1 phase. As discussed in more detail below, *S. cerevisiae* RAS proteins are required for stimulation of adenylyl cyclase, encoded by the *CYR1* gene.[97] The production of cAMP is required for progression through the G1 phase of the cell cycle.[58-61] It is not yet known if the interaction of RAS with adenylyl cyclase is direct or mediated through other proteins. The effects of cAMP are mediated through the cAMP-dependent protein kinase (cAPK), which in yeast, as in most other organisms, consists of two molecules of a regulatory subunit and two of a catalytic subunit.[42] In *S. cerevisiae* there is one gene, *BCY1*, encoding the regulatory subunit, and three genes, *TPK1, 2,* and *3* each encoding catalytic subunits.[94,95] It is likely that RAS has other effector functions in *S. cerevisiae*,[93,106] and I will discuss the evidence for this later. The nature of these alternate functions are not currently understood.

   S. cerevisiae RAS, like mammalian ras, bind GTP and DGP.[86] The bound nucleotide exchanges slowly on the isolated protein, and exchange is accelerated by auxiliary proteins. Yeast RAS have an intrinsic GTP hydrolysis activity which is also accelerated by auxiliary proteins.[89,90] Yeast RAS, as mammalian RAS, is active when in its GTP-bound state.[28,29] Yeast RAS protein, like mammalian RAS protein, undergoes an essential carboxy terminal processing event during maturation that is dependent upon its terminal signal sequences, Cys-A-A-X, where A is any aliphatic amino acid and X is the terminal amino acid.[20,76] These processing events, which are similar for mammals and yeast, are required for the proper membrane localization of RAS but will not be discussed further.

   Point mutations can activate yeast RAS, as they do for mammalian ras. Thus, strains carrying the mutant $RAS2^{val19}$ allele, equivalent to the mutation at codon 12 that activates H-*ras*, have a clearly mutant phenotype.[47,80] Such strains fail to respond properly to nutrient deprivation: they fail to arrest in G1; they fail to accumulate carbohydrate stores; they fail to become heat shock resistant; and they die upon prolonged starvation. Similar phenotypes are seen upon disruption of the *BCY1* gene, which leads to unregulated activity of the cAPK catalytic subunits.[95,97] Thus, intact RAS- and cAMP-regulated pathways appear required for maintenance of normal responses to nutrient deprivation.

   The regulation of RAS is quite complex, and several molecules that regulate it are known. The product of the *CDC25* gene is proposed to catalyze guanine nucleotide exchange on RAS[10,77] and homologs of *CDC25* have been found in other yeasts and higher eukaryotes.[17,35,43,81] Other exchange catalysts may yet be found. The products of the IRA1 and IRA2 genes negatively regulate RAS by promoting hydrolysis of bound GTP.[87-90] The IRA proteins are homologous to sar1, found in the fission yeast *Schizosaccharomyces pombe*,[103] and to GAP and NF-1, present in mammalian cells. RAS activity is also regulated by a vigorous feedback control mediated through the action of the cAPK.[71] Glucose feeding is the only well-documented stimulus for cAMP production in yeast,[91] and this response requires both RAS and CDC25.[62,63,101]

III. AN OVERVIEW OF RAS IN
Schizosaccharomyces pombe

   The fission yeast *S. pombe* contains *ras1*, the single close homolog of oncogenic mammalian RAS.[32] There are several important physiological and biochemical differences between RAS in the two yeasts. In *S. pombe* ras1 is not essential for viability, but rather it is required for sexual activity and the maintenance of an elongated rather than spherical morphology.[34,68] Almost certainly ras1 does not mediate its effects through adenylyl cyclase. Mammalian RAS can substitute for *S. pombe* ras1, as it can in *S. cerevisiae*.

In *S. pombe*, sexual activity is a function of mating type, and this subject has been reviewed elsewhere.[24] As in *S. cerevisiae*, there are two mating types in *S. pombe*. Only opposite mating types conjugate, and the resulting diploids almost always immediately commence sporulation. Conjugation and sporulation both require at least two signals: mating pheromone, mediated through transmembrane receptors of the serpentine class and their associated heterotrimeric G proteins,[73] and nutrient deprivation, which is probably mediated through adenylyl cyclase.[49] In the absence of adenylyl cyclase, conjugation and sporulation occur even without starvation. In the absence of ras1 function, neither mating nor sporulation occur.

S. pombe strains with the *ras1*[val17]-activated allele display abnormalities in sexual behavior.[34,68] The *ras1*[val17] cells conjugate with very low efficiency, and in the presence of cells of the opposite mating type develop long mating tubes. Sporulation is normal. It is likely that the conjugal abnormalities result from an excess of ras1 activity, since overexpression of the *sar1* gene product, which encodes an inhibitor of ras1 and has structural and functional homology to the *S. cerevisiae* IRA proteins and mammalian GAP, restores normal conjugation levels in strains expressing activated ras1.[103] Cells respond to mating signals and nutritional deprivation, even in the presence of constitutively activated ras1.

The regulation of ras1 in *S. pombe* shows similarities to the regulation of RAS in *S. cerevisiae*. The product of the *ste6* gene has structural homology to the *S. cerevisiae CDC25* gene product, and genetic analysis suggests it plays a similar role.[43] There is, however, another *S. pombe* gene, *ral2*, that shows no homology to other known genes of the RAS system, but by genetic analysis appears required for the activation of ras1.[33,35] As mentioned above, the sar1 protein is homologous to the IRA proteins, and genetic analysis also suggests it plays a similar role in the regulation of RAS activity. There is no evidence that sar1 or the IRA proteins have any RAS-effector function.

The effector(s) of ras1 are not known. Two protein kinases, bry1 and byr2, appear to act downstream of ras1.[66,104] These protein kinases are most homologous to the protein kinases encoded by the STE11 and STE7 genes of *S. cerevisiae*. STE11 and STE7 are known to be essential to the mating pheromone response pathway of that yeast.[55] I speculate later on the relationship of these kinases to the RAS-dependent signaling pathways in mammals.

IV. THE REGULATION OF RAS IN YEASTS BY GUANINE NUCLEOTIDE EXCHANGE

The first evidence that RAS is itself regulated derived from the study of the *S. cerevisiae CDC25* gene, which encodes a protein predicted to be 185 kDa.[10] Temperature-sensitive mutations of this gene behaved like temperature-sensitive mutations of the *CYR1* gene, indicating that *CDC25* might encode a component of the RAS pathway. Further genetic analysis indicated that although disruption of *CDC25* was lethal in wild-type yeast, activated alleles of *RAS2* suppressed this lethality. Subsequently, dominant negative (interfering) mutations of RAS were discovered that had alterations in their guanine nucleotide-binding domains.[77] The interference caused by these mutant genes could be overcome by overexpression of the *CDC25* gene. This result suggested that the dominant negative RAS product formed an ineffective complex with the CDC25 protein, thus, blocking its ability to interact with and activate wild-type proteins. This model has been confirmed by recent biochemical experiments demonstrating the physical interaction of CDC25 and RAS.[45,65] This interaction occurs preferentially when RAS is in its GDP-bound state.[65]

Such studies have led to the hypothesis that CDC25 directly catalyzes the exchange of the bound guanine nucleotide on RAS for free nucleotide. Since the concentration of GTP inside the cell is higher than the concentration of GDP, and the majority of RAS in cells is bound to GDP, the net result of catalyzing exchange should be the activation of RAS. Direct support for this idea comes from biochemical studies of the interactions between RAS and the product of the *SCD25* gene.[17] *SCD25* was isolated as a gene that could suppress the temperature-sensitive defects of a *cdc25^{ts}* mutant strain.[8] It encodes a protein with clear structural homology to the CDC25 protein. The SCD25 protein, purified from an *Escherichia coli* expression system, does catalyze the exchange of guanine nucleotide bound to purified RAS protein. No one has yet convincingly demonstrated this property for recombinant CDC25 protein, presumably because it has not been synthesized in a native form from a bacterial expression system, or because it requires auxiliary factors to become activated.

It is quite possible that CDC25 is itself regulated. First, mutation of CDC25 can activate its activity.[10] Moreover, as mentioned before, feeding glucose to yeast starved for a fermentable carbon source leads to a rapid and transient rise in cAMP. The transient elevation in this response requires both RAS and CDC25. Only the carboxy terminal third of CDC25 is required for this response.[101] The cAMP transient is not seen in cells that lack wild-type RAS but contain activated RAS2^{val19}, which does not require CDC25 for function, and is not seen in cells lacking both CDC25 and IRA1. The rapid rise in cAMP levels during glucose feeding is not attributable to alterations in the regulatory action of the IRA proteins, although the rapid decrease in cAMP levels that ensues is probably the consequence of the IRA proteins.[90] At the moment we lack direct proof that glucose feeding affects CDC25 function. An alternate hypothesis, worthy of consideration, is that a heterotrimeric G protein induces the cAMP transient brought about by glucose feeding. If correct, this hypothesis would increase the parallelism of the *S. cerevisiae* RAS-effector system with that of *S. pombe* (see section below). In fact, the product of *GPA2*, a G_α protein, appears to be involved in the cAMP transient.[69] Its mode of action is unclear, and it is unknown if homologs of *GPA2* will be found.

It is highly likely that RAS/CDC25 interactions have been conserved in the evolution of RAS systems. The yeast *S. pombe* has a homolog of *S. cerevisiae* CDC25, called ste6, which likewise participates in the ras1 pathways.[43] In fact, there is a family of proteins occurring throughout evolution that have structural homology to the *S. cerevisiae* CDC25 product. Among these are *BUD5*[11,74] that together with another *ras*-like protein, called *BUD1*, participates in a pathway in *S. cerevisiae* involved in the determination of cell polarity. More recently, a gene in *Drosophila melanogaster*, called son of sevenless, *(SOS)* has been found that encodes a member of this family.[82] *SOS* is likely to interact with the closest *Drosophila* homolog of RAS.

Mammalian RAS, containing the mutations in the guanine nucleotide binding domain that are analogous to the mutations found in the *S. cerevisiae* dominant negative mutant RAS, also interfere with wild-type RAS function in *S. cerevisiae*.[77] It is likely, therefore, that mammalian RAS can interact with yeast CDC25. Moreover, such mutant mammalian *ras* genes are dominant lethals when expressed in mammalian cells, and so it is likely that molecules like CDC25 are needed in mammalian cells to maintain RAS in its active state.[27] It is not certain that nucleotide exchange will be a control point in mammalian cells, subject to various intracellular and extracellular signals, but this seems likely.

CDC25-like and GAP-like molecules are not necessarily the only regulatory entities that interact with RAS. Two genes, one in *S. cerevisiae, RPI,*[51] and one in *S. pombe, ral2,*[33,35] have been described that have the characteristics of genes that encode regulators of RAS. These genes encode products that are not structurally related to either CDC25 or GAP nor do they appear to act through these types of agents. It is likely that other exchange proteins will be found in *S. pombe*. The *S. pombe* ste6 gene product that is homologus to *CDC25* is needed only for ras1 function during conjugation. By contrast, cells lacking *ral2* behave like cells lacking *ras1*: they are deficient in conjugation and sporulation and are spherical. It is not known if *ral2* encodes an exchange factor, or if it acts upstream of one. Some molecules might catalyze the guanine nucleotide exchange of RAS, yet not be structurally related to the CDC25-like family. There is precedence for this within the larger ras superfamily.[41] Some molecules may regulate RAS nucleotide exchange by blocking the exchange, and precedent for this mechanism is also found within the RAS superfamily.[57] Our knowledge of molecules capable of interacting with and regulating RAS must be considered incomplete, and further research in this area is clearly warranted.

V. THE REGULATION OF RAS BY GTPase-ACTIVATING PROTEINS

RAS proteins slowly hydrolyze GTP *in vitro*, and Trahey and McCormick noted that this rate of hydrolysis did not match the rate of hydrolysis of GTP by RAS protein microinjected into oocytes.[98] This observation led to the discovery of a GTPase-activating activity, which has subsequently been found ubiquitously in eukaryotic cells. The first GTPase-activating proteins, or GAPs, were isolated from human and bovine sources.[39,99] Mammalian GAPs are described in detail in other chapters of this book.

Subsequent to the discovery of GAP, a locus in *S. cerevisiae* was found that appeared required to regulate wild-type RAS. Disruptions of this gene, called IRA1 (inhibitor of *ras*), resulted in the same set of phenotypes seen in cells with activating mutations of RAS2.[88] Furthermore, disruption of the *CDC25* gene in *iral*⁻ cells restored a wild-type phenotype. The IRA1 gene was cloned by complementation screening, and sequence analysis revealed that it encoded a very large protein, greater than 300 kDa, containing a domain with significant homology to mammalian GAP. The IRA1 protein indeed possesses RAS GTPase-stimulating activity.[87] A second, highly related gene, called IRA2, was discovered by hybridization screening with an IRA1 probe.[90] It encodes a protein with a complementary function. Expression of mammalian GAP can complement the loss of IRA function in yeast, further evidence for the conservation of the system.[3]

We do not know if or how the IRA proteins are controlled in *S. cerevisiae*. Their large size certainly suggests that they may be a focal point of regulation. One hint of their physiological role is the abnormal cAMP response to glucose feeding seen in IRA-deficient cells.[88] Instead of displaying the normal biphasic response, with a rapid rise in cAMP followed by a rapid decline to normal levels in the continued presence of glucose, *iral*⁻ cells fed glucose have a rapid rise in cAMP which does not subsequently decline. Thus, the IRA genes may function in the stringent feedback control of RAS that is known to be the result of activating the cAPK.[71]

The *sarl* gene of *S. pombe* also encodes a member of the GAP family of proteins.[103] The *sarl* gene was discovered in a screen for multicopy genes capable of reversing the effects of *rasl* activation. Disruption of *sarl* leads to a phenotype virtually indistinguishable from the phenotype that results from the activation of *rasl*. Expression of sarl in *S. cerevisiae* complements the loss of the *IRA* genes, and, similarly, expression of mammalian GAP in *S. pombe* complements the loss of the *sarl* gene. *sarl* encodes a protein of about 70 kDa, appreciably smaller than the IRA proteins, and closer in size to mammalian GAP, containing a domain with significant homology to GAP. GAP and sarl are about equally diverged from each other and the IRA proteins.

The gene for the von Recklinghausen's neurofibromatosis disease locus, *NF-1*, encodes a GAP-like molecule.[108] Humans with von Recklinghausen's neurofibromatosis have a predisposition to developmental and proliferative abnormalities in cells of neuroepithelial origin. The NF1 protein stimulates the GTPase activity of mammalian RAS and complements both IRA and sarl deficiency in the respective yeasts.[2] NF1 inhibits mammalian RAS activity when both are expressed in *S. cerevisiae*. Like sarl, but unlike the IRA proteins or GAP, the NF-1 protein also appears to inhibit the activated form of RAS. The homology between NF-1 and the IRA proteins is especially close and significantly extends beyond the domain that is considered merely necessary for catalytic activity. In fact, NF-1 is closer in structure to the yeast IRA proteins than it is to mammalian GAP. This extended homology implies additional functional homology will be found, and that duplication of *GAP*-like genes occurred very early in evolution.

All of these studies confirm the ubiquity and functional conservation of GAP-like molecules and their role in RAS regulation. Several scientists have speculated that GAP-like molecules may have RAS-dependent effector functions as well.[1] Although there are reasons for entertaining this hypothesis for mammalian cells,[105] the weight of evidence is against this hypothesis in yeast. Disruption of *sar1* in *S. pombe*, or the IRA1 and IRA2 genes of *S. cerevisiae*, does not appear to block RAS-effector function. However, we cannot completely rule out the GAP-effector hypothesis in yeast. First, it is by no means clear that we have identified all the GAP-like molecules in yeasts. Given their diversity, it is likely we have not. If such molecules had redundant RAS-effector function, disruption of one or two of them would not necessarily inhibit RAS function. Second, we do not fully understand RAS function, and RAS might have multiple effectors, some of which might be GAP-like. Perhaps a more useful question is whether RAS effectors will have GAP-like activity. The answer is not yet in.

VI. EFFECTORS OF RAS IN *S. cerevisiae*

The first known effector target for RAS is the adenylyl cyclase protein of *S. cerevisiae*.[97] Mammalian and yeast RAS both can activate an adenylyl cyclase complex *in vitro*, although, activation of adenylyl cyclase is not likely to be the function of RAS in vertebrates[4,5] or even in the yeast *S. pombe*.[34] Effects on adenylyl cyclase explains most of the effects of RAS on *S. cerevisiae*, yet stimulation of adenylyl cyclase is not its only function.[106]

The genetic and biochemical evidence that RAS is required for stimulating adenylyl cyclase in *S. cerevisiae* is overwhelming. Cells containing mutant and activated *ras* (e.g., RAS2[val19]) display the same phenotypes as cells with constitutively activated cAMP-dependent protein kinase catalytic subunits: heat shock sensitivity, failure to survive a prolonged nitrogen starvation, failure to accumulate carbohydrates upon nutrient limitation, failure to arrest properly in G1 upon starvation, and failure to sporulate.[47,80,97] The phenotypes of cells carrying RAS2[val19] can be restored to wild-type by overexpressing the *S. cerevisiae PDE* genes, which encode cAMP phosphodiesterases.[72,80] Cells with attenuated cAMP-dependent protein kinase regulatory subunits are not RAS-dependent.[95,97] Cells that lack RAS are rescued from lethality if either the *CYR1* gene, encoding adenylyl cyclase, or the *TRK* genes, encoding the catalytic subunits of the cAPK, dependent protein kinase subunit, are overexpressed.[46,94]

The cAMP levels in cells are dependent upon RAS.[97] Cells with activated RAS have two- to four-fold elevated cAMP levels. Cells lacking RAS have immeasurable levels of cAMP. These effects are dramatically enhanced in cells that have attenuated cAMP-dependent protein kinase catalytic subunits and, thereby, lack feedback control of cAMP levels.[71] Up to 1000-fold differences in cAMP levels can then be seen in cells that have RAS vs. cells that lack RAS or that lack CDC25. The biphasic cAMP response to glucose feeding is not seen in cells containing the constitutively activated *RAS2*[val19] allele.[62]

Both yeast and mammalian RAS proteins stimulate the Mg^{2+}-dependent activity of adenylyl cyclase purified from yeast.[9] Stimulation can be 50-fold over basal levels, and is completely dependent upon RAS being in its GTP-bound state.[29]

The *S. cerevisiae CYR1* was the first eukaryotic gene encoding adenylyl cyclase to be cloned.[46] It encodes a 220 kDa protein that bears virtually no homology to the adenylyl cyclases which have since been characterized from higher eukaryotes.[53] In particular, yeast adenylyl cyclases do not contain membrane-spanning domains. The *S. pombe* adenylyl cyclase resembles the *S. cerevisiae* form, both in structure[109] and to some degree in regulation,[49,50] although it is not regulated by RAS. The catalytic domain of the *S. cerevisiae* CYR1 comprises the terminal 40 kDa. The sequence is otherwise noteworthy for its large (60 kDa) leucine-rich repeat domain. This domain contains approximately 25 units of a 23 amino acid repeat with the consensus sequence PXXαXXLXXLXXLXLXXNXαXXα (where P represents proline, N asparagine, L leucine; α represents any aliphatic amino acid; and X represents any amino acid). This repeat motif is found in many apparently unrelated proteins encoded in genomes from yeasts to insects to vertebrates.[31] It is not yet known if vertebrates contain a homolog of this form of adenylyl cyclase, but this appears likely (see below).

The domains of adenylyl cyclase that are required for RAS responsiveness have been mapped.[14] The N terminal third of the molecule is not required, up to within 100 amino acids of the repeat domain. The structural integrity of the repeat domain is itself stringently required, and almost any insertion or deletion within this domain destroys RAS responsiveness. Overexpression of the repeat domain alone suffices to inhibit the RAS responsiveness of wild-type adenylyl cyclase.[31] Inhibition could be due to the formation of ineffective adenylyl cyclase complexes if adenylyl cyclase normally dimerizes, or to the ineffective sequestering of some limiting component needed for wild-type RAS/adenylyl cyclase interaction.

RAS interacts directly with an adenylyl cyclase complex purified from yeast,[27] but it is not known if RAS interacts directly with adenylyl cyclase. RAS does not appear to stimulate adenylyl cyclase made in *E. coli*.[46] (There is one contrary report in the literature[100] that is widely discredited.) Allele-specific interactions have not answered this question definitively, but are consistent with the idea that the interaction is direct; suppressors of attenuated *ras* alleles have been found and these have arisen by mutations of adenylyl cyclase that map between the repeat and catalytic domains.[22,56] Unfortunately, these mutations in adenylyl cyclase all create a hypersensitive protein, rather than a protein that is responsive only to the specific mutant RAS protein. Thus, this issue has not been resolved by genetics or by biochemistry.

It is possible that RAS acts indirectly upon adenylyl cyclase or requires auxiliary factors. A highly purified adenylyl cyclase complex can be purified from yeast that retains its RAS responsiveness. This complex contains other proteins. One of these, called CAP (adenylyl cyclase associated protein) has an apparent mobility in SDS-polyacrylamide gels of 70 kDa.[30] The protein was purified and the gene encoding it cloned. The N terminus of CAP (the first 168 amino acids out of 526) is required for RAS stimulation of adenylyl cyclase and for full cellular responsiveness to RAS.[37]

It is not yet clear if CAP is required to maintain adenylyl cyclase in a RAS-responsive configuration or whether the effects of RAS are mediated through CAP. There is a CAP homolog in *S. pombe*, and it is required for *S. pombe* adenylyl cyclase to be responsive to its regulatory controls.[50] Although the adenylyl cyclase of *S. pombe* is not RAS-responsive, these combined facts do not resolve the issue, since either CAP or adenylyl cyclase could have evolved an altered regulatory control. Unfortunately, the N terminal domains of CAP do not efficiently cross-complement in the other organism, which is consistent with either hypothesis. If CAP is a RAS-responsive element, it cannot be the only one in *S. cerevisiae* since cells lacking CAP still show some response to activated RAS.[38]

CAP is itself a very interesting molecule in that it is bifunctional.[37] We have just discussed the functions of its N terminus. In *S. cerevisiae*, the C terminus (the last 158 amino acids out of 526) is required for a wide variety of functions which at first appear unrelated to RAS or to adenylyl cyclase. Cells lacking the C terminus of CAP are temperature-sensitive, sensitive to nutrient extremes, large, round, multinucleated, have disrupted actin filaments, and bud randomly from the cell surface.[102] Most of these phenotypes can be effectively suppressed by the overexpression of the gene encoding profilin,[102] an actin, polyproline and phosphatidylinositol-binding protein. Moreover, disruption of profilin leads to a set of phenotypes similar to those observed upon the disruption of the C terminus of CAP. These results suggest that the C terminus of CAP may effect both phospholipid metabolism and cytoarchitecture. The *S. pombe* and *S. cerevisiae* CAP C terminal domains cross-complement very efficiently.[50]

The C and N terminus of CAP are connected by the middle third of the protein, containing a polyproline segment. Cells expressing CAP with just this middle third domain deleted are nearly normal for all CAP-related phenotypes, although they are somewhat large and round.[37] Curiously, this middle region is very highly conserved in evolution.[50] It is likely that current phenotypic assays in yeast do not reflect the functional importance of the middle region of CAP. The most reasonable hypothesis is that this domain of CAP coordinates components of signal transduction pathways with the organization of the cytoskeleton.

Another important line of work indicates the complexity and similarity of RAS-effector systems in yeasts and higher cells. In a search for dominant negative mutants, Michaeli et al.[64] found that overexpression of a mutant human H-*ras* protein incapable of carboxy terminal processing suppressed the abnormal phenotype of RAS2^{val19} cells. This suppression was not overcome by the coexpression of adenylyl cyclase. Unprocessed H-*ras* remains cytoplasmic, but does not interfere (to a first approximation) with the localization or processing of yeast RAS2. Additional studies demonstrate that only cytoplasmic H-*ras* in its GTP-bound state can suppress the RAS2^{val19} phenotype.[64] Similar observations have been made for unprocessed H-*ras* in the cells of more complex eukaryotes.[110] The surprising and unexplained inference is that cytoplasmic RAS interferes with the function of wild-type RAS, perhaps by sequestering an important mediator of its actions to the wrong cellular sub compartment.

The effects of RAS upon adenylyl cyclase are not sufficient to explain all its effects upon *S. cerevisiae*. Several different observations suggest that RAS has other functions. First, haploid spores with disruptions of the adenylyl cyclase or cAMP-dependent protein kinase genes are not uniformly incapable of germination or vegetative growth, while haploid spores with disruptions in both *ras* genes are completely incapable of vegetative growth.[93,94] Second, multicopy genes that efficiently suppress the growth defects of cells lacking adenylyl cyclase do not as effectively suppress the growth defects of cells lacking RAS.[93] Finally, certain multicopy suppressors of the phenotypes associated with the loss of the C terminus of CAP can suppress these phenotypes in cells lacking all of CAP only if such cells also contain an activated RAS.[38] Cells with an activated cAMP-signaling pathway are not responsive to these suppressors. The importance of this conclusion is two-fold. First, an alternative function of RAS in *S. cerevisiae* may be found that will represent a function conserved in evolution. Second, RAS may have multiple independent functions in the same cell. This prospect must be given serious consideration in attempts to understand the role of RAS in higher cells.

S. cerevisiae may also be used to hunt for mammalian effectors of RAS in a more direct way: cDNA expression screening. As mentioned above, fragments of adenylyl cyclase can inhibit RAS function. It is reasonable to expect, therefore, that expression of mammalian cDNAs encoding RAS effectors might inhibit RAS function in yeast, and form the basis of a genetic screen. Such screens have, in fact, yielded new mammalian genes.[15] The function of these genes is still under study.

VII. EFFECTORS OF RAS IN *S. pombe*

To understand *ras1* function in *S. pombe* it is necessary to understand the sexual life cycle of the fission yeast. This subject has been reviewed elsewhere,[24] and a brief synopsis was given in an earlier section. About a dozen sterile (*ste*) loci have been identified that when mutated render *S. pombe* cells unable to mate. *ste6* is one of these, and, as was discussed previously, is a homolog of *S. cerevisiae CDC25*. *ras1* is *ste5*.[67] *byr1* and *byr2*, discussed briefly above, are *ste1* and *ste8*, respectively.[67,104]

S. pombe cells only mate when starved, and then mate with cells of the opposite mating type in response to a soluble mating pheromone. The presence of the activated *ras1*^{val17} allele renders cells hypersensitive to mating factor, and exposure to pheromone under conditions of starvation induce exaggerated morphological response and agglutinability.[34] Despite this, or rather because of it, *ras1*^{val17} cells conjugate poorly. The hypersensitivity of *ras1*^{val17} cells to mating factor is dependent upon starvation. From these results we can conclude that activation of *ras1* does not abrogate the response of the cell to two essential external stimuli for mating: starvation and mating pheromone.

The relationship between *ras1* and the two well-defined signaling pathways, mating pheromone detection and starvation, can be inferred from the existing literature. Starvation appears to induce many of the components required for pheromone detection, including ste6, the mating pheromones and their receptors, and ste11, which is directly required for the production of these inducible components.[24] Starvation signaling may be mediated through adenylyl cyclase. Cells lacking adenylyl cyclase will mate even in the absence of starvation.[49] However, cells lacking adenylyl cyclase do not display the hypersensitivity to pheromones seen in *ras1*^{val17} cells. On the other hand, cells with disruptions in proteases postulated to degrade mating factor do show the *ras1*^{val17} phenotype.[44] Moreover, cells with activated alleles of *gpa1*, which encodes the G_α subunit that mediates the recognition of the pheromone by its receptor, also display this phenotype.[73] Activation of *gpa1* cannot bypass the loss of ras1; and activation of *ras1* cannot bypass the need for mating pheromone.[73] This set of relationships indicates that ras1 and the pheromone receptor apparatus jointly control sexual activity. We will return to the important biochemical implications of this shortly.

Candidate effectors for ras1 can be found by selecting for genes on multicopy plasmids that can suppress the defects of *ras1*$^-$ cells. Two such genes, *byr1* and *byr2*, have been described in the literature.[66,104] Each gene, when on a multicopy plasmid, can suppress the sporulation defects of *ras1*$^-$ cells. Disruption of *byr1* or *byr2* leads to absolute defects in mating and conjunction. *byr1* on a multicopy plasmid can suppress the sporulation defects of *byr2*$^-$ cells, but not the reverse.[104] These relationships suggest that ras1 acts, directly or indirectly, upon byr2 which acts, directly or indirectly, upon byr1. This is by no means proven, and many other models are compatible with the data.

byr1 and *byr2* each are predicted to encode protein kinases. *byr1* is most homologous to the product of the *S. cerevisiae* STE7 gene, and byr2 is most homologous to the product of the *S. cerevisiae* STE11 gene. Both STE7 and STE11 are kinases that function on the mating pheromone pathway in *S. cerevisiae*.[55] This suggests that the pheromone-dependent mating pathways of these two yeasts are at least partially conserved, although it has not yet been shown that byr1 and byr2 are regulated by the mating pheromone apparatus in *S. pombe*.

It is not known if the byr proteins represent direct targets for ras1 interaction. In any event, there must be other functions of ras1 in *S. pombe* cells, since *ras1*$^-$ cells are round rather than rod shaped. Neither *byr* gene in multiple copy suppresses this phenotype, nor does disruption of either *byr* gene cause this phenotype.[67,104]

VIII. TWO SPECULATIONS ABOUT RAS

I bring this review near to a close with two speculations. First, analogies between *S. pombe*, *S. cerevisiae*, and mammalian cells suggest a possible conservation of the biochemical function of RAS in *S. pombe* and mammals. In both *S. pombe* and *S. cerevisiae* there are other kinases required for sexual function: in *S. pombe*, spk1[96] and in *S. cerevisiae*, FUS3 and KSS1.[16,25] These kinases are most closely related to the ERK/MAP protein kinases that have been identified in mammalian cells as response elements in extracellular signaling mediated through receptor protein tyrosine kinases.[6,7,12,13] In *S. cerevisiae*, evidence suggests that the FUS3 and KSS1 kinases act downstream of STE7 and STE11.[36,85] Therefore, let us assume that this same relationship will hold between spk1 and byr1. Let us further make a larger assumption: the cascade of kinases common in *S. pombe* and *S. cerevisiae* that regulate ERK/MAP-like protein kinases has a parallel in mammalian cells, with an ERK/MAP protein kinase Kinase and an ERK/MAP protein kinase kinase Kinase. In mammalian cells, the phosphorylation and activation of the ERK/MAP protein kinases is both RAS-dependent and growth-factor dependent.[54,91,107] A set of homologies between the yeast *S. pombe* and mammalian cells may then be readily imagined in which a cascade of kinases under the dual control of RAS and an extracellular factor describe signal transduction pathways in both organisms. This hypothesis naturally suggests that the biochemical function of *S. pombe* ras1 and mammalian oncogenic RAS are closely related, if not identical. A further implication of this model is that the ERK/MAP protein kinases will be regulated by kinases homologous to STE7 and byr1, which will in turn be regulated by kinases homologous to STE11 and byr2. Obviously, this model is constructed from fragmentary data, but its far reaching predictions are testable.

The second speculation is a generalization about the physiological role of RAS and GAP-like molecules in the regulation of signaling by extracellular factors. We have reasonably inferred that one role of the *S. cerevisiae* IRA proteins is the feedback regulation of RAS activity. In the absence of these proteins, cells respond to glucose feeding by an unregulated elevation of cAMP. In the absence of cAMP-dependent protein kinase activity the same end results. We can surmise a similar role of sar1 in *S. pombe*. In the absence of this regulator of ras1, cells become hypersensitive to mating factor and mating becomes very inefficient. The same result ensues if the ability of the cell to degrade mating pheromone is blocked. With regard to pheromone hypersensitivity, we can infer that either it is inefficient for a cell to commit prematurely to conjugation, or that an exaggerated response itself blocks conjugation, or both. In any case, it is vital for the cell to be able to desensitize and "tune" its response to the extracellular factor. One can immediately recognize in this a general problem in cellular signal recognition. From the study of RAS in metazoans it is apparent that, as in *S. pombe*, many signaling pathways are codependent upon RAS.[21,27,52,79,83,84] Even in *S. cerevisiae* there is suggestive evidence that another

G protein is involved in the stimulation of cAMP by glucose feeding.[69] I propose that, in general, the codependence of a signaling pathway upon RAS serves at least one purpose, enabling the cell to tune its signal response through a feedback mechanism mediated by GAP-like molecules.

IX. CONCLUSION

From the study of yeast RAS we have learned much about its regulation: the existence of exchange proteins, the role of GTPase-activating proteins, and the importance of the latter in feedback inhibition. We will undoubtedly uncover other components involved in regulating RAS, and many of these and their interactions may be conserved. We have also learned much about RAS-effector functions: RAS may have multiple functions within the same cell, RAS is active in its GTP-bound state, and interesting and conserved auxiliary proteins have been defined.

We stand to learn much more about RAS-effector functions from yeasts, including a complete biochemical picture of how RAS activates adenylyl cyclase, a definition of its alternate function in *S. cerevisiae*, and a definition of its primary function in *S. pombe*. There is reason to believe that RAS function will be conserved between *S. pombe* and vertebrates. The study of the *S. pombe* RAS-dependent sexual differentiation kinase cascade should enable the discovery of a similar cascade involved in growth and developmental signaling in mammalian cells. Finally, both fission and budding yeasts can be used in genetic screens to identify new components of mammalian signal transduction pathways.

ACKNOWLEDGMENTS

I wish to thank Stevan Marcus, Scott Powers, Linda Van Aelst, and many others for useful discussions and helpful guidance to the literature and also Patricia Bird for her help in preparing this review. This work was supported by the National Cancer Institute of the National Institutes of Health and the American Cancer Society. Michael Wigler is an American Cancer Society Research Professor.

REFERENCES

1. **Adari, H., Lowy, D., Willumsen, B., Der, C., and McCormick, F.,** *Science,* 240, 518, 1988.
2. **Ballester, R., Marchuk, D., Boguski, M., Saulino, A., Letcher, R., Wigler, M., and Collins, F.,** *Cell,* 63, 851, 1990.
3. **Ballester, R., Michaeli, T., Ferguson, K., Xu, H.-P., McCormick, F., and Wigler, M.,** *Cell,* 59, 681, 1989.

4. Beckner, S., Hattori, S., and Shih, T., *Nature*, 317, 71, 1985.
5. Birchmeier, C., Broek, D., and Wigler, M., *Cell*, 43, 615-621.
6. Boulton, T., Gregory, J., Jong, S.-M., Wang, L.-H., Ellis, J., and Cobb, M., *J. Biol. Chem.*, 265, 2713, 1990a.
7. Boulton, T., Nye, S., Robbins, D., Ip, N., Radiziejewska, E., Morgenbesser, S., DePinho, R., Panayotatos, N., Cobb, M., and Yancopoulos, G., *Cell*, 65, 663, 1990.
8. Boy-Marcotte, E., Damak, P., Camonis, J., Garreau, H., and Jacquet, M., *Gene*, 77, 21, 1989.
9. Broek, D., Samiy, N., Fasano, O., Fujiyama, A., Tamanoi, F., Northup, J., and Wigler, M., *Cell*, 41, 763, 1985.
10. Broek, D., Toda, T., Michaeli, T., Levin, L., Birchmeier, C., Zoller, M., Powers, S., and Wigler, M., *Cell*, 48, 789, 1987.
11. Chant, J., Corrado, K., Pringle, J., and Herskowtiz, I., *Cell*, 65, 1213, 1991.
12. Chao, M., *Cell*, 68, 995, 1992.
13. Cobb, M., Boulton, T., and Robbins, D., *Cell Reg.*, 2, 965, 1991.
14. Colicelli, J., Field, J., Ballester, R., Chester, N., Young, D., and Wigler, M., *Mol. Cell. Biol.*, 10, 2539, 1990.
15. Colicelli, J., Nicolette, C., Birschmeier, C., Rodgers, L., Riggs, M., and Wigler, M., *Proc. Natl. Acad. Sci. U.S.A.*, 88, 2913, 1991.
16. Courchesne, W., Kunisawa, R., and Thorner, J., *Cell*, 58, 1107, 1989.
17. Damak, D., Boy-Marcotte, E., Le-Roscouet, D., Guilbaud, R., and Jacquet, M., *Mol. Cell. Biol.*, 11, 202, 1991.
18. DeFeo Jones, D., Scolnick, E., Koller, R., and Dhar, R., *Nature*, 306, 707, 1983.
19. DeFeo Jones, D., Tatchell, K., Robinson, L., Sigal, I., Vass, W., Lowy, D., and Scolnick, E., *Science*, 228, 179, 1985.
20. Der, C. and Cox, A., *Cancer Cells*, 3, 331, 1991.
21. Deshpande, A. and Kung, H., *Mol. Cell. Biol.*, 7, 1285, 1987.
22. DeVendittis, E., Vitelli, A., Zahn, R., and Fasano, O., *EMBO J.*, 5, 3657, 1986.
23. Dhar, R., Nieto, A., Koller, R., DeFeo-Jones, D., and Scolnick, E., *Nucleic Acids Res.*, 12, 3611, 1984.
24. Egel, R., Nielsen, O., and Weilguny, D., *Trends Genet.*, 6, 369, 1990.
25. Elion, E., Grisafi, P., and Fink, G., *Cell*, 60, 649, 1990.
26. Ellis, R., DeFeo, D., Shih, T., Gonda, M., Young, H., Tsuchida, N., Lowy, D., and Scolnick, E., *Nature*, 292, 506, 1981.
27. Feig, L. and Cooper, G., *Mol. Cell. Biol.*, 8, 3235, 1988.
28. Field, J., Broek, D., Kataoka, T., and Wigler, M., *Mol. Cell. Biol.*, 7, 2128, 1987.
29. Field, J., Nikawa, J., Broek, D., MacDonald, B., Rodgers, L., Wilson, I., Lerner, R., and Wigler, M., *Mol. Cell. Biol.*, 8, 2159, 1988.
30. Field, J., Vojtek, A., Ballester, R., Bolger, G., Colicelli, J., Ferguson, K., Gerst, J., Kataoka, T., Michaeli, T., Powers, S. et al., *Cell*, 61, 319, 1990.
31. Field, J., Xu, H.-P., Michaeli, T., Ballester, R., Sass, P., Wigler, M., and Colicelli, J., *Science*, 247, 464, 1990.
32. Fukui, Y. and Kaziro, Y., *EMBO J.*, 4, 687, 1985.
33. Fukui, Y. and Yamamoto, M., *Mol. Gen. Genet.*, 215, 26, 1988.
34. Fukui, Y., Kosasa, T., Kaziro, Y., Takeda, T., and Yamamoto, M., *Cell*, 44, 329, 1986.
35. Fukui, Y., Miyake, S., Satoh, M., and Yamamoto, M., *Mol. Cell. Biol.*, 9, 5617, 1989.
36. Gartner, A., Nasmyth, K., and Ammerer, G., *Genes Dev.*, 6, 1280, 1992.
37. Gerst, J., Ferguson, K., Vojtek, A., Wigler, M., and Field, J., *Mol. Cell. Biol.*, 11, 1248, 1991.
38. Gerst, J., Rodgers, L., Riggs, M., and Wigler, M., *Proc. Natl. Acad. Sci. U.S.A.*, 89, 4338, 1992.
39. Gibbs, J., Schaber, M., Allard, W., Sigal, I., and Scolnick, E., *Proc. Natl. Acad. Sci. U.S.A.*, 85, 5026, 1988.

40. Han, M. and Sternberg, P., *Cell*, 63, 921, 1990.
41. Hart, M., Eva, A., Evans, T., Aaronson, S., and Cerione, R., *Nature*, 354, 311, 1991.
42. Hixson, C. and Krebs, E., *J. Biol. Chem.*, 225, 2137, 1980.
43. Hughes, D., Fukui, Y., and Yamamoto, M., *Nature*, 344, 355, 1990.
44. Imai, Y. and Yamamoto, M., *Mol. Cell. Biol.*, 12, 1827, 1992.
45. Jones, S., Vignais, M.-L., and Broach, J., *Mol. Cell. Biol.*, 11, 2641, 1991.
46. Kataoka, T., Broek, D., and Wigler, M., *Cell*, 43, 493, 1985.
47. Kataoka, T., Powers, S., McGill, C., Fasano, O., Strathern, J., Broach, J., and Wigler, M., *Cell*, 37, 437, 1985.
48. Kataoka, T., Powers, S., Cameron, S., Fasano, O., Goldfarb, M., Broach, H., and Wigler, M., *Cell*, 40, 19, 1985.
49. Kawamukai, M., Ferguson, K., Wigler, M., and Young, D., *Cell Reg.*, 2, 155, 1991.
50. Kawamukai, M., Gerst, J., Field, J., Riggs, M., Rodgers, L., Wigler, M., and Young, D., *Mol. Biol. Cell*, 3, 167, 1992.
51. Kim, J.-H. and Powers, S., *Mol. Cell. Biol.*, 11, 3894, 1991.
52. Korn, L., Siebel, C., McCormick, F., and Roth, R., *Science*, 236, 840, 1987.
53. Krupinski, J., Coussen, F., Bakalyar, H., Tang, W.-J., Feinstein, G., Orth, K., Slaughter, C., Reed, R., and Gilman, A., *Science*, 244, 1558, 1989.
54. Leevers, S. and Marshall, C., *EMBO J.*, 11, 569, 1992.
55. Marsh, L., *Annu. Rev. Cell Biol.*, 7, 699, 1991.
56. Marshall, M., Gibbs, J., Scolnick, E., and Sigal, I., *Mol. Cell. Biol.*, 8, 52, 1988.
57. Matsui, Y., Kikuchi, A., Araki, S., Hata, Y., Kondo, J., Teranishi, Y., and Takai, Y., *Mol. Cell. Biol.*, 10, 4116, 1990.
58. Matsumoto, K., Uno, I., and Ishikawa, T., *Exp. Cell Res.*, 146, 151, 1983.
59. Matsumoto, K., Uno, I., and Ishikawa, T., *J. Bact.*, 157, 277, 1984.
60. Matsumoto, K., Uno, I., and Ishikawa, T., *Yeast*, 1, 15, 1985.
61. Matsumoto, K., Uno, I., Oshima, Y., Ishikawa, T., *Proc. Natl. Acad. Sci. U.S.A.*, 79, 2355, 1982.
62. Mbonyi, K., Beullens, M., Detremerie, K., Geerts, L., and Thevelein, J., *Mol. Cell. Biol.*, 8, 3051, 1988.
63. Mbonyi, K., Van Aelst, L., Arguelles, J., Jans, A., and Thevelein, J., *Mol. Cell. Biol.*, 10, 4518, 1990.
64. Michaeli, T., Field, J., Ballester, R., O'Neill, K., and Wigler, M., *EMBO J.*, 8, 3039, 1989.
65. Munder, T. and Furst, P., *Mol. Cell. Biol.*, 12, 2091, 1992.
66. Nadin-Davis, S. and Nasim, A., *EMBO J.*, 7, 985, 1988.
67. Nadin-Davis, S. and Nasim, A., *Mol. Cell. Biol.*, 10, 549, 1990.
68. Nadin-Davis, S., Nasim, A., and Beach, D., *EMBO J.*, 5, 2963, 1986.
69. Nakafuku, M., Obara, T., Kaibuchi, K., Miyajima, I., Miyajima, A., Itoh, H., Nakamura, S., Arai, K.-I., Matsumoto, K., and Kaziro, Y., *Proc. Natl. Acad. Sci. U.S.A.*, 85, 1374, 1988.
70. Neuman-Silberberg, F., Schejter, E., Hoffmann, F., and Shilo, B.-Z., *Cell*, 37, 1027, 1984.
71. Nikawa, J., Cameron, S., Toda, T., Ferguson, K., and Wigler, M., *Genes Dev.*, 1, 931, 1987.
72. Nikawa, J., Sass, P., and Wigler, M., *Mol. Cell. Biol.*, 7, 3629, 1987.
73. Obara, T., Nakafuku, M., Yamamoto, M., and Kaziro, Y., *Proc. Natl. Acad. Sci. U.S.A.*, 88, 5877, 1991.
74. Powers, S., Gonzales, E., Christensen, T., Cubert, J., and Broek, D., *Cell*, 65, 1225, 1991.
75. Powers, S., Kataoka, T., Fasano, O., Goldfarb, M., Strathern, J., Broach, J., and Wigler, M., *Cell*, 36, 607, 1984.
76. Powers, S., Michaelis, S., Broek, D., Santa Anna, S., Field, J., Herskowitz, I., and Wigler, M., *Cell*, 47, 413, 1986.

77. Powers, S., O'Neill, K., and Wigler, M., *Mol. Cell. Biol.*, 9, 390, 1989.
78. Reymond, C., Gomer, R., Mehdy, M., and Firtel, R., *Cell*, 39, 141, 1984.
79. Rubin, G., *Trends Genet.*, 7, 372, 1991.
80. Sass, P., Field, J., Nikawa, J., Toda, T., and Wigler, M., *Proc. Natl. Acad. Sci. U.S.A.*, 83, 9303, 1986.
81. Shou, C., Farnsworth, C., Neel, B., and Feig, L., *Nature*, 358, 51, 1992.
82. Simon, M., Bowtell, D., Dodson, G., Laverty, T., and Rubin, G., *Cell*, 67, 701, 1991.
83. Smith, M., DeGudicibus, S., and Stacey, D., *Nature*, 320, 540, 1986.
84. Sternberg, P. and Horvitz, R., *Trends Genet.*, 7, 366, 1991.
85. Stevenson, B., Rhodes, N., Errede, B., and Sprague, G., Jr., *Genes Dev.*, 6, 1293, 1992.
86. Tamanoi, F., Samiy, N., Rao, M., and Walsh, M., *Cancer Cells III: Growth Factors and Transformation*, Feramisco, J., Ozanne, B., and Stiles, L., Eds., Cold Spring Harbor Laboratory, Cold Spring Harbor, NY, 251, 1985.
87. Tanaka, K., Lin, B., Wood, D., and Tamanoi, *Proc. Natl. Acad. Sci. U.S.A.*, 88, 468, 1991.
88. Tanaka, K., Matsumoto, K., and Toh-E, A., *Mol. Cell. Biol.*, 9, 757, 1989.
89. Tanaka, K., Nakafuku, M., Satoh, T., Marshall, M., Gibbs, J., Matsumoto, K., Kaziro, Y., and Toh-E, A., *Cell*, 60, 803, 1990.
90. Tanaka, K., Nakafuku, M., Tamanoi, F., Kaziro, Y., Matsumoto, K., and Toh-E, A., *Mol. Cell. Biol.*, 10, 4303, 1990.
91. Thevelein, B. and Beullens, M., *J. Gen. Microbiol.*, 131, 3199, 1985.
92. Thomas, S., DeMarco, M., D'Arcangelo, G., Halegous, S., and Brugge, J., *Cell*, 68, 1031, 1992.
93. Toda, T., Broek, D., Field, J., Michaeli, T., Cameron, S., Nikawa, J., Sass, P., Birchmeier, C., Powers, S., and Wigler, M., *Oncogene and Cancer*, Aaronson, S. A. et al., Eds., Japan Science Society Press, Tokyo, 1987, 253.
94. Toda, T., Cameron, S., Sass, P., and Wigler, M., *Cell*, 50, 277, 1987.
95. Toda, T., Cameron, S., Sass, P., Zoller, M., Scott, J., McMullen, B., Murwitz, M., Krebs, E., and Wigler, M., *Mol. Cell. Biol.*, 7, 1371, 1987.
96. Toda, T., Shimanuki, M., and Yanagida, M., *Genes Dev.*, 5, 60, 1991.
97. Toda, T., Uno, I., Ishikawa, T., Powers, S., Kataoka, T., Broek, D., Cameron, S., Broach, J., Matsumoto, K., and Wigler, M., *Cell*, 40, 27, 1985.
98. Trahey, M. and McCormick, F., *Science*, 238, 542, 1987.
99. Trahey, M., Wong, G., Halenbeck, R., Rubinfeld, B., Martin, G. A., Ladner, M., Long, M., Crosier, W., Watt, K., Koths, K., and McCormick, F., *Science*, 242, 1697, 1988.
100. Uno, I., Mitsuzawa, H., Matsumoto, K., Tanaka, K., Oshima, T., and Ishikawa, T., *Proc. Natl. Acad. Sci. U.S.A.*, 82, 7855, 1985.
101. Van Aelst, L., Boy-Marcotte, E., Camonis, J., Thevelein, J., and Jacquet, M., *Eur. J. Biochem.*, 193, 675, 1990.
102. Vojtek, A., Haarer, B., Field, J., Gerst, J., Pollard, T., Brown, S., and Wigler, M., *Cell*, 66, 497, 1991.
103. Wang, Y., Boguski, M., Riggs, M., Rodgers, L., and Wigler, M., *Cell Reg.*, 2, 453, 1991.
104. Wang, Y., Xu, H.-P., Riggs, M., Rodgers, L., and Wigler, M., *Mol. Cell. Biol.*, 11, 3554, 1991.
105. Wigler, M., *Nature*, 346, 696, 1990.
106. Wigler, M., Field, J., Powers, S., Broek, D., Toda, T., Camerson, S., Nikawa, J., Michaeli, T., Colicelli, J., and Ferguson, K., *Cold Spring Harb. Symp. Quant. Biol.*, 53, 649, 1988.
107. Wood, K., Sarnecki, C., Roberts, T., and Blenis, J., *Cell*, 68, 1041, 1992.
108. Xu, G., O'Connell, P., Viskochil, D., Cawthon, R., Robertson, M., Culver, M., Dunn, D., Stevens, J., Gesteland, R., White, R., and Weiss, R., *Cell*, 62, 599, 1990.

109. **Young, D., Riggs, M., Field, J., Vojtek, A., Broek, D., and Wigler, M.,** *Proc. Natl. Acad. Sci. U.S.A.,* 86, 7989, 1989.
110. **Yu, C., Tsai, M., and Stacey, D.,** *Mol. Cell. Biol.,* 10, 6683, 1990.
111. **Zoller, M., Yonemoto, W., Taylor, S., and Johnson, K.,** *Gene,* 99, 171, 1991.

Chapter 8

REGULATION AND FUNCTION OF TWO DEVELOPMENTALLY REGULATED *ras* GENES IN *Dictyostelium*

R. Keith Esch, Gerald Weeks, and Richard A. Firtel

TABLE OF CONTENTS

0-8493-5214-2/93/$0.00 + $.50
© 1993 by CRC Press, Inc.

I. INTRODUCTION

Genes encoding guanine nucleotide-binding *ras* proteins have now been identified in all eukaryotic organisms examined, ranging from yeast to humans. These proteins display a high degree of evolutionary conservation, implicating *ras* in a fundamentally essential cellular role in all eukaryotes. The downstream effect(s) of *ras* in metazoans remains unclear, as does the extent of functional redundancy in organisms harboring more than one *ras* protein. The cellular slime mold *Dictyostelium discoideum* is one such organism. Two very similar *ras* genes with distinct patterns of expression have been identified, Dd *ras* and Dd *ras*G.

Dictyostelium discoideum is a relatively simple eukaryote having an unusual life cycle that renders it attractive for studies on proliferation and differentiation in general and the involvement of *ras* genes with these processes in particular. The processes of cell division and cell differentiation are largely distinct during the life cycle of *Dictyostelium*.[1-3] Vegetative *D. discoideum* cells exist as independent amoebae as long as ample nutrients are available. Upon nutrient deprivation, however, growth and cell division cease and the previously free living cells initiate an interactive developmental program. Approximately 3 h after starvation, a cell initiates a cAMP-mediated response/relay cascade by secreting cAMP into its surroundings. Nearby amoebae detect the cAMP signal via cell surface cAMP receptors that they stimulate adenylate cyclase through a G-protein-mediated process. The cAMP, thus produced, is then released, causing the signal to be relayed. In addition to stimulation of adenylate cyclase, cAMP binding to the receptor results in chemotaxis up the cAMP gradient — a process mediated through the activation of guanylate cyclase and phospholipase C. At ~6 min intervals, pulses of extracellular cAMP are detected and relayed and thus, a directed signaling system is established. As with many receptor-mediated processes, the transient, pulsatile activation of these signaling responses is controlled by sequential activation and adaptation of these pathways. Removal of cAMP by extracellular phosphodiesterase permits the pathways to deadapt in preparation for the next cAMP signal. As responding cells chemotax toward the source of cAMP pulses, a mound-shaped multicellular aggregate is formed. The aggregation process begins at ~5 h and typically requires ~2 h for completion. By ~12 h after starvation, a single tip, which continues to act as a cAMP oscillator and functions as an organizing center for morphogenesis, can clearly be distinguished on the mound. During this stage, specialization of cell function, which initiates late in aggregation, continues, and the initial spatial pattern of two distinct cell types (prestalk and prespore) is established (Figure 1). The tipped mound gradually extends vertically and then falls to the substratum forming a migrating slug or pseudoplasmodium by ~16 h of development. In the slug, the prestalk and prespore cells are largely segregated: the anterior 15% of the slug is composed of prestalk cells and the posterior 85% is predominantly prespore cells with the posterior-most 3 to 5% being prestalk cells that will

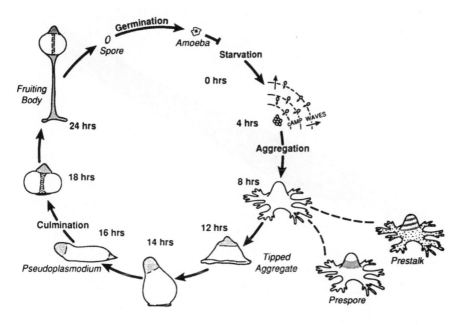

FIGURE 1. *Dictyostelium discoideum* developmental cycle. Drawings depict morphology of multicellular structures at the indicated times after starvation. Shading represents prestalk and stalk regions. In the late aggregate stage (8 h), the time when prestalk and prespore cells first appear, separate cartoons are used to depict the spatial distribution of the prestalk cells and the prespore cells as determined using Dd *ras/lacZ*[9] and *Sp60/lacZ*,[33] respectively. (*SP60* encodes a prespore coat protein). At later stages, the shaded areas depict the spatial patterns of the Dd *ras* prestalk gene. The unshaded areas express *SP60*.

eventually form the basal disk of the fruiting body (Figure 2). When a suitable environment is reached, the migrating slug reestablishes firm contact with the substratum and undergoes culmination. During this process, prestalk cells lift the prespore cell mass by forming a vertical stalk through the spore mass and extending beneath it. Terminal differentiation occurs for both cell types: the prestalk cells vacuolize to form the stalk and prespore cells are encapsulated by a spore coat to become spores. By ~25 h after starvation, the process is complete and a mature fruiting body results. The life cycle starts anew when the spore mass or sorocarp erupts to release spores which, after germination, give rise to vegetative amebae.

II. *Dictyostelium ras* SEQUENCE ANALYSES

Predicted amino acid sequences of both *Dictyostelium ras* proteins, Dd *ras*, and Dd *rasG*, as well as human Ha-*ras*-1, human *rap*, and *Dictyostelium rap* proteins are presented in Figure 3. The *Dictyostelium ras* proteins are 82%

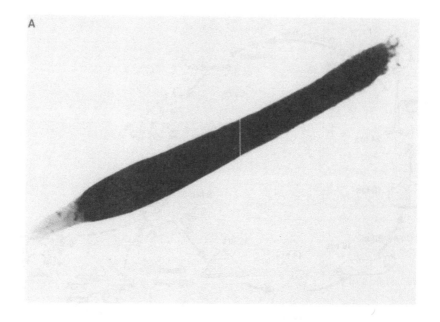

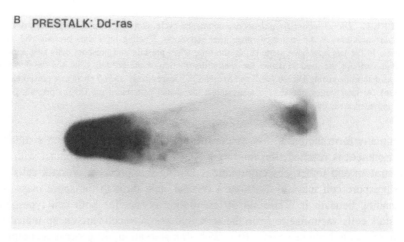

FIGURE 2. Spatial expression of prestalk and prespore genes at slug stage. Cells transformed with *SP50/lacZ* (A) and Dd *ras/lacZ* (B) were allowed to develop to slug stage and were then fixed and stained for β-galactosidase activity.[9,33]

conserved relative to one another and each is between 60 and 69% conserved relative to those of human, *Drosophila*, and yeast. Sequence similarity is extremely high in the amino terminal 81 amino acids, thus, confirming the strong evolutionary conservation of this region. For these first 81 amino acids, Dd *ras* and Dd *rasG* are 98% conserved relative to one another and each is 93% conserved relative to *ras* proteins of the other species.

The highly conserved, functionally defined domains identified in other *ras* proteins are also present in both Dd *ras* and Dd *ras*G. The four small domains implicated in guanine nucleotide-binding/GTPase activity are found in the *Dictyostelium ras* proteins as is the region of *ras* that interacts with its downstream target. In mammals, this effector domain coincides with a site of interaction with the GTPase-activating protein (GAP), suggesting that the *Dictyostelium ras* proteins may interact with a GAP-like molecule.[4] The carboxyl terminal CAAX box required for posttranslational farnesylation, localization to the cell membrane, and hence, proper function, is conserved in the *Dictyostelium ras* proteins. Consistent with this, *ras* protein in *Dictyostelium* has been shown to be predominantly membrane-associated in an acetylated form.[5] However, the additional cysteine residue (at position 181 or 184), which for mammalian N-*ras* and Ha-*ras* is palmitoylated, is not present in either Dd *ras* or Dd *ras*G. In this respect, the *Dictyostelium ras* proteins are more similar to Ki-*ras*, which has a region rich in basic amino acids substituted for the additional modified cysteine.

III. EXPRESSION OF *ras* GENES DURING *Dictyostelium* GROWTH AND DEVELOPMENT

The Dd *ras* and Dd *ras*G genes are differentially expressed during *Dictyostelium* growth and development. Both genes express a 1.2-kb mRNA during vegetative growth; however, the Dd *ras*G transcript is present at much higher levels than is the Dd *ras* transcript.[6,7] In wild-type parental strains, Dd *ras*G mRNA levels increase approximately two-fold during the first 2 to 3 h of development and, subsequently, decline to very low levels by 6 h (aggregate stage). The Dd *ras* 1.2 kb mRNA is not expressed during early development but is reinduced during late aggregation followed slightly later by two smaller messages (~1 kb), which are expressed from distal promoters and are indistinguishable by RNA blot analysis.[9,10] Both the 1.2 and 1-kb transcript levels increase through the tipped aggregate and slug stages (8 to 18 h). All three then decline late during culmination as the fruiting body is formed (20 to 24 h).[7]

IV. SYNTHESIS OF *ras* PROTEIN

Both polyclonal and monoclonal *ras*-specific antibodies have been used to detect *ras* protein in extracts from growing and developing *Dictyostelium*.[7,11] Immunoprecipitation studies using the common *ras*-specific monoclonal antibody (Y13-259) show that *ras* protein synthesis is maximal during vegetative growth and the first 2 h after starvation and then continuously drops throughout the remainder of early development. Synthesis then increases during slug stage and finally drops to negligible levels by culmination. Major and minor *ras* protein species have been observed and have been designated p23 [Dd ras] and p24[Dd ras] respectively. In growing cells, the p24 [Dd ras] protein is a very small fraction of the total material immunoprecipitated by the *ras*-specific monoclonal anti-

```
              130                  ● ● ●      ○          150
DdrasG   G E G Q D L A K S  F G S P   F L E T S  A K I R V N V E E A F Y S
Ddras    N - - E - - - - -  G - N C   - M - - S  - I - - I - - - - I - -
Ha-ras1  R Q A - - R - Y -  I H - Y   I - - T -  Q G - - D - T
Ddrap1   E Q - E E - R K -  D C Y -   A - N K -  - Q I - N
rap1A    E Q - - N - R Q W  C N C A   - S - S K  I - N - H - D

              160               170              180            189
DdrasG   L V R E I R K D L K G D S K P E K G K K K R P L K A C T L L
Ddras    - - - - - - - - E - - - - - Q S S G - A Q - K K Q - L I -
Ha-ras1  - - - - - - Q H K L R K L N - P D E S G P G C M S C K - V - S
Ddrap1   - - Q - - N R K N P V G P P S K A K S - - K P K K K - A - -
rap1A    - - Q - - N R K T P V E K K K P K K K S - - - - - - - L - -
```

FIGURE 3. Comparison of the derived amino acid sequences of the *Dictyostelium ras* and *rap* proteins with the human *ras* (Ha-*ras*) and *rap* proteins (*rap1A*). Amino acids are identified by the single-letter code. The Dd *rasG* protein sequence is numbered and the other sequences are aligned relative to it with gaps inserted to maximize homology. The circles in the region 137 to 145 indicate amino acids conserved between Dd *rasG* and human *ras* proteins but diverged in the Dd *ras* protein. Solid symbols indicate identity, while the open circle indicates chemical similarity.

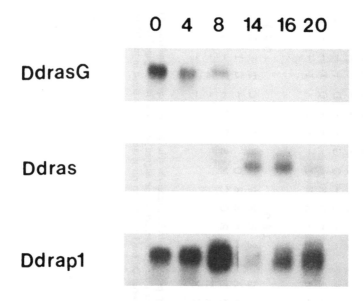

FIGURE 4. Developmental expression of Dd *ras*G, Dd *ras*, and Dd *rap1*. At the indicated times (hours) after starvation, total RNA was isolated from developing cells. Subsequent isolation of polyadenylated RNA was performed by oligo (dT) cellulose chromatography. For each time point, 5 μg of poly(A) was size fractionated on 1.5% formaldehyde-agarose gels. Resulting Northern blots were probed with Dd *ras*G, Dd *ras*, and Dd *rap1* cDNAs as indicated.

body (≤5%); however, in cells of the pseudoplasmodial stage, the fraction of p24$^{Dd\ ras}$ is significantly greater (~25%).[11] Since the Y13-259 antibody recognizes an epitope of *ras* proteins that is completely conserved in both *Dictyostelium ras* proteins, it would be expected to immunoprecipitate both Dd *ras* and Dd *ras*G gene products. The pattern of *ras* protein synthesis observed through development is consistent with the combined expression of Dd *ras* and Dd *ras*G mRNAs.[12] Evidence indicates that p24$^{Dd\ ras}$ is not a precursor to p23$^{Dd\ ras}$, but the nature of the relationship between the two proteins and specific correspondence to Dd *ras* and Dd *ras*G remains unknown.[13]

V. SPATIAL AND CELL-TYPE EXPRESSION OF Dd *ras*

Since Dd *ras* is expressed during multicellular stages as differentiation occurs, cell-type and spatially restricted expression have been determined. *In situ* staining for β-galactosidase in transformants expressing *lacZ* driven from the Dd *ras* promoter (Dd *ras*/*lacZ*) has been used to monitor Dd *ras* expression through development (Figure 5). At the aggregation stage, a fairly random speckled pattern of β-gal-staining cells is seen. The staining cells, which comprise ~10% of the population, are seen both within the central region of the aggregate, which will eventually give rise predominantly to prespore cells, and at the

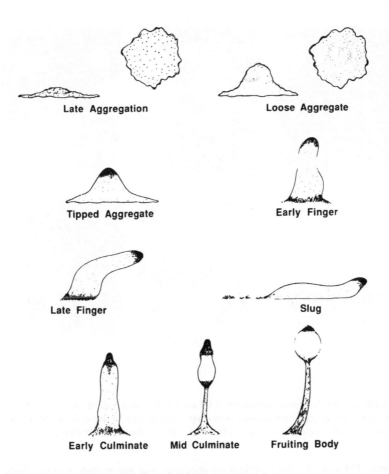

FIGURE 5. Dd *ras* spatial expression summary cartoon. Drawings were made from both published and unpublished results of β-gal expression studies from developmental stages indicated. For the late aggregation and loose aggregate stages, both top (right) and side (left) views are presented. (From Esch, R. K. and Firtel, R. A., *Genes Dev.,* 5, 9, 1991. With permission.)

periphery where cells are still migrating inward. As a tip forms on the mound, Dd *ras*-expressing cells appear to sort to this region, possibly in response to cAMP signaling, and become localized in this region. Staining of mounds during tip formation shows a spiral pattern of Dd *ras/lacZ*-expressing cells, suggesting the movement of these cells toward the tip is in response to proposed spiral cAMP waves emitted from the tip. After the tipped mound extends vertically, Dd *ras*-expressing cells are observed primarily at the anterior tip but also at the base of the structure. In addition, a small number of Dd *ras*-expressing cells are found scattered within the prespore domain and probably represent anterior-like cells.[14] At intervals during slug migration, many of the β-gal-staining posterior basal cells are sloughed off. The cells that then become the rear of the slug now

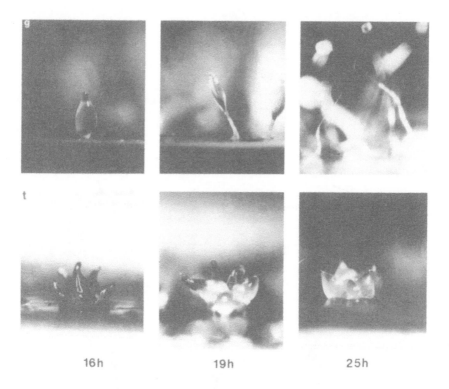

16h   19h   25h

FIGURE 6. Developmental phenotypes of Dd *ras* transformants. Dd *ras*-Gly12(g) and Dd *ras*-Thr12(t) expressing cells were allowed to develop synchronously on nitrocellulose filters and individual aggregates were photographed at the times indicated (hours into development).

express Dd *ras*. As the culmination proceeds and the mature fruiting body is formed, Dd *ras*-expressing cells comprise the caps located at the top and the base of the sorocarp and in the stalk and basal disk,[9] similar to what is observed for differentiation inducing factor (DIF)-induced prestalk genes.[15,16]

The above Dd *ras*-directed spatial pattern of staining clearly revealed its prestalk cell-type-specific expression observed by others. A closer examination of the Dd *ras/lacZ* staining in slugs and early culminants resulted in identification of a core region at the anterior that shows a lower level of staining. This lighter-staining region closely resembles a domain of prestalk cells designated prestalk B. The pattern of Dd *ras/lacZ* cells that do stain in the anterior suggests that Dd *ras* is expressed in a subpopulation of prestalk cells very similar to those designated prestalk A and is similar to that observed for the DIF-inducible prestalk gene *ecmA*.[9,15]

We examined expression of Dd *ras/lacZ* in vegetatively growing cells when the 1.2-kb mRNA is expressed, and observed that ~10% of the cells stained. A Dd *ras* promoter deletion was constructed that removes the 1.2-kb mRNA start site and results in a complete lack of expression of the 1.2-kb message.

When this deleted Dd *ras* promoter is fused to *lacZ* and its resultant staining pattern analyzed in *Dictyostelium*, it is found to have a few notable differences to that seen using the parental construct: there is no expression detected in vegetatively growing cells, staining intensity during development is reduced, and a much lower relative level of expression is observed in basal cells compared to that seen in anterior cells.[9]

VI. EFFECTS OF cAMP ON *ras* GENE EXPRESSION

Because of the fundamentally important role of cAMP in aggregation and during multicellular development,[2,17] and the known induction of prespore and some prestalk genes by cAMP,[18] regulation of *ras* expression by cAMP has been examined. After many hours in shaking cultures without nutrients, Dd *ras*G mRNA levels do not decline as they do following ~3 h of nutrient deprivation on a solid substratum. Mimicking the environment of early development during aggregation by supplying pulses of cAMP to shaking cultures, produces both the initial increase and subsequent decrease in Dd *ras*G levels. This decline can be prevented by including caffeine which, in *D. discoideum*, inhibits the activation of adenylate cyclase[19] and, thus, blocks the cAMP relay response.[8] Addition of a high, constant level of cAMP, however, does not result in a decline in Dd *ras*G mRNA. These results suggest that the decline of Dd *ras*G expression during early aggregation is dependent on the cellular responses to cAMP pulses mediated through the same signal transduction pathways controlling aggregation.[2]

In contrast, Dd *ras* expression is induced by constant high levels of cAMP in fast shaking cultures. Dd *ras* mRNA is observed only after cultures are shaken for 4 to 6 h to allow the expression of cAMP receptors, signal transduction components, and required exogenous cAMP and is, thus, similar to prestalk-enriched, cAMP-inducible genes such as pst-cath/CP2.[20-22] (Addition of a high level of cAMP earlier than this represses early pulse-induced responses necessary to establish cells competent to induce cell-type-specific genes.[2,3] This is thought to be caused by an adaptation and downregulation of aggregation-stage cAMP receptors.) Although levels of the 1.2- and 1-kb messages increase, the 1.2-kb transcript of Dd *ras* is more responsive to cAMP and is induced to levels five-fold higher than seen during multicellular development. Cells from dissociated aggregates shaken without cAMP rapidly lose Dd *ras* mRNA, but subsequently reinduce Dd *ras* expression upon cAMP addition.[7] These findings suggest that high levels of cAMP present in the multicellular structure from the late aggregate through the slug stage likely contribute to the activation of Dd *ras* expression.

VII. FUNCTION OF *ras* IN *Dictyostelium*

Although the roles of the two *ras* genes in *Dictyostelium* are not yet known, various studies have lead to suggestions on function that are worthy of consideration. Perhaps the simplest suggestion is that since Dd *ras*G and Dd *ras* are differentially expressed, Dd *ras*G function may be limited to vegetative growth and very early development, while Dd *ras* may function primarily during multicellular development with a specific role in the anterior tip and/or prestalk cell differentiation. It is worth noting here that a member of the *ras*-related *rap* family of genes has been identified in *Dictyostelium*. Expression of this gene, Dd *rap1*, is strongest at aggregation when the expression of either Dd *ras*G or Dd *ras* is minimal[23] (Figure 4). Due to this relative expression pattern and the finding that in mammalian cells the *rap* gene can suppress the transformed phenotype induced by an activated *ras* product,[24] it has been suggested that the *ras* and *rap* products of *Dictyostelium* may also have antagonistic roles.[23]

The basis of the observation that at any one time only 10% of the vegetative cells detectably express Dd *ras*/*lacZ* is not well understood. It is not known whether a small subset of cells constantly express Dd *ras* or a larger number of cells transiently express Dd *ras* during different intervals within the cell cycle. Either case invokes a degree of discriminatory regulation that suggests Dd *ras* may have a specific function in growing cells. One possibility is that Dd *ras* function is required for a certain period of the cell cycle and that expression is restricted to that time. Since cells early in the cell cycle have a propensity to differentiate into prestalk cells,[25-28] it is possible that Dd *ras* may have a role in regulating this process.

Further suggestions concerning the function of Dd *ras* in development arise from experiments with cells expressing a mutant Dd *ras* protein and from the combination of these studies with the analysis of Dd *ras* spatial expression. *Dictyostelium* cells expressing Dd *ras* with the putative dominant activating mutation G12T ($Gly_{12} \rightarrow Thy_{12}$) do not develop normally. The expression of the activated *ras* protein causes reduced ability to chemotax in response to cAMP and altered morphogenesis resulting in multiply tipped aggregates and the inability to proceed through development to form fruiting bodies.[29] The impaired response to cAMP, evident during both chemotaxis and the emergence of the organizing center at the apical tip, suggests a deficiency in cAMP signal transduction. When assayed in the Dd *ras* G12T mutant and compared to wild-type cells, the adenylate cyclase activation pathway is unaffected,[29] but the guanylate cyclase pathway displays a marked increase in desensitization.[30] Moreover, recent results show small changes in phosphotidylinositol levels in Dd *ras* G12T vs. control cells.[31] It, therefore, seems likely that the abnormal phenotypes displayed by Dd *ras* G12T expressing cells result, at least in part, from perturbation of these signaling pathways mediating the chemotactic response. This involvement of Dd *ras* in chemotaxis, combined with the apparent sorting of Dd *ras*-expressing cells to the tip of developing aggregates,[32] suggests Dd *ras* plays a role in establishing the organizing center that is crucial for proper morphogenesis during the multicellular stages.

Expression of Dd *ras*G containing the G12T mutation results in more restricted effects on development. Aggregation is aberrant when cells develop submerged on a plastic surface; however, both aggregation and tip formation are normal when cells are developed on filters. Thus, the perturbation is confined to early development, the time at which Dd *rasG* is maximally expressed in wild-type cells.[6]

Although the suggestions presented here do not define specific cellular functions for products of *Dictyostelium ras* genes, it is clear that the developmental context supplied by the uncommon features of the *Dictyostelium* life cycle offers productive avenues toward delineation of these functions. Differentially restricted developmental expression and differential relationships with extracellular cAMP signaling should prove especially useful in addressing functional redundancy in these cells expressing more than one *ras* gene.

REFERENCES

1. **Loomis, W. F., Ed.**, *The Development of Dictyostelium discoideum*, Academic Press, New York, 1982.
2. **Firtel, R. A., van Haastert, P. J. M., Kimmel, A. R., and Devreotes, P.**, G-protein linked signal transduction pathways in development: *Dictyostelium* as an experimental system, *Cell*, 58, 235, 1989.
3. **Mann, S. K. O. and Firtel, R. A.**, Regulation of gene expression and cell-type differentiation via signal transduction processes in *Dictyostelium*, in *Cell Activation: Genetic Approaches, Vol. II.*, in *Advanced Regulation of Cell Growth*, Mond, J., Weiss, A., and Camber, J., Eds., Raven Press, New York, 1991, 9.
4. **McCormick, F.**, *ras* GTPase activating protein: signal transmitter and signal terminator, *Cell*, 56, 5, 1989.
5. **Weeks, G., Lime, A. F., and Pawson, T.**, A *ras* encoded protein in *Dictyostelium discoideum* is acylated and membrane associated, *Mol. Microbiol.*, 1, 347, 1987.
6. **Robbins, S. M., Williams, J. G., Jermyn, K. A., Spiegelman, G. B., and Weeks, G.**, Growing and developing *Dictyostelium* cells express different *ras* genes, *Proc. Natl. Acad. Sci. U.S.A.*, 86, 938, 1989.
7. **Reymond, C. D., Gomer, R. H., Medhy, M., and Firtel, R. A.**, Developmental regulation of a *Dictyostelium* gene encoding a protein homologous to mammalian *ras* protein, *Cell*, 39, 141, 1984.
8. **Khosla, M., Robbins, S. M., Spiegelman, G. B., and Weeks, G.**, The regulation of DdrasG gene expression during *Dictyostelium* development, *Mol. Cell. Biol.*, 10, 918, 1990.
9. **Esch, R. K. and Firtel, R. A.**, cAMP and cell sorting control the spatial expression of a developmentally essential cell-type-specific *ras* gene in *Dictyostelium*, *Genes Dev.*, 5, 9, 1991.
10. **Reymond, C. D. and Thompson, N. A.**, Analysis of multiple transcripts of the Dd *ras* gene during *Dictyostelium discoideum* development, *Dev. Genet.*, 12, 139, 1991.
11. **Weeks, G. and Pawson, T.**, The synthesis and degradation of *ras*-related gene products during growth and differentiation in *Dictyostelium discoideum*, *Differentiation*, 33, 207, 1987.
12. **Pawson, T., Amiel, T., Hinze, E., Auersberg, N., Neave, N., Sobolewski, A., and Weeks, G.**, Regulation of a *ras*-related protein during development of *Dictyostelium discoideum*, *Mol. Cell. Biol.*, 5, 33, 1985.

13. **Weeks, G. and Pawson, T.**, The synthesis and degradation of *ras*-related gene products during growth and differentiation in *Dictyostelium discoideum*, *Differentiation*, 33, 207, 1987.

14. **Sternfeld, J. and David, C. N.**, Fate and regulation of anterior-like cells in *Dictyostelium* slugs, *Dev. Biol.*, 93, 111, 1982.

15. **Williams, J. G., Duffy, K. T., Lane, D. P., McRobbie, S. J., Harwood, D., Traynor, D., Kay, R. R., and Jermyn, K. A.**, Origins of the prestalk-prespore pattern in *Dictyostelium* development, *Cell*, 59, 1157, 1989.

16. **Jermyn, K. A. and Williams, J. G.**, An analysis of culmination in *Dictyostelium* using prestalk and stalk-specific cell autonomous markers, *Development*, 111, 779, 1991.

17. **Devreotes, P.**, *Dictyostelium discoideum*: a model system for cell-cell interactions in development, *Science*, 245, 1054, 1989.

18. **Mehdy, M. C. and Firtel, R. A.**, A secreted factor and cAMP jointly regulate cell-type-specific expression in *Dictyostelium*, *Mol. Cell. Biol.*, 5, 705, 1985.

19. **Brenner, M. and Thoms, S. D.**, Caffeine blocks activation of cAMP synthesis in *Dictyostelium discoideum*, *Dev. Biol.*, 101, 136, 1984.

20. **Mehdy, M. D., Ratner, D., and Firtel, R. A.**, Induction and modulation of cell-type-specific gene expression in *Dictyostelium*, *Cell*, 32, 761, 1983.

21. **Pears, C. J., Mahbubani, H. M., and Williams, J. G.**, Characterization of two highly diverged but developmental co-regulated cysteine proteinase genes in *Dictyostelium discoideum*, *Nucleic Acids Res.*, 13, 8853, 1985.

22. **Datta, S., Gomer, R. H., and Firtel, R. A.**, Spatial and temporal regulation of a foreign gene by a prestalk-specific promoter in transformed *Dictyostelium discoideum* cells, *Mol. Cell. Biol.*, 6, 811, 1986.

23. **Robbins, S. M., Suttorp, V. V., Weeks, G., and Spiegelman, G. B.**, A *ras* related gene from the lower eukaryote *Dictyostelium* that is highly conserved relative to the human *rap* genes, *Nucleic Acids Res.*, 18, 5265, 1990.

24. **Kitayama, H., Sugimoto, Y., Matsuzaki, T., Ikawa, Y., and Noda, M.**, A *ras*-related gene with transformation suppressor activity, *Cell*, 56, 77, 1989.

25. **van Lookeren Campagne, M. M., Duschl, G., and David, N. C.**, Dependence of cell type proportioning and sorting on cell cycle phase in *Dictyostelium discoideum*, *J. Cell. Sci.*, 70, 133, 1984.

26. **Weijer, C. J., Duschl, G., and David, C. N.**, Dependence of cell-type proportioning and sorting on cell cycle phase in *Dictyostelium discoideum* slugs: evidence that cyclic AMP is the morphogenetic signal for prespore differentiation, *Development*, 103, 611, 1984.

27. **McDonald, S. A. and Durnston, A. J.**, The cell cycle and sorting behavior in *Dictyostelium discoideum*, *J. Cell Sci.*, 66, 195, 1984.

28. **Gomer, R. H. and Firtel, R. A.**, Cell-autonomous determination of cell-type choice in *Dictyostelium* development by cell-cycle phase, *Science*, 237, 758, 1987.

29. **Reymond, C. D., Gomer, R. H., Nellen, W., Theibert, A., Devreotes, P., and Firtel, R. A.**, Phenotypic changes induced by a mutated *ras* gene during the development of *Dictyostelium* transformants, *Nature*, 323, 340, 1986.

30. **van Haastert, P. J. M., Kesbeke, F., Reymond, C. D., Firtel, R. A., Luderus, E., and van Driel, R.**, Aberrant transmembrane signal transduction in *Dictyostelium* cells expressing a mutated *ras* gene, *Proc. Natl. Acad. Sci. U.S.A.*, 84, 4905, 1987.

31. **van der Kaay, J., Draijer, R., and van Haastert, P. J. M.**, Increased conversion of phosphatidylinositol to phosphatidylinositol phosphate in *Dictyostelium* cells expressing a mutated *ras* gene, *Proc. Natl. Acad. Sci. U.S.A.*, 87, 9197, 1990.

32. **Sternfeld, J. and David, C. N.**, Cell sorting during pattern formation in *Dictyostelium*, *Differentiation*, 20, 10, 1981.

33. **Haberstroh, L. and Firtel, R. A.**, A spatial gradient of expression of a cAMP-regulated prespore cell-type-specific gene in *Dictyostelium*, *Genes Dev.*, 4, 596, 1990.

Chapter 9

ras GENES IN *Drosophila melanogaster*

Zeev Lev

TABLE OF CONTENTS

I. GENERAL PROPERTIES

A. THE *Drosophila ras* HOMOLOGS

The first indication of the presence of *Drosophila* genes homologous to vertebrate *ras* genes was presented by Shilo and Weinberg who used a v-Ha-*ras* DNA probe to detect three bands in genomic *Drosophila melanogaster* DNA blots.[1] Later three *ras*-like genes — *Ras1, Ras2,* and *Ras3*, were cloned and characterized.[2-5] Recently, a *rab*- and an *arf*-like gene were also identified and cloned.[6,37]

Ras1,[7] also termed *Dras1* and *Dmras85D*,[2,4] was cloned independently by two groups.[2,4] It is located at region 85D on the polytene chromosome of the larval salivary gland.[2] The gene contains two introns, 150 and 75 bp long, respectively, and codes for a protein 189 amino acids long.[4,8,9] *Ras2*,[7] also termed *Dras2* and *Dmras64B*,[2-4] was independently cloned three times, and mapped to region 64B.[2-4] The gene contains two introns, about 630 and 56 bp long, respectively, and codes for a 191 amino acid long protein.[4,10,11] The third member of the group, *Ras3*,[7] also termed *Dras3* and *Rap1*,[5,12] is mapped to 62B.[5] Its cDNA sequence is colinear with the genomic DNA sequence suggesting that *Ras3* is an intronless gene. The Ras3 protein contains 182 amino acids.[12]

Analysis of sequence homologies among the three *Drosophila Ras* genes and other *ras*-like genes indicated that they are related to different families in the *Ras* branch of the *ras* supergene superfamily.[13] The *Drosophila Ras1* belongs to the Ras family which includes the three human transforming *ras* genes and the yeast RAS genes. *Ras2* belongs to another gene family which contains the human nontransforming R-*ras* gene and the TC21 gene. *Ras3* is most similar to the *Rap* family which contains the *rap1*/Ki-*rev* gene and other *rap* and *smg* genes. The *Drosophila Ras* genes are more similar to their human counterparts than to each other. For example, in the region spanning the amino acids at positions 3 to 162 there is only 65% similarity between the *Drosophila* Ras1 and Ras2 proteins, but 86 and 72% similarity between Ras1 and the human Ha-*ras* and Ras2 and the human R-*ras* proteins, respectively. Apparently, these genes have diverged before the divergence of the invertebrate and the vertebrate radiations.

The presence of all three genes in other species of the genus *Drosophila* has been proved indirectly by hybridizing poly(A) RNA blots from these species with *D. melanogaster Ras* DNA probes.[14] Thus, genes homologous to *Ras1, Ras2,* and *Ras3* are found in *D. simulans* and *D. yakuba*, which are sibling species in the subgroup Melanogaster; in *D. ananassae*, which represents another subgroup in the group Melanogaster, subgenus Sophophora; and in *D. virilis*, which represents another subgenus, Drosophila.

rab3A is a small neuronal GTP-binding protein specifically localized to synaptic vesicles. The *Drosophila rab3* homolog, localized to 47B, reveals high evolutionary conservation of *rab3A* (76%) and *rab3B* (78%), including their carboxy-terminal Cys-X-Cys sequence.[38] Although only distantly related to the *ras* genes, another *Drosophila* gene which should be mentioned briefly is *arl*. This gene was isolated during a genome walk in region 72AB for the purpose of cloning the *brahma* gene.[6] The gene was located by virtue of a 1.05-kb transcript identified immediately proximal to the *brahma* locus. Genomic sequence analysis and cDNA showed that the gene contains two introns, 96 and 278 nucleotides long, respectively. The predicted product is a protein containing 180 amino acids and of a molecular weight of 20,250 kDa. The protein sequence is very similar to the ADP-ribosylation factors (ARFs).[15] A lethal mutation in this locus, *arl*,[6] can be completely rescued with cloned *arl* genomic sequences, thus, implying that the *arl* gene product has an essential function in *D. melanogaster*.

B. EXPRESSION DURING DEVELOPMENT

Each of the three *Drosophila Ras* genes codes for one major larger transcript, and in addition for one or two shorter transcripts.[10,14,16,17] *Ras1* codes for 2.0 and 1.3 kb transcripts, *Ras2* codes for 1.8 and 1.4 kb transcripts, and *Ras3* for 2.9, 1.9, and 1.5 kb transcripts.[14] Two larger transcripts assigned to *Ras2*,[3,4] are actually the product of another gene termed *CS1/Rop*, located near *Ras2*.[18] The basis for the differences in length among the transcripts of each gene has not been determined. However, since less than 600 bases are required for coding a Ras protein, then all these transcripts, including the shorter ones, are large enough to code for a full-length protein. Interestingly, the sizes of almost all transcripts of each *Ras* gene were conserved in four other *Drosophila* species: *D. simulans, D. yakuba, D. ananassae*, and *D. virilis*, except for the larger *Ras2* transcript which is slightly longer in *D. ananassae* and *D. virilis*.[14]

The pattern of expression during development of all three *D. melanogaster* ras genes is similar. The larger transcript of each gene is expressed constitutively during all developmental stages, including unfertilized eggs, embryos, larvae, pupae, and adult flies. The shorter transcripts are found only in unfertilized eggs and early embryos.[14] Presumably, they accumulate as maternal RNA in the developing oocytes, and are degraded following egg deposition during embryogenesis. However, a short period of *de novo* zygotic expression cannot be excluded. Consequently, the shorter *Ras1* and *Ras3* transcripts may be detected in adult females but not in males.[17] Quantitative analysis of transcript concentration revealed that the percentages of *Ras1, Ras2*, and *Ras3* transcripts in egg poly(A) RNA were 0.05, 0.05, and 0.17%, respectively.[14] Generally the percentage of individual transcripts during development was in the range of 0.01 to 0.03% of poly(A) RNA.

The transcriptional patterns of *Ras1* and *Ras3* were also examined in the Schneider 2 cell line, derived from embryonic cells,[19] and in cells derived from the neoplastic tumors found in brains of *lethal(2)giant larvae [l(2)gl]* mutant larvae. These cells represent undifferentiated neuroblasts.[20] In the Schneider 2 tissue culture cells all transcripts were detected.[17] Apparently, the control apparatus which regulates the *Ras1* and *Ras3* gene expression in these cells is of the maternal/embryonic type. On the other hand, only the larger, constitutive transcripts were detected in the neoplastic neuroblasts. Thus, in these cells, in spite of their embryonic, undifferentiated nature, the pattern of *Ras* gene expression is of the postembryonic type.

C. TISSUE AND ORGAN SPECIFICITY

Two different approaches have been taken in the study of the distribution of *Ras* gene expression in *Drosophila* tissues: *in situ* hybridization and transcription fusions with a reporter gene. Using the first approach, the spatial distribution of the transcripts of the three *Ras* genes was found to be very similar.[16] In the embryo, a homogenous, uniform pattern of hybridization was found. It should be noted that in this experiment both the maternal and the *de novo* zygotic transcripts are simultaneously detected, therefore, it is still possible that the uniform maternal contribution is actually masking a more specific pattern of the zygotic *Ras* gene expression. In the larva no hybridization could be observed in polytenized tissues such as salivary glands or fat cells. However, strong hybridization was observed in the imaginal disks, including the eye-antenna disks, the wing disks, the genital disks, and the anlagen of the testis and ovaries. Within the disks the pattern of hybridization was uniform. In addition, *Ras* transcripts were detected in the cortex of the larval brain, but not in its interior. Finally, in adult flies specific hybridization was found in the brain cortex and in the thoracic and abdominal ganglia, in the flight muscles, and in the ovaries.

In situ hybridization detects the distribution of the native transcripts and, therefore, accurately reflects the actual spatial regulation of *ras* gene expression. However, in most cases its resolution is limited to organs and tissues rather than single cells. Higher resolution, down to the single cell level, may be obtained with reporter genes, usually the bacterial *lacZ* gene. In preliminary experiments using the *Ras2* gene promoter fused to *lacZ*, a pattern similar but not identical to the pattern described above was obtained.[21] It should be noted that this approach requires cloning of a reporter gene downstream to the native promoter. Therefore, any transcription control elements which might be located in the deleted sequences are eliminated.

D. STUDIES IN HETEROLOGOUS SYSTEMS

In two different experiments human and *Drosophila* systems were intermingled. In one case a fusion between the N terminus of the activated c-Ha-*ras* oncogene and the C terminus of the *Ras3* gene was constructed.[5] The chimeric gene, driven by the promoter of the native c-Ha-*ras* protooncogene, was used to transfect Rat-1 cells. Transformed foci were obtained although the transformation efficiency was about 10% as compared with transfections in which intact activated c-Ha-*ras* DNA was used. A similar construct, but with the C terminus of *Ras1* did not yield any transformed foci unless a viral promoter was added. In the latter case the transformation efficiency was 2% as compared with intact activated c-Ha-*ras* DNA. Since the C termini of the *Drosophila* Ras proteins differ greatly from the human *ras* sequence, it appears that the only contribution of the *Drosophila* sequences is to provide the required anchoring to the cell membrane, and to serve as a spacer between the membrane and the active N terminus.

In order to overexpress activated c-Ha-*ras* oncogene in *Drosophila* Schneider 2 cells, it was fused to the strong, constitutive *copia* promoter.[22] Surprisingly, only transient transfections yielded high amounts of Ha-*ras* p21 protein. Trials to obtain stably transformed lines failed, since the neomycin-resistant lines obtained never contain the human gene. However, when the *copia* promoter was replaced with the inducible metallothionein (MT) promoter it was possible to isolate stably transformed lines which did contain integrated Ha-*ras* genes. Following induction with cadmium chloride, high amounts of Ha-*ras* p21 protein were produced. Apparently the lengthy overexpression of this exogenous protein is lethal to the Schneider 2 cells, probably due to an interaction such as inhibition or competition with components of one of the endogenous *Ras* pathways.

II. FUNCTIONAL STUDIES

A. *Ras1*

The product of the *sevenless (sev)* gene is a receptor tyrosine kinase required for correct differentiation of the R7 photoreceptor in the eye of *D. melanogaster*.[23] The ligand of this receptor is the product of the *bride of sevenless (boss)* gene produced in the adjacent R8 photoreceptor. To find additional genes participating in this signal transduction pathway downstream to *sevenless*, a genetic screen for mutations which decrease the effectiveness of signaling by this receptor was carried out.[8] Seven genes were identified in the screen. The normal products of these genes are probably required for proper signaling by the *sevenless* receptor tyrosine kinase. Two mutations in one gene were located to a chromosomal region where the *Ras1* gene is found. In a complementation experiment using a genomic DNA fragment containing the intact *Ras1* gene, the mutated phenotype was completely rescued, suggesting that the gene in question is indeed *Ras1*. The *Ras1* locus in the two mutated chromosomes was sequenced and changes in the amino acid sequence of the putative Ras1 protein were discovered — in one case it was a change of aspartate to asparagine in position 38, and in the other case the glutamate at position 62 was changed into lysine. Apparently, the Ras1 protein is part of the pathway which transmits a signal from the Sevenless receptor to promote the differentiation of the R7 photoreceptor.

Another gene identified in this screen, and in another study,[39] was *Son of sevenless (Sos)*. A mutated dominant allele of *Sos* was previously isolated by virtue of its ability to suppress a particular mutant allele of *sevenless*.[24] A genomic DNA fragment of the *Sos* gene was isolated after an extensive chromosomal walk. Subsequently, *Sos* cDNA clones were isolated and sequenced. A considerable homology was found between the Sos protein and the yeast *Saccharomyces cerevisiae CDC25* and *SCD25* genes. The homology with these genes is 28 and 22%, respectively, in a region of 380 amino acids. *CDC25* and *SCD25* are guanine nucleotide exchange factors which catalyze the activation of the yeast Ras2 protein by converting it from the GDP-bound form to the GTP-bound form.[25,26] Thus *Sos* is an exchange factor, activating a *D. melanogaster* ras protein, probably *Ras1*. In summary, the *sevenless*-mediated signal transduction pathway includes the activation of the sevenless receptor by its ligand, the Boss protein. The receptor transmits a signal, directly or indirectly, to the Sos exchange factor which activates the Ras1 protein. Ras1 then transmits the signal to its unknown effector target. Interestingly, both *Ras1* and *Sos* participate in the transduction of another signal mediated by the epidermal growth factor (EGF) receptor tyrosine kinase.[8]

B. *Ras2*

The signal transduction pathways in which *Ras2* is involved are still unknown, and none of the additional genes presumably participating in these pathways have been described. However, several indirect evidences suggest that the *Rop* gene is a reasonable candidate to interact with *Ras2*.

During the isolation and characterization of the *Drosophila Ras2* promoter region it was shown that in addition to the *Ras2* gene this promoter regulates another gene, in the opposite polarity relative to the *Ras2* gene.[17] Apparently, the two genes are regulated by a bidirectional promoter. The new gene was initially termed *CS1* and is now termed *Rop* for *Ras opposite*. The transcription start sites of *Ras2* and *Rop* were localized and found to be merely 94 nucleotides apart. Deletion analysis proved that certain *cis*-acting elements within the promoter region are required for full transcriptal activity of both genes simultaneously.[27] Two of these sites are protected from DNaseI digestion by *trans*-acting factors.[28] Clearly, the *Drosophila Ras2* promoter is an authentic bidirectional promoter. In many bidirectional promoters isolated thus far there is some kind of interaction between the two divergently transcribed genes, either regulative or functional. Similarly, *Ras2* and *Rop* can interact with each other. A prerequisite for such an interaction is coexpression of the two genes in the same tissue.

To determine if *Rop* is expressed simultaneously in the same tissue as *Ras2*, the spatial distribution of their transcripts during development was determined by *in situ* hybridization and by constructing lines of transgenic flies containing transcription fusions of the *Rop* or *Ras2* gene promoter and the bacterial *lacZ* gene.[21] Interestingly, identical patterns were obtained for *Rop* and *Ras2* in several tissues. Clearly, if *Rop* is some factor associated with *Ras2*, this peculiar mode of regulation by a bidirectional promoter is most adventitious for the two genes. For example, the Rop protein is available in the same tissue as Ras2, ready to interact with it when required. From the nucleotide sequence of several cloned cDNAs the sequence of the putative Rop protein, 597 amino acids long, was deduced. However, no proteins with significant sequence similarity with *Rop* were found.[29] Thus, the identity of *Rop* and how it is related to *Ras2* remained to be discovered.

To study the impact of expressing activated *Ras* in *Drosophila*, the glycine at position 14 of *Ras2* which corresponds to glycine at position 12 of the human Ha-*ras*, was mutated *in vitro* into valine.[10] The activated *Ras2* gene was fused to the inducible promoter of the heat shock 70 gene (hsp70) and introduced into the *D. melanogaster* germ line by P-element-mediated transduction. Even without heat induction, the ectopic expression of the activated *Ras2* gene driven by the basal transcription activity of the hsp70 promoter, was sufficient to produce a large number of phenotypic changes. Most stocks had low fertility and showed developmental disturbances in their wings and eyes. Bristles were of variable size and placement, particularly on the anterior margin of the wing. In many cases, sternopleural bristles, and humeral bristles of the thorax were bent or forked. Higher levels of the activated Ras2 protein induced by heat-shocking of the flies generated additional effects. When pulse-heated for 1 to 2 h during third instar and pupal stages, bloated wings were obtained due to failure in the fusing of the dorsal and ventral surfaces. In addition, wings were smaller, distorted, and sometimes missing on one side of the fly. If heat-shocked earlier, the flies rarely had wing defects. Strikingly, dorsal-to-ventral scars developed in the adult eye. These scars consisted of unpigmented, irregular, and fused ommatidia. Two hours of heating during the third instar larval stage resulted in 60 to 80% lethality. Onset of pupariation appeared to be a highly sensitive step since heating a few hours before or after this event left no survivors. Blackened masses developed in these animals, particularly in the larval mid-intestine and gastric ceca, and in various sites within the pupa. Other organs, including the wings, head, abdomen, and legs were aborted and distorted.

The pleiotropic effects induced by the activated *Ras2* hamper genetic analysis of the mutant gene. An alternative approach has recently been taken by Bishop and Corces who constructed transgenic flies which carry activated *Ras2* driven by the promoter of the flight muscle actin 88F gene.[30] Viable flies have been obtained with visible phenotypes, mostly disturbances in wing development.

C. *Ras3/Rap1*

The *Roughened (R)* mutation of *D. melanogaster* is a dominant mutation that disrupts eye development. Flies heterozygous for the mutation have rough eyes due to irregular spacing and orientation between adjacent ommatidia. In addition, the R7 photoreceptor is missing in most of ommatidia. The eyes of homozygous flies are even rougher. The identical chromosomal location of *Roughened* and the *Ras3* gene initiated a set of experiments designed to test the possibility that they are eventually the same gene.[12] Sequence analysis of the *Ras3* gene in wild-type and *Roughened* chromosomes, and in chromosomes containing revertants of *Roughened*, proved this assumption. *Ras3* appears to be the *Drosophila* homolog of the human *Rap1A* and *Rap1B* genes. Out of the 184 amino acids of the rap1A protein, 87% are identical and 93% are conserved in

Ras3. The *Roughened* mutation is a point mutation that changes a phenylalanine at position 157 (position 156 of human Ha-*ras*) to leucine, and the *Roughened* revertants are stop codons within the *Ras3/Rap1* protein. The phenylalanine at 157 is one of the few residues conserved outside the GTP-binding region in all members of the ras superfamily.

Homozygous null *Ras3/Rap1⁻* mutants are lethal at the late larval stages.[12] Since the gene is expressed maternally, it is likely that its product may be also required in embryogenesis.[14] The *Roughened* allele keeps the vital properties of the wild-type protein since flies which carry it over a deletion are viable. On the other hand, this mutation is not just an overproducer since flies that carry the *Roughened* allele and a wild-type allele have less severe phenotype than flies which carry the mutated allele over a deletion. Thus, this allele produces a protein with some kind of a novel, or altered function. The nature of this function is not known, but considering the role of the *Ras1* gene product in establishing the identity of the R7 photoreceptor, it is possible that *Ras1* and *Ras3/Rap1* may have an antagonistic role in R7 determination. It is interesting to note that the human *rap1A* gene may interact in a similar way with the human *Ki-ras* gene, since overexpressed rap1A protein can suppress the activity of a transforming *Ki-ras* gene.[31]

D. THE INTERACTION BETWEEN THE *awd* AND THE *prune* GENES

The phenotype of the recessive *prune (pn)* mutation of *D. melanogaster* is brown eye color. This is due to an alteration in the activity of GTP cyclohydrolase, the enzyme that catalyzes the first step in the biosynthesis of pteridines, which determine the eye color in *Drosophila*. The *abnormal wing disc (awd)* is another mutation of *D. melanogaster* which has a larval, recessive lethal phenotype. In homozygous *awd* null mutants the wing, leg, and eye-antenna imaginal disks in third instar larvae differentiate with varying degrees of aberration and cell death. The ovaries, larval brain, and proventriculus are often abnormal and these larvae die as prepupae.[32] *Killer of prune (awd^{K-pn})* is a conditional dominant lethal allele of *awd*. awd^{K-pn} flies are normal, and have no apparent mutated phenotype on their own. However, in combination with *prune*, namely in the double mutants *pn awd^{K-pn}*, this allele is dominant lethal and they die as second or third instar larvae. Recent molecular cloning of both *awd* and *prune* genes has revealed the identity of these genes and helped to explain the lethal nature of the *prune/awd^{K-pn}* interaction. According to this explanation, it is very likely that a putative *ras*-like gene is involved in the interaction.

The *awd* gene product appears to be a nucleoside diphosphate kinase (NDP kinase), which catalyzes transphosphorylation of both free and bound GDP to GTP.[33] In homozygous *awd⁻* embryos and larvae the absence of a specific NDP kinase may affect the free GTP/GDP ratio in the cells and consequently abolish normal development. Alternatively, the phosphorylation of bound GDP to GTP may be hampered. NDP kinases are involved in the phosphorylation of GDP bound to tubulin. Furthermore, recent studies carried out with an ARF, a small GTP-binding protein, have shown that three human, bovine, and mouse NDP kinases were able, in the absence of nucleotide exchange activity, to convert ARF-GDP directly into ARF-GTP.[34] The *prune* gene product has a significant sequence homology with the catalytic domain of mammalian GAPs.[35] Thus, Prune-GAP may downregulate the activity of a *Drosophila* RAS-like protein by converting its active GTP-bound form to the GDP-bound form. In homozygous *prune* flies this GAP activity is missing. The higher concentration of the GTP-bound form transduces a stronger signal resultign in abortive eye-color production. The *prune/awd*$^{K-pn}$ lethal interaction suggests that the GDP-bound form of the Ras-like protein regulated by the Prune-GAP is subjected to direct transphosphorylation by the awd-NDP kinase. In *awd*$^{K-pn}$ mutants this activity is probably enhanced compared with the wild-type level, but still without apparent side effects due to the compensating GTPase activity accelerated by the prune-GAP. However, in *prune awd*$^{K-pn}$ double mutants, the combination of the overreactive NDP kinase encoded by the *awd*$^{K-pn}$ allele and the lack of the prune-GAP, results in an excessive, lethal signal. The identity of the putative *ras*-like gene which sits in the core of this interaction remains to be seen.

III. CONCLUDING REMARKS

Only three *ras* genes and one *rab* gene have been isolated thus far in *Drosophila*. In comparison with the intensive research carried out on the different aspects of the biology of *Drosophila*, and the large number of *ras* and *ras*-like genes isolated in other systems, this number is surprisingly low. In addition, besides *Sos*, which might be an exchange factor of *Ras1*, no other *ras* regulator has been isolated. (See Note Added in Proof, at end of chapter, concerning *Gap1*.)

Research on these genes at the biochemical level has just begun and only preliminary data concerning the biochemical properties of *Drosophila ras* gene is available.[9,11] On the other hand, studies on the expression of these genes during development are quite detailed, providing a most comprehensive description on the expression of a *ras* gene during the life cycle of a multicellular organism. Apparently, the three *Drosophila Ras* genes are expressed constitutively during all developmental stages, and they are moderately abundant in poly(A) RNA. However, studies bearing on the distribution of *Ras* gene expression in *Drosophila* tissues indicated that this expression is spatially regulated and is localized in a relatively small number of tissues. It is difficult to speculate on the role of the *Ras* genes in those tissues among which are organs such as the imaginal disks which contain undifferentiated, proliferating cells, and terminally differentiated tissues such as the adult brain. In any case, the nature of their function in a given tissue needs to be determined by tissue-specific inducers and/or effectors. Interestingly, the *Drosophila Ras1* and *Ras3* genes may be expressed together during eye differentiation, and be active in parallel signal transduction pathways or even in the same pathway.

Genetic methods are very efficient in several invertebrates, particularly nematodes and fruit flies. For example, the first definitive indication for any role of a *ras* gene in a developmental pathway has been discovered in the nematode.[36,37] A genetic approach using the fruit fly may also contribute invaluable data concerning the functions of the *ras* genes and details of the signal transduction pathway in which they participate. For example, second-site mutations were induced which enhance the wing phenotype of flies expressing activated *Ras2* under the control of the flight muscle actin 88f gene promoter.[30] Recently, seven components associated with signals transduced by the *sevenless* and EGF receptor tyrosine kinases were identified.[8,12,24] Two of them are genes encoding a Ras protein and a guanine nucleotide exchange factor. The identity of the other five elusive genes remains to be determined.

NOTE ADDED IN PROOF

A. *Ras1*

Loss-of-function mutations in the *Ras1* gene result in downregulation of the signal transmitted from the *sevenless* receptor tyrosine kinase.[8] Therefore it is reasonable to assume that a gain-function-mutation in *Ras1* would increase this signal. To test this hypothesis transgenic flies carrying an activated *Ras1* gene (a Gly12 was changed into Val12), under the control of the *sevenless* promoter, were constructed.[40] Indeed, experiments with these strains showed that the activated Ras1 protein induced the formation of supernumerary photoreceptor R7 cells and rescued R7 cells from transformation into cone cells in *sevenless* null mutants. Thus, the activated Ras1 protein bypasses the requirement for the *sevenless* gene product. Similar constructs containing activated *Ras2* rather than *Ras1* did not show any interactions with the *sevenless* gene function, suggesting that *Ras2* is not part of this pathway.

Two additional genes associated with the *sevenless*/*Ras1* signal transduction pathway have been recently identified. The first one is *Gap1*, a putative GTPase activated protein of *Ras1*.[41] *Gap1* was isolated in an enhancer trap screen for enhancers that induce specific expression of a reporter gene in the eye imaginal disk. It was also isolated in a gentic screen for mutations that enhance the phenotype of a temperature-sensitive *sevenless* allele, and independently, in a genetic screen for viable recessive mutations that show rough eye phenotype. The predicted Gap1 protein has significant homology with the catalytic domain of known GAPs. It is most similar to rasGAP and NF1 (39 and 31% identity along 167 amino acids, respectively). Furthermore, out of 15 residues conserved in the catalytic domains of 5 mammalian and yeast GAPs, 14 residues are conserved in the *Drosophila* Gap1. Genetic analysis showed that reduction in the *Gap1* gene activity increased the effectiveness of signalling by *sevenless*. Complete inactivation of *Gap1* induced supernumerary R7 cells and could rescue the lack of the *sevenless* function. Thus, *Gap1* is located downstream to *sevenless* in the signal transduction pathway that induces R7 differentiation. It is a negative regulator of this pathway, probably by enhancing the GTPase activity of *Ras1*. Since the phenotype of loss-of-function mutations in *Gap1* is opposite to the phenotype observed in loss-of-function mutations in *Ras1*, the Gap1 protein may not serve as a positive effector target of Ras1.

The identification of an additional step in the *sevenless* signal transduction pathway emerged from studies on *Raf*, the *Drosophila* homolog of the mammalian Raf-1 gene. The *Raf* locus is identical to *l(1)polehole*, an early embryonic recessive lethal mutation. In mammals this serine-threonine kinase is acting downstream to *ras* in many pathways induced by receptor tyrosine kinases. In *Drosophila*, *Raf* functions downstream to the *torso* receptor tyrosine kinase in determining the embryonic terminal patterns. A recent study suggests that it also participates in the pathway triggered by *sevenless*.[42] Weak *Raf* alleles suppressed gain-of-function mutations of *sevenless*. Activated Raf protein, obtained by fusing the Raf kinase domain to the signal sequence and the extracellular and transmembrane domains of the Torso protein, was put under the control of the *sevenless* promoter. When this protein was expressed in transgenic flies, supernumerary R7 cells were produced, similarly to the effect of activated Ras1 protein on the same pathway. The activated Raf protein transmits a signal that results in the differentiation of the R7 cells even if the function of *sevenless* is completely absent. Weak *Raf* alleles decrease the activity of the activated *Ras1* allele, but reduced dosage of the Ras1 protein does not impair the signalling from *Raf*. Apparently *Raf* acts downstream to *Ras1*, and the Ras1 protein is rate limiting in the signal transmission from *sevenless*, but not from *Raf*.

B. *Ras2*

Recent data regarding the localization of the *Rop* and *Ras2* gene transcripts during embryogenesis, and the homology found between the Rop protein and the products of a novel *S. cerevisiae* gene family, suggest that these two genes may be involved in vesicle trafficking among *Drosophila* cellular membranes.[43]

In situ hybridization to whole-mount embryos detected, during germ band retraction, specific *Ras2* transcripts in the garland cells. *Rop* is also detected in the garland cells and, in addition, in the CNS. During all later stages *Ras2* transcripts are limited to a specific set of cells in the CNS, and to the garland cells. *Rop* transcripts are fairly abundant in all cells of the CNS. They are less abundant in the peripheral nervous system, but appear in all neurogenic clusters. Distinct co-expression of both genes in the antennal-maxillary complex is also evident.

A search of protein databases revealed significant homology between the Rop protein and the yeast proteins Sly1p, Sec1p, and Vps33p(Slp1p); all of them are associated with protein traffic among yeast cellular membranes.[44] In addition, Rop and these proteins share a conserved motif with β-COP, a coat protein isolated from rat Golgi-bound nonclathrin vesicles. This homology suggests that the Rop protein may also function in vesicle trafficking among membranes of *Drosophila* cells. The tissue-specific expression of the *Rop* gene during embryogenesis in the CNS, a tissue that is highly active in membrane recycling, and particularly in the garland cells, is in concert with this suggestion. The garland cells, also termed wreath cells, surround the esophagus near its joining with the proventriculus. The cells are already differentiated in the embryo, and they persist to the adult stage. Studies bearing on the ultrastructure of these cells revealed extensive invaginations of the plasma membrane, which are the origin of a lot of coated pits and coated vesicles. The garland cells are considered as accessory cells to the open blood system and have been implicated in the removal of waste materials from the hemolymph by endocytosis.

Our finding that the Rop protein is homologous to the products of the *S. cerevisiae SLY1* gene is very stimulating, since the SLY1-20 allele was isolated in a screen designed to identify suppressors of the loss of the *YPT1* gene function.[45] *YPT1* is a *ras*-like gene essential for ER-to-Golgi protein transport and possibly also participates in intra-Golgi vesicle movements. The *SLY1-20* allele which allows *YPT1*-independent growth contains a single point mutation in the wild-type *SLY1* gene. A possible mechanism of the suppression is that *SLY1* encodes a protein acting downstream of *YPT1*. Thus, *SLY1* may interact, directly or indirectly, with *YPT1*.

The unusually close proximity of the *Ras2* and *Rop* genes and their joint regulative mechanism could be due solely to the need for their products in the same tissue. However, since Sly1p may interact with a small GTPase protein of the *ras* superfamily, it is conceivable that the Rop and Ras2 proteins may also interact functionally with one another in these tissues. Considering the relatively high expression of *Ras2* in the garland cells, and its putative interaction with *Rop*, it is possible that *Ras2* is also involved in endocytic processes and/or other transport pathways mediated by vesicle trafficking.

ACKNOWLEDGMENTS

I am grateful to V. Corces, J. Bishop, and O. Segev for providing unpublished results. This work was supported by grants from the Israel-U.S. Binational Science Foundation, the Basic Research Foundation administered by the Israel Academy of Science and Humanities, and the Israel Cancer Research Fund.

REFERENCES

1. **Shilo, B.-Z. and Weinberg, R. A.,** DNA sequence homologous to vertebrate oncogenes are conserved in *Drosophila melanogaster*, *Proc. Natl. Acad. Sci. U.S.A.*, 78, 6789, 1991.
2. **Neuman Silberberg, F. S., Schejter, E., Hoffmann, F. M., and Shilo, B. Z.,** The *Drosophila ras* oncogenes: structure and nucleotide sequence, *Cell*, 37, 1027, 1984.
3. **Mozer, B., Marlor, R., Parkhurst, S., and Corces, V.,** Characterization and developmental expression of a *Drosophila ras* oncogene, *Mol. Cell Biol.*, 5, 885, 1985.

4. **Brock, H. W.**, Sequence and genomic structure of *ras* homologues Dmras85D and Dmras64B of *Drosophila melanogaster, Gene*, 51, 129, 1987.

5. **Schejter, E. D. and Shilo, B. Z.**, Characterization of functional domains of p21 ras by use of chimeric genes, *EMBO J.*, 4, 407, 1985.

6. **Tamkun, J. W., Kahn, R. A., Kissinger, M., Brizuela, B. J., Rulka, C., Scott, M. P., and Kennison, J. A.**, The arflike gene encodes an essential GTP-binding protein in *Drosophila, Proc. Natl. Acad. Sci. U.S.A.*, 88, 3120, 1991.

7. **Lindsley, D. L. and Zimm, G.**, The genome of *Drosophila melanogaster*, Dros. Info. Service 68, 1990.

8. **Simon, M. A., Bowtell, D. D. L., Dodson, G. S., Laverty, T. R., and Rubin, G. M.**, Ras1 and a putative guanine nucleotide exchange factor perform crucial steps in signaling by the sevenless protein tyrosine kinase, *Cell*, 67, 701, 1991.

9. **Sahar, D. and Lev, Z.**, unpublished results.

10. **Bishop, J. G. and Corces, V. G.**, Expression of an activated *ras* gene causes developmental abnormalities in transgenic *Drosophila melanogaster, Genes Dev.*, 2, 567, 1988.

11. **Halachmi, N. and Lev, Z.**, unpublished results.

12. **Hariharan, I. K., Carthew, R. W., and Rubin, G. M.**, The Drosophila Roughened mutation: Activation of a rap homolog disrupts eye development and interferes with cell determination, *Cell*, 67, 717, 1991.

13. **Chardin, P.**, Small GTP-binding proteins of the ras family: a conserved functional mechanism?, *Cancer Cells*, 3, 117, 1991.

14. **Lev, Z., Kimchie, Z., Hessel, R., and Segev, O.**, Expression of *ras* cellular oncogenes during development of *Drosophila melanogaster, Mol. Cell Biol.*, 5, 1540, 1985.

15. **Sewell, J. L. and Kahn, R. A.**, Sequences of the bovine and yeast ADP-ribosylation factor and comparison to other GTP-binding proteins, *Proc. Natl. Acad. Sci. U.S.A.*, 85, 4620, 1988.

16. **Segal, D. and Shilo, B. Z.**, Tissue localization of *Drosophila melanogaster ras* transcripts during development, *Mol. Cell Biol.*, 6, 2241, 1986.

17. **Kimchie, Z., Segev, O., and Lev, Z.**, Maternal and embryonic transcripts of *Drosophila* protooncogenes are expressed in Schneider 2 culture cells but not in *l(2)gl* transformed neuroblasts, *Cell Differ.*, 26, 79, 1989.

18. **Cohen, N., Salzberg, A., and Lev, Z.**, A bidirectional promoter is regulating the *Drosophila ras2* gene, *Oncogene*, 3, 137, 1988.

19. **Schneider, I. and Blumenthal, A. B.**, *Drosophila* cell and tissue culture, in *The genetics and biology of Drosophila*, Ashburner, M. and Wright, T. R. F., Eds., Academic Press, New York, 1979, 265.

20. **Gateff, E. and Mechler, B. M.**, Tumor-suppressor genes of *Drosophila melanogaster, Crit. Rev. Oncol.*, 1, 221, 1989.

21. **Cohen, N., Kimchie, Z., and Lev, Z.**, unpublished results.

22. **Johansen, H., Van der Straten, A., Sweet, R., Otto, E., Maroni, G., and Rosenberg, M.**, Regulated expression at high copy number allows production of a growth-inhibitory oncogene product in *Drosophila* Schneider cells, *Genes Dev.*, 3, 882, 1989.

23. **Rubin, G. M.**, Signal transduction and the fate of the R7 photoreceptor in Drosophila, *Trends Genet.*, 7, 372, 1991.

24. **Rogge, R. D., Karlovich, C. A., and Banerjee, U.**, Genetic dissection of a neurodevelopmental pathway: *Son of sevenless* functions downstream of the sevenless and EGF receptor tyrosine kinases, *Cell*, 64, 39, 1991.

25. **Jones, S., Vignais, M.-L., and Broach, J. R.**, The CDC25 protein of *Saccharomyces cerevisiae* promotes exchange of guanine nucleotides bound to ras, *Mol. Cell Biol.*, 11, 2641, 1990.

26. **Crechet, J. B., Poullet, P., Camonis, J., Jacquet, M., and Parmeggiani, A.**, Different kinetic properties of the two mutants, RAS2Ile152 and RAS2Val19, that suppress CDC25 requirement in RAS/adenylate cyclase pathway in *Saccharomyces cerevisiae, J. Biol. Chem.*, 265, 1563, 1991.

27. **Lev, Z., Segev, O., Cohen, N., Salzberg, A., and Shemer, R.,** Structure of the *Drosophila Ras2* bidirectional promoter, in *Ras Oncogenes,* Spandidos, D., Ed., Plenum, New York, 1989, 75.

28. **Segev, O.,** personal communication.

29. **Salzberg, A. and Lev, Z.,** unpublished results.

30. **Bishop, J. G. and Corces, V. G.,** personal communication.

31. **Kitayama, H., Sugimoto, Y., Matsuzaki, T., Ikawa, Y., and Noda, M. A.,** *ras*-related gene with transformation suppressor activity, *Cell,* 56, 77, 1989.

32. **Dearolf, C. R., Hersperger, E., and Shearn, A.,** Developmental consequences of awdb3, a cell-autonomous lethal mutation of Drosophila induced by hybrid dysgenesis, *Dev. Biol.,* 129, 159, 1988.

33. **Biggs, J., Hersperger, E., Steeg, P. S., Liotta, L. A., and Shearn, A.,** A Drosophila gene that is homologous to a mammalian gene associated with tumor metastasis codes for a nucleoside diphosphate kinase, *Cell,* 63, 933, 1990.

34. **Randazzo, P. A., Northup, J. K., and Kahn, R. A.,** Activation of a small GTP-binding protein by nucleoside diphosphate kinase, *Science,* 254, 850, 1991.

35. **Teng, D. H., Engele, C. M., and Venkatesh, T. R.,** A product of the prune locus of Drosophila is similar to mammalian GTPase-activating protein, *Nature,* 353, 437, 1991.

36. **Han, M. and Sternberg, P. W.,** *let-60,* a gene that specifies cell fates during *C. elegans* vulval induction, encodes a ras protein, *Cell,* 63, 921, 1990.

37. **Beitel, G. J., Clark, S. G., and Horvitz, H. R.,** *Caenorhabditis elegans* ras gene *let-60* acts as a switch in the pathway of vulval induction, *Nature,* 348, 503, 1990.

38. **Johnston, P. A., Archer, B. T., Robinson, K., Mignery, G. A., Jahn, R., and Sudhof, T. C.,** rab3A attachment to the synaptic vesicle membrane mediated by a conserved poly-isoprenylated carboxy-terminal sequence, *Neuron,* 7, 101, 1991.

39. **Bonfini, L., Karlovich, C. A., Dasgupta, C., and Banerjee, U.,** The Son of sevenless gene product: a putative activator of Ras, *Science,* 255, 603, 1992.

40. **Fortini, M. E., Simon, M. A., and Rubin, G. R.,** *Nature,* 355, 559, 1992.

41. **Gaul, U., Mardon, G., and Rubin, G. M.,** *Cell,* 68, 1007, 1992.

42. **Dickson, B., Sprenger, F., Morrison, D., and Hafen, E.,** *Nature,* 360, 600, 1992.

43. **Salzberg, A., Cohen, N., Halachmi, N., Kimchie, Z., and Lev, Z.,** in press.

44. **Aalto, M. K., Keranen, S., and Ronne, H.,** *Cell,* 68, 181, 1992.

45. **Dascher, C., Ossig, R., Gallwitz, D., and Schmitt, H. D.,** *Mol. Cell Biol.,* 11, 872, 1991.

SECTION II
The *ras* Superfamily

Chapter 10

ras HOMOLOGS: A COMPARISON OF PRIMARY STRUCTURES

Pierre Chardin

TABLE OF CONTENTS

I. INTRODUCTION: THE DISCOVERY OF
ras-RELATED GENES

The discovery of H-*ras*, K-*ras*, and N-*ras* genes has been described in the first part of this book. In 1983, the same year that N-*ras* was discovered, an open reading frame located between the actin and tubulin genes of *Saccharomyces cerevisiae* was shown to encode a protein, YPT, sharing ~30% identity with mammalian *ras* proteins. This YPT protein was first considered as the yeast homolog of mammalian *ras*, but the RAS1 and RAS2 genes were soon discovered in *S. cerevisiae* and shown to encode proteins much more closely related to mammalian *ras* (see Chapter 7). It was, thus, clear that, at least in yeast, some proteins distantly related to *ras* were present. One year later the *rho* genes were discovered, first in *Aplysia*, then in human, rho proteins also shared ~30% identity with ras or YPT proteins. This second discovery of a protein distantly related to *ras* gave additional credit to the idea that ras proteins belonged to a family and the fact that these first two discoveries had been fortuitous strongly suggested that it was a large family. It, thus, seemed reasonable to use more systematic approaches to isolate new members of this family.

Two kinds of homology probing have been most instrumental in the discovery of new *ras*-related genes: the use of degenerated oligonucleotide mixes corresponding to conserved regions (usually the DTAGQE sequence around position 61), or low stringency hybridization with already isolated probes, sometimes in distant organisms. The other important strategy has been to isolate small G proteins from various tissues by biochemical methods, on the basis of their GTP/GDP binding ability. Sequencing of short peptides followed by cDNA cloning with corresponding oligonucleotide probes enabled the full sequence to be obtained. Several of these proteins turned out to be identical to the ones predicted by molecular genetics approaches.

Several yeast mutants were also shown to encode *ras*-related proteins, such as SEC4, a secretion mutant, CDC42, a cell division control mutant, or RSR1, a suppressor of CDC24; some *Dictyostelium* genes that have been isolated by their expression at a specific time in development, such as Dd *ras* or *sas*1 and *sas*2, are also related to *ras* and YPT genes respectively. And a *Caenorhabditis elegans* mutant of vulval induction (let-60) was found to encode a *ras* protein (see the first section of this book).

These approaches turn out to be so efficient that the primary structure of more than 60 *ras*-related proteins is now available from cloned cDNA sequences.[1-50]

The purpose of this chapter is to list the various sequences already published and to compare the primary structures of most of them. Several sequences recently determined in *Dictyostelium*,[46] in nematodes,[47] in *Mucor racemosus*,[48] in *Neurospora crassa*,[49] or in the electric ray[23] were not included since they encoded proteins very similar to the ones already isolated from other organisms. A *ras*-related gene has also been found in *Aplysia*;[50] partial sequences suggest that it might represent a new member of the *ras* branch, however, more sequence data would be needed to clarify this point and it was not included in the alignment.

Comparative analysis allows us to define conserved and variable regions that can be located on the three-dimensional model of the *ras* protein, providing important clues to the major functional regions and opening new ways to analyze their functions. A detailed analysis of the functional role of conserved amino acids has been published elsewhere,[51] and will only be briefly summarized here. Sequence comparisons also enable the classiiati elate pteins into four main branches and various subbranches allowing the prediction of functional specificities for each subclass of this family.

II. SEQUENCE COMPARISONS

A first alignment of closely related proteins in the *ras* family easily shows that *ras*-related proteins might be subdivided into four main parts: (1) The N terminal end, before the first conserved residue (K5) is highly variable in length and sequence, ranging from four amino acids in most *ras* proteins, *rap1*, CDC42 to 30 amino acids in R-*ras*; (2) amino acids 5 to 83 represent a first region of high conservation; (3) amino acids 84 to 164 are also conserved but to a lower extent, with two main boxes of homology; and (4) the C terminal part, from amino acid 165 to the C terminus is highly variable, even among closely related proteins such as H-*ras*/K-*ras*/N-*ras*, except for the last four amino acids that always include a cysteine and often fit the CAAX consensus for farnesylation or geranylgeranylation, cleavage and carboxymethylation (see Chapter 3).

The three-dimensional structure of the H-*ras* protein has been determined (see Chapter 2) and shows a single domain around the GTP/GDP binding site, with two contiguous regions representing the two "lobes" of a "heart-like" structure. Interestingly, the two main central regions of high and low conservation are corresponding to these two distinct parts of the molecule. The first one corresponds to the highly conserved region from amino acid 5 to 83 and is mainly the phosphate-binding site (P-site) with two main boxes of homology around positions 12 and 61, it includes two regions in position 32 to 36 and 61 to 68 that are the most mobile parts of the protein core, where the major changes from the GDP-bound to the GTP-bound conformations are observed. The second one from amino acid 84 to 164, conserved to a lower extent, has two main boxes of homology corresponding to the binding site of the guanine ring (G-site). In this region there are no major conformational changes from the GDP-bound state to the GTP-bound state. The structure of the C terminal end is not known, because it appears as a disordered part in the crystal, suggesting that this C terminal tail flips out of the major globular G-binding domain. This subdivision in four parts has been conserved in Figure 1 alignments.

A. THE N TERMINAL EXTENSION

In Figure 1A are listed the names of the genes, C in H-*ras*-C means Chicken H-*ras*, similarly Dd means *Dictyostelium*, D is for *Drosophila*, Y is for

GENE	Reference		N°		N-terminal extension
Hras	Capon et al.	1983	1	HRAS	MTEY
HrasC	Westaway et al.	1986	2	HRASC	MTEY
Kras	McGrath et al.	1983	3	KRAS	MTEY
Nras	Taparowsky et al.	1983	4	NRAS	MTEY
Dras1	Neuman-Silberberg	1984	5	DRAS1	MTEY
Dd rasG	Robbins et al.	1989	6	DdRASG	MTEY
Ddras	Reymond et al.	1984	7	DdRAS	MTEY
RASY1	DeFeo-Jones et al.	1983	8	RASY1	MQGNKSTMTEY
RASY2	Powers et al.	1984	9	RASY2	MPLNKSNIREY
RASSP	Fukui et al.	1985	10	RASSP	MRSTYLREY
TC21	Drivas et al.	1990	11	TC21	MAAAAGRLRQEKY
Dras2	Mozer et al.	1985	12	DRAS2	MQMQTY
Rras	Lowe et al.	1987	13	RRAS	MSSGAASGTGRGRPRGGGPGFGDPPPSETH
rap1A	Pizon et al.	1988	14	RAP1A	MREY
Drap1	Hariharan et al.	1991	16	DRAP1	MREY
rap1B	Pizon et al.	1988	15	RAP1B	MREY
rap2A	Pizon et al.	1988	14	RAP2A	MREY
rap2B	Ohmstede et al.	1990	17	RAP2B	MREY
RSR1	Bender and Pringle	1989	18	RSR1	MRDY
ralA	Chardin & Tavitian	1986	29	RALA	MAANKPKGQNSLALH
ralB	Chardin & Tavitian	1989	20	RALB	MAANKSKGQSSLALH

FIGURE 1. Sequence alignment of the *ras* superfamily of GTPases. Sequences are grouped as the members of each branch.

rhoA	Yeramian et al.	1987	21	RHOA	MAAIRK
rhoB	Chardin et al.	1988	22	RHOB	MAAIRK
rhoC	Chardin et al.	1988	22	RHOC	MAAIRK
rho-O	Ngsee et al.	1991	23	RHO-O	MAAIRK
rhoApl	Madaule and Axel	1985	24	RHOAP	MAAIRK
RHOY1	Madaule et al.	1987	25	RHOY1	MSQQVGNSIRR
rac1	Didsbury et al.	1989	26	RAC1	MQAI
rac2	Didsbury et al.	1989	26	RAC2	MQAI
G25KA	Shinjo et al.	1990	27	G25KA	MQTI
G25KB	Munemitsu et al.	1990	28	G25KB	MQTI
CDC42	Johnson & Pringle	1990	29	CDC42	MQTL
TC10	Drivas et al.	1990	11	TC10	MPGAGRSSMAHGPGALML
RHOY2	Madaule et al.	1987	25	RHOY2	MSEKAVRR

FIGURE 1 (continued).

rab1A	Touchot et al.	1987	30	RAB1A	MSSMNPEYDYLF
rab1B	Vielh et al.	1989	31	RAB1B	MNPEYDYLF
YPTm	Palme et al.	1988	32	YPTM	MSNEFDYLF
YPT1	Gallwitz et al.	1983	33	YPT1	MNSEYDYLF
YPTSP	Fawell et al.	1989	34	YPTSP	MNPEYDYLF
SAS1	Saxe and Kimmel	1988	35	SAS1	MTSPATNKSAAYDYLI
SAS2	Saxe and Kimmel	1988	35	SAS2	MTSPATNKPAAYDFLV
YPT2	Haubruck et al.	1990	36	YPT2	MSTKSYDYLI
SEC4	Salminen and Novick	1987	37	SEC4	MSGLRTVSASSGNGKSYDSIM
rab8	Chavrier et al.	1990	38	RAB8	MAKTYDYLF
rab10	Chavrier et al.	1990	38	RAB10	MAKKTYDLLF
rab3A	Zahraoui et al.	1988	39	RAB3A	MASATDSRYGQKESSDQNFDYMF
rab3B	Zahraoui et al.	1989	40	RAB3B	MASVTDGKHGVKDASDQNFDYMF
rab3C	Matsui et al.	1988	41	RAB3C	MRHEAPMQMASAQDARYGQKDSSDQNFDYMF
rab2	Touchot et al.	1987	30	RAB2	MAYAYLF
rab4A	Zahraoui et al.	1988	39	RAB4A	MSETYDFLF
rab4B	Chavrier et al.	1990	38	RAB4B	MAETYDFLF
YPT3	Miyake and Yamamoto	1990	42	YPT3	MCQEDEYDYLF
rab11	Chavrier et al.	1990	38	RAB11	MGTRDDEYDYLF
Ara	Matsui et al.	1989	43	Ara	MSSDDEGREEYLF
rab6	Zahraoui et al.	1989	40	RAB6	MSTYGGDFGNPLRKF
RYH1	Hengst et al.	1990	44	RYH1	MSENYSFSLRKF
rab5	Zahraoui et al.	1989	40	RAB5	MASRGATRPNGPNTGNKICQF
rab7	Bucci et al.	1988	45	RAB7	MLL
rab9	Chavrier et al.	1990	38	RAB9	?
ran/TC4	Drivas et al.	1990	11	ran/TC4	MAAQGEPQVQF

	5	15	25	35	44	54	64	74
HRAS	KLVVVGAGGV	GKSALTIQLI	QNHFVDEYDP	TIEDSYR-KQ	VVIDGETCLL	DILDTAGQEE	YSAMRDQYMR	TGEGFLCVFA
HRASC	KLVVVGAGGV	GKSALTIQLI	QNHFVDEYDP	TIEDSYR-KQ	VVIDGETCLL	DILDTAGQEE	YSAMRDQYMR	TGEGFLCVFA
KRAS	KLVVVGAGGV	GKSALTIQLI	QNHFVDEYDP	TIEDSYR-KQ	VVIDGETCLL	DILDTAGQEE	YSAMRDQYMR	TGEGFLCVFA
NRAS	KLVVVGAGGV	GKSALTIQLI	QNHFVDEYDP	TIEDSYR-KQ	VVIDGETCLL	DILDTAGQEE	YSAMRDQYMR	TGEGFLCVFA
DRAS1	KLVVVGAGGV	GKSALTIQLI	QNHFVDEYDP	TIEDSYR-KQ	VVIDGETCLL	DILDTAGQEE	YSAMRDQYMR	TGEGFLLVFA
DdRASG	KLVIVGGGGV	GKSALTIQLI	QNHFIDEYDP	TIEDSYR-KQ	VTIDGETCLL	DILDTAGQEE	YSAMRDQYMR	TQQGFLCVYS
DdRAS	KLVIVGGGGV	GKSALTIQLI	QNHFIDEYDP	TIEDSYR-KQ	VSIDDETCLL	DILDTAGQEE	YSAMRDQYMR	TQQGFLCVYS
RASY1	KIVIVGGGGV	GKSALTIQFI	QSYFVDEYDP	TIEDSYR-KQ	VVIDDKVSIL	DILDTAGQEE	YSAMREQYMR	TGEGFLLVYS
RASY2	KLVVVGGGGV	GKSALTIQLT	QSHFVDEYDP	TIEDSYR-KQ	VVIDDEVSIL	DILDTAGQEE	YSAMREQYMR	NGEGFLLVYS
RASSP	KLVVVGDGGV	GKSALTIQLI	QSHFVTDYDP	TIEDSYT-KK	CEIDGEGALL	DVLDTAGQEE	YSAMREQYMR	TGEGFLLVYN
TC21	RLVVVGGGGV	GKSAITIQFI	QSYFVTDYDP	TIEDSYT-KQ	CVIDGRAARL	DILDTAGQEE	FGAMREQYMR	TGEGFLLVFS
DRAS2	KLVVVGGGGV	GKSALTIQFI	QSYFVTDYDP	TIEDSYT-KQ	CNIDVPAKL	DILDTAGQEE	FSAMREQYMR	SGEGFLLVFA
RRAS	KLVVVGGGGV	GKSALTIQFI	QSYFVSDYDP	TIEDSYT-KI	CSVDGIPARL	DILDTAGQEE	FGAMREQYMR	AGHGFLLVFA
RAP1A	KLVVLGSGGV	GKSALTVQFV	QGIFVEKYDP	TIEDSYR-KQ	VEVDCQQCML	EILDTAGTEQ	FTAMRDLYMK	NGQGFALVYS
DRAP1	KIVVLGSGGV	GKSALTVQFV	QCIFVEKYDP	TIEDSYR-KQ	VEVDGQQCML	EILDTAGTEQ	FTAMRDLYMK	NGQGFVLVYS
RAP1B	KLVVLGSGGV	GKSALTVQFV	QGIFVEKYDP	TIEDSYR-KQ	VEVDAQQCML	EILDTAGTEQ	FTAMRDLYMK	NGQGFALVYS
RAP2A	KVVVLGSGGV	GKSALTVQFV	TGTFIEKYDP	TIEDFYR-KE	IEVDSSPSVL	EILDTAGTEQ	FASMRDLYIK	NGQGFILVYS
RAP2B	KVVVLGSGGV	GKSALTVQFV	TGSFIEKYDP	TIEDFYR-KE	IEVDSSPSVL	EILDTAGTEQ	FASMRDLYIK	NGQGFILVYS
RSR1	KLVVLGAGGV	GKSCLTVQFV	QGVYLDTYDP	TIEDSYR-KT	IEIINKVFDL	EILDTAGIAQ	FTAMRELYIK	SGMGFLLVYS
RALA	KVIMVGSGGV	GKSALTLQFM	YDEFVEDYEP	TKADSYR-KK	VVLDGEEVQI	DILDTAGQED	YAAIRDNYFR	SGEGFLCVFS
RALB	KVIMVGSGGV	GKSALTLQFM	YDEFVEDYEP	TKADSYR-KK	VVLDGEEVQI	DILDTAGQED	YAAIRDNYFR	SGEGFLLVFS

FIGURE 1 (continued).

```
RHOA   KLVIVGDGAC GKTCLLIVFS KDQFPEVYVP TVFENYV-AD IEVDGKQVEL ALWDTAGQED YDRLRPLSYP DTDVILMCFS
RHOB   KLVVVGDGAC GKTCLLIVFS KDEFPEVYVP TVFENYV-AD IEVDGKQVEL ALWDTAGQED YDRLRPLSYP DTDVILMCFS
RHOC   KLVIVGDGAC GKTCLLIVFS KDQFPEVYVP TVFENYI-AD IEVDGKQVEL ALWDTAGQED YDRLRPLSYP DTDVILMCFS
RHO-O  KLVIVGDGAC GKTCLLIVFS KDQFPEVYVP TVFENYV-AD IEVDGKQVEL ALWDTAGQED YDRLRPLSYP DTDVILMCFS
RHOAP  KLVIVGDGAC GKTCLLIVFS KDQFPEVYVP TVFENYV-AD IEVDGKQVEL ALWDTAGQED YDRLRPLSYP DTDVILMCFS
RHOY1  KLVIVGDGAC GKTCLLIVFS KGQFPEVYVP TVFENYV-AD VEVDGRRVEL ALWDTAGQED YDRLRPLSYP DSNVVLICFS
RAC1   KCVVVGDGAV GKTCLLISYT TNAFPGEYIP TVFDNYS-AN VMVDGKPVNL GLWDTAGQED YDRLRPLSYP QTDVFLICFS
RAC2   KCVVVGDGAV GKTCLLISYT TNAFPGEYIP TVFDNYS-AN VMVDSKPVNL GLWDTAGQED YDRLRPLSYP QTDVFLICFS
G25KA  KCVVVGDGAV GKTCLLISYT TNKFPSEYVP TVFDNYA-VT VMIGGEPYTL GLFDTAGQED YDRLRPLSYP QTDVFLVCFS
G25KB  KCVVVGDGAV GKTCLLISYT TNKFPSEYVP TVFDNYA-VT VMIGGEPYTL GLFDTAGQED YDRLRPLSYP QTDVFLVCFS
CDC42  KCVVVGDGAV GKTCLLISYT TNQFPADYVP TVFDNYA-VT VMIGDEPYTL GLFDTAGQED YDRLRPLSYP STDVFLVCFS
TC10   KCVVVGDGAV GKTCLLMSYA NDAFPEEYVP TVFDHYA-VS VTVGGKQYLL GLYDTAGQED YDRLRPLSYP MTDVFLICFS
RHOY2  KLVIIGDGAC GKTSLLVFT  LGKFPEQYHP TVFENYV-TD CRVDGIKVSL TLWDTAGQEE YERLRPFSYS KADILIGFA
```

```
RAB1A     KLLLIGDSGV GKSCLLLRFA DDTYTESYIS TIGVDFKIRT IELDGKTIKL QIWDTAGQER FRTITSSYYR GAHGIIVVYD
RAB1B     KLLLIGDSGV GKSCLLLRFA DDTYTESYIS TIGVDFKIRT IELDGKTIKL QIWDTAGQER FRTVTSSYYR GAHGIIVVYD
YPTM      KLLLIGDSSV GKSCFLLRFA DDSYVDSYIS TIGVDFKIRT VEVEGKTVKL QIWDTAGQER FRTITSSYYR GAHGIIIVYD
YPT1      KLLLIGNSGV GKSCLLLRFS DDTYTNDYIS TIGVDFKIKT VELDGKTVKL QIWDTAGQER FRTITSSYYR GSHGIIIVYD
YPTSP     KLLLIGDSGV GKSCLLLRFA DDTYTESYIS TIGVDFKIRT FELEGKTVKL QIWDTAGQER FRTITSSYYR GAHGIIIVYD
SAS1      KLLLIGDSGV GKSCLLLRFS EDSFTPSFIT TIGIDFKIRT IELBGKRIKL QIWDTAGQER FRTITTAYYR GAMGILLVYD
SAS2      KLLLIGDSGV GKSCLLLRFS DGSFTPSFIA TIGIDFKIRT IELEGKRIKL QIWDTAGQER FRTITTAYYR GAMGILLVYD
YPT2      KLLLIGDSGV GKSCLLLRFS EDSFTPSFIT TIGIDFKIRT IELDGKRIKL QIWDTAGQER FRTITTAYYR GAMGILLVYD
SEC4      KILLIGDSGV GKSCLLVRFV EDKFNPSFIT TIGIDFKIKT VDINGKKVKL QLWDTAGQER FRTITTAYYT GAMGIILVYD
RAB8      KLLLIGDSGV GKTCVLFRFS EDAFNSTFIS TIGIDFKIRT IELDGKRIKL QIWDTAGQER FRTITTAYYR GAMGIMLVYD
RAB10     KLLLIGDSGV GKTCVLFRFS DDAFNTFIS  TIGIDFKIKT VELAGKKIKL QIWDTAGQER FHTITTSYYR GAMGIMLVYD
RAB3A     KILIIGNSSV GKTSFLFRYA DDSFTPAFVS TVGIDFKVKT IYRNDKRIKL QIWDTAGQER YRTITTAYYR GAMGFILMYD
RAB3B     KLLIIGNSSV GKTSFLLRYA DDTFTPAFVS TVGIDFKVKT VYRHEKRVGL QIWDTAGQER YRTITTAYYR GAMGFILMYD
RAB3C     KLLIIGNSSV GKTSFLFRYA DDSFTSAFVS TVGIDFKVKT VFKNEKRIKL QIWDTAGQER YRTITTAYYR GAMGFILMYD
RAB2      KYIIIGDTGV GKSCLLLQFT DKRFQPVHDL TIGVEFGARI ITIDGKQIKL LIWDTAGQES FRWMTRSYYR GAAGALLVYD
RAB4A     KFLVIGNAGT GKSCLLHQFI EKKFKDDSNH TIGVEFGSKI INVGGKYVKL QIWDTAGQER FRSVTRSYYR GAAGALLVYD
RAB4B     KFLVIGSAGT GKSCLLHQFI ENKFKQDSNH TIGVEFGTRS IQVDGKTIKA QIWDTAGQER YRAITSAYYR GAVGALLVYD
YPT3      KTVLIGDSGV GKSNLLMRFT RNEFNIESKS TIGVEFATRN IVLDNKKIKA QIWDTAGQER YRAITSAYYR GAVGALIVYD
RAB11     KVVLIGDSGV GKSNLLSRFT RNEFNLESKS TIGVEFATRS IQVDGKTIKA QIWDTAGQER YRAITSAYYR GAVGALLVYD
Ara       KIVVIGDSAV GKSNLLDRYA RNEFSANSKA TIGVEFQTQS MEIBKEVKA  QIWDTAGQER FRAVTSAYYR GAVGALVVYD
RAB6      KLVFLGEQSV GKTSLITTRFM YDSFINTYQA TIGIDFLSKT MYLEDRTVRL QLWDTAGQER FRSLIPSYIR DSTVAVVVYD
RYH1      KLVFLGEQSV GKTSLIITRFM YDQFINTYQA TIGIDFLSKT MYLEDRTVRL QLWDTAGQER FRSLIPSYIR DSSVAIIVYD
RAB5      KLVLLGESAV GKSSLVLRFV KGQFHEFQES TIGAAFLTQT VCLIDTTVSL NIWDTAGQEG YHSLAPMYYR GAQAAIVVYD
RAB7      KVIILGDSGV GKTSLMNQYV NKKFSNQYKA TIGADFLTKE VMVDDRLVTM QIWDTAGQER FQSLGVAFYR GADCCVLVFD
RAB9                 NKFDIQLFH TIGVEFLNKD LEVDGHFVTM QIWDTAGQER FRSLRTPFYR GSDCCLLTFS
ran/TC4   KLVLVGDGGT GKTTFVKRHL TGEFEKKYVA TLGVEVHPLV FHTNRGPIKF NVWDTAGQEK FGGLRDGYYI QAQCAIIMFD
```

FIGURE 1 (continued).

```
84         94         104        114        123        133        142        152        162
INNTKSFEDI HQYREQIKRV KDSDDVPMVL VGNKCDLAA- RTVESRQAQD LARSYG-IPY IETSAKTRQG VEDAFYTLVR EIR
INNTKSFEDI HQYREQIKRV KDSDDVPMVL VGNKCDLPA- RTVETRQAQD LARSYG-IPY IETSAKTRQG VEDAFYTLVR EIR
INNTKSFEDI HHYREQIKRV KDSEDVPMVL VGNKCDLPS- RTVDTKQAQD LARSYG-IPF IETSAKTRQG VDDAFYTLVR EIR
INNSKSFADI NLYREQIKRV KDSDDVPMVL VGNKCDLPT- RTVDTKQAHE LAKSYG-IPF IETSAKTRQG VEDAFYTLVR EIR
VNSAKSFEDI GTYREQIKRV KDAEEVPMVL VGNKCDLAS- WNVNNEQARE VAKQYG-IPY IETSAKTRMG VDDAFYTLVR EIR
ITSRSSFDEI ASFREQILRV KDKDRVPMIV VGNKCDLESD RQVTTGEGQD LAKSFG-SPF LETSAKIRVN VEEAFYSLVR EIR
ITSRSSYDEI ASFREQILRV KDKDRVPLIL VGNKADLDHE RQVSVNEGQE LAKD-S-LSF HESSAKSRIN VEEAFYSLVR EIR
VTSRNSFDEL LSYQQIQRV  KDSDYIPVVV VGNKLDLENE RQVSYEDGLR LAKQLN-APF LETSAKQAIN VDEAFYSLIR LVR
ITSKSSLDEL MTYYQQILRV KDTDYVPIVW VGNKSDLENE KQVSYQDGLN MAKQMN-APF LETSAKQAIN VEEAFYTLAR LVR
ITSRSSFDEI STFYQQILRV KDKDTFPVVL VANKCDLEAE RVVSRAEGEQ LAKSMH-CLY VETSAKLRLN VEEAFYSLVR TIR
VTDRGSFEEI YKFQRQILRV KDRDEFPMIL IGNKADLDHQ RQVTQEEGQQ LARQLK-VTY MEASAKIRMN VDQAFHELVR VIR
LNDHSSFDEI PKFQRQILRV KDRDEFPMLM VGNKCDLKHQ QQVSLEEAQN TSRNLM-IPY IECSAKLRVN VDQAFHELVR IVR
INDRQSFNEV GKLFTQILRV KDRDDFPVVL VGNKADLESQ RQVPRSEASA FGASHM-VAY FEASAKLRLN VDEAFEQLVR AVR
ITAQSTFNDL QDLRBQILRV KDTEDVPMIL VGNKCDLEDE RVVGKEQGQN LARQMN-CAF LESSAKSKIN VNEIFYDLVR QIN
ITAQSTFNDL QDLRBQILRV KDTDDVPMVL VGNKCDLEEE RVVGKELGKN LAITQFN-CAF METSAKAKVN VNDIFYDLVR QIN
ITAQSTFNDL QDLRBQILRV KDTDDVPMIL VGNKCDLEDE RVVGKEQGQN LARQMN-CAF LESSAKSKIN VNEIFYDLVR QIN
LVNQQSFQDI KPMRDQIIRV KRYEKVPVIL VGNKVDLESE REVSSSEGRA LAEEMG-CPF METSAKSKTM VDELFAEIVR QMN
LVNQQSSQDI KPMRDQIIRV KRYERVPMIL VGNKVDLEGE REVSYGEGKA LAEEMS-CPF METSAKNRAS VDELFAEIVR QMN
VTDRQSLEEL MELRBQVLRI KDSDRVPMVL IGNKADLINE RVISVEEGIE VSSKWGRVPF YETSALLRSN VDEVFVDLVR QII
ITEMESFAAT ADFRBQILRV KEDENVPFLL VGNKSDLEDK RQVSVEEAKN RADQWN-VNY VETSAKTRAN VDKVFFDLMR EIR
ITEHESFTAT AEFRBQILRV KEEDKIPLLV VGNKSDLEER RQVPVEEARS KAEEMG-VQY VETSAKTRAN VDKVFFDLMR EIR
```

```
IDSPDSLENI PEKWTPEVK- HFCPNVPIILL VGNKKDLRND EPVKPEEGRD MANRIGAFGY MECSAKTKDG VREVFEMATR AAL
VDSPDSLENI PEKWVPEVK- HFCPNVPIILL VANKKDLRSD EPVRTDDGRA MAVRIQAYDY LECSAKTKEG VREVFETATR AAL
IDSPDSLENI PEKWTPEVK- HFCPNVPIILL VGNKKDLRQD EPVRSEEGRD MANRISAFGY LECSAKTKEG VREVFEMATR AGL
IDSPDSLENI PEKWTPEVK- HFCPNIPIILL VGNKKTAGDD EPVKPDDAKE MGSRIKAFGY LECSAKTKEG VREVFELASR AAL
IDSPDSLENI PEKWTPEVR- HFCPNVPIILL VGNKKDLRND EPVRPEDGRA MAEKINAYSY LECSAKTKEG VRDVFETATR AAL
IDLPDSLENV QEKWIAEVL- HFCQGVPIILL VGCKVDLRND QPVTSQEGQS VADQIGATGY YECSAKTGYG VREVFEAATR ASL
LVSPASFENV RAKWYPEVR- HHCPNTPIILL VGTKLDLRDD TPITYPQGLA MAKEIGAVKY LECSALTQRG LKTVFDEAIR AVL
LVSPASYENV RAKWFPEVR- HHCPSTPIILL VGTKLDLRDD APITYPQGLA LAKEIDSVKY LECSALTQRG LKTVFDEAIR AVL
VVSPSSFENV KEKWVPEIT- HHCPKTPFLL VGTQIDLRDD KPITPETAEK LARDLKAVKY VECSALTQRG LKNVFDEAIL AAL
VVSPSSFENV KEKWVPEIT- HHCPKTPFLL VGTQIDLRDD KPITPETAEK LARDLKAVKY VECSALTQRG LKNVFDEAIL AAL
VISPPSFENV KEKWFPEVH- HHCPGVPCLV VGTQIDLRDD RPITSEQCSR LARELKAVKY VECSALTQRG LKNVFDEAIV AAL
VVNPASFQNV KEEWVPELK- EYAPNVPFLL IGTQIDLRDD KPICVEQQQK LAKEIGACCY VECSALTQKG LKTVFDEAII AIL
VDNFESLINA RTKWNADEAL- RYCPDAPIVL VGLKKDLRQE EMVPIEDAKQ VARAIGAKKY MECSALTGEG VDDVFEVATR TSL
```

FIGURE 1 (continued).

```
VTDQESFNNV K-QVVLQEIDR YASENVNKLL VGNKCDLTTK KVVDYTTAKE FADSLG-IPF LETSAKNATN VEQSFMTMAA EIK
VTDQESYANV K-QVVLQEIDR YASENVNKLL VGNKSDLTTK KVVDNTTAKE FADSLG-VPF LETSAKNATN VEQAFMTMAA EIK
ITDMESFNNV K-QVVLDEIDR YANDSVRNVL VGNKCDLAEN RAVDTSVAQA YAQEVG-IPF LETSAKESIN VEEAFLAMSA AIK
VTDQESFNGV K-MVVLQEIDR YATSTVLKLL VGNKCDLKOK RVVEYDVAKE FADANK-MPF LETSALDSTN VEDAFLTMAR QIK
VTDQDSFNNV K-QVVLQEIDR YAVEGVNRLL VGNKSDMVDK KVVEYSVAKE FADSLN-IPF LETSAKDSTN VEQAFLTMSR QIK
VTDEKSFGNI R-NMIRNIEQ HATDSVNKML IGNKCDMAEK KVVDSSRGKS LADEYG-IKF LETSAKNSIN VEEAFISLAK DIK
VTDEKSFGSI R-NMIRNIEQ HASDSVNRML IGNKCDMTEK KVVDSSRGKS LADEYG-IKF LETSAKNSVN VEEAFIGLAK DIK
VTDKKSFDNV R-TWFSNVEQ HASENVYKIL IGNKCDCEDQ RQVSFEQGQA LADELG-VKF LEAASKTNVN VDEAFFTLAR EIK
VTDERIFTNI K-QVVFKTVNE HANDEAQLLL VGNKSDM-ET RVVTADQGEA LAKQLG-IPF IESSAKNDDN VNEIFFTLAK LIQ
ITNEKSFDNI R-NMIRNIEE HASADVEKMI LGNKCDVNDK RQVSKERGEK LALDYG-IKF METSAKANIN VENAFFTLAR DIK
ITNGKSFENI S-KVVLRNIDE HANEDVERML LGNKCDMDDK RVVPKGKGEQ IAREHG-IRF FETSAKVNIN IEKAFLTLAE DIL
ITNEESFNAV Q-DWSTQIKT YSWDNAQVLL VGNKCDMEDE RVVSSERGRQ LADHLG-FEF FEASAKDNIN VKQTFERLVD VIC
ITNEESFNAV Q-DWATQIKT YSWDNAQVIL VGNKCDMEEE RVVPTEKGQL LAEQLG-FDF FEASAKENIS VRQAFERLVD AIC
ITNEESFNAV Q-DWSTQIKT YSWDNAQVIL VGNKCDMEDE RVISTERGQH LGEQLG-FEF FETSAKDNIN VKQTFERLVD IIC
ITRRDTFNHL T-TWLEDARQ HSNSNMVIML IGIKSDLESR REVKEEGEA FAREHG-LMF METSAKTASN VEEAFINTAK EIY
ITSRETYNAL T-NMLTDARM LASQNIVIIL CGNKKOLDAD REVTFLEASR FAQENE-LMF LETSALTGED VEEAFVQCAR KIL
IAKHLTYENV E-RWLKELRT LASPNIVVIL CGNKKDLDPE REVTFLEASR FAQENE-LMF LETSALTGEN VEEAFLKCAR TIL
ITKQSSFDNV G-RWLKELRE HADSNIVIML VGNKTDLLHL RAVSTEEAQA FAE-NN-LSF IETSAMDASN VEEAFQTVLT EIF
IAKHLTYENV E-RWLKELRD HADSNIVIML VGNKSDLRHL RAVPTDEARA FAEKNG-LSF IETSALDSTN VEAAFQTLIT EIY
ITRRITFESV G-RWLDELKI HSDTTVARML VGNKCDLENI RAVSVEEGKA LAEEBG-LFF VETSALDSTN VKTAFEMVIL DIY
ITNVNSFQQT T-KWIDDVRT ERGSDVIIML VGNKTDLADK RQVSIEEGER KAKELN-VMF IETSAKAGYN VKQLFRRVAA ALP
ITNHNSFVNT E-KWIEDVRA ERGDDVIVL VGNKTDLADK RQVTQEEGEK KAKELK-IMH METSAKAGHN VKLLFRKIAQ MLP
ITNEESFARA K-NMVKELQR QASPNIVIAL SGNKADLANK RAVDFQEAQS YADDNS-LLF METSAKTSMN VNEIFMAIAK KLP
VTAPNTFKTL D-SMRDEFLI QASFNFPFVV LGNKIDLENR QVATKRAQAW CYSKNN-IPY FETSAKEAIN VEQAFQTIAR NAL
VDDSQSFQNL S-NMKKEFIY YADVSFPFVI LGNKIDISER QVSTEEAQAW CRDNGD-YPY FETSAKDATN VAAAFEEAVR RVL
VTSRVTYKNV P-NMHRDLVR VC-ENIPIVL CGNKVDIKDS KVKAKSIVFH RKKNLQ-YY- -DISAKSNYN FEKPFLWLAR KLI
```

```
HRAS                 QHKLRKLNPPDESGPGCMSCKCVLS
HRASC                QHKLRKLNPPDESGPGCMNCKCVLS
KRAS (A)             QYRLKKISKEKTPGCVKIKKCIIM
NRAS                 QYRMKKLNSSDDGTQGCMGLPCVVM
DRAS1                KDKDNKGRRGRKMNKPNCRFKCKML
DdRASG               KDLKGDSKPEKGKKKRPLKACTLL
DdRAS                KELKGDQSSGKAQKRKKQCLIL
RASY1     110 a.a.   RSKQSAEPQKNSSANARKEYSGGCIIC
RASY2     120 a.a.   NAKQARKQQAAPGGNTSEASKSGSGGCIIS
RASSP                RYNKSEEKGFQNKQAVOCVIC
TC21                 KFQEQECPPSPEPTRKEKDKKGCHCVIF
DRAS2                KFQIAQRPF-IEQDYKKKGKRKCCLM
RRAS                 KYQEQELPPSPSSAPRKKGGGCPCVLL
RAP1A                RKTP-VEKKKPKKKSCLLL
DRAP1                KKSPEKKQKKPKKSLCVLL
RAP1B                RKTPVPGKARKKSSQQLL
RAP2A                YAAQPDKDDPCCSACNIQ
RAP2B                YAAQSNGDEGCCSACVIL
RSR1    80a.a.       SNRTGISATSQQKKKKNASTCTTL
RALA                 ARKMEDSKEKNGKKKRKSLAKRIRERCCIL
RALB                 TKKMSENKDKNGKKSSKNKKSFERCCLL
```

(B)   KHKEKMSKDGKKKKKKSKTKCVIM

FIGURE 1 (continued).

RHOA	QARRGKKKSGCLVL
RHOB	QKRYGSQNGCINCCKVL
RHOC	QVRKNKRRRGCPIL
RHO-O	QAKTKSKSPCLLL
RHOAP	QVKKKKGGCVVL
RHOY1	MGKSKTNGKAKKNTEKKKKCVLL
RAC1	CPPPVKKRKRKCLLL
RAC2	CPQPTRQQKFACSLL
G25KA	EPPEPKKSRRCVLL
G25KB	EPPETQPKRKCCIF
CDC42	EPPVIKKSKKCTIL
TC10	TPKKHTVKKRIGSRCINCCLIT
RHOY2	LMKKEPGANCCIIL

RAB1A	KRMGPGATAGAEKSNVKIQSTPVKQSGGGCC
RAB1B	KRMGPGAASGG-ERPNLKIDSTPVKSASGGCC
YPTM	KSKAGSQAALERKPSNVVQMKGRPIQQEQQKSSRCCST
YPT1	QSMSQQNLNETTQKKEDKGNVNLKGQSLTNTGGGCC
YPTSP	ERMGNNTFASSNAKSSVKVGQGTNVSQSSSNCC
SAS1	KRMIDTPNEQPQVVQPGTNLGANNNKKKACC
SAS2	KRMIDTPND-P----DHTICITPNNKRNTCC
YPT2	KQKIDAENEFSNQANNVDLGNDRTVKRCC
SEC4	EKIDSNKLVGVGNGKEGNISINSGSNSSKSNCC
RAB8	AKMDKKLEGNSPQGSNQGVKITTPDQQKRSSFFRCVLL
RAB10	RKTPVKEPNSENVDISSGGGVTGWKSKCC
RAB3A	EKMSESLDTADPAVTGAKGPQLSDQQVPPHQDCAC
RAB3B	DKMSDSLDT-DPSMLGSSKNTRLSDTPPLLQQNCSC
RAB3C	DKMSESLET-DPAITAAKQNTRLKETPPPPQPNCGC
RAB2	EIYEKIQEGVFDINNEANGIKIGFQHAATNATHAGNQGQQAGGGCC
RAB4A	NKIESGELDPERMGSGIQYGDAALRQLRSPRRTQAPNAQECGC
RAB4B	NKIDSGELDPERMGSGIQYGDASLRQLRQPRSAQAVAPQPCGC
YPT3	RIVSNRSLEAGIDGVHPTAGQTLNIAPTMNDLNKKKSSSQCC
RAB11	RIVSQKQMSDRRRENDMSPSNNVVPIHVPPTTENKPKVQQCQNI
Ara	NNVSRKQLNSDTYKDELTVNRVSLVKDINSASKQSSGFSOCSST
RAB6	GMESTQDRSREDMIDIKLEKPQEQPVSEQGCSC
RYH1	GMENVETQSTQMIDVSIQPNENESSQNC
RAB5	KNEPQNPGANSARGGGVDLTEPTQPTRNQOCSN
RAB7	KQETEVELYNEFPEPIKLDKNERAKASAESCSC
RAB9	ATEDRSDHLIQTDTVSLHRKPKPSSSCC
ran/TC4	GDPNLEFVAMPLSPHQKLSWTQLWQHSMSTT

FIGURE 1 (continued).

Saccharomyces cerevisiae, SP for *Schizosaccharomyces pombe*, genes written in capital letters are from *S. cerevisiae*. After the name of the gene is listed the reference of its first description. On the right side appears the sequence of the N terminal extension of the protein, before K5, the first strictly conserved residue. Unless otherwise stated, all numberings refer to H-*ras*. In *ras* there are only four residues from the N terminal end to the first highly conserved amino acid (K5), however, this short tail sticks out of the major globular domain, so the structure can accommodate the much longer N terminal parts found in some other *ras*-related proteins such as R-*ras*, *rab3*, *rab5*, or SEC4 where 20 to 30 amino acid extensions can probably form small additional domains that might be involved in the interaction with specific proteins.

B. THE PHOSPHATE-BINDING PART

In Figure 1B are aligned the sequences of the first highly conserved part, from residues 5 to 83, by blocks of 10 amino acids. The GXXXXGKS/T motif around position 15 (where S/T means serine or threonine at this position) is found in all *ras*-related proteins, in ARFs, Sar1p, α-subunits of heterotrimeric G proteins, elongation factors, SRP receptor, and in many other nucleotide-binding proteins. This region adopts a very rigid loop structure with the side chain of lysine (K16) folding back to interact with the main chain carbonyl groups of G10 and A11. This seems to be an essential motif of many phosphate-binding sites. In normal *ras*, glycines are always found in position 12 and 13, and replacement of G12 by any other amino acid except proline leads to deep structural changes and the acquisition of a transforming potential for *ras* proteins (see the first section of this book). However, the Ψ_2 angles of these G12 and G13 are not very unusual. We found that some *rab* proteins such as *rab3*, that have neither G12 nor G13, still possess a GTPase activity comparable to p21 *ras* GTPase, *rho* proteins that have an alanine insteady of glycine 13 also possess a GTPase comparable or even higher than *ras*. This provides further arguments that G12 and G13 are not absolute requirements for the GTPase activity and are not directly involved in the γ-phosphate hydrolysis. The reason why substitution of G12 by any other amino acid except proline results in lowered GTPase and transforming activity in *ras* is probably that *ras* cannot tolerate large-side chains at this position without affecting the whole structure of the P-site. Amino acids 25 to 27 are at the beginning of loop L2, F28 stabilizes the binding of the guanine through hydrophobic interactions, the aromatic ring being perpendicular to the plane of the guanine ring, a conformation often found in other nucleotide-binding proteins. Amino acids 30 to 34 represent the most detached loop of the structure with all side chains going outside; it is, thus, difficult to explain why Y/F 32 is strictly conserved. It might come close to the magnesium ion and β-phosphate in the GDP-bound conformation. In the GTP-bound form, the hydroxyl group of the T35 side chain is close to an oxygen of the γ-phosphate and is also involved in the coordination of the Mg^{2+} ion. The I36 side chain is going outside, a surprising feature regarding its hydrophobicity. Then amino acids 37 to 45 represent a β-sheet with most side chains sticking outside. Residues with side chains going outside are usually the most variable. It is noteworthy that the only exception to this correlation between "variable" and "external" regions are regions 36 to 48 and 61 to 68, where most of the side chains stick outside in both the GDP and GTP structures of p21 *ras*, but where a rather high conservation is found, especially among the diverse members of each branch. The three-dimensional model of p21 *ras* cannot explain this conservation, it might be that the interaction of p21 with its target induces an important conformational change in this region that would be important for p21 function, but could not be predicted from the structure of the isolated protein. Since region 36 to 46 represents the site of GAP interaction, and is rather conserved, there is probably a family of GAP proteins, all of them having similarities in the sites of interaction with the various proteins of the *ras* family. However, these structural similarities of GAP-like proteins might be difficult to recognize at the sequence level. The site for ADP-ribosylation of *rho* proteins by Botulinum exo-enzyme C3 (Asn41 of *rho*) is at position 39 in *ras* numbering, this modification results in a dramatic effect on *rho* function *in vivo* as demonstrated by the phenotypic changes; however, no effects of ADP-ribosylation on biochemical properties have been found, again suggesting the major importance of this region *in vivo*, that is not understood by *in vitro* studies of isolated proteins.

Amino acids 46 to 49 form a sharp turn with acidic side chains interacting with the basic side chains of positions 161 and 164. In this region a protein that would recognize the C terminal part could displace loop 46 to 49 and the conformational change might propagate to the P-site. The second highly conserved region is the DTAGQE motif around position 60. This region is located close to the β- and γ-phosphates of GTP. D57 is coordinated to magnesium in the GDP-bound form, and to a water molecule that contacts magnesium in the GTP-bound form. G60 is hydrogen-bounded to the γ phosphate of GTP and participates in the conformational change after hydrolysis of this γ phosphate.

The side chain of Q61 is able to adopt two different conformations, one where the side chain is sticking outside, and the other where the side chain is coming close to a water molecule contacting the γ-phosphate. Thus, in this conformation the side chain of glutamine 61 is likely to play an important role in the intrinsic GTPase. The *rap1* and *rap2* proteins possess a T insteady of this Q61 and have a lower GTPase activity than *ras*, so it seems that the replacement of this glutamine by a threonine might significantly affect the GTPase rate. However, the hydrolysis of GTP in the cell is several orders of magnitude quicker than in isolated proteins and involves GAP-like proteins, suggesting that the rate of the intrinsic GTPase might be of little physiological significance. The hydrolytic mechanism in the presence of GAP is unknown. Heterotrimeric G proteins also have a Q in the position corresponding to *ras* Q61 and it has been found that the substitution of this glutamine by an arginine significantly reduces the GTPase rate, too.

Region 61 to 68 is a loop and beginning of an α-helix with several side chains going outside. We have recently shown that a mutation in this region impairs GAP stimulation of GTP hydrolysis, but not GAP interaction *in vitro*, suggesting that this region is important for the hydrolytic process induced by GAP. It is also the major "switch region" changing in the GDP- vs. GTP-bound conformations, and thus, a potential interaction site for proteins specifically recognizing the GDP- or the GTP-bound forms.

C. THE GUANINE-BINDING PART

In Figure 1C are listed the sequences of the second conserved region from residues 84 to 164, also by blocks of 10. Some insertions have been omitted for clarity: one insertion in *ralB* between amino acids 104 and 105, a 12 or 13 amino acid insertion is found in all *rho* proteins between positions 122 and 123 (see second paragraph of discussion), a four amino acid insertion in *rab7* between positions 107 and 108. There are two highly conserved regions, both involved in the binding of the guanine ring. The LVGNKXDL sequence: N116 side chain interacts both with the hydroxyl group of T144 side chain and with the N7 atom of the guanine ring. K117 interacts with an oxygen of the ribose and D119 side chain interacts both with S145 side chain, N1 atom, and NH_2 group of the guanine ring. This motif is conserved in most G proteins and seems to be essential for the specificity of guanine recognition.

The ETSAK motif is highly conserved in all proteins of the *ras* family. E143 makes a hydrogen bond with Y141, and a salt bridge with R123, T144 side chain OH interacts with Asn116, and it might be noted that the other amino acids found in this position are only serine, also with an OH group or cysteine with a SH group. S145 interacts with the side chain of D119. The K147 long hydrophobic side chain is staking on one site of the guanine ring and probably stabilizes its binding through hydrophobic interactions. Consistent with this interpretation, in some proteins it is replaced by leucine that also has a long hydrophobic side chain. The ETSAK motif is not conserved in the other known G proteins.

D. THE C TERMINAL EXTENSION AND CAAX MOTIF

In Figure 1D are listed the C terminal extensions, after the last highly conserved residue at position 164. These sequences were arbitrarily aligned to match the CAAX boxes or were aligned on the last cysteine when the consensus CAAX could not be found (*rab* and YPT proteins).

The C terminal extensions are highly variable in size and sequence and probably act as a flexible spacer arm from the membrane-bound C terminus to the beginning of the globular domain, in the cytosol. Although their sequences are variable, they are mainly homogeneous in length in each of the three branches, with the exception of *S. cerevisiae* proteins. The shortest C terminal extensions are found in the *rho* branch (14 to 17 amino acids), and thus, in these proteins the major globular domain is expected to be closer to the membrane. The proteins of the *ras* branch have extensions of 18 to 30 amino acids while in the *rab* branch their length extends from 27 to 47 amino acids. It is clear than an extension of 47 amino acids can represent a small additional domain of the protein. In *S. cerevisiae ras* the C terminal domains are as large as the guanine nucleotide-binding domain. It is believed that this C terminal part that comes close to the membrane and is the most characteristic of each protein provides the specificity for the interaction with an integral membrane protein (or with other transiently membrane-bound proteins) that can have an "exchange factor" function.

In *ras* proteins ending with a consensus CAAX, the cysteine is farnesylated, the last three residues are clipped, and the now C terminal cysteine is carboxymethylated. These modifications help in anchoring the C terminal in the membrane (see Chapter 3). Among the many proteins ending with a CAAX consensus, those where X, the C terminal residue, is Met, Ser, or Cys are farnesylated, while those where X is Leu (or Phe or Asn) are geranylgeranylated. The *rab* proteins ending with CC or CXC also seem to be geranylgeranylated. However, the GGGCC sequence found in yeast YPT is conserved in its mammalian homolog *rab1*, and the CNC sequence found in RYH1 is CSC in its mammalian homolog *rab6*, this evolutionary conservation suggests that additional sequence requirements not yet understood can provide some kind of specificity. The *ran*/TC4 protein does not end with a cysteine motif and is not membrane-bound but has a nuclear localization.

III. DISCUSSION

A. INTERNAL/EXTERNAL RESIDUES AND POTENTIAL TARGETS FOR INTERACTING PROTEINS

Figure 2 summarizes the alignments and defines a consensus sequence. The upper line is the sequence of H-*ras* as a reference. Strictly conserved positions are indicated by a star underneath. Positions where only two or three amino acids can account for the residue found in most proteins are also indicated, for instance, in position 6: L, V, or C are found in most proteins (compare Figure 2 and the alignments in Figure 1B). Dashes indicate more variable positions. The four main conserved regions involved in GDP/GTP binding might easily be seen on this figure, while a few other amino acids such as K5, Y71, and F156 are strictly conserved. The aromatic rings of Y71 and F156 are going inside and stabilize the structure through many hydrophobic interactions. In many other positions where side chains are going inside, the conservation is not strict, but only amino acids with hydrophobic side chains are found. The reason why K5 is conserved is not understood. Besides these, two additional regions in positions 35 to 48 and 64 to 85 are also conserved, although to a lower extent; higher degrees of conservation are found in each branch, suggesting that all proteins in a given branch might recognize related targets interacting with these conserved positions. Regions 25 to 43, 47 to 50, 61 to 68, 85 to 88, 104 to 108, 121 to 124, 126 to 129, 131 to 133, 135 to 138, 148 to 150, and 161 to 164 are the main regions where most of the side chains stick outside. They might provide the specificity for the interaction of each protein with its specific regulators and/or targets. In *ras*, it is already known that the two main switch regions in positions 32 to 36 and 61 to 68 are involved in GAP interaction, we can already predict that proteins specifically interacting with the GDP-bound form will also interact with these regions since they are the only ones where conformational differences between the GDP- or GTP-bound forms might be seen. The other external regions and N or C terminal extensions are potential targets for the exchange factor or other uncharacterized proteins.

B. INSERTIONS AND DELETIONS

Among all insertions and deletions, several are branch-specific providing additional arguments for the existence of four well-defined branches.

ras **branch** — a deletion between positions 122 and 123 appeared in the *ras* sub-branch before *Drosophila*, it is not found in yeast *ras*, *Dictyostelium ras* or in other sub-branches of the *ras* branch, so it has appeared after the divergence of the phylums leading to yeast and insects/vertebrates. This also strongly suggests that mammalian H-*ras*, K-*ras*, and N-*ras* appeared by duplications of a single ancestral vertebrate gene, followed by some divergence.

rho **branch** — a single deletion is found in position 103 and one insertion between amino acids 138 and 139. A 12 amino acid insertion is found in all proteins of the *rho* branch between positions 122 and 123. The sequences of these insertions are given here:

```
RHOA     RND  EHTRRELAKMKQ  EPV
RHOB     RSD  EHVRTELARMKQ  EPV
RHOC     RQD  EHTRRELAKMKQ  EPV
RHO-0    AGD  EHTRRELAKMKQ  EPV
RHOAP    RND  ESTKRELMKMKQ  EPV
RHOY1    RND  PQTIEQLRQBGQ  QPV
RAC1     RDD  KDTIEKLKEKKL  TPT
RAC2     RDD  KDTIEKLKEKKL  API
G25KA    RDD  PSTIEKLAKNKQ  KPI
G25KB    RDD  PSTIEKLAKNKQ  KPI
CDC42    RDD  KVIIEKLQRQRL  RPI
TC10     RDD  PKTLARLNDMKE  KPI
RHOY2    RQE  AHFKENATDEM-  VPI
```

At least the size of this insertion is conserved, suggesting that it has appeared in a common ancestral gene. This insertion occurs in loop 116 to 126 where it is not expected to modify the overall structure of the molecule but certainly modifies the structure of this loop and the following α-helix.

rab **branch** — one deletion is found in position 95 and one insertion between amino acids 41 and 42. Some insertions or deletions are only found in a single member of a sub-branch: one deletion in SEC4 in position 120 and one insertion in *ralB* between amino acids 104 and 105. A four amino acids insertion in *rab7* between positions 107 and 108.

ran **branch** — a two amino acids deletion is found between positions 140 and 143.

```
KLVVVGAGGVGKSALTIQLIQNHFVDEYDPTIEDSYRKQVVIDGETCLLDILDTAGQEE
 5      10   15 17              28   32  35   40         47      53 55 57        63
KLVVVG-GGVGKSCLL--L----F---Y--TIE-SY-K-V-ID-----L-IWDTAGQE-
   VLIL  SAC  TAFT  F    Y   F   VG NF R I VE    I LL      T
   CILI  ST   S     Y        S   F  D  A C LN    A VF
   *      *    **                 *                 **** *


YSAMRDQYMRTGEGFLCVFAINNTKSFEDIHQYREQIKRVKDSDDVPMVLVGNKCDLAA
64   68   71            82 84              97                113 115   119
Y--LR--YYR-G-GFL-VFSI----SF------W--------------LVGNK-DL--
   F    MT   MP T VII CYDV    TL     R              VI TQ   M
   I       IK A CA  M AL      Y      Y              IL
           *                                        *      *


RTVESRQAQDLARSYGIPYIETSAKTRQGVEDAFYTLVREIR
125       134        141 143   147          156       164
--V----G---A------Y-ETSAK---NV---F--------
   I      A    G      F  C  L   GL
                      S
                      * **           *
```

FIGURE 2. Summary of alignments: consensus sequence for the *ras* superfamily of GTPases. See text for details.

C. CONSTRUCTION OF A PHYLOGENETIC TREE

Since the N and C terminal parts are of variable length and sequence, they cannot be aligned. Thus, only the core region from amino acid 5 to 164 constituting the GTP/GDP binding domain has been taken into account for the alignments. The methods used have been described in Reference 51. The consensus phylogenetic tree deduced from these results is represented in Figure 3. The names of the genes and their references are identical to those described in Figure 1A. It should be noted that while the existence of the four distinct branches *ras, rho, rab,* and *ran*/TC4 was clear from several different approaches using different programs; the details in some sub-branches have been found by only one calculation method and can only be considered as indicative.

The *ras* and *rho* branches are the most homogeneous with proteins always sharing more than 50% identity, the *rab* branch that now includes the highest number of proteins is also the most heterogeneous, some members of the *rab* branch sharing only 40% identity with other members of the same branch. The *ran*/TC4 protein is slightly more closely related to *rab* proteins, however, it has several unusual characteristics that place it as the first representant of a new branch.

D. AN ESTIMATE OF THE NUMBER OF *ras*-RELATED PROTEINS IN MAMMALS

There has been a rapid increase in the number of newly discovered *ras*-related proteins from 1985 to 1989. There now seems to be a ralentando, and different groups working independently have frequently isolated identical proteins by different approaches. As an example, a large fraction of the small G

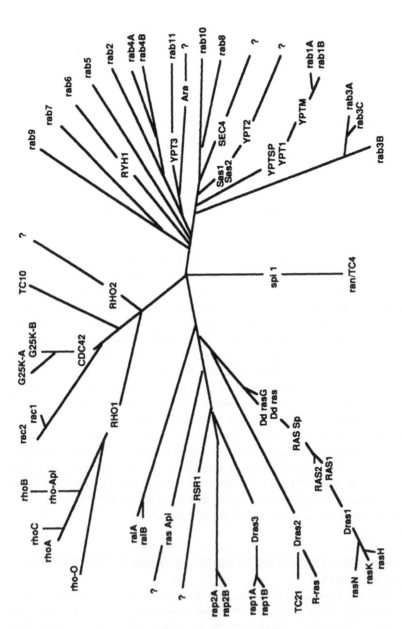

FIGURE 3. Phylogenetic tree of *ras*-related GTPases showing the four main branches: *ras* branch, *rho* branch, *rab* branch, and *ran* branch.

proteins isolated on biochemical criteria turned out to be identical to those predicted from the cDNAs isolated by homology probing. Another striking example is the *rap1A* protein that has first been isolated by low stringency hybridization with a D-*ras3* probe, then as a cDNA able to revert v-K-*ras* transformation: K-*rev*-1 and as a cDNA for a protein first supposed to be a botulinum toxin D substrate. A small G protein found in brain, neutrophil, and platelets (*smg* p21) also appeared to be identical to *rap1* and a similar protein was isolated from platelet cytosol. It also appeared that a small G protein that was phosphorylated and translocated in response to Iloprost was *rap1B* (see Chapter 11).

Thus, the fact that isolation of small G proteins by various approaches and from diverse cell types leads us more and more frequently to rediscover already known proteins strongly suggests that a major fraction of this family has now been discovered, and we might reasonably assume that at least 50, but probably not much more than 100 proteins, of the ras family exist in humans or other mammals.

E. RELATION TO OTHER G PROTEIN FAMILIES

A large number of GTP/GDP binding proteins are present in all eukaryotic cells. These G proteins are members of at least six distinct families: the α-subunits of heterotrimeric G proteins directly involved in signal transduction, the elongation factors of protein biosynthesis, the signal recognition particle (SRP) receptor, the sar1p family, the ADP-ribosylation factor (ARF) family, and the *ras* family.

The consensus sequence of *ras*-related proteins was aligned to the sequences of other G proteins representing the different families: ARF, sar1p, Era, α-subunits, SRP receptor and elongation factor Tu. The GX4GKS/T motif could easily be recognized in all of them, always followed by Y or F in the next 12 amino acids. The DXXG motif was also easily found in most of them, except in the SRP receptor where two different regions matched this consensus. A T was always present at approximately 20 amino acids upstream of this motif, corresponding to *ras* T35. The LVGNKXDL motif was also unambiguously found in all of them, but the following ETSAK motif was much more difficult to recognize, and its localization is only tentative for several proteins. This is not so surprising since only K147 of *ras* is directly interacting with the guanine ring, the other amino acids interacting only with side chains that form the guanine-binding pocket. It is interesting to note that the few strictly conserved amino acids of the ras family are probably representing absolute requirements for GTP/GDP binding since most of them are also present in distantly related G proteins, the only exception being this ETSAK sequence. Thus, the ras family cannot be defined on the absolute criteria of strictly conserved amino acids but rather by a high conservation of the consensus sequence described in Figure 2.

F. A CONSERVED FUNCTIONAL MECHANISM? THE STRUCTURAL POINT OF VIEW

The functions of these proteins are discussed in several other chapters of this book. Sequence comparisons enable us to define external regions that are potential targets for specific interacting proteins; since these regions are highly variable, a large variety of target proteins are also expected. Nevertheless, the general structures of all *ras*-related proteins are highly analogous, suggesting a conserved functional mechanism. *Rab* proteins are very likely involved in vesicular traffic, *rho* proteins seem to be implicated in the interaction of membranes with cytoskeletal components, and preliminary results suggest that *rap* and *ral* proteins too are associated with vesicles and interact with cytoskeletal components. Furthermore, the ARF and sar1p proteins, that are only distantly related to *ras* in sequence, but might be analogous in structure, also seem to be involved in cellular compartmentation and traffic. Therefore, except the RAS1, RAS2, and RSR1 proteins of *S. cerevisiae*, all small G proteins are very homologous in structure and most of them are involved in traffic of intracellular membrane components. Since *ras* proteins too have a very similar structure, we suggest that *ras* performs a similar "shuttle" function at the plasma membrane, to control lateral segregation of multiprotein complexes involved in signal transduction.[52]

ACKNOWLEDGMENTS

P. C. is supported by INSERM.

REFERENCES

1. **Capon, D. J., Chen, E. Y., Levinson, A. D., Seeburg, P. H., and Goeddel, D. V.,** Complete nucleotide sequences of the T24 human bladder carcinoma oncogene and its normal homologue, *Nature,* 302, 33, 1983.
2. **Westaway, D., Papkoff, J., Moscovici, C., and Varmus, H. E.,** Identification of a provirally activated c-Ha-*ras* oncogene in an avian nephroblastoma via a novel procedure: cDNA cloning of a chimaeric viral-host transcript, *EMBO J.,* 5, 301, 1986.
3. **McGrath, J. P., Capon, D. J., Smith, D. H., Chen, E. Y., Seeburg, P. H., Goeddel, D. V., and Levinson, A. D.,** Structure and organization of the human Ki-*ras* protooncogene and a related processed pseudogene, *Nature,* 304, 501, 1983.
4. **Taparowsky, E., Shimizu, K., Goldfarb, M., and Wigler, M.,** Structure and activation of the human N-*ras* gene, *Cell,* 34, 581, 1983.
5. **Neuman-Silberberg, F. S., Schejter, E., Hoffman, F. M., and Shilo, B. Z.,** The drosophila *ras* oncogenes: structure and nucleotide sequence, *Cell,* 37, 1027, 1984.
6. **Robbins, S. M., Williams, J. G., Jermyn, K. A., Spiegelman, G. B., and Weeks, G.,** Growing and developing *Dictyolstelium* cells express different *ras* genes, *Proc. Natl. Acad. Sci. U.S.A.,* 86, 938, 1989.
7. **Reymond, C. D., Gomer, R. H., Mehdy, M. C., and Firtel, R. A.,** Developmental regulation of a *Dictyostelium* gene encoding a protein homologous to mammalian *ras* proteins, *Cell,* 39, 141, 1984.

8. **Defeo-Jones, D., Scolnick, E. M., Koller, R., and Dhar, R.**, *ras*-related gene sequences identified and isolated from *Saccharomyces cerevisiae, Nature,* 306, 707, 1983.

9. **Powers, S., Kataoka, T., Fasano, O., Goldfarb, M., Strathem, J., Broach, J., and Wigler, M.**, Genes in *S. cerevisiae* encoding proteins with domains homologous to the mammalian *ras* proteins, *Cell,* 36, 607, 1984.

10. **Fukui, Y. and Kaziro, Y.**, Molecular cloning and sequence analysis of a *ras* gene from *Schizosaccharomyces pombe, EMBO J.,* 4, 687, 1985.

11. **Drivas, G. T., Shih, A., Coutavas, E., Rush, M. G., and D'Eustachio, P.**, Characterization of four novel ras-like genes expressed in a human teratocarcinoma cell line, *Mol. Cell. Biol.,* 10, 1793, 1990.

12. **Mozer, B., Marlor, R., Parkhurst, S., and Corces, V.**, Characterization and developmental expression of a Drosophila *ras* oncogene, *Mol. Cell. Biol.,* 5, 885, 1985.

13. **Lowe, D., Capon, D., Delwart, E., Sakaguchi, A., Naylor, S., and Goeddel, D.**, Structure of the human and murine R-*ras* genes, novel genes closely related to *ras* proto-oncogenes, *Cell,* 48, 137, 1987.

14. **Pizon, V., Chardin, P., Lerosey, I., Olofsson, I., and Tavitian, A.**, Human rap1 and Rap2 cDNAs, homologous to the Drosophila melanogaster gene D*ras* 3, encode proteins closely related to *ras* in the effector region, *Oncogene,* 3, 210, 1988.

15. **Pizon, V., Lerosey, I., Chardin, P., and Tavitian, A.**, Nucleotide sequence of a human cDNA encoding a *ras*-related protein (rap1B), *Nucleic Acids Res.,* 16, 7719, 1988.

16. **Hariharan, I. K., Carthew, R. W., and Rubin, G. M.**, The Drosophila *Roughened* mutation: activation of a rap homolog disrupts eye development and interferes with cell determination, *Cell,* 67, 717, 1991.

17. **Ohmstede, C.-A., Farell, F. X., Reep, B. R., Clemetson, K. J., and Lapetina, E. G.**, Rap2B: a ras related GTP-binding protein from platelets, *Proc. Natl. Acad. Sci. U.S.A.,* 87, 6527, 1990.

18. **Bender, A. and Pringle, J. R.**, Multicopy suppression of the *cdc24* budding defect in yeast by *CDC42* and three newly identified genes including the *ras*-related gene *RSR1, Proc. Natl. Acad. Sci. U.S.A.,* 86, 9976, 1989.

19. **Chardin, P. and Tavitian, A.**, The *ral* gene: a new *ras*-related gene isolated by the use of a synthetic probe, *EMBO J.,* 5(9), 2203, 1986.

20. **Chardin, P. and Tavitian, A.**, Coding sequences of human *ral*A and *ral*B cDNAs, *Nucleic Acids Res.,* 17, 4380, 1989.

21. **Yeramian, P., Chardin, P., Madaule, P., and Tavitian, A.**, Nucleotide sequence of human *rho* cDNA clone 12, *Nucleic Acids Res.,* 4, 1869, 1987.

22. **Chardin, P., Madaule, P., and Tavitian, A.**, Coding sequence of human *rho* cDNA clone 6 and clone 9, *Nucleic Acids Res.,* 16, 2717, 1988.

23. **Ngsee, J. K., Elferink, L. A., and Scheller, R. H.**, A family of ras-like GTP binding proteins expressed in electromotor neurons, *J. Biol. Chem.,* 266, 2675, 1991.

24. **Madaule, P. and Axel, R.**, A novel *ras*-related gene family, *Cell,* 41, 31, 1985.

25. **Madaule, P., Axel, R., and Myers, A.**, Characterization of two members of the *rho* gene family from the yeast *Saccharomyces cerevisiae, Proc. Natl. Acad. Sci. U.S.A.,* 84, 779, 1987.

26. **Didsbury, J., Weber, R. F., Bokoch, G. M., Evans, T., and Snyderman, R.**, *rac*, a novel *ras*-related family of proteins that are botulinum toxin substances, *J. Biol. Chem.,* 264, 16378, 1989.

27. **Shinjo, K., Koland, J. G., Hart, M. J., Narasimhan, V., Johnson, D. I., Evans, T., and Cerione, R. A.**, Molecular cloning of the gene for the human placental GTP-binding protein Gp (G25K): identification of this GTP-binding protein as the human homolog of the yeast cell division cycle protein CDC42, *Proc. Natl. Acad. Sci. U.S.A.,* 87, 9853, 1990.

28. **Munemitsu, S., McInnis, M. A., Clark, R., McCormick, F., Ullrich, A., and Polakis, P.**, Molecular cloning and expression of a G25K cDNA, the human homolog of the yeast cell cycle gene CDC42, *Mol. Cell. Biol.,* 10, 5977, 1990.

29. **Johnson, D. I. and Pringle, J. R.,** Molecular characterization of CDC42, a *Saccharomyces cerevisiae* gene involved in the development of cell polarity, *J. Cell. Biol.,* 111, 143, 1990.

30. **Touchot, N., Chardin, P., and Tavitian, A.,** Four additional members of the *ras* gene superfamily isolated by an oligonucleotide strategy: molecular cloning of YPT related cDNAs from rat brain, *Proc. Natl. Acad. Sci. U.S.A.,* 84, 8210, 1987.

31. **Vielh, E., Touchot, N., Zahraoui, A., and Tavitian, A.,** Nucleotide sequence of a rat cDNA: *RAB*1B, encoding *RAB*1-YPT related protein, *Nucleic Acids Res.,* 17, 1770, 1989.

32. **Palme, K., Diefenthal, T., Sander, C., Vingron, M., and Schell, J.,** Identification of guanine-nucleotide binding proteins in plants: structural analysis and evolutionary comparison of the ras-related YPT gene family from *Zea maïs,* in *The Guaninine Nucleotide Binding Proteins,* Bosch, L., Ed., Plenum Press, NATO ASI series, New York, 273, 1988.

33. **Gallwitz, D., Donath, C., and Sander, C.,** A yeast gene encoding a protein homologous to the human c-*has/bas* proto-oncogene product, *Nature,* 306, 704, 1983.

34. **Fawell, E., Hook, S., and Armstrong, J.,** Nucleotide sequence of a gene encoding a YPT1-related protein from *Schizosaccharomyces pombe, Nucleic Acids Res.,* 17, 4383, 1989.

35. **Saxe, S. A. and Kimmel, A. R.,** Genes encoding novel GTP-binding proteins in *Dictyostelium, Dev. Genet.,* 9, 259, 1988.

36. **Haubruck, H., Engelke, U., Mertins, P., and Gallwitz, D.,** Structural and functional analysis of YPT2, an essential ras-related gene in the fission yeast *Schizosaccharomyces pombe* encoding a sec4 protein homologue, *EMBO J.,* 9, 1959, 1990.

37. **Salminen, A. and Novick, P.,** A *ras*-like protein is required for a post golgi event in yeast secretion, *Cell,* 49, 527, 1987.

38. **Chavrier, P., Vingron, M., Sander, C., Simons, K., and Zerial, M.,** Molecular cloning of YPT/SEC4-related cDNAs from an epithelial cell line, *Mol. Cell. Biol.,* 10, 6578, 1990.

39. **Zahraoui, A., Touchot, N., Chardin, P., and Tavitian, A.,** Complete coding sequences of the *ras* related rab3 and 4 cDNAs, *Nucleic Acids Res.,* 16, 1204, 1988.

40. **Zahraoui, A., Touchot, N., Chardin, P., and Tavitian, A.,** The human rab genes encode a family of GTP binding proteins related to the Yeast YPT1 and SEC4 products involved in secretion, *J. Biol. Chem.,* 264, 12394, 1989.

41. **Matsui, Y., Kikuchi, A., Kondo, J., Hishida, T., Teranishi, Y., and Takai, Y.,** Nucleotide and deduced amino acid sequence of a GTP-binding protein family with molecular weights of 25,000 from bovine brain, *J. Biol. Chem.,* 263, 11071, 1988.

42. **Miyake, S. and Yamamoto, M.,** Identification of ras-related, YPT family genes in *Schizosaccharomyces pombe, EMBO J.,* 9, 1417, 1990.

43. **Matsui, M., Sasamoto, S., Kunieda, T., Nomura, N., and Ishizaki, R.,** Cloning of *ara,* a putative *Arabidopsis thaliana* gene homologous to the *ras*-related gene family, *Gene,* 76, 313, 1989.

44. **Hengst, L., Lehmeier, T., and Gallwitz, D.,** The ryh1 gene in the fission yeast *Schizosaccharomyces pombe* encoding a GTP-binding protein related to ras, rho and YPT: structure, expression and identification of its human homologue, *EMBO J.,* 9, 1949, 1990.

45. **Bucci, C., Frunzio, R., Chiariotti, L., Brown, A., Rechler, M., and Bruni, C.,** A new member of the ras gene superfamily identified in a rat liver cell line, *Nucleic Acids Res.,* 16, 9979, 1988.

46. **Robbins, S. M., Suttorp, V. V., Weeks, G., and Spiegelman, G. B.,** A ras-related gene from the lower eucaryote *Dictyolstelium* that is highly conserved relative to the human rap genes, *Nucleic Acids Res.,* 18, 5265, 1990.

47. **Han, M. and Sternberg, P. W.,** let-60, a gene that specifies cell fates during *C. elegans* vulval induction, encodes a ras protein, *Cell,* 63, 921, 1990.

48. **Casale, W. L., McConell, D. G., Wang, S.-Y., Lee, Y.-J., and Linz, J. E.,** Expression of a gene family in the dimorphic fungus *Mucor racemosus* which exhibits striking similarity to human ras genes, *Mol. Cell. Biol.,* 10, 6654, 1990.

49. **Altschuler, D. L., Muro, A., Schijman, A., Bravo Almonacid, F., and Torres, H. N.,** *Neurospora crassa* cDNA clones coding for a new member of the ras protein family, *FEBS Lett.,* 273, 103, 1990.

50. **Swanson, M., Elste, A., Greenberg, S., Schwartz, J., Aldrich, T., and Furth, M.,** Abundant expression of rat proteins in Aplysia neurons, *J. Cell Biol.,* 103, 485, 1986.
51. **Valencia, A., Chardin, P., Wittinghofer, A., and Sander, C.,** The ras protein family: evolutionary tree and role of conserved amino acids, *Biochemistry,* 30, 4637, 1991.
52. **Chardin, P.,** Small GTP-binding proteins of the ras family, a conserved functional mechanism?, *Cancer Cells,* 3, 117, 1991.

Chapter 11

Krev-1 AND THE RELATED GENES

Hitoshi Kitayama and Makoto Noda

TABLE OF CONTENTS

0-8493-5214-2/93/$0.00 + $.50
© 1993 by CRC Press, Inc.

I. INTRODUCTION

The *ras*-related Krev-1 gene was isolated as a transformation suppressor gene against *ras*-transformed "DT" cells in our laboratory. The same gene and its product were discovered independently by other groups through different approaches and named *rap1A* and *smg* p21A, respectively. Thus far, three Krev-1-related genes (*rap1B/smg* p21B, *rap2*, and *rap2B*) have been identified in mammalian cells. In this chapter we shall mainly describe our genetic analyses of the Krev-1 gene and discuss about possible mechanisms of action of the Krev-1 protein.

II. ISOLATION OF THE GENES

We tried to isolate transformation suppressor genes using the expression cloning method shown in Figure 1.[1,2] The DT cell is a transformed derivative of the mouse fibroblast cell line, NIH 3T3, containing two copies of v-K-*ras* gene.[3] Frequency of spontaneous reversion is very low in DT cells. DT cells were transfected with a normal human fibroblast cDNA expression library followed by various types of enrichment procedures designed to selectively eliminate malignant cells in the cell population. Final selection was carried out by direct microscopic observation, and seven flat revertants were obtained in a series of experiments. To recover the transfected plasmids, total DNA extracted from revertants was digested with *Sal*I, which has a unique cutting site within the vector, ligated at a low DNA concentration and transformed into *Escherichia coli*. After Ampicillin selection of the bacteria, plasmid was extracted from each bacterial clone, and its revertant-inducing activity was assessed by transfection assay on DT cells. A cDNA clone with transformation-suppressor activity was obtained from one of the revertants and named Krev-1 (Kirsten-*ras*-revertant-1). Krev-1 could also suppress the transforming activities of H-*ras* and N-*ras* genes, but not of other oncogenes tested (i.e., *src, mos, raf, fos*).

Two other groups isolated the same gene through independent approaches. Tavitian's group screened a human cDNA library with a Drosophila *Dras3* gene[4] as a probe at low stringency condition and obtained two *ras*-related genes, *rap1* (=*rap1A*) and *rap2*.[5] Later, they also isolated another cDNA clone closely related to *rap1* and named it *rap1B*.[6] As a result, the original *rap1* is now called *rap1A*. Takai et al.[7] purified a series of novel small molecular weight GTP-binding proteins (smgs) from bovine brain membranes, and one of the proteins with a 21 kDa molecular weight was named *smg* p21. The cDNA encoding *smg* p21 (=*smg* p21A) was isolated from a bovine brain cDNA library by using an oligonucleotide probe designed from the partial amino acid sequence.[7] The nucleotide sequence of the human Krev-1 was found to be the same as that of *rap1A*, and deduced amino acid sequence encoded by these cDNAs was identical to the amino acid sequence of the bovine *smg* p21A.

Another related GTP-binding protein named *smg* p21B was purified from bovine brain and human platelets by Takai's group, which turned out to be

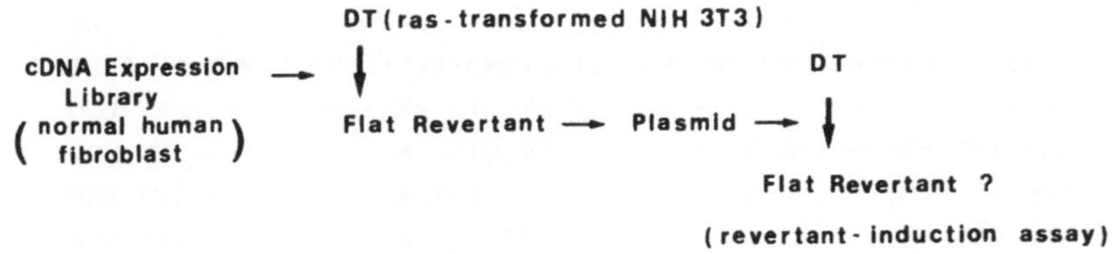

DT (ras - transformed NIH 3T3)

cDNA Expression → ↓ DT
Library
(normal human)   Flat Revertant → Plasmid → ↓
  fibroblast

Flat Revertant ?

(revertant - induction assay)

FIGURE 1. Experimental outline.

identical to the product of the *rap1B* gene.[8] The gene *rap2B* was found by Lapetina's group during the expression screening of a platelet cDNA library with the monoclonal antibody M90 recognizing an epitope on *ras* p21s (residues 107 to 130).[9]

III. STRUCTURE OF THE Krev-1 AND THE RELATED PROTEINS

The amino acid sequence of the Krev-1 protein shares around 50% similarity with ras proteins (Figure 2). Similarities are especially high in the functional domains[10] known in the H-ras protein, such as GDP/GTP-binding domains, effector domain, and C terminal CAAX (C, cysteine residue; A, aliphatic residue; and X, any residue) motif that is essential for posttranslational modification.

Rap1B/*smg* p21B is closely related to the Krev-1/rap1A/*smg* p21A with only nine amino acids difference. rap2B is 90% identical to rap2. Residues 21 to 24, 30, 31, and 61 to 64 are conserved among the proteins encoded by K*rev-1*, *rap1B, rap2, rap2B,* and D*ras3*, but they are diverged in ras proteins.

The Saccharomyces *RSR1* gene (also called *BUD1*[11]) was isolated from the yeast genomic library as a multicopy suppressor for a *cdc24* mutation.[12] Although overall structure of the *RSR1*-encoded protein is similar to that of Krev-1, residues 30, 31, and 62 are divergent from the other Krev-1-related proteins.

IV. TISSUE DISTRIBUTION AND SUBCELLULAR LOCALIZATION OF THE Krev-1-RELATED PROTEINS

K*rev*-1 gene is expressed in many tissues.[2] The RNA expression was detected in brain, thymus, lung, spleen, liver, kidney, colon, uterus, and ovary of the rat. Relatively higher levels of mRNA were detected in the brain and colon when β-actin mRNA was used as a normalization standard.

Takai et al. examined the protein expression in various rat tissues with antiserum recognizing both *smg* 21A and *smg* 21B.[13] The *smg* p21s were detected in cerebrum, cerebellum, adrenal gland, thymus, lung, heart, liver, small intestine, kidney, and testis. Among these tissues, *smg* p21s were found in the cerebellum at the highest level and were also abundant in cerebrum, lung, and testis.

```
              10        20        30        40        50
H-Ras   MTEYKLVVVGAGGVGKSALTIQLIQNHFVDEYDPTIEDSYRKQVVIDGET
Krev-1  -R------L-S---------V-FV-GI--EK-------------EV-CQQ
Rap 1B  -R------L-S---------V-FV-GI--EK-------------EV-AQQ
Rap 2   -R---V--L-S---------V-FVTGT-IEK-----------EIEV-SSP
Rap 2B  -R---V--L-S---------V-FVTGS-IEK-----------EIEV-SSP
Dras3   -R---I--L-S---------V-FV-CI--EK-------------KVNDRQ
RSR 1   -RD-----L--------C--V-FV-GVYL-T-----------TIE--NKV

              60        70        80        90        100
H-Ras   CLLDILDTAGQEEYSAMRDQYMRTGEGFLCVFAINNTKSFEDIHQYREQI
Krev-1  -M-E------T-QFT----L--KN-Q--AL-YS-TAQST-N-LQDL----
Rap 1B  -M-E------T-QFT----L--KN-Q--AL-YS-TAQST-N-LQDL----
Rap 2   SV-E------T-QFAS---L-IKN-Q--IL-YSLV-QQ--Q--KPM-D--
Rap 2B  SV-E------T-QFAS---L-IKN-Q--IL-YSLV-QQ-SQ--KPM-D--
Dras3   -M-E-VN---T-QFT---NL--KN- SDS-WSTRSRRNRRLT-CRT----
RSR 1   FD-E------IAQFT---EL-IKS-M---L-YSVTDRQ-L-ELMEL---V

              110       120       130       140
H-Ras   KRVKDSDDVPMVLVGNKCDLA ARTVESRQAQDLARSY GIPYIETSAKT
Krev-1  L----TE----I--------EDE-V-GKE-G-N---QWCNCAFL-S---S
Rap 1B  L----T-----I--------EDE-V-GKE-G-N---QWNNCAFL-S---S
Rap 2   I---RYEK--VI-----V--ESE-E-S-SEGRA--EEW -C-FM-----S
Rap 2B  I---RYER---I-----V--EGE-E-SYGEGKA--EEW SC-FM-----N
Dras3   L----T-------------E E-V-GKELGKN--TQF NCAFM-----A
RSR 1   L-I----R-----I---A--INE-VISVEEGIEVSSKWGRV-FY----LL

        150       160       170       180       189
H-Ras   RQGVEDAFYTLVREI RQHKLRKLNPPDESGDGCMSCKCVLS (189aa)
Krev-1  KIN-NEI--D---Q-N-KTPVE-KK-KKK-         -L-L (184aa)
Rap 1B  KIN-NEI--D---Q-N-KTPVPGKARKKS-         -LQL (184aa)
Rap 2   KTM-DEL-AEI--QMNYAAQPD-DD-CCSA         -NIQ (183aa)
Rap 2B  KAS-DEL-AEI--QMNYAAQSNGDEGCCSA          --IL (183aa)
Dras3   KVN-N-I--DWSGRST-SRPR-NRRSRKVP          ---L (182aa)
RSR 1   -SN-DEV-VD---Q-I-(92aa)KKKKKNAST        -TIL (272aa)
```

FIGURE 2. Comparison of the amino acid sequences.

Light microscopic immunocytochemical studies showed that *smg* p21s in brain were distributed abundantly in the cytoplasmic region of neuronal cell bodies and moderately in neuropil, whereas c-*ras* p21s were more abundant in neuropil than in the cytoplasmic region of neuronal cell bodies.[13] Subcellular fractionation analyses of cerebrum also revealed partly distinct localization of *smg* p21s and c-*ras* p21s.[13] Indirect immunofluorescence and cell fractionation analyses using antibodies recognizing residues 121 to 137 of the rap1A and

TABLE 1
Effects of Point Mutations on the Revertant-Inducing
Activity of K*rev*-1

Type[a]	Amino acid position	Predicted role in H-*ras*	Normal	Mutant	Reversion flat/neo[R] (%)	Ratio
A	12	Phosphate-binding	Gly	Val	4.4	1.6
B	17	Mg^{2+}-binding	Ser	Asp	<0.5	<0.2
B	38	Effector	Asp	Ala	1.2	0.4
B	38		Asp	Asn	<0.4	<0.2
A	59	Phosphate-binding	Ala	Thr	13.7	4.9
C	61	(Highly mobile loop)	Thr	Gln	1.1	0.4
D	61		Thr	Lys	<0.2	<0.1
C	63		Gln	Glu	12.7	4.5
B	116	Guanine-binding	Asn	His	0.7	0.3
C	160	(C terminal α-helix)	Asp	Thr	5.5	2.0
B	167		Arg	Gly	1.8	0.6
B	181	Membrane association	Cys	Ser	<0.5	<0.2
Wild					2.8	1.0

[a] Mutations fall into four types: A, from normal *ras*-type to activated *ras*-type; B, from normal *ras*-type to inactivated *ras*-type; C, from K*rev*-1-specific-type to normal *ras*-type; D, from K*rev*-1-specific-type to activated *ras*-type.

rap1B proteins suggested that ras and Krev-1 proteins reside in distinct subcellular compartments, and that the rap1 proteins are associated with the Golgi complex.[14] Takai's group has purified smg 21B from human platelet membrane,[8] while Nozawa's group has purified *smg* p21A/Krev-1 from human platelet cytosol.[8]

V. BIOLOGICAL ACTIVITIES OF THE MUTANT K*rev*-1 GENES

To understand the functional organization of the protein, we introduced a series of single amino acid substitutions into the K*rev*-1 gene. Mutant genes were transfected into DT cells and their revertant-inducing activities were tested (Table 1).[15] The mutations we tested can be classified into four types. The first class (Type A) includes the substitutions from c-*ras*-type to activated *ras*-type. This type of mutation includes amino acid substitutions within the putative phosphate-binding region. Both Val^{12} and Thr^{59} mutations activated the trans-formation-suppressing activity of the K*rev*-1 gene. When NIH 3T3 cells were transfected with these mutants, no transformed cells could be observed. The data suggest that the Krev-1 protein and the ras protein have opposing functions despite their structural similarities. The second class of mutations (Type B) include the ones corresponding to the inactivating mutations previously found in *ras* oncogenes such as a dominant negative mutation at the Mg^{2+}-binding site (Asp^{17}),[16] effector region mutations, (Ala^{38} and Asn^{38}), a guanine-binding region mutation, (His[116]), and a C terminal CAAX motif mutation (Ser[181]). All these mutations resulted in the decrease in the transformation-suppressor activity, indicating that these regions are essential for the function of the Krev-1 protein as they are for the functions of ras proteins. The data are also consistent with a model that the Krev-1 protein is regulated by the GTP/GDP exchange mechanism like other G proteins and is able to interact with the effector molecule for ras. The fourth cysteine residue from the COOH terminus in ras is farnesylated and the farnesylation is followed by proteolytic cleavage of the C terminal three residues and carboxylmethylation of the cysteine.[17] In smg 21B, it has been shown that Cys[181] is geranylgeranylated,[18] and the removal of the three amino acids as well as carboxymethylation of the cysteine residue probably also occur. We also introduced mutations in some of the Krev-1-specific amino acid residues changing to either c-*ras*-type (Type C) or oncogenic *ras*-type (Type D). Codon 61 was changed in both ways, and in both cases the mutants exhibited reduced suppressor activity. Therefore, Thr[61] seems to be important for the transformation-suppressor activity of Krev-1. The glutamic acid to glutamine mutation at amino acid 63 (Type C) resulted in an augmented revertant-inducing activity. We also tested the effects of other substitutions in codon 63, and the data indicated that charged amino acids (Glu, Asp, Arg) at this position potentiate the revertant-inducing activity of the Krev-1 protein.[15] The fact that the region from codon 61 to 64 (Thr-Glu-Gln-Phe) is conserved among all the Krev-1-related proteins suggests that this region probably plays an important regulatory role in these proteins.

VI. SUPPRESSOR ACTIVITY ON HT 1080 CELLS

Our experiments described thus far were carried out by using artificially transformed murine cell lines, DT. To assess biological activities of the Krev-1 in human tumor-derived cell lines, we employed a human fibrosarcoma cell line, HT 1080, which contains an activated N-*ras* gene.[19]

The wild-type K*rev*-1 gene or some of its mutants were transfected into HT 1080 cells, and the growth properties and the tumorigenicity in nude mice of the transfectants were examined (Table 2).[15] The cells harboring the wild-type K*rev*-1 gene remained tumorigenic, although, the growth of the transfectants was slower than that of control cells. The K*rev*-1 gene carrying a mutation (Asn[38]) in the putative effector-binding region could not slow the growth of HT 1080 cells. The CAAX box mutant (Ser[181]) showed a residual activity. The activating mutants of Val[12] and Glu[63] showed strong tumor-suppressor activities in HT 1080 cells. The transfectants showed lower growth rate and the colony-forming efficiencies in soft agar than did the control cells. In nude mice assay, the transfectants only formed small nodules early on, but the nodules were regressed afterward without establishing tumors.

TABLE 2
Growth Properties of the HT 1080 Cells Transfected with Wild-Type or Mutant K*rev*-1

DNA	Soft agar colonies (%)	Doubling time (h)	Tumorigenicity[a] Day 15	Day 32	Day 45
Vector	69	23	3/3 (SSS)	3/3 (LMS)	3/3 (LLM)
WT	76	34	2/3 (MS)	3/3 (LLM)	3/3 (LLL)
Val[12]	10	32	0/3	1/3 (S)	0/3
Asn[38]	75	23	3/3 (MSS)	3/3 (LLL)	3/3 (LLL)
Thr[59]	41	36	3/3 (SSS)	3/3 (LLM)	3/3 (LLL)
Glu[63]	28	42	2/3 (SS)	1/3 (S)	0/3
Ser[181]	62	35	1/3 (M)	2/3 (M)	3/3 (LMS)

[a] Approximate size of each tumor(v) was estimated by multiplying the width(w), the length(l), and the height(h) of the tumor[$v = w \times l \times h$, in millimeters] and is indicated as follows: S, $v < 300$; M, $300 < v < 1000$; L, $1000 < v$.

VII. TRANSFORMATION SUPPRESSOR ACTIVITIES OF THE K*rev*-1/H-*ras* CHIMERIC GENES

We have mentioned that Krev-1 and ras exhibit mutually counteracting biological activities despite their strong structural similarity. To map the region responsible for the transformation-suppressor activity, we took advantage of the structural similarity and constructed the K*rev*-1/H-*ras* chimeric genes. As parental genes we used activated Val[12] mutants of human K*rev*-1 and H-*ras* in hopes of obtaining clear results. The coding regions of both genes were divided into three portions. Portion I (N terminus — residue 59) contains putative phosphate- and ribose-binding regions as well as the effector region, portion II (residues 60 to 111) the codons 61 to 64 characteristic of the K*rev*-1 family, and portion III (residue 112 — C terminus) the divergent region as well as C terminal CAAX motif.

The chimeric genes were transfected into DT cells and the revertant-inducing activities were tested (Figure 3A).[20] As expected, the parental K*rev*-1 (Val[12]) gene did, and the other parent, H-*ras* (Val[12]), did not show the transformation-suppressor activity. When the portion I or the portion I plus II of the parental H-*ras* (Val[12]) genes were replaced by the corresponding portion(s) of the K*rev*-1 gene (C4 and C7), the chimeric genes acquired the revertant-inducing activity. On the other hand, reciprocal chimeric genes, C1 and C6, exhibited reduced suppressor activity. These data suggest that the portion I of the K*rev*-1 gene is responsible for its transformation-suppressor activity. The fact that chimeric gene C2, whose portion I is derived from H-*ras* and portion II and III are from K*rev*-1 with activated Glu[63] mutation, did not show the revertant-inducing activity, further supports this idea.

Zhang and Lowy constructed extensive collection of *ras*/K*rev*-1 chimeric genes (Figure 3B).[21] They divided the portion I into three parts (residues 5 to

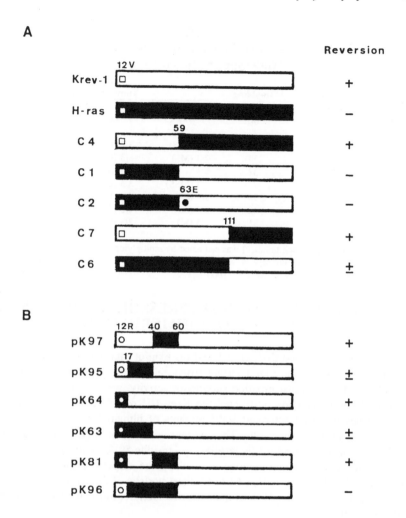

FIGURE 3.   Revertant-inducing activities of the H-*ras*/K*rev*-1 chimeric genes.

17, 18 to 40, and 41 to 60). Transfection experiments showed that the chimeric genes with amino acids 18 to 40 derived from the K*rev*-1 gene (pK81 and pK97) could induce flat revertants. Therefore, the region responsible for the transformation-suppressor activity must be located within the residues 18 to 40.

VIII. IDENTIFICATION OF THE AMINO ACIDS CRITICAL FOR THE TRANSFORMATION-SUPPRESSOR ACTIVITY

There are seven divergent amino acids within the residues 18 to 40 between H-ras and Krev-1 (Table 3). To find which residue(s) among these seven

TABLE 3
Transformation and Revertant-Inducing Activities
of H-*ras* Mutants

			On NIH 3T3		On DT	
			Soft agar colonies (%, size)		Reversion flat/neo^R (%)	
H-*ras* (Val^{12})	21	31	+ + 25	(L)	− 0/568 (<0.18)	
Mutants:	−IQLIQNHFVDE−					
(Krev-1-type)						
H1	V−FV−GI−−EK		− <1.5		± 5/704 (0.71)	
H2	V−FV−−−−−−−		+ + 14	(L)	− 0/318 (<0.31)	
H3	−−−−−GI−−EK		− <1.5		+ 10/806 (1.24)	
H4	−−−−−GI−−−−		± 6.8	(S)	− 0/171 (<0.58)	
H5	−−−−−−−−−EK		− <1.5		+ 11/628 (1.75)	

divergent amino acids is important for the transformation-suppressor activity, we constructed a series of H-*ras* mutants. All or part of the divergent residues were substituted to the K*rev*-1 type, and the resulting mutants were transfected into DT and NIH 3T3 cells (Table 3). Some mutants (H3, H5, and possibly H1) acquired significant levels of revertant-inducing activity. Since common mutations among these positive mutants were Glu^{30} and Lys^{31}, these residues seem to be important for the transformation-suppressor activity.

We also constructed a series of K*rev*-1 mutants carrying H-*ras*-type amino acids in this region. The biological activity of those mutants are shown in Table 4. The mutants retaining the K*rev*-1-type 30 and 31 residues (K2 and K4) exhibited the revertant-inducing activity. On the other hand, the K*rev*-1 mutants harboring H-*ras*-type Asp^{30} and Glu^{31} residues (K1, K3, and K5) exhibited decreased transformation-suppressor activity. These results support the above conclusion that residues 30 and 31 are important for the transformation-suppressor activity of the Krev-1 protein.

To further explore the significance of residues 30 and 31, single amino acid substitutions were introduced into these two positions of the H-*ras* and the K*rev*-1 genes. H-*ras* has acidic amino acids at both positions 30 and 31, whereas Krev-1 has an acidic residue at position 30 and a basic residue at position 31. H-*ras* mutants with single K*rev*-1-type mutations (H6 and H7) failed to show strong enough suppressor activity, although H7 (codon 31 mutation) showed a weak activity (Table 5). Mutant H13 with a basic amino acid, Arg, at residue 31, exhibited significant revertant-inducing activity, whereas mutant H14 with a neutral Gly residue at position 31 failed to show transformation suppressor activity. These results suggest the importance of the basic amino acid at codon 31 for the biological activity of Krev-1 protein. Parallel experiments with K*rev*-1 mutants supported this idea. The K*rev*-1 mutants with basic amino acids at residue 31 (K6 and K13) retained the revertant-inducing activity. By contrast,

TABLE 4
Transformation and Revertant-Inducing Activities
of K*rev*-1 Mutants

			On NIH 3T3	On DT
			Soft agar colonies (%, size)	Reversion flat/neo^R (%)
Krev-1 (Val¹²)	21 31		− <2.0	+ 23/1209 (1.9)
Mutants:	-VQFVQGIFVEK-			
(*ras* type)				
K1	I-LI-NH--DE		− <2.0	± 2/303 (0.66)
K2	I-LI--------		− <2.0	+ 2/124 (1.6)
K3	-----NH--DE		+ + 34 (M)	± 2/278 (0.72)
K4	-----NH----		− <2.0	+ 2/ 53 (3.8)
K5	---------DE		− <2.0	± 2/844 (0.24)

TABLE 5
Effects of Mutations in Codons 30 and 31 on the Biological
Activities of H-*ras* (Val¹²) and K*rev*-1 (Val¹²)

	Codon		On NIH 3T3	On DT
	30	31	Soft agar colonies (%, size)	Reversion flat/neo^R (%)
Parent:				
H-*ras*(Val¹²)	[Asp]	[Glu]	15 (L)	− 0/568 (<0.18)
Mutants:				
H6	[Glu]	[—]	23 (L)	− 0/244 (<0.41)
H7	[—]	{Lys}	4.8 (S)	+ 2/204 (0.98)
H12	{Lys}	{Lys}	0	± 1/504 (0.20)
H13	[—]	{Arg}	2.3 (S)	+ 7/375 (1.9)
H14	[—]	Gly	6.3 (S)	± 2/366 (0.55)
Parent:				
K*rev*-1 (Val¹²)	[Glu]	{Lys}	0	+ 23/1209 (1.9)
Mutants:				
K6	[Asp]	{—}	0	+ 4/276 (1.4)
K7	[—]	[Glu]	0	− 0/116 (<0.86)
K12	{Lys}	{—}	2.6 (S)	± 2/260 (0.77)
K13	[—]	{Arg}	1.9 (S)	+ 4/224 (1.8)
K14	[—]	Gly	0	− 0/178 (<0.56)

[ ]:acidic, { }:basic.

substitutions to an acidic (K7) or a neutral amino acid (K14) abolished the revertant-inducing activity. Based on these findings, we concluded that both residues 30 and 31, conserved among the K*rev*-1 family, are essential for the effective transformation-suppressor activity, although the contribution of basic Lys[31] seemed to be greater than that of Glu[30] (Kitayama et al., unpublished observations).

IX. OTHER FEATURES

The counteracting activity of the rap1B protein, the close relative of Krev-1, against H-ras proteins was observed in the *Xenopus* oocyte system. Lapetina et al. investigated the effects of microinjected rap1B or rap2B on the H-ras (Val[12])-induced germinal vesicle breakdown in *Xenopus* oocytes.[22] They reported that rap1B efficiently (more than 90%) inhibited the germinal vesicle breakdown while rap2B showed rather weak effects.

The Krev-1 protein can efficiently bind to the *ras* p21 GTPase-activating protein (GAP) *in vitro* and inhibit the ras/GAP interaction. GAP was first discovered in *Xenopus* oocytes by McCormick et al.[23] Later, it was shown that ras interacts with GAP through its effector-binding domain.[24] The question of whether GAP is able to interact with the protein product of the K*rev*-1 gene and stimulate its intrinsic GTPase activity was examined.[25,26] Although Krev-1 could bind to GAP, its GTPase activity was not stimulated. Moreover, the affinity between Krev-1 and GAP was even greater than that between ras and GAP: in one report a 100-fold difference was recorded.[25] As a consequence, GAP-mediated ras GTPase activation was inhibited by the addition of Krev-1. Physiological relevance of this finding remains to be evaluated. Another set of GAPs specific for Krev-1 have been found[27] and cDNAs for one of them isolated and characterized.[28] Interestingly, the "rap1GAP" shows no structural similarity with "rasGAP", and there is evidence indicating that Krev-1 interacts with the rap1GAP through domains different from the "effector" domain.

The release of the GDP molecule followed by the binding with GTP is believed to be a critical step for ras protein activation. Factors involved in the release of GDP have been partially purified and characterized.[29,30] A similar GDP dissociation stimulator (GDS) for *smg* 21B was purified and its cDNA was isolated.[31] This GDS has a rather broad substrate specificity: effective for *smg* p21s, K-ras, and rhoA, but not for H-ras, rhoB, and smg p25A/rab3A (see Chapter 27).

Phosphorylation *in vitro* as well as *in vivo* of the Krev-1 and related proteins has bene studied to some detail, and the evidence indicates that these proteins serve as substrates for cAMP-dependent protein kinase (PKA). The identity has been established between rap1B/*smg* p21B and a major PKA-substrate of 22 kDa in platelet microsomes previously termed thrombolamban.[32-34] Purified smg 21B can be phosphorylated *in vitro* by PKA at Ser[179].[35] The phosphorylated form of *smg* p21B shows increased affinity to GDS and decreased affinity to membrane.[35] It was also shown that GDS has an ability to induce the dissociation of the *smg*

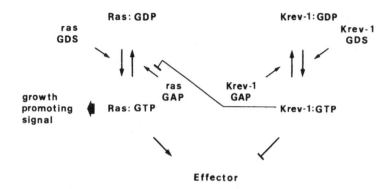

FIGURE 4. A model for the Krev-1 action.

p21B prebound to the membrane.[36] Therefore, it was speculated that GDS may be responsible for the cAMP-induced release of the phosphorylated form of *smg* p21s from the platelet microsomes to the cytoplasm.[37]

In addition to the membrane-bound rap1B/*smg* p21B, platelets also contain rap1A/*smg* p21A/Krev-1 in the cytosol.[38] Interestingly, the molecules seem to be already phosphorylated in the platelet cytosol, since the purified protein could be phosphorylated by PKA *in vitro* only after the treatment with alkaline phosphatase.[39]

In neutrophil, a major low molecular weight GTP-binding protein associated with the membrane fraction was found to be rap1A.[40] Neutrophil rap1A is a good substrate for PKA *in vitro* and, probably, *in vivo*. The target site for this phosphorylation was identified as Ser[180]. Effects of this phosphorylation on the neutrophil functions are yet to be elucidated (see below).

Krev-1 was copurified with the NADPH oxidase cytochrome b_{558} light chain prepared from stimulated neutrophils.[41] In cell-free NADPH oxidase activation system, two kinds of G proteins are required. One is a pertussis toxin-sensitive heterotrimeric G protein and the other is a pertussis toxin-insensitive GTP-binding protein. Gabig et al. reported the data suggesting Krev-1 or a related protein as a candidate for the latter protein.[42] More recently, however, another member of the ras protein family, rac 2, has also been suggested as a feasible candidate.[43,44] Since it was found that rap1A/Krev-1 is the most abundant low molecular weight G protein in neutrophil membrane,[40] elucidation of the role of this protein in the above-mentioned or some other neutrophil function remains as an interesting subject.

X. MODEL OF ACTION OF THE Krev-1 PROTEIN

One of the models for the mechanism of Krev-1-mediated growth suppression is shown in Figure 4.

The proteins ras and Krev-1 are regulated by GDP/GTP exchange reaction. In both proteins the GTP-bound form is active and the GDP-bound form is inactive. GTPase-activating proteins, specific for each protein, stimulate intrinsic GTPase activity and GDP-bound proteins are accumulated. GDP dissociation stimulators promote the release of GDP and accelerate the GDP/GTP exchange reaction. Since Krev-1 protein shares a common structure with ras protein in its effector target domain, it is likely that Krev-1 and ras interact with the same target molecule. Formation of the growth-promoting ras/target complex might be inhibited by overexpressed Krev-1 protein. Although the Krev-1 product were found to be able to interact with *ras*-GAP, whether GAP is the target protein or not remains to be explored. At present we cannot exclude the possibility that Krev-1/target complex generates a growth inhibitory signal independently from the ras-mediated growth promoting signal. However, our finding that the H-*ras* gene could be converted to a transformation-suppressor by only two mutations, raises the possibility that Krev-1 represents an evolutionarily generated, dominant negative antagonist for ras proteins.

REFERENCES

1. **Noda, M., Kitayama, H., Matsuzaki, T., Sugimoto, Y., Okayama, H., Bassin, R. H., and Ikawa, Y.,** Detection of genes with a potential for suppressing the transformed phenotype associated with activated *ras* genes, *Proc. Natl. Acad. Sci. U.S.A.*, 86, 162, 1989.
2. **Kitayama, H., Sugimoto, Y., Matsuzaki, T., Ikawa, Y., and Noda, M.,** A *ras*-related gene with transformation suppressor activity, *Cell*, 56, 77, 1989.
3. **Noda, M., Selinger, Z., Scolnick, E. M., and Bassin, R. H.,** Flat revertants isolated from Kirsten sarcoma virus-transformed cells are resistant to the action of specific oncogenes, *Proc. Natl. Acad. Sci. U.S.A.*, 80, 5602, 1983.
4. **Schejter, E. D. and Shilo, B.,** Characterization of functional domains of p21 *ras* by use of chimeric genes, *EMBO J.*, 4, 407, 1985.
5. **Pizon, V., Chardin, P., Lerosey, I., Olofsson, B., and Tavitian, A.,** Human cDNAs *rap* 1 and *rap* 2 homologous to the Drosophila gene D*ras* 3 encode proteins closely related to *ras* in the "effector" region, *Oncogene*, 3, 201, 1988.
6. **Pizon, V., Chardin, P., and Tavitian, A.,** Nucleotide sequence of a human cDNA encoding a *ras*-related protein (*rap* 1B), *Nucleic Acids Res.*, 16, 7719, 1988.
7. **Kawata, M., Matsui, Y., Kondo, J., Hishida, T., Teranishi, Y., and Takai, Y.,** A novel small molecular weight GTP-binding protein with the same putative effector domain as the *ras* proteins in bovine brain, *J. Biol. Chem.*, 263, 18965, 1988.
8. **Matsui, Y., Kikuchi, A., Kawata, M., Kondo, J., Teranishi, Y., and Takai, Y.,** Molecular cloning of *smg* p21B and identification of *smg* p21 purified from bovine brain and human platelets as *smg* p21B, *Biochem. Biophys. Res. Commun.*, 166, 1010, 1990.
9. **Ohmstede, C.-A., Farrell, F. X., Reep, B. R., Clemetson, K. J., and Lapetina, E. G.,** *RAP* 2B:A *RAS*-related GTP-binding protein from platelets, *Proc. Natl. Acad. Sci. U.S.A.*, 87, 6527, 1990.
10. **Barbacid, M.,** *ras* genes, *Annu. Rev. Biochem.*, 56, 779, 1987.
11. **Chant, J. and Herskowitz, I.,** Genetic control of bud site selection in yeast by a set of gene products that constitute a morphogenetic pathway, *Cell*, 65, 1203, 1991.
12. **Bender, A. and Pringle, J. R.,** Multicopy suppression of the *cdc24* budding defect in yeast by *CDC42* and three newly identified genes including the *ras*-related gene *RSR1*, *Proc. Natl. Acad. Sci. U.S.A.*, 86, 9976, 1989.

13. **Kim, S., Mizoguchi, A., Kikuchi, A., and Takai, Y.**, Tissue and subcellular distributions of the *smg-21/rap* 1/Krev-1 proteins which are partly distinct from those of c-*ras* p21s, *Mol. Cell. Biol.*, 10, 2645, 1990.

14. **Beranger, F., Goud, B., Tavitian, A., and de Gunzburg, J.**, Association of the *Ras*-antagonistic Rap 1/Krev-1 proteins with the Golgi complex, *Proc. Natl. Acad. Sci. U.S.A.*, 88, 1606, 1991.

15. **Kitayama, H., Matsuzaki, T., Ikawa, Y., and Noda, M.**, Genetic analysis of the Kirsten-*ras*-revertant 1 gene:Potentiation of its tumor suppressor activity by specific point mutations, *Proc. Natl. Acad. Sci. U.S.A.*, 87, 4284, 1990.

16. **Feig, L. A. and Cooper, G. M.**, Inhibition of NIH3T3 cell proliferation by a mutant *ras* protein with preferential affinity for GDP, *Mol. Cell. Biol.*, 8, 3235, 1988.

17. **Casey, P. J., Solski, P. A., Der, C. J., and Buss, J. E.**, p21^{ras} is modified by a farnesyl isoprenoid, *Proc. Natl. Acad. Sci. U.S.A.*, 86, 8323, 1989.

18. **Kawata, M., Farnsworth, C. C., Yoshida, Y., Gelb, M. H., Glomset, J. A., and Takai, Y.**, Posttranslationally processed structure of the human platelet protein *smg* p21B:Evidence for geranylgeranylation and carboxyl methylation of the C terminal cysteine, *Proc. Natl. Acad. Sci. U.S.A.*, 87, 8960, 1990.

19. **Paterson, H., Reeves, B., Brown, R., Hall, A., Furth, M., Bos, J., Jones, P., and Marshall, C. J.**, Activated N-*ras* controls the transformed phenotype of HT 1080 human fibrosarcoma cells, *Cell*, 51, 803, 1987.

20. **Kitayama, H., Matsuzaki, T., Ikawa, Y., and Noda, M.**, A domain responsible for the transformation suppressor activity in Krev-1 protein, *Jpn. J. Cancer Res.*, 81, 445, 1990.

21. **Zhang, K., Noda, M., Vass, W. C., Papageorge, A. G., and Lowy, D. R.**, Identification of small clusters of divergent amino acids that mediate the opposing effects of *ras* and Krev-1, *Science*, 249, 162, 1990.

22. **Campa, M. J., Chang, K.-J., Vedia, L. M., Reep, B. R., and Lapetina, E. G.**, Inhibition of *ras*-induced germinal vesicle breakdown in *Xenopus* oocytes by *rap*-1B, *Biochem. Biophys. Res. Commun.*, 174, 1, 1991.

23. **Trahey, M. and McCormick, F.**, A cytoplasmic protein stimulates normal N-*ras* p21 GTPase, but does not affect oncogenic mutants, *Science*, 238, 542, 1987.

24. **Cales, C., Hancock, J. F., Marshall, C. J., and Hall, A.**, The cytoplasmic protein GAP is implicated as the target for regulation by the *ras* gene product, *Nature*, 332, 548, 1988.

25. **Frech, M., John, J., Pizon, V., Chardin, P., Tavitian, A., Clark, R., McCormick, F., and Wittinghofer, A.**, Inhibition of GTPase activating protein stimulation of *Ras*-p21 GTPase by the K*rev*-1 gene product, *Science*, 249, 169, 1990.

26. **Hata, Y., Kikuchi, A., Sasaki, T., Schaber, M. D., Gibbs, J. B., and Takai, Y.**, Inhibition of the ras p21 GTPase-activating protein-stimulated GTPase activity of c-Ha-*ras* p21 by *smg* p21 having the same putative effector domain as *ras* p21s, *J. Biol. Chem.*, 265, 7104, 1990.

27. **Kikuchi, A., Sasaki, T., Araki, S., Hata, Y., and Takai, Y.**, Purification and characterization from bovine brain cytosol of two GTPase-activating proteins specific for *smg* p21, a GTP-binding protein having the same effector domain as c-*ras* p21s, *J. Biol. Chem.*, 264, 9133, 1989.

28. **Rubinfeld, B., Munemitsu, S., Clark, R., Conroy, L., Watt, K., Crosier, W. J., McCormick, F., and Polakis, P.**, Molecular cloning of a GTPase activating protein specific for the Krev-1 protein P21^{rap1}, *Cell*, 65, 1033, 1991.

29. **Wolfman, A. and Macara, I. G.**, A cytosolic protein catalyzes the release of GDP from p21^{ras}, *Science*, 248, 67, 1990.

30. **Downward, J., Riehl, R., Wu, L., and Weinberg, R. A.**, Identification of a nucleotide exchange-promoting activity for p21^{ras}, *Proc. Natl. Acad. Sci. U.S.A.*, 87, 5998, 1990.

31. **Kaibuchi, K., Mizuno, T., Fujioka, H., Yamamoto, T., Kishi, K., Fukumoto, Y., Hori, Y., and Takai, Y.**, Molecular cloning of the cDNA for simulatory GDP/GTP exchange protein for *smg* p21s (*ras* p21-like small GTP-binding proteins) and characterization of stimulatory GDP/GTP exchange protein, *Mol. Cell. Biol.*, 11, 2873, 1991.

32. **Fischer, T. H. and White, G. C.**, Partial purification of and characterization of thrombolamban, a 22,000 dalton cAMP-dependent protein kinase substrate in platelets, *Biochem. Biophys. Res. Commun.*, 149, 700, 1987.

33. **Kawata, M., Kikuchi, A., Hoshijima, M., Yamamoto, K., Hashimoto, E., Tamamura, H., and Takai, Y.,** Phosphorylation of *smg* p21, a *ras* p21-like GTP-binding protein, by cyclic AMP-dependent protein kinase in a cell-free system and in response to prostaglandin E_1 in intact human platelets, *J. Biol. Chem.,* 264, 15688, 1989.

34. **White, T. E., Lacal, J. C., Reep, B., Fischer, T. H., Lapetina, E. G., and White, G. C., II,** Thrombolamban, the 22-kDa platelet substrate of cyclic AMP-dependent protein kinase, is immunologically homologous with the Ras family of GTP-binding proteins, *Proc. Natl. Acad. Sci. U.S.A.,* 87, 758, 1990.

35. **Hata, Y., Kaibuchi, K., Kawamura, S., Hiroyoshi, M., Shirataki, H., and Takai, Y.,** Enhancement of the action of *smg* p21 GDP/GTP exchange protein by the protein kinase A-catalyzed phosphorylation of *smg* p21, *J. Biol. Chem.,* 266, 6571, 1991.

36. **Kawamura, S., Kaibuchi, K., Hiroyoshi, M., Hata, Y., and Takai, Y.,** Stoichiometric interaction of *smg* p21 with its GDP/GTP exchange protein and its novel action to regulate the translocation of *smg* p21 between membrane and cytoplasm, *Biochem. Biophys. Res. Commun.,* 174, 1095, 1990.

37. **Lapetina, E. G., Lacal, J. C., Reep, B. R., and Vedia, L. M.,** A *ras*-related protein is phosphorylated and translocated by agonist that increase cAMP levels in human platelets, *Proc. Natl. Acad. Sci. U.S.A.,* 86, 3131, 1989.

38. **Nagata, K., Itoh, H., Katada, T., Takenaka, K., Ui, M., Kaziro, Y., and Nozawa, Y.,** Purification, identification and characterization of two GTP-binding proteins with molecular weights of 25,000 and 21,000 in human platelet cytosol, *J. Biol. Chem.,* 264, 17000, 1989.

39. **Nagata, K., Nagao, S., and Nozawa, Y.,** Low Mr GTP-binding proteins in human platelets:cyclic AMP-dependent protein kinase phosphorylates m22KG(I) in membrane but not c21KG in cytosol, *Biochem. Biophys. Res. Commun.,* 160, 235, 1989.

40. **Quilliam, L. A., Mueller, H., Bohl, B. P., Prossnitz, V., Sklar, L. A., Der, C. J., and Bokoch, G. M.,** Rap 1A is a substrate for cyclic AMP-dependent protein kinase in human neutrophils, *J. Immunol.,* 147, 1628, 1991.

41. **Quinn, M. T., Parkos, C. A., Walker, L., Orkin, S. H., Dinauer, M. C., and Jesaitis, A. J.,** Association of a *Ras*-related protein with cytochrome *b* of human neutrophils, *Nature,* 342, 198, 1989.

42. **Eklund, E. A., Marshall, M., Gibbs, J. B., Crean, C. D., and Gabig, T. G.,** Resolution of a low molecular weight G protein in neutrophil cytosol required for NADPH oxidase activation and reconstitution by recombinant K*rev*-1 protein, *J. Biol. Chem.,* 266, 13964, 1991.

43. **Abo, A., Pick, E., Hall, A., Totty, N., Teahan, C. G., and Segal, A.,** Activation of the NADPH oxidase involves the small GTP-binding protein p21^{rac1}, *Nature,* 353, 668, 1991.

44. **Knaus, U., Heyworth, P. G., Evans, T., Curnutte, J. T., and Bokoch, G. M.,** Regulation of phagocyte oxygen radical production by the GTP-binding protein Rac 2, *Science,* 254, 1512, 1991.

Chapter 12

ral GENE PRODUCTS AND THEIR REGULATION

Larry A. Feig and Renee Emkey

TABLE OF CONTENTS

0-8493-5214-2/93/$0.00 + $.50
© 1993 by CRC Press, Inc.

I. INTRODUCTION

The *ralA* and *ralB* genes belong to the ras superfamily. Like G proteins, the proteins encoded by members of this superfamily bind GTP and GDP with high affinity and cycle between the active GTP-bound and inactive GDP-bound states (for review see Reference 1). The former is promoted by the replacement of GDP with GTP, the latter by hydrolysis of GTP to GDP. The Ras proteins can be distinguished from the G protein family by their smaller size (20 to 30 kDa compared to 40 to 50 kDa). They also appear to exist in cells as monomers, in contrast to the heterotrimeric G proteins that contain β- and γ-subunits. Finally, the intrinsic GTPase activity of the Ras superfamily members can be stimulated by specific GTPase-activating proteins (GAPs).

The Ras superfamily has been divided into subfamilies based upon amino acid sequence similarities between their encoded proteins.[2] The *ras* gene subfamily contains the H-*ras*, N-*ras*, and K-*ras* protooncogenes, the *ral*A and *ralB* genes, the *rap1* and *rap2* genes and the R-*ras* gene. Other subfamilies, *rho, rab, arf,* and *ran(TC4),* display less sequence similarity to *ras* genes.

Like many members of the *ras* superfamily, the *ral* genes were originally detected by sequence similarity to *ras,* rather than by functional assay. Specifically, an oligonucleotide probe, based on a region of ras protein involved in GTP hydrolysis (amino acids 57 to 63), identified highly similar *ralA* and *ralB* genes in cDNA libraries from simian, human, and mouse cells.[3,4] Homologs of *ral* have been found in *Xenopus laevis,* but not in *Saccharomyces cerevisiae*[5] (L. A. Feig, unpublished observation). The predicted amino acid sequence of these proteins showed that in addition to the region of the probe, all other segments of ras protein that are known to be involved in interactions with guanine nucleotides are highly conserved in these two newly identified proteins. This suggested, and it was subsequently demonstrated, that Ral proteins bind and hydrolyze GTP.[6]

By analogy to better characterize GTP-binding proteins, it is likely that Ral proteins also function as molecular switches. They are expected to be active in altering the activity of a downstream target protein when bound to GTP and inactive when bound to GDP. Although this suggests how Ral activity is regulated, it does not suggest which cellular systems Ral proteins influence. In fact, the physiological role of Ral proteins is not known. The possibilities keep growing as new information implicates GTP-binding proteins in the regulation of a wide variety of cellular processes. These include signal transduction across the plasma membrane, vesicle and protein targeting, actin filament formation, coupling stages of the cell cycle, and protein translation.[17] The task of identifying a function for Ral may not be as difficult as it seems, since research on Ras and other GTP-binding proteins not only establishes paradigms for how these proteins may function, but it also suggests many experimental strategies to reveal the physiological function of Ral proteins. In this review, we will summarize how our knowledge of other GTP-binding proteins is being exploited to reveal the function of the ras-related Ral proteins.

II. BIOCHEMISTRY OF Ral PROTEINS

The genes *ralA* and ral*B* encode extremely similar (85% identity) proteins of 206 amino acids with predicted molecular masses of ~24 kDa (p24 Ral).[3,4] Ral proteins share 55% sequence identity with p21 Ras. The difference in molecular weight is due primarily to an additional 11 amino acids on the N terminus and an insertion of five amino acids near the C terminus of Ral. In the four major segments involved in binding and hydrolyzing guanine nucleotides, the two families are almost indistinguishable. This sequence similarity is reflected in similar biochemical properties: (1) both p24 Ral and p21Ras bind GTP and GDP with affinities ~10^{-11} M; (2) both show a dramatic increase in nucleotide dissociation rates when Mg^{2+} is chelated with EDTA; (3) both hydrolyze GTP very slowly (~0.05 min^{-1} at 37°C); and (4) both display decreased GTPase activity when the same critical amino acids are mutated.[6] Moreover, NMR analysis reveals that the environment around the phosphates of guanine nucleotides is extremely similar in the two proteins; however, subtle differences have been detected. For example, the drop in GTPase activity induced by a Leu substitution is greater in Ras (position 61Gln) than in the comparable mutation in Ral (position 72Gln). Moreover, the rate of GDP dissociation from Ral is greater than that from Ras, particularly when Mg^{2+} is not present.[6] Overall, however, Ras and Ral display very similar interactions with guanine nucleotides. The physiological significance of this may be that the activities of Ral and Ras are regulated in similar ways. This contrasts with another subfamily member, Rap, which has an amino acid difference at a position comparable to 61Gln in Ras. Rap proteins, therefore, have an intrinsic GTPase activity that is comparable to a GTPase-defective Ras mutant.[8]

Also contributing to the high sequence similarity between Ras and Ral are additional regions of the proteins whose functions have not yet been determined. These include Ras sequences 75 to 82 and 97 to 104 that reside between the second and third nucleotide-binding domains.

Ras and Ral sequences do diverge in other regions that reside on the outside of the proteins and potentially interact with other molecules. For example, amino acids from position 20 to 50 and 62 to 70 of Ras are involved in binding GAPs and have been implicated in affecting downstream target proteins.[9] Ral proteins display some sequence similarity to Ras in this region, however clear differences exist. This is consistent with Ras and Ral having distinct GAPs.[10] Moreover, Ral is involved in a regulatory pathway distinct from Ras, since constitutively activated Ral proteins do not oncogenically transform cells (R. Emkey and L. A. Feig, unpublished observation).

III. CELLULAR AND SUBCELLULAR LOCALIZATION OF Ral PROTEINS

Ral proteins likely serve a function used in all cells, since *ralA* gene expression has been detected in all cell types examined. Particularly high levels have

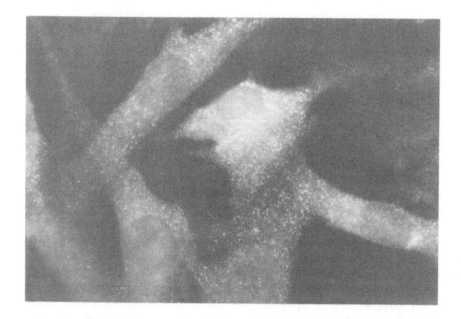

FIGURE 1. Immunofluorescence localization of p24 ral in NIH 3T3 cells. NIH 3T3 cells were fixed with 3.7% formalin and then permeabilized with 0.5% saponin. The cells were then incubated with affinity-purified anti-ralA antibody and stained with Texas Red conjugated anti-rabbit IgG.

been found in brain, testes, and platelets.[11,12] In some cases, the subcellular localization of Ras superfamily members has given a strong indication of cellular function. For example, p21 Ras exists primarily at the inner surface of the plasma membrane consistent with its role in transducing signals emanating from outside the cell. Members of the *rab* family are localized to specific vesicular compartments. The discovery of Rab5 primarily in membranes of early endosomes led to experiments that documented a role for this protein in targeting vesicle fusion associated with endocytosis.[13]

Affinity-purified antibodies to recombinant RalA have shown that like Ras, p24 Ral is found almost exclusively in membrane fractions of cell lysates.[14] In the marine ray, *Discopyge ammata*, Ral has been detected in isolated synaptic vesicles and appears to be concentrated in presynaptic terminals.[15] Since these antibodies were made against the entire RalA protein, they likely cross-react with the highly similar RalB. Thus, it is not known at present which ral family member is being observed. We used these same antibodies (kindly provided by P. Chardin) in immunofluorescence experiments with NIH 3T3 cells. Figure 1 shows punctate staining throughout the cytoplasm consistent with the presence of p24 Ral in vesicles. The staining pattern was not consistent with vesicles associated with Golgi, the subcellular site of Rap, Arf, and some Rab family members. The pattern was more similar to vesicles of the endocytic pathway such as clathrin-coated vesicles and smooth endosomes.

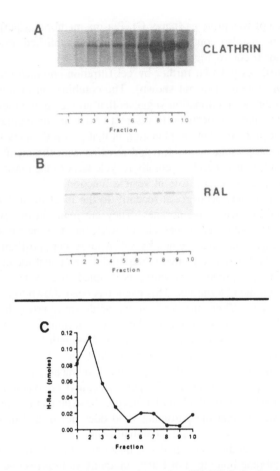

FIGURE 2. Fractionation of calf brain membranes on a Ficoll/D₂O density gradient. Calf brains were homogenized and crude membrane fractions were isolated by centrifugation. They were then fractionated on a Ficoll/D_0 gradient. Ten fractions were collected and an equal amount of each were assayed as described below. Panel A shows the distribution of clathrin-heavy chains in fractions of the gradient by Coomassie staining of SDS-PAGE gels. As expected the majority of membranes containing clathrin were found at the bottom of the gradient. Panel B shows ral protein in these fractions by Western blotting using anti-ralA antibodies. Panel C shows H-*ras* protein in these fractions. It has a distinctly different distribution from ral in this gradient. Consistent with its known localization in plasma membranes, H-*ras* was found exclusively in light vesicle fractions. H-*ras* p21 was immunoprecipitated with anti-ras monoclonal YA6-172 and then quantitated by ^{32}P[GTP]-binding of immunoprecipitates.

To investigate this further, membrane fractions of calf brain were separated on a Ficoll/D₂O density gradient and the presence of Ral protein was assessed by Western blotting.[16] As shown in Figure 2, panel B, p24 Ral was detected in all gradient fractions. These included dense membrane fractions containing clathrin (see panel A) and light fractions containing plasma membranes, as evidenced by the presence of Ras proteins (panel C). Plasma membranes purified by more conventional means also contained Ral proteins (unpublished observations, L. Feig; and personal communication, P. Chardin).

Clathrin vesicles purified further by gel filtration chromatography retained significant Ral protein (data not shown). The combination of immunofluorescence and membrane fractionation suggests that unlike some other small GTP-binding proteins that are localized to discreet vesicular compartments, p24 Ral is a component of many. One explanation is that p24 Ral enters the endocytic pathway from the plasma membrane and remains with the membrane fraction as clathrin vesicles uncoat and eventually recycle back to the plasma membrane. Thus, Ral may play a general role in vesicle function.

Much attention has been given recently to the mechanism by which Ras-related proteins localize to membranes. They are made in the cytoplasm and become associated with membranes via a CAAX motif at the carboxy terminus and another recognition sequence nearby.[17,18] A three-step, posttranslational process involving the CAAX sequence is required for efficient membrane binding of p21 Ras. First, a prenoid derivative is coupled to a cysteine residue four amino acids from the C terminus. This event has been shown to be required for Ras function *in vivo*. Then, the AAX amino acids are removed by proteolysis, and finally, the α-carboxyl group of the now C terminal cysteine residue is methylated. Ras proteins are prenylated at this Cys with C_{15} farnesyl, and the signal for this particular lipid appears to be the Met at the X position of CAAX.[19] Ral proteins, like many other Ras superfamily members, have a Leu at this position, which appears to specify C_{20} geranylgeranyl prenylation of Cys. It is not yet clear what role individual lipid molecules have on membrane interactions. It is clear, however, that the particular lipid does not in and of itself specify localization to a specific membrane compartment.

In p21 Ras, an additional signal just upstream of the CAAX box also contributes to membrane binding. For Ras[H], another Cys four residues upstream is a palmitoylation site. For Ras[K], a polybasic stretch of amino acids is involved.[18] Alteration of these sites results in poor membrane binding, yet interestingly, the protein still functions! Ral proteins are more like K-Ras in that they have some basic amino acids at a comparable position, but their influence, as well as that of prenylation, on specific membrane localization has not yet been studied.

IV. Ral-GAP

The activity of Ras superfamily members is thought to be controlled by at least two types of proteins. Guanine nucleotide-releasing factors (GRFs) that activate function by promoting the exchange of GTP for bound GDP, and GAPs that inhibit function by accelerating the hydrolysis of GTP.[9] GRFs, which may function like hormone receptors that activate G proteins, have been identified for some Ras-related proteins, however, their regulatory role is only now being investigated. In contrast, a growing body of evidence supports the idea that GAPs for Ras (Ras-GAP and NF-1) act as both negative regulators and as components of the downstream target of Ras. For example, a negative regulatory role is consistent with the observation that the GTPase-stimulating activity of GAP decreases after stimulation of T cell by antigen, which is concomitant with an increase in the proportion of Ras bound to GTP.[20] Moreover, GAP associates with and becomes tyrosine-phosphorylated by some growth factor receptors.[21,22] In response to receptor activation, GAP also associates with two other proteins whose functions are not yet known.[23]

A second function of GAP, a component of a downstream target, is consistent with the observation that mutations in Ras that decrease Ras-GAP interaction invariably decrease downstream coupling efficiency. More directly, Ras and GAP are both required for carbachol inhibition of K+ fluxes in isolated atrial membranes *in vitro*.[24] GAP is also necessary for H1-kinase activation in *Xenopus* oocytes by microinjected Ras.[25] Many more details of this potential dual role for GAP need to be revealed, however, these studies illuminate the key role this class of protein plays in mediating Ras function.

For these reasons, characterization of GAPs for Ral will undoubtedly reveal important clues to help identify the regulatory systems influenced by p24 Ral. Ral-GAP was identified in and partially purified from cytosolic extracts of rat and bovine brain and testes.[10] The biochemical properties of this protein distinguish it from GAPs for other Ras superfamily members including those for Ras, Rho, and Rap. Its most striking biochemical feature is its large size as assessed by gel filtration (M_r greater than 10^6). This may be an overestimate since sucrose gradients indicate it is between 150 and 443 kDa. Although this may reflect nonspecific aggregation, a GAP for Ras, NF-1, is also very large (~ 300 kDa) as predicted from cDNA sequence. Given the fact that Ras-GAP (120 kDa) requires only its C terminal 30 kDa to influence Ras, the large size of Ral-GAP may indicate its activity is regulated by complex mechanisms.[26] Alternatively, like the Bcr protein, which is both a GAP (for the GTP-binding protein, rac) and a protein kinase, another biochemical property may be encoded in these additional sequences.[27] Since Ras-GAP is likely a downstream target of Ras, proteins that interact with these additional sequences of Ral-GAP may represent downstream targets of Ral. In fact, the large size of Ral-GAP suggests that these potential target proteins will be purified in a complex with Ral-GAP.

The Ras paradigm may be appropriate for predicting some aspects of Ral-GAP function, because just as Ral is similar in many ways to Ras, our limited knowledge of Ral-GAP suggests it is similar to Ras-GAP. For example, the K_m of Ral-GAP for Ral is similar to that of Ras-GAP for Ras.[10] Furthermore, Ras-GAP activity has been shown to be inhibited by a set of lipid molecules, some of which are produced in cells in response to mitogens.[28,29] Ral-GAP activity is inhibited by the same set (R. Emkey and L. A. Feig, unpublished observations), which differs from that shown to effect Rho-GAP and R-Ras. Finally, Ral-GAP failed to promote the GTPase activity of mutant Ral proteins containing amino acid substitutions that in Ras lead to GAP-insensitive proteins.[10]

It must be kept in mind, however, that Ral may function in a manner that is very different than the signal-transducing Ras. Ral may function in vectorial transport like protein translation factors and other vesicle-localized Ras family members that are involved in vesicle sorting. The processes controlled by these proteins require the continuous cycling from the GTP- to GDP-bound state to maintain activity. In this type of system a GAP protein could actually function as a positive regulator of function, since the intrinsic GTPase activity of this family of proteins is too slow for significant cycling of the GTP-binding protein.

V. OVEREXPRESSION OF NORMAL AND MUTANT Ral PROTEINS

A very useful approach for assessing the function of a protein is to study the consequence of increasing its activity in cells by transfection of normal or constitutively active forms. For Ras, oncogenic mutations that decrease intrinsic GTPase activity and destroy responsiveness to GAP, yield highly active proteins that are locked predominantly in the GTP state in cells. Expression of these mutants in some cells leads to oncogenic transformation and in others, differentiation. As previously described, similar mutations in *ral* yield proteins that are GTPase-deficient and fail to respond to Ral-GAP, implying they too will be constitutively active in cells. However, no striking phenotypes have been detected to date after transfection of these mutant genes.

An alternative approach, that is potentially more revealing, is to observe the consequence of inhibiting endogenous Ral function in cells by transfecting a dominant inhibitory form of p24 Ral. Such interfering *ras* mutants have been constructed and used extensively to identify regulatory pathways in mammalian cells that involve Ras function.[30-35] Because of the similarity between p21 Ras and p24 Ral, it is likely that similar strategies can be accomplished for the latter. One such mutation is at Ser17 of Ras, one of the key amino acids involved in binding Mg^{2+} associated with bound nucleotide.[30,36] By analyzing a series of mutations at this site, it is clear that a unique characteristic of this type of dominant inhibitory mutant is that binding specificity is changed from equal affinity for GTP and GDP to preferential affinity for GDP.[37] It also has been shown that these mutants are constitutively inactive, such that even when bound to GTP they cannot activate downstream targets. A variety of experiments suggest that these mutants interfere with normal Ras function by competing for a guanine nucleotide releasing factor that normally activates Ras.[38]

A comparable mutation (Ser to Asn at position 28) has been made in Ral and the mutant's interactions with guanine nucleotides were compared to wild-type ral. One way to assess the affects of a mutation on the affinity of these proteins for GTP and GDP is to measure the rate of nucleotide dissociation. As shown in Figure 3, the dissociation rates of GTP and GDP from normal Ral are similar, consistent with their similar affinities for the protein. In contrast, S28N RalA shows an increase in GDP dissociation and an even greater GTP-dissociation rate. This indicates that, as expected, the mutant Ral protein has preferential affinity for GDP. Thus, one can be reasonably confident, given the tight correlation between this unique binding specificity and Ras inhibitory phenotype,

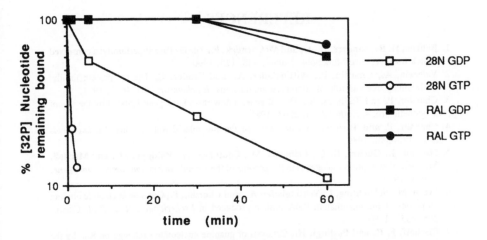

FIGURE 3. Nucleotide binding properties of normal and 28N Ral proteins. Normal and mutant Ral proteins, purified from an *E. coli* expression system were allowed to bind either $5 \times 10^{-8} M$ [g-^{32}P]GTP or [^{35}S]GDP and then the rate of dissociation of labeled nucleotide was measured by introducing $10^{-4} M$ unlabeled GTP or GDP. The percent of labeled protein was measured as a function of time by passing the sample through nitrocellulose filters. Two experiments were averaged in the figure.

that the mutant Ral protein is also locked in an inactive state. Thus, it is likely that when expressed in cells, the mutant will interfere with the activation of endogenous Ral. The phenotype of such cells is presently under investigation.

As described earlier, the continuous cycling of Ral may be required to perform its function. Therefore, Ral mutants that are defective in GTPase activity and unresponsive to GAP may actually interfere with normal Ral function in cells. In this case, these mutants will likely compete for the downstream target of Ral, rather than its upstream activator.

VI. CONCLUSION

It is clear that much work needs to be done to define the function of Ral proteins. The knowledge of the structure and function of other GTP-binding proteins will undoubtedly make this task earlier. In particular, the purification and cloning of Ral-GAP and the use of dominant inhibitory forms of Ral to interfere with normal Ral function are sure to be enlightening.

REFERENCES

1. **Bourne, H. R., Sanders, D. A., and McCormick, F.,** The GTPase superfamily: a conserved switch for diverse cell functions, *Nature,* 348, 125, 1990.
2. **Valencia, A., Chardin, P., Wittinghofer, A., and Sander, C,** The ras protein family: evolutionary tree and role of conserved amino acids, *Biochemistry,* 30, 4637, 1991.
3. **Chardin, P. and Tavitian, A.,** The ral gene: a new ras-related gene isolated by the use of a synthetic probe, *EMBO J.,* 5, 2203, 1986.
4. **Chardin, P. and Tavitian, A.,** Sequence of the Ras-related Ral proteins, *Nucleic Acids Res.,* 17, 4380, 1989.
5. **Moreau, J., Guellec, R. L., Leibovici, M., Couturier, A., Philippe, M., and Mechali, M.,** Detection of proto-oncogenes in the genome of the amphibian *Xenopus laevis, Oncogene,* 4, 443, 1989.
6. **Frech, M., Schlichting, I., Wittinghofer, A., and Chardin, P.,** Guanine nucleotide binding properties of the mammalian RalA protein produced in *Escherichia coli, J. Biol. Chem.,* 265, 6353, 1990.
7. **Bischoff, F. R. and Postingl, H.,** Catalysis of guanine nucleotide exchange on Ran by the mitotic regulator RCC1, *Nature,* 354, 80, 1991.
8. **Quilliam, L. A., Der, C. J., Clark, R., O'Rourke, E. C., Zhang, K., McCormick, F., and Bockoch, G. M.,** Biochemical characterization of baculovirus-expressed rap1A/Krev-1 and its regulation by GTPase-activating proteins, *Mol. Cell. Biol.,* 10, 2901, 1990.
9. **Bollag, G. and McCormick, F.,** Regulators and effectors of ras proteins, *Annu. Rev. Cell Biol.,* 7, 601, 1991.
10. **Emkey, R., Freedman, S., and Feig, L. A.,** Characterization of a GTPase activating protein for the Ras-related Ral protein, *J. Biol. Chem.,* 266, 9703, 1991.
11. **Olofsson, B., Chardin, P., Touchot, N., Zahraoui, N., and Tavitian, A.,** Expression of the ras-related ralA and rab genes in adult mouse tissues, *Oncogene,* 3, 231, 1988.
12. **Polakis, P. G., Weber, B. N., Didsbury, J. R., Evans, T., and Snyderman, R.,** Identification of the ral and rac gene products, low molecular mass GTP-binding proteins from human platelets, *J. Biol Chem.,* 264, 163, 1989.
13. **Gorvel, J.-P., Chavrier, P., Zerial, M., and Gruenberg, J.,** rab5 controls early endosome fusion in vitro, *Cell,* 64, 915, 1991.
14. **Bhullar, R. P., Chardin, P., and Haslam, R. J.,** Identification of multiple ral gene products in human platelets that account for some but not all of the platelet Gn-proteins, *FEBS Lett.,* 260, 48, 1990.
15. **Ngsee, J. K., Elferink, L. A., and Scheller, R. H.,** A family of ras-like GTP-binding proteins expressed in electromotor neurons, *J. Biol. Chem.,* 266, 2675, 1991.
16. **Forgac, M. and Cantley, L.,** Characterization of the ATP-dependent proton pump of clathrin coated vesicles, *J. Biol. Chem.,* 259, 8101, 1984.
17. **Gibbs, J.,** Ras-C-terminal processing enzymes — new drug targets?, *Cell,* 65, 1, 1991.
18. **Hancock, J. F., Paterson, H., and Marshall, C. J.,** A polybasic domain or palmitoylation is required in addition to the CAAX motif to localize p21Ras to the plasma membrane, *Cell,* 63, 133, 1990.
19. **Maltese, W. A.,** Posttranslational modification of proteins by isopreinoids in mammalian cells, *FASEB J.,* 4, 3319, 1990.
20. **Downward, J., Graves, J. D., Warne, P. H., Rayter, S., and Cantrell, D. A.,** Stimulation of p21 ras upon T-cell activation, *Nature,* 346, 719, 1990.
21. **Ellis, C., Moran, M., McCormick, F., and Pawson, T.,** Phosphorylation of GAP and GAP-associated proteins by transforming and mitogenic tyrosine kinases, *Nature,* 343, 377, 1990.
22. **Kaplan, D. R., Morrison, D. K., Wong, G., McCormick, F., and Williams, L. T.,** PDGF b-Receptor stimulates tyrosine phophorylation of GAP and association of GAP with a signalling complex, *Cell,* 61, 125, 1990.

23. **Moran, M. F., Polakis, P., McCormick, F., Pawson, T., and Ellis, C.,** Protein-tyrosine kinases regulate the phosphorylation, protein interactions, subcellular distribution and activity of p21Ras GTPase-activating protein, *Mol. Cell. Biol.,* 11, 1804, 1991.

24. **Yatani, A., Okabe, K., Polakis, P., Halenbeck, R., McCormick, F., and Brown, A. M.,** ras p21 and GAP inhibit coupling of muscarinic receptors to atrial K + channels, *Cell,* 61, 769, 1990.

25. **Dominguez, I., Marshall, M. S., Gibbs, J. B., Garcia de Herreros, A., Cornet, M. E., Graziani, G., Diaz-Meco, M. T., Johansen, T., McCormick, F., and Moscat, J.,** Role of GTPase activating protein in mitogenic signalling through phosphatidylcholine-hydrolyzing phospholipase C, *EMBO J.,* 10, 3215, 1991.

26. **Marshall, M. S., Hill, W. S., Ng, A. S., Vogel, U. S., Schaber, M. D., Scolnick, E. M., Dixon, R. A. F., Sigal, I. S., and Gibbs, J. B.,** A C-terminal domain of GAP is sufficient to stimulate ras p21 GTPase activity, *EMBO J.,* 8, 1105, 1989.

27. **Diekmann, D., Brill, S., Garrett, M. D., Totty, N., Hsuan, J., Monfries, C., Hall, C., Lim, L., and Hall, A.,** Bcr encodes a GTPase-activating protein for p21Rac, *Nature,* 351, 400, 1991.

28. **Tsai, M. H., Yu, C. L., Wei, F. S., and Stacey, D. S.,** The effect of GTPase activating protein upon Ras is inhibited by mitogenically responsive lipids, *Science,* 243, 522, 1989.

29. **Yu, C. L., Tsai, M. H., and Stacey, D. W.,** Serum stimulation of NIH 3T3 cells induces the production of lipids able to inhibit GTPase-activating protein activity, *Mol. Cell. Biol.,* 10, 6683, 1990.

30. **Feig, L. A. and Cooper, G. M.,** Inhibition of NIH 3T3 cell proliferation by a mutant ras protein with preferential affinity for GDP, *Mol. Cell. Biol.,* 8, 3235, 1988.

31. **Cai, H., Szeberenyi, J., and Cooper, G. M.,** Effect of a dominant inhibitory Ha-ras mutation on mitogenic signal transduction in NIH 3T3 cells, *Mol. Cell. Biol.,* 10, 5314, 1990.

32. **Szeberenyi, J., Cai, H., and Cooper, G. M.,** Effect of a dominant inhibitory Ha-ras mutation on neuronal differentiation of PC12 cells, *Mol. Cell. Biol.,* 10, 5324, 1990.

33. **Stacey, D. W., Roudebush, M., Day, R., Mosser, S. D., Gibbs, J. B., and Feig, L. A.,** Dominant inhibitory Ras mutants demonstrate the requirement for Ras activity in the action of tyrosine kinase oncogenes, *Oncogene,* 1991.

34. **Gupta, S. K., Callego, C., and Johnson, G. L.,** Mitogenic pathways regulated by G protein oncogenes, *Mol. Biol. Cell,* 3, 123, 1992.

35. **Medema, R. H., Wubbolts, R., and Bos, J. L.,** Two dominant inhibitory mutants of p21 ras interfere with insulin-induced gene expression, *Mol. Cell. Biol.,* 11, 5963, 1991.

36. **Pai, E. F., Krengel, U., Petsko, G. A., Goody, R. S., Kabsch, W., and Wittinghofer, A.,** Refined crystal structure of the triphosphate conformation of H-ras p21 at 1.35 A resolution: implications for the mechanism of GTP hydrolysis, *EMBO J.,* 9, 2351, 1990.

37. **Farnsworth, C. and Feig, L. A.,** Dominant inhibitory mutations in the Mg^{2+} binding site of Ras blocks its activation by GTP, *Mol. Cell. Biol.,* 11, 4822, 1991.

38. **Stacey, D. W., Feig, L. A., and Gibbs, J. G.,** Dominant inhibitory Ras mutants selectively inhibit the activity of either cellular or oncogenic Ras, *Mol. Cell. Biol.,* 11, 4053, 1991.

Chapter 13

THE *rho* GENE FAMILY

Rosario Perona, Rafael P. Ballestero, and Juan Carlos Lacal

TABLE OF CONTENTS

I. INTRODUCTION

Since the discovery of the implications of the family of *ras* genes in human malignancy, several families of small monomeric GTPases have been identified with a molecular mass of 20 to 25 kDa. Almost 100 distinct proteins of this superfamily have now been characterized. Based on the degree of conservation of their respective primary sequences, different families have been defined and grouped into three major branches. Although for most of these proteins little more than their primary structure is known, this provisional assignment may provide some light into the understanding of their biological functions.

The *ras* polypeptides are GTP-binding proteins with a weak GTPase activity. Their intrinsic GTPase activity is regulated *in vivo* by at least two factors, designated as GAP[1] (for GTPase-activating protein) and neurofibromin, the product of the NF-1 gene,[2] present in a large variety of tissues and established cell lines. Other factors regulating the GTP/GDP cycle have been also identified, including a *ras*-guanine nucleotide exchange factor, rGEF,[3] a *ras*-guanine nucleotide releasing factor, *ras*-GRF,[4] and a *ras*-exchange promoting factor, REP.[5]

Among the extensive superfamily of *ras*-related genes, the *rho* family (*ras* homolog) comprise a group of genes highly conserved in evolution, with detected members in yeast, *Aplysia* (marine snail), *Discopyge ommata* (marine ray), rat, and humans. In this chapter we will summarize only the information available on *rho* genes and their products. Other members of this family including CDC42, G25K, TC10, and *rac,* are discussed on separate chapters of this book.

II. IDENTIFICATION AND CHARACTERIZATION OF *rho* GENES

The *rho* genes were first identified during the analysis of a cDNA library isolated from the abdominal ganglia of the mollusc *Aplysia californica,* while searching for clones homologous to mammalian peptide hormones.[6] Low stringency screening resulted in the isolation of a cDNA fragment that shared weak homology with the α-subunit of human chorionic gonadotropin (hCG).

The nucleotide sequence of this fragment revealed some degree of homology to the hCG at the amino acid level. This sequence was embedded in a single open reading frame of 157 nonhomologous amino acids. Comparison to known proteins in the database uncovered that the protein codified by this gene was 35% identical to the Harvey *ras*-p21 protein. In fact, it also codified for a 21,000 Da protein, similar to all known members of the *ras* family.

When the expression of this newly isolated gene was investigated using poly(A)[+] RNA from the *Aplysia* abdominal ganglion, two mRNAs corresponding to 4.5 and 3.3 kb were detected. Hybridization of cDNA libraries from human peripheral T cell and brain with the *Aplysia rho* cDNA led to the identification of a set of clones that hybridized with the *Aplysia* probe, but not with the N- or K-*ras* probes utilized.[7] Differential hybridization of the isolated human cDNA clones with discrete regions of the *Aplysia* cDNA clone made it possible to discern three classes of human *rho* genes. They were classified as clones that hybridized strongly with an N terminal probe but weakly with the C terminal probe of the *Aplysia rho* cDNA (class I, clone H12, or *rho* A);[6,8] clones that hybridized strongly with the C terminus (class II, clone H6, or *rho* B),[6,9] and finally, clones that hybridized strongly with both the C and N termini (class III, clone H9, or *rho* C).[6]

The three human *rho* genes are 85% homologous to the *Aplysia* gene, being highly homologous to each other (Figure 1). Maximal divergences reside in two regions located in the carboxy half of the protein, as previously observed among the members of the *ras* gene family.[10] Genes corresponding to *rho*A, B, and C have also been found in other mammals[6] with an almost complete identity in their protein sequences and *rho*C is ubiquitously expressed in mouse tissues with higher levels in ovary and testis.[11]

Jähner and Hunter independently isolated the *rho*B gene as an immediate early gene by differential screening of quiescent fibroblasts stimulated with growth factors or after activation of a temperature-sensitive mutant of a retroviral protein tyrosine kinase.[12] The *rho*B gene is transiently activated at the transcriptional level by v-*fps* and by the epidermal growth factor (EGF).[12] Its labile mRNA is inducible in the presence of cycloheximide but not actinomycin D. Serum has a weak effect on *rho*B induction and no effect was observed with phorbol esters. These data suggest that *rho*B can be induced by activation of protein tyrosine kinases through a pathway independent of protein kinase C (PKC).

Two different members of the *rho* gene family have been isolated from *Saccharomyces cerevisiae* designated as RHO1 and RHO2, which are 53% homologous to each other.[13] RHO1 is also 70% identical to the *Aplysia* gene. Genetic experiments demonstrate that RHO1 is an essential gene in yeast while RHO2 is not required for cell viability. Overexpression of the catalytic subunit of the cAMP-dependent protein kinase or RAS-Val[19], both capable of suppressing lethality of RAS deletions in yeast, does not suppress the lethality of a RHO1 mutation suggesting that *ras* and *rho* genes function in independent pathways in yeast.

Recently, a new member of this family has been isolated from the marine ray *Discopyge ommata*, and designated as *rho*-O.[14] By cellular fractionation, the *rho*-O product, which is 93% identical to the *Aplysia californica rho* protein, has been localized to the presynaptic terminals.[14]

III. PRIMARY STRUCTURE OF *rho* PROTEINS

When the amino acid sequence encoded by the *Aplysia rho* gene was compared with that of the human H-*ras* gene, the two sequences aligned with 35% homology and three gaps (Figure 1). The *rho* genes encoded polypeptides of 191 amino acids, this is 2 or 3 residues longer than the 188 or 189 amino acid

```
                    10              20              30            40
H-ras         M T E Y K L V V V G A G G V G K S A L T I Q L I Q N H F V D E Y D P T I E D
Aplysia-rho   M A A I R K K L V I V G D G A C G K T C L L I V F S K D Q F P E V Y V P T V F E
rho A         M A A I R K K L V I V G D G A C G K T C L L I V F S K D Q F P E V Y V P T V F E
rho C         M A A I R K K L V I V G D G A C G K T C L L I V F S K D Q F P E V Y V P T V F E
rho B         M A A I R K K L V V V G D G A C G K T C L L I V F S K D E F P E V Y V P T V F E

                    50              60              70            80
H-ras         S Y R K Q V V I D G E T C L L D I L D T A G Q E E Y S A M R D Q Y M R T G E G F
Aplysia-rho   N Y V A D I E V D G K Q V E L A L W D T A G Q E D Y D R L R P L S Y P D T D V I
rho A         N Y V A D I E V D G K Q V E L A L W D T A G Q E D Y D R L R P L S Y P D T D V H
rho C         N Y I A D I E V D G K Q V E L A L W D T A G Q E D Y D R L R P L S Y P D T D V H
rho B         N Y V A D I E V D G K Q V E L A L W D T A G Q E D Y D R L R P L S Y P D T D V I

                    90             100             110           120
H-ras         L C V F A I N N T K S F E D I H Q Y R E Q I K R V K D S D D V P M V L V G N K C
Aplysia-rho   L M C F S I D S P D S L E N I P E K W T P E V R H F C P N V P I I L V G N K K
rho A         L M C F S I D S P D S L E N I P E K W T P E V K H F C P N V P I I L V G N K K
rho C         L M C F S I D S P D S L E N I P E K W T P E V K H F C P N V P I I L V G N K K
rho B         L M C F S V D S P D S L E N I P E K W V P E V K H F C P N V P I I L V A N K K

                   130             140             150           160
H-ras         D L A A R T V E S R Q A Q D L A R S Y G I P Y I E T
Aplysia-rho   D L R N D E S T K R E L M K M Q E P V R P E D G R A M A E K I N A Y S Y
rho A         D L R N D E H T R R E L A K M K Q E P V K P E E G R D M A N R I G A F G Y
rho C         D L R Q D E H T R R E L A K M K Q E P V R S E E G R D M A N R I S A F G Y
rho B         D L R S D E H V R T E L A R M K Q E P V R T D D G R A M A V R I Q A Y D Y
```

```
                              170          180          190           200
H-ras              SAKTRQGVEDAFYTLVREIRQHKL RKLNPPDESGPG
Aplysia-rho   LECSAKTKEGVRDVFETATRAALQVKK KKKGG
rho A         MECSAKTKDGVREVFEMATRAALQARRGKKKSG
rho C         LECSAKTKEGVREVFEMATRAGLQVRKNKRRRG
rho B         LECSAKTKEGVREVFETATRAALQKRRYGSQNGCINC

H-ras         CMSCKCVLS
Aplysia-rho   CVVL
rho A         CLVL
rho C         CPIL
rho B         CKVL
```

FIGURE 1. Comparison of primary structures of human *rho* genes to the *Aplysia californica rho* gene and the human H-ras gene.

long *ras* proteins.[6] Although the *Aplysia rho* gene shares 35% homology at the amino acid level with the H-*ras* gene product, the homology is clustered revealing regions of complete divergence and four regions of strong conservation. Mammaliam *rho*A, B, and C proteins also display about 35% amino acid identity with *ras* proteins, mainly clustered in the same four highly homologous internal regions corresponding to the GTP-binding site. As described before, *rho* genes have also been identified in the baking yeast, *S. cerevisiae*. Their sequence homology to both human *rho* genes and the H-*ras* gene are schematically depicted in Figure 2.

Maximal divergence between *ras* and *rho* proteins extends from residues 121 to 140 (122 to 155 for *rho*), where no homology is observed. This region, which coincides with the first variable domain of the classical *ras* gene family, encompasses a gap in the alignment because of the insertion of a 14 amino acid block in all *rho* proteins not present in *ras*-p21 proteins. The last 21 C terminal amino acids of *rho* correspond to the second variable domain of the classical *ras* gene family. This domain is 12 amino acids shorter in *rho* than in *ras* proteins and conserves only two residues. Thus, *rho* proteins diverge most in regions in which the individual *ras* genes are also divergent. This comparison proved that interposition of constant and variable domains is retained not only among members of a given *ras* gene family, but also among different families.

Within the most conserved domains, strong homology is observed in four discreet regions which have been suggested are involved in GTP binding or hydrolysis (see also Chapter 2).

1. Between residues 5 to 17 (7 to 19 for *rho*) *ras* and *rho* proteins share 67% homology. Of relevance is to note that in this region *rho* proteins retain a Gly[14], equivalent to position 12 in *ras* proteins, a residue that is altered in several transforming *ras* genes.[10] However, at least one significant difference is found in this region: *rho* proteins have an Ala[15] corresponding to Gly[13] in *ras*, suggesting that they might have different biochemical properties.

2. Residues 57 to 64 (59 to 66 for *rho*) show even stronger conservation, with 7 out of 8 identical amino acids and one conservative change. Interestingly, these two conserved regions (residues 5 to 17 and 57 to 64) include the specific residues of cellular *ras* genes, which if mutated, result in genes with transforming potential.[15]

3. Residues 113 to 120 (114 to 121 for *rho*) show a 7 out of 8 amino acid conservation. This region of *ras* and *rho* share sequence homology to the GTP-binding protein of *Escherichia coli*, EF-Tu.[16]

4. Residues 141 to 158 (156 to 173 for *rho*) show 11 out of 18 homologous amino acids with mostly conservative changes. These two regions (residues 113 to 120 and 141 to 158) have been related to the GTP-binding activity of *ras*-p21 proteins, and may be related specifically to the GDP/GTP interchange.

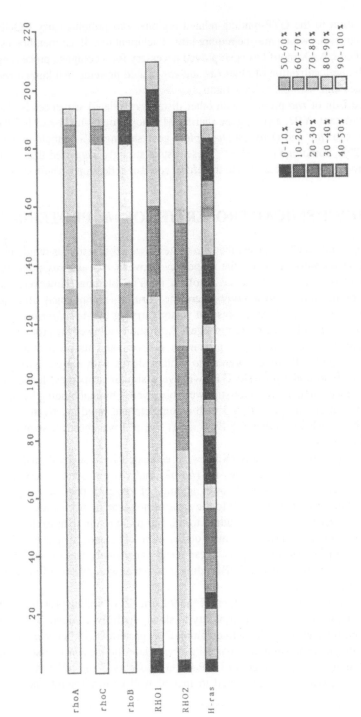

FIGURE 2. Schematical representation of regions of homology among several *rho* proteins.

In addition to the GTP-binding-related regions, *rho* proteins carry COOH terminal sequences which may be required for attachment to cellular membranes. The so-called CAAX motif has been proved necessary for a complete processing and membrane localization of other *ras* and *ras*-related proteins. All known *rho* polypeptides end with the CAAX motif.

Comparison of *rho* proteins with other distant *ras*-related sequences is also highly significant. YP2, a yeast gene sharing 34% homology with H-*ras*, is 32% homologous to *rho*, and D*ras*2, a *Drosophila* gene sharing 50% homology with H-*ras*, is 29% homologous to *rho*.[17] These results strongly suggested that these genes share a common ancestor, and therefore, could encode proteins with similar functions.

IV. BIOCHEMICAL PROPERTIES OF *rho* PROTEINS

Madaule and Axel[6] proposed that *rho* proteins would function as regulatory GTP-binding proteins based on the homology between *rho* and *ras* proteins, especially at regions thought to be involved in GTP binding. However, the definitive proof came from a study where the purified, recombinant, *Aplysia rho* protein was utilized.[18] An expression vector carrying the *A. californica rho* gene under the control of the bacteriophage λ P_L promoter allowed the production of large amounts of the *rho* p21 protein in *E. coli,* and its purification to approximately 90%. Purified *Aplysia rho* protein shares with *ras*-p21 proteins its ability to bind and hydrolyze GTP with very similar properties; however, some differences in their characteristics were observed. Optimal binding by *rho*-p21 occurred at around pH 7 to 8.5 with a complete and rapid inactivation for both proteins at pHs lower than 6.5 By contrast, *ras*-p21 was more resistant than *rho* proteins to alkali treatment.

When dissociation constants (Kds) were estimated by Scatchard analysis, differences in the binding activities for *rho*-p21 and *ras*-p21 were observed.[19] *rho*-p21 showed a 5- to 10-fold lower affinity of *rho*-p21 than of *ras*-p21. Consistent with this finding, a 100-fold excess of nonguanine nucleotides competed more efficiently for GTP binding to the *rho*-p21 protein than to the *ras*-p21 protein. Similarly, guanosine nucleotides more efficiently displaced GTP binding to *rho*-p21 than to *ras*-p21. These results may imply that *rho*-p21 had a lower specificity for GTP and GDP binding, consistent with different functions for *rho*- and *ras*-p21 proteins.

One of the most important features of monomeric GTP-binding proteins with known regulatory functions is their ability to slowly hydrolyze bound GTP at a rate of about 0.01 min^{-1}. The active complex (G protein-GTP) becomes inactivated by this mechanism and remains inactive while the protein is associated with GDP, until the replacement of GDP for GTP occurs.[20] *rho*-p21 has a very low GTPase activity when compared to that of well-known heterotrimeric G proteins.

The products of the wild-type and a Val[14] mutant of the human *rho*A gene have also been expressed in *E. coli* and their products purified.[21] In contrast to the results obtained for the *rho*-p21 from *Aplysia*,[18] GTP hydrolysis for the wild-type *rho*A protein was two-fold faster than for the normal *ras* protein. Substitution of Gly[14] to Val[14] is equivalent to the activating mutation Val[12] in *ras*-p21, which reduces its intrinsic GTPase activity by 5- to 10-fold. Therefore, a decreased GTPase activity was expected in this mutated *rho* protein. Indeed, Val[14] *rho*A showed a decreased intrinsic GTPase activity similar to the Val[12] mutation in *ras*.[21] Finally, Takai and co-workers have purified mammalian *rho* proteins from bovine brain and demonstrated GTP-binding and GTPase properties slightly different to those described with recombinant proteins.[22-24] Although there is no satisfactory explanation for these differences, it may well be explained by the different procedures utilized for purification.

Guanine nucleotide exchange for normal *rho*-p21 and R-*ras*-p23 are very similar.[25] Substitution of Gly[12] for Val[12] in the *ras*-p21 protein results in a small decrease (2- to 3-fold) in the nucleotide exchange rate.[19] A similar effect was observed on the exchange rate of *rho*-p21 with a Val[14] substitution, with a net increased half-life for GTP exchange to greater than 100 min.[25] This alteration in the *in vitro* exchange rate may not reflect the intracellular behavior of the protein, since there are several factors involved in the regulation of this critical step. As it has been shown for other well-characterized GTPases, regulation of GDP/GTP exchange is essential for regulation of *rho* protein function, as we will discuss in the following sections.

V. CELLULAR LOCALIZATION OF *rho* PROTEINS

Besides the conservation of relevant regions of their primary structure and their ability to bind and hydrolyze GTP, two additional properties of the *ras* proteins are also found in *rho*-p21. These two regions are most likely involved in their cellular localization and, therefore, their biological function: the CAAX motif, and a stretch of charged residues at the C terminal end (see Chapter 3).

The *ras* proteins are synthesized as soluble precursors[26] that migrate to the inner leaflet of the cell membrane while they acquire tightly bound lipids.[27] All known *ras* proteins, although highly divergent at the last 30 residues at their C termini, end with the CAAX motif, where C is a cysteine, A is an aliphatic amino acid, and X any amino acid. This region seems to be the recognition sequence involved in a complex series of events leading to farnesylation, proteolysis, and methylesterification. All *ras* proteins, except K-*ras*[4B], are also palmitoylated at a close Cys residue, strengthening its interaction with the membrane. An alternative mechanism is the display of a polybasic sequence close to the CAAX motif, which efficiently substitutes palmitoylation, as in the K-*ras*[4B] protein.

Deletion of the C terminal amino acids on v-H-*ras* prevents lipid attachment and membrane binding, and abolishes its transforming potential.[27] Other proteins such as the γ-subunit of the classical G proteins terminate with the CAAX sequence as well,[28] with Leu as the final residue (X), indicative of geranylgeranylation instead of farnesylation, found in proteins where X is any other residue. As with *ras* proteins, the CAAX motif seems to be responsible for membrane attachment of the heterotrimeric complex.[29]

The C terminal sequence of *rho* (Cys-Val-Val-Leu) suggests is may also associate with the cytoplasmic face of the cell membrane by a similar mechanism, also involving geranylgeranylation. Just upstream of the putative membrane-binding site, most *rho* proteins contain a stretch of lysines and/or arginines, similar to the one found in the *Drosophila* protein Dras2[30] and in the human K-*ras*[4B] protein.[7,31] As indicated before, these basic residues presumably facilitate or stabilize membrane binding by interaction with acidic lipids (see also Chapter 3).

There are reports finding significant amounts of *rho* proteins in the cytoplasm,[32] synaptosomal membranes,[33] presynaptic vesicles,[14] Golgi, and membranes of the rod outer segments.[34,35] Thus, the products of the *rho* genes probably have a preferential membrane association, although the exact intracellular location awaits further investigation. A deeper look at the biogenesis and cellular localization of *rho* proteins is needed before reaching a final conclusion. This research will be important to clarify some contradictory results on the biological function of *rho* proteins.

VI. GAP FACTORS FOR *rho* PROTEINS

A cytoplasmic protein capable of increasing the intrinsic GTPase activity of normal but not oncogenic *ras* proteins, has been identified in *Xenopus* oocytes and a wide range of mammalian cells. This factor, designated as *ras*-GAP, is thoroughly discussed in Section III of this book. *ras*-GAP, purified from bovine brain and human placenta, is a monomeric 125 kDa protein.[2] GAP activity has also been found for other small GTP-binding proteins such as *rap, rac, ral,* and *rho* (see Section III of this book).

The GAP-like activity for *rho*A has been detected on crude cytoplasmic extracts of human spleen tissue, as a monomeric 29 kDa protein.[23] This *rho*-GAP factor specifically increases the *in vitro* GTPase activity of normal rho A protein, but has no effect on *ras* or R-*ras* proteins. As is the case with oncogenic *ras* mutants, the Val[14] *rho*A protein does not respond to *rho*-GAP, showing an identical behavior to the effect of a Val[12] oncogenic mutation in *ras*-p21. It is interesting to investigate whether all three mammalian *rho* proteins would interact with the same purified 29 kDa *rho*-GAP, or whether there is a different GAP molecule for each family member. It has been reported that interaction of *ras*-GAP with *ras*-p21 and *rho*-GAP with *rho*-p21 can be inhibited by phospholipids,[36] but the physiological significance of this interaction remains to be elucidated.

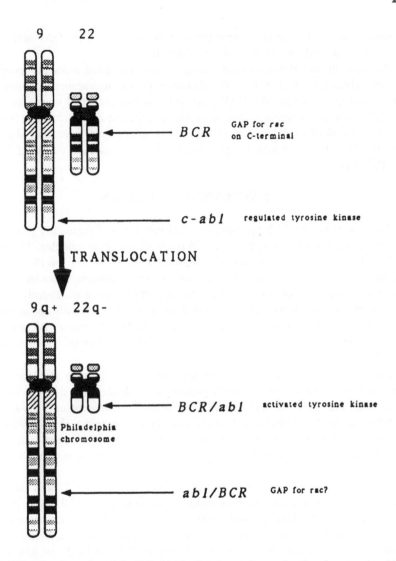

FIGURE 3. Generation of the Philadelphia chromosome by translocations between the *abl* and BCR genes.

Partial sequencing of *rho*-GAP revealed significant homology with the product of the bcr (breakpoint cluster region) gene, which is an important region involved in the chromosomal translocations present in chronic myeloid leukemias and acute lymphoid leukemias of Philadelphia chromosome positive patients (Figure 3). A bcr-related protein, n-chimerin, also has significant homology with the *rho* A-GAP peptide.[37] Out of 31 amino acids in the *rho*A-GAP peptide, bcr and n-chimerin show 50 and 43% identity, respectively.[38] The carboxyl terminal domains of bcr and n-chimerin gene products both have *in vitro* GAP activity for *rac* proteins and are members of the *rho* branch.

Although the initial characterization suggested an estimated molecular weight for purified rhoA-GAP of 29 kDa, preliminary evidence from sequence obtained from partial clones of the *rho*-GAP gene suggests that its actual size is larger (A. Hall, personal communication). Recently, a GAP-like activity has also been described for *rho*B by Yamamoto et al.[39] It has an apparent molecular weight of 37,000 Da although it may complex to an estimated molecular weight of 150 to 200,000 Da.

VII. EXCHANGE FACTORS

In addition to *rho*-GAP proteins, two different types of regulatory proteins have been described for *rho* proteins: GDI (GDP dissociation inhibitor)[40,41] and GDS (GDP dissociation stimulator).[42] GDI specific for *rho*A-p21 and *rho*B-p20 has been purified from bovine brain cytosol. *rho*-GDI inhibits dissociation of GDP from *rho*B-p21 but does not affect binding of GTPγS, a nonhydrolyzable GTP analog, to the guanine nucleotide-free form. Evidence has also been presented that *rho*-GDI makes a complex with the GDP-bound form of *rho* protein with a molar ratio of 1:1, but not with the GTPγS-bound or the guanine nucleotide-free form.[43]

Previous kinetic studies have revealed that the rate-limiting step for the GDP/GTP exchange reaction of *rho*B protein is the dissociation of GDP from it[44] as described for other well-known G proteins.[45] In this context, *rho*-GDI interacts with the GDP-bound form of *rho* and, thereby, regulates the GDP/GTP exchange reaction by inhibiting the dissociation of GDP from, and the subsequent binding of GTP to the protein.[46,47] Thus, *rho*-GDI appears to serve as an inhibitory regulatory protein.

The *rho*-GDI protein has been purified to near homogeneity[46] and a cDNA clone has been isolated from a bovine brain cDNA library.[40] *rho*-GDI is a single polypeptide of 27,000 Da expressed in most tissues in rat.[41] Sequence comparison analysis revealed that *rho*-GDI is a novel protein and shares a weak amino acid sequence homology with the yeast CDC25 protein.[48] The CDC25 protein has been suggested to interact with the GDP-bound inactive form of RAS2 protein and convert it to the GTP-bound active form.[48] The region of the CDC25 protein homologous to *rho*-GDI has been proposed to be essential for viability of yeast cells and for the activation of RAS2 protein. The *rho*-GDI cDNA has been expressed in *E. coli* and COS7 cells. Once purified from *E. coli*, this *rho*-GDI protein showed the predicted activity on purified *rho* proteins.[40]

More recently, a third type of regulatory proteins, GDS (for GDP dissociation stimulator) has been described for *rho*A and *rho*B proteins.[42] The protein has been partially purified from bovine brain cytosol and has an estimated molecular weight of 50,000 Da. *rho*-GDS stimulates the dissociation of GDP from *rho*A and *rho*B proteins but were inactive for c-Ha-*ras*-p21.

Figure 4 summarizes the role of the three regulatory activities acting on *rho* proteins: *rho*-GAP stimulates the GTPase activity without affecting the GDP/GTP exchange reaction and *rho*-GDI inhibits the GDP/GTP exchange reaction without affecting the GTPase activity.[43] It seems that the *rho*-GDI/*rho*-p21 complex becomes soluble, indicating an alternative mechanism to inactivate *rho*-p21.[43] Since *rho*-GDS stimulates the GDP/GTP exchange reaction without affecting the GTPase activity, the conversion of the *rho* proteins from the GDP-bound (inactive) form to the GTP-bound (active) form is regulated by both GDS and GDI and the reverse reaction is regulated by *rho*-GAP.

VIII. BIOLOGICAL FUNCTION

As with other members of the superfamily, *rho* proteins have been investigated for their ability to duplicate known biological functions of *ras* proteins. Therefore, investigations have focused on their putative activity as transforming agents on murine fibroblasts, differentiation effects on PC12 cells, and induction of maturation in *Xenopus laevis* oocytes. As a peculiarity, the biological function of *rho* proteins has been investigated mostly in an indirect way rather than using a direct approach. This has been possible thanks to the finding that *rho* proteins are good substrates of a bacterial toxin, the botulinum C3 ADP-ribosyltransferase.[21,49-56] However, the interpretation is not as clean as desired since there are other intracellular substrates distinct from *rho* proteins which could hamper the results. This includes closely related proteins such as *rac*, and a protein (G21K) found in thymocytes.[51]

Bacterial enzymes that ADP-ribosylate proteins have proved invaluable for studying the functions of their eukaryotic targets, particularly G proteins. Exoenzyme C3 is a newly described ADP-ribosyltransferase secreted by C and D strains of *Clostridium botulinum* but distinct from the classical C1 and D neurotoxins.[55,57] In some eukaryotic cells, C3 has a single protein substrate which is found mostly in the soluble fraction,[52] although some reports also revealed membrane-associated substrates.[53] The size of its intracellular mayor substrate is about 21 kDa, and its interaction with GTP led to the suggestion that it was a member of the *ras* superfamily of proteins.[32]

When introduced into mammalian cells, purified exoenzyme C3 ADP-ribosylated intracellular proteins and the cells underwent temporary drastic morphological alterations (Figure 5). Treated NIH 3T3 cells round profoundly, become refractile and project dendritic processes attached to the plate. Similar effects were observed using CV-1, Hep-2 and HeLa cells.[58] Treatment of Vero cells with C3 caused the disappearance of microfilaments and induced actino-

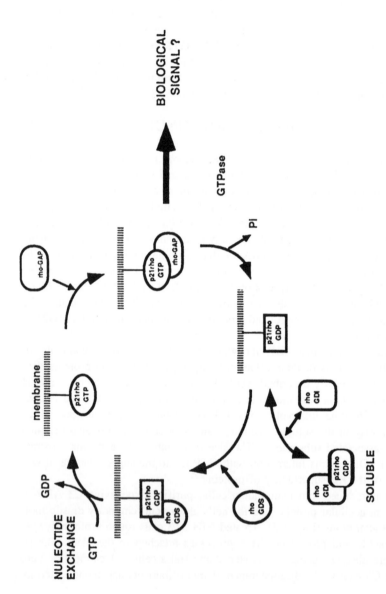

FIGURE 4. Mechanism of regulation of *rho* proteins by interchange of guanine nucleotides and GTP hydrolysis. All known components such as *rho*-GAP, *rho*-GDI, and *rho*-GDS are depicted. For details on the function of each particular factor, see the text.

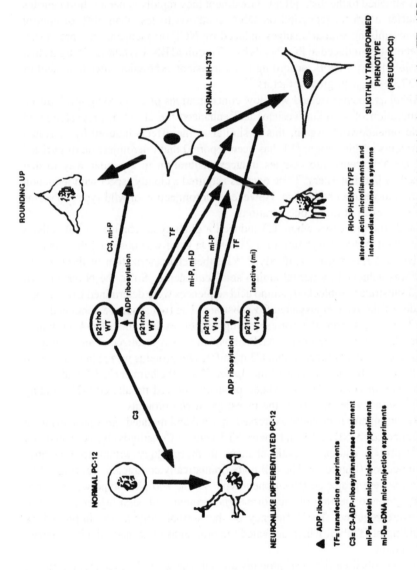

FIGURE 5. Schematic representation of the biological activity of *rho* proteins analyzed either by transfection (TF), ADP-ribosylation (C3), microinjection of purified proteins (mi-P), or DNA (mi-D).

morphic shape changes without any apparent direct effect upon actin.[54] Although transformation induced by *ras* proteins also promotes morphological alterations,[59] they seem different than the changes observed after C3 treatment.

PC12 cells treated with the exotoxin C3 responded with a different shape change as compared to fibroblasts. While PC12 cells are normally round and poorly attached to the dish, after C3 treatment they rapidly generate short neurites and suffer a drastic reduction on DNA synthesis to less than 30% of control cells, resembling similar changes induced by NGF or oncogenic *ras* proteins.[58] Differentiation induced in PC12 cells by C3 required RNA synthesis,[58] suggesting a possible role for *rho*B, according to its transient expression after treatment of NIH 3T3 cells with growth factors.[12]

Xenopus oocytes treated with low concentrations of C3 (100 ng/ml) induced the migration of germinal vesicles and enhanced the effect of progesterone in vesicle breakdown.[60] Again, this is similar to the effects induced by activated *ras* proteins in this system.[61] It has been reported that microinjection of purified Val[14] *rho*A proteins into oocytes induces maturation in a similar way to that induced by phorbol esters.[60] The oocytes acquired a cloudy aspect with induction of germinal vesicle breakdown. However, microinjection of wild-type *rho* seems to be inert even at high concentrations.

In all the processes where C3 induced a phenotypic change on the cells, a concomitant ADP-ribosylation of a 21,000-Da band was observed that was not reactive to a *ras*-specific antibody.[54,58] Further characterization of the C3 substrate was achieved by partial amino acid sequencing. Since the N terminal of the C3 substrate was blocked, amino acid sequences were determined from tryptic peptides of the ADP-ribosylated C3 substrate. The two sequences obtained covered 23 residues from the corresponding sequences in human *rho*A protein.[57] The modification of the *in vivo* ADP-ribosylation was localized at an asparagine 41 residue.[62] Direct evidence that C3 modifies *rho* proteins was obtained *in vitro* with the recombinant products of the *Aplysia*[56] and the human *rho*A[57] and *rho*C[54] genes. The *in vitro* ADP-ribosylated products showed no altered GTP-binding or GTPase activities in any of the investigated *rho* proteins.

Recombinant, bacterially expressed, normal and mutated *rho*A proteins were microinjected in NIH 3T3 and Swiss 3T3 cells.[63] Constitutively, activated *rho* (Val[14]) protein causes dramatic changes in morphology within 15 min after microinjection (Figure 5). The cell body dramatically contracted, but finger-like processes were projected (''*rho* phenotype''). The *rho* phenotype was also obtained by microinjection of the normal *rho* protein, although higher concentrations were required. The efficiency of the normal protein was improved by prebinding of GTPγS, which activated the protein as the constitutively activated Val[14] mutant.

As described earlier, *rho* proteins are substrates of C3 exoenzyme, but no change in their biochemical properties were observed *in vitro*. When cells were injected with ribosylated normal *rho* protein, the cells rounded up in an analogous fashion to that observed after injecting purified C3 transferase.[63] However, the morphology alteration is different that the typical *rho* phenotype. When ribosylated Val[14] mutant was injected, there was no apparent effect on the morphology of the cells[63] (Figure 5). Since ribosylation takes place near the putative effector domain,[62] it could impede *rho* proteins in triggering an intracellular signal. These results suggest that ADP-ribosylation of *rho* proteins by C3 inactivates its biological function.

An explanation can be that injection of ribosylated normal *rho*, which is most likely in the GDP form due to its intrinsic GTPase activity, blocked endogenous *rho* proteins by competing for an upstream factor,[43] but cannot itself produce an effect since it is ribosylated. Ribosylated Val[14] *rho* cannot bind to this putative factor since it is permanently in the GTP form. Since ribosylated normal *rho* still interacts with *rho*-GAP,[63] it will be maintained in the GDP form. The latter evidence claims for a role for *rho* in maintaining some aspects of the organization of the cytoskeletal network that affects cell shape.

As described earlier, Hunter and co-workers have demonstrated that *rho*B is expressed transiently after growth factor stimulation of quiescent cells or activation of a temperature-sensitive v-*fps* retroviral oncogene.[12] These results raise the possibility that *rho*B may function as a transcription activator gene or at least that it responds as an immediate early gene with still unknown function. If *rho*A and *rho*B are confirmed in their respective functions as regulators of cell shape and as an early gene, it would be a single case among the *ras* superfamily of proteins. Although both genes are very similar at the protein level, they would perform quite distinct cellular functions. Further investigations on the role of *rho*B after induction are required to reach such a surprising conclusion.

IX. TRANSFORMING ACTIVITY OF *rho* GENES

Based on the dramatic changes observed after microinjection of purified *rho*A proteins into mammalian cells, one could speculate that *rho* proteins were involved in the regulation of cell shape. Thus, overexpression of these genes in normal cells could be detrimental, or at least interfere, with normal growth.

Cell lines overexpressing *rho*A proteins were shown to have different properties than expected, contradicting the above argument. They showed reduced dependency of serum and loss of some degree of contact inhibition,[64,65] two well-accepted landmarks of cellular transformation. Indeed, they were also shown to be tumorigenic when inoculated into nude mice, although tumorigenicity was low compared to *ras* genes. However, no increase in any of these effects were observed using mutant genes equivalent to activated *ras* mutants. Finally, *rho*A was not able to cooperate with *myc* in transformation of rat embryo fibroblast even at high levels of expression.[64,65]

We have recently investigated the biological function of the *Aplysia rho* gene by heterologous expression into NIH 3T3 cells.[66] Both the wild-type gene

TABLE 1
Tumorigenicity of Cell Lines Overexpressing *rho* Proteins

Cell line	Gene transfected	Tumorigenicity in nude mice	
		Latency	Incidence
Control (neo^r)	pSV-2-neo	—	0/3
WT-6	*Aplysia rho* wild type pSV-2WT	7w	2/2
Val-14.5	*Aplysia rho* Val-14 pSV-2-Val^{14}	3w	3/3

Note: The wild-type *Aplysia californica rho* gene was a generous gift of Dr. R. Axel.[6] Mutated Val[14] was generated by PCR (polymerase chain reaction) using synthetic linkers (data not shown). Plasmids carrying either no insert (pSV-2-neo), or the wild-type (pSV-2WT) or mutated Val[14] (pSV-2-Val[14]) *rho* gene were generated and introduced into NIH 3T3 cells by standard transfection assays. Clones overexpressing each of the desired gene were selected by Northern blot analysis and grown under appropriate conditions. Athymic nude mice were inoculated with 1×10^6 cells per injection and animals followed up for tumor development for three months. Latency is indicated in weeks.

as well as a Val[14] mutant were introduced into the pZIP-neo eukaryotic vector and then transfected into NIH 3T3 cells. Cell lines expressing the wild-type gene were tumorigenic with a long latency period of 2 months (Table 1). The latency period decreased to three weeks when cell lines expressing a Val[14] mutant were injected.

We have also investigated serum-independent growth of isolated cell lines overexpressing wild-type and mutated *Aplysia rho* genes. In good agreement with the above results, the wild-type *rho* gene was able to induce serum-independent growth, but this effect was less pronounced than that induced by the Val[14] mutant (Figure 6). These results suggest an activation of the oncogenic potential of the *Aplysia rho* gene with a single amino acid substitution, analogous to that found in activated *ras*-p21 proteins. Thus, the *Aplysia rho*-p21 and mammalian *ras*-p21 proteins have comparable biological properties, and may be involved in the regulation of critical steps of cell proliferation.

X. CONCLUSIONS

The above results bring about an interesting discussion: can *rho* proteins be considered real signal transducers, in the light as Gs, Gi, or *ras* proteins are considered? Or on the contrary, are *rho* proteins similar to structural proteins such as tubulin, which also bind GTP but has no transducing function? The fact that several *rho*-specific regulatory factors, such as *rho*-GAP, GDI, and GDS have been found, makes them better candidates for the transducer family than for the structural one. *ras*-GAP has been shown to be phosphorylated by several protein tyrosine kinases. Thus, perhaps the analogous *rho*-GAP, which is distinct from *ras*-GAP, is subject to similar modifications which would provide a connection between *rho* and the protein tyrosine kinase pathway.

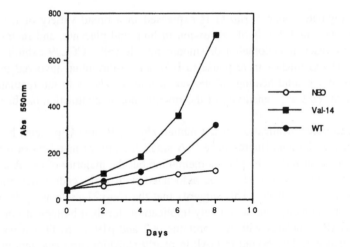

FIGURE 6. Growth of cell lines overexpressing either normal (WT-6) or mutated (Val[14.5]) *rho* proteins under conditions of serum depletion. Cells were grown in 24-well miniwell plates in 0.5% fetal calf serum. At times indicated, the cell number was estimated by crystal violet staining as described elsewhere.[67]

Further support to the transducer hypothesis for *rho* proteins has been provided from the evidence that when the human *rho*A and *Aplysia rho* proteins are overexpressed in NIH 3T3 cells, they confer growth factor independence. Moreover, cell lines carrying retroviral vectors expressing normal and mutated proteins were tumorigenic when injected into nude mice. Research on the regulation of all factors described above which interact with *rho* proteins, as well as deciphering of the intracellular alterations induced by overexpression of individual *rho* genes, are needed to finally understand their biological function.

NOTE ADDED IN PROOF

During the last few months great progress has been made in the understanding of some of the biological functions of *rho* proteins. While a definitive picture has not been reached yet, strong evidence has been generated to strengthen the three major roles so far associated to *rho*. Thus, *rho* proteins have been involved in transient transcriptional activation and regulation of cell shape changes induced by growth factors, and in tumor development.

A new member of the *rho* family has been cloned from hamster CCL39 fibroblasts and human HeLa cells and designated as *rho* G (*ras* homolog growth related).[68] A phylogenetic analysis reveals that *rho* G belongs to the group comprising yeast CDC42 and human CDC42Hs, *rac*1, *rac*2, and TC10, *rho* G mRNA accumulates 8 to 10 h after serum addition as a result of transient transcriptional activation. In agreement with these observations, *rho* B[12] was isolated as an immediate early gene and *rac*2 has been described also to be a growth regulated gene.[69] *rho* G is expressed in a broad variety of tissues in humans, with high levels of expression in lung and placenta, and seems to be related to vascular endothelial or smooth muscle cells. CCL39 exhibit a high level of *rho* G mRNA in response to FGF and α-thrombin, involved in angiogenesis and wound healing. Moreover, a role for *rho* A in the regulation of GTP-enhanced Ca^{2+} sensitivity of the smooth muscle contraction has been proposed.[70]

Intracellular localization of the human *rho* A, B, and C proteins has been visualized by immunofluorescence.[71] A small fraction of members of all three proteins is localized to the plasma membrane but the majority of *rho* A and *rho* C is cytosolic, whereas *rho* B is associated with early endosomes and prelysosomal compartments. This suggests that the *rho*-p21 proteins cycle on and off the plasma membrane with likely implications for their biological functions.

*ras*GAP associates with two proteins p62 and p190.[72] p190 is a phosphoprotein that is tightly bound to GAP in nearly stoichiometric amounts in mitogenically stimulated and tyrosine kinase transformed cells.[73,74] The observation that p190 association with GAP impairs the GTPase activity of GAP,[74] implicates p190 in the *ras* signal transduction pathway. Sequence analysis reveals that the C-terminus of the protein contains a domain with sequences very similar with those found in the bcr gene product, n-chimerin, and *rho*-GAP.[72] Each of these three proteins exhibit GAP activity for at least one member of the *rho* family. Furthermore, recombinant p190 protein shows *in vitro* GAP-like activity specific for members of the *rho* family.[75] Thus, formation of a complex between *ras*-GAP and p190 following growth factor stimulation may provide a mechanism to couple signals mediated by the *ras* and *rho* GTPases.

An elegant study from Hall's group has provided some essential clues to further determine the biological function of the *rho* and *rac* products.[76,77] *rho* A has been shown to be an essential element for the formation of stress fibers and focal adhesion while *rac* may be related to membrane ruffling in response to serum and growth factors. Stress fiber formation induced by serum was due to a lysophospholipid, most likely lysophosphatidic acid bound to serum albumin.

All these results, along with those described in previous sections, suggest that different members of the *rho* family may be involved in different aspects of the response of mammalian cells to growth stimuli.

ACKNOWLEDGMENTS

The *Aplysia rho* gene was a generous gift from R. Axel. This work was supported by grants from Dirección General de Investigación Científica y Técnica (DGICYT), Fundación Ramón Areces, and Laboratorios Serono R. Perona is Titulado Superior (CSIC); R. P. Ballestero is a Fellow from the Spanish Department for Education and Science; J. C. Lacal is Investigador Científico (CSIC) and Honorary Professor of the Universidad Autónoma de Madrid.

REFERENCES

1. **Trahey, M. and McCormick, F. A.**, Cytoplasmic protein stimulates natural N-ras p21 GT pase, but does not affect oncogenic mutants, *Science*, 238, 542, 1987.
2. **Xu, G., O'Connell, P., Viskochil, D., Cawthon, R., Robertson, M., Culver, M., Dunn, D., Stevens, J., Gesteland, R., White, R., and Weiss, R.**, The neurofibromatosis type 1 gene encodes a protein related to GAP, *Cell*, 62, 599, 1990.
3. **Huang, H. C., Kung, H. F., and Kamata, T.**, Purification of a factor capable of stimulating the guanine nucleotide exchange reaction of ras proteins and its effect on ras related small molecular mas G proteins, *Proc. Natl. Acad. Sci. U.S.A.*, 87, 8008, 1990.
4. **Wolfman, A. and Macara, J.**, A cytosolic protein catalizes the release of GDP from p21-ras, *Science*, 248, 67, 1990.
5. **Downward, J., Rielhl, R., Wu, L., and Weinberg, R. A.**, Identification of a nucleotide exchange promoting activity for p21-ras, *Proc. Natl. Acad. Sci. U.S.A.*, 87, 5998, 1990.
6. **Madaule, P. and Axel, R.**, A novel ras-related gene family, *Cell*, 41, 31, 1985.
7. **McGrath, J. P., Capon, D. J., Smith, D. H., Chenne, Y., Soeburg, P. H., Goeddel, D. V., and Levinson, A. D.**, Structure and organization of the human Ki-ras proto-oncogene and a related processed pseudogene, *Nature*, 304, 501, 1983.
8. **Yeramian, P., Chardin, P., Madaule, P., and Tavitian, A.**, Nucleotide sequence of human rho cDNA clone 12, *Nucleic Acid. Res.*, 15, 1869, 1987.
9. **Chardin, P., Madaule, P., and Tavitian, A.**, Coding sequence of human *rho* cDNAs clone 6 and clone 9, *Nucleic Acid. Res.*, 16, 2717, 1988.
10. **Downward, J.**, The ras superfamily of small GTP-binding proteins, *TIBS*, 15, 469, 1990.
11. **Olofsson, B., Chardin, P., Touchot, N., Zahraoui, A., and Tavitian, A.**, Expression of the ras-related ralA, rho12 and rab genes in adult mouse tissues, *Oncogene*, 3, 231, 1988.
12. **Jähner, D. and Hunter, T.**, The *ras*-related gene *rho*B is an immediate-early gene inducible by v-*fps*, epidermal growth factor, and platelet-derived growth factor in rat fibroblasts, *Mol. Cell. Biol.*, 11, 3682, 1991.
13. **Madaule, P., Axel, R., and Myers, A. M.**, Characterization of two members of the *rho* gene family from the yeast *Saccharomyces cerevisiae*, *Proc. Natl. Acad. Sci. U.S.A.*, 84, 779, 1987.
14. **Ngsee, J. K., Elferink, L. A., and Scheller, R. H.**, A family of *ras*-like GTP-binding proteins expressed in electromotor neurons, *J. Biol. Chem.*, 266, 2675, 1991.
15. **Fasano, O., Aldrich, T., Tamano, F., Taparowsky, B., Furth, M., and Wigler, M.**, Analysis of the transforming potential of the human H-*ras* gene by random mutagenesis, *Proc. Natl. Acad. Sci. U.S.A.*, 81, 4000, 1984.
16. **Leberman, R. and Egner, V. C.**, Homologies in the primary structure of GTP binding proteins: the nucleotide binding site of EF-Tu and p21, *EMBO J.*, 3, 339, 1984.
17. **Gallwitz, D., Donath, C., and Sander, C.**, A yeast gene encoding a protein homologous to the human C-*ras/bas* proto-oncogene product, *Nature*, 306, 704, 1983.
18. **Anderson, P. S. and Lacal, J. C.**, Expression of the *Aplysia californica rho* gene in *Escherichia coli*: purification and characterization of its encoded p21 product, *Mol. Cell. Biol.*, 7, 3620, 1987.
19. **Lacal, J. C. and Aaronson, S. A.**, Activation of *ras* p21 transforming properties associated with an increase in the release rate of bound guanine nucleotide, *Mol. Cell. Biol.*, 6, 4211, 1986.
20. **Bourne, H. R., Sanders, D. A., and McCormick, F.**, The GTPase superfamily: conserved structure and molecular mechanism, *Nature*, 348, 125, 1990.
21. **Aktories, K., Braun, U., Rösener, S., Just, I., and Hall, A.**, The *rho* gene product expressed in *E. coli* is a substrate of botulinum ADP-ribosyl transferase C3, *Biochem. Biophys. Res. Commun.*, 158, 209, 1989.

22. **Hoshijima, M., Kondo, J., Kikuchi, A., Yamamoto, K., and Takai, Y.**, Purification and characterization from bovine brain membranes of a GTP-binding protein with a M, of 21,000, ADP-ribosylated by an ADP-ribosyltransferase contaminated in botulinum toxin type C1. Identification as the *rho* A gene product, *Mol. Brain Res., 7, 9, 1990.*

23. **Yamamoto, K., Kondo, J., Hishida, T., Teranishi, Y., and Takai, Y.**, Purification and characterization of a GTP-binding protein with a molecular weight of 20,000 in bovine brain membranes, *J. Biol. Chem., 263, 9926, 1988.*

24. **Morii, N., Sekine, A., Ohashi, Y., Nakao, K., Imura, H., Fijuwara, M., and Narumiya, S.**, Purification and properties of the cytosolic substrate for botulinum ADP-ribosyltransferase, *J. Biol. Chem., 263, 12,420, 1988.*

25. **Garret, M. D., Self, A. J., van Oers, C., and Hall, A.**, Identification of distinct cytoplasmic targets for *ras*/R-*ras* and *rho* regulatory proteins, *J. Biol. Chem., 264, 10, 1989.*

26. **Lowy, D. R., Zhang, I. C., Declue, J. C., and Willumsen, B. M.**, Regulation of p21^{ras} activity, *TIGS.* 346, 351, 1991.

27. **Willumsen, B. M., Christensen, A., Hubbert, N. L., Papageorge, A. G., and Lowy, D. R.**, The p21-*ras* C-terminus is required for transformation and membrane association, *Nature,* 310, 583, 1984.

28. **Hurley, J. B., Fong, H., Teplow, D. B., Dreyer, W. J., and Simon, M. I.**, Isolation and characterization of a cDNA clone for the ganma subunit of a Bovine retinal transducin, *Proc. Natl. Acad. Sci. U.S.A.,* 81, 6948, 1984.

29. **Hancock, J. F., Paterson, H., and Marshall, C. J.**, A polybasic domain or palmitoylation is required in addition to the CAAX motif to localize p21^{ras} to the plasma membrane, *Cell,* 63, 133, 1990.

30. **Newman-Silberberg, S., Schejter, E., Hoffman, F. M., and Shilo, B.**, The Drosophila *ras* oncogenes: structure and nucleotide sequences, *Cell,* 37, 1027, 1984.

31. **Shimizu, K., Birnbaum, D., Ruley, M., Fasano, O., Suard, Y., Edlund, L., Taparowsky, T., Goldfarb, M., and Wigler, M.**, The structure of Ki-*ras* gene of the human lung carcinoma cell line Calu-1, *Nature,* 304, 497, 1983.

32. **Braun, U., Habermann, B., Just, I., Aktories, K., and Vandekerckhove, J.**, Purification of the 22 kDa protein substrate of botulinum ADP-ribosyltransferase C3 from porcine brain cytosol and its characterization as a GTP-binding protein highly homologous to the *rho* gene product, *FEBS Lett.,* 243, 70, 1989.

33. **Shigeskuni, K., Kikuchi, A., and Takai, Y.**, Intrasynaptosomal distribution of the *ras, rho* and *smg*-25A GTP-binding proteins in bovine brain, *Mol. Brain Res.,* 6, 167, 1989.

34. **Toki, C., Oda, K., and Ikehara, Y.**, Demonstration of GTP-binding proteins and ADP-ribosylated proteins in rat liver golgi fraction, *Biochem. Biophys. Res. Commun.,* 164, 333, 1989.

35. **Wielund, T., Ulibarri, I., Gierschik, P., Hall, A., Aktories, K., and Jakobs, K. H.**, Interaction of recombinant *rho* A GTP-binding proteins with photoexcited rhodopsin, *FEBS Lett.,* 274, 111, 1990.

36. **Tsai, M., Hall, A., and Stacey, D. W.**, Inhibition by phospholipids of the interaction between R-*ras, rho,* and their GTPase-activating proteins, *Mol. Cell. Biol.,* 9, 5260, 1989.

37. **Diekmann, D., Brill, S., Garret, M. D., Totty, N., Hsuan, J., Monfries, C., Hall, C., Lim, L., and Hall, A.**, Bcr encodes a GTPase-activating protein for p21^{rac} *Nature,* 351, 400, 1991.

38. **Hall, C., Monfries, C., Smith, P., Lim, H. H., Kozma, R., Ahmed, S., Vanniasingham, V., Leung, T., and Lim, L.**, Novel human brain cDNA encoding a 34000 M^r protein n-Chimaerin, related to both the regulatory domain of protein kinase C and BCR, the product of the breakpoint cluster region gene, *J. Mol. Biol.,* 211, 11, 16, 1990.

39. **Yamamoto, J., Kikuchi, A., Ueda, T., Ohga, N., and Takai, Y.**, A GTPase-activating protein for *rho*B p20, a *ras* p21-like GTP-binding protein — partial purification, characterization and subcellular distribution in rat brain, *Mol. Brain Res.,* 8, 105, 1990.

40. **Fukumoto, Y., Kaibuchi, K., Hori, Y., Fujioka, H., Araki, S., Ueda, T., Kikuchi, A., and Takai, Y.,** Molecular cloning and characterization of a novel type of regulatory protein (GDI) for the *rho* proteins, *ras* p21-like small GTP-binding proteins, *Oncogene*, 5, 1321, 1990.

41. **Shimizu, K., Kaibuchi, K., Nonaka, H., Yamamoto, J., and Takai, Y.,** Tissue and subcellular distributions of an inhibitory GDP/GTP exchange protein (GDI) for the *rho* proteins by use of its specific antibody, *Biochem. Biophys. Res. Commun.*, 175, 199, 1999.

42. **Isomura, M., Kaibuchi, K., Yamamoto, T., Kawamura, S., Katayama, M., and Takai, Y.,** Partial purification and characterization of GDP dissociation stimulator (GDS) for the *rho* proteins from bovine brain cytosol, *Biochem. Biophys. Res. Commun.*, 169, 652, 1990.

43. **Ohga, N., Kikuchi, A., Ueda, T., Yamamoto, J., and Takai, Y.,** Rabbit intestine contains a protein that inhibits the dissociation of GDP from and the subsequent binding of GTP to *rho* B p20, a *ras* p21-like GTP-binding protein, *Biochem. Biophys. Res. Commun.*, 163, 1523, 1989.

44. **Kuroda, S., Kikuchi, A., and Takai, Y.,** Kinetic analysis of the binding of guanine nucleotides to bovine brain *rho* B p20, a *ras* p21-like GTP-binding protein, *Biochem. Biophys. Res. Commun.*, 163, 674, 1989.

45. **Hall, A. and Self, A. J.,** The effect of Mg^{2+} on the guanine nucleotide exchange rate of p21 N-*ras*, *J. Biol. Chem.*, 261, 10,963, 1986.

46. **Ueda, T., Kikuchi, A., Ohga, N., Yamamoto, J., and Takai, Y.,** Purification and characterization from bovine brain cytosol of a novel regulatory protein inhibiting the dissociation of GDP from and the subsequent binding of GTP to *rho* B p20, a *ras* p21-like GTP-binding protein, *J. Biol. Chem.*, 265, 9373, 1990.

47. **Isomura, M., Kikuchi, A., Ohga, N., and Takai, Y.,** Regulation of binding of *rho* B p20 to membranes by its specific regulatory protein, GDP dissociation inhibitor, *Oncogene*, 6, 119, 1991.

48. **Broek, D., Toda, T., Michaeli, T., Levin, L., Birchmeier, C., Zoller, M., Powers, S., and Wigler, M.,** The *S. cerevisiae* CDC25 gene product regulates the *ras*/Adenylate cyclase pathway, *Cell*, 48, 789, 1987.

49. **Kikuchi, A., Yamamoto, K., Fujita, T., and Takai, Y.,** ADP-ribosylation of the bovine brain *rho* protein by botulinum toxin type C1, *J. Biol. Chem.*, 263, 16,303, 1988.

50. **Narumiya, S., Sekine, A., and Fujiwara, M.,** Substrate for botulinum ADP-ribosyltransferase Gb, has an amino acid sequence homologous to a putative *rho* gene product, *J. Biol. Chem.*, 263, 17,255, 1988.

51. **Didsbury, J., Weber, R. F., Bokoch, G. M., Evans, T., and Snyderman, R.,** *rac*, a novel *ras*-related family of proteins that are botulinum toxin substrates, *J. Biol. Chem.*, 264, 16,378, 1989.

52. **Bokoch, G. M., Parkos, C. A., and Mumbys, S. M.,** Purification and characterization of the 22,000-Dalton GTP-binding protein substrate for ADP-ribosylation by botulinum toxin, G_{22K}, *J. Biol. Chem.*, 263, 16,744, 1988.

53. **Ohashi, Y. and Narumiya, S.,** ADP-ribosylation of a M^r 21,000 membrane protein by type D botulinum toxin, *J. Biol. Chem.*, 262, 1430, 1987.

54. **Chardin, P., Boquet, P., Madaule, P., Popoff, M. R., Rubin, E. J., and Gill, D. M.,** The mammalian G protein *rho* C is ADP-ribosylated by *Clostridium botulinum* exoenzyme C3 and affects actin microfilaments in Vero cells, *EMBO J.*, 8, 1087, 1989.

55. **Rubin, E. J., Gill, D. M., Boquet, P., Popoff, M. R.,** Functional modification of a 21-Kilodalton G protein when ADP-ribosylated by exoenzyme C3 of *Clostridium botulinum*, *Mol. Cell. Biol.*, 8, 418, 1988.

56. **Quilliam, L. A., Lacal, J. C., and Bokoch, G. M.,** Identification of *rho* as a substrate for botulinum toxin C^3 catalyzed ADP-ribosylation, *FEBS Lett.*, 247, 221, 1989.

57. **Aktories, K., Weller, U., and Chatwal, G. S.,** *Clostridium botulinum* type C produced a novel ADP-ribosyltransferase distinct from botulinum C2 toxin, *FEBS Lett.*, 212, 109, 1987.

58. **Nishiki, T., Narumiya, S., Morii, N., Yamamoto, M., Fujiwara, M., Kamata, Y., Sakaguchi, G., and Kozaki, S.,** ADP-ribosylation of the *rho/rac* proteins induces growth inhibition, neurite outgrowth and acetilcholine esterase in cultured pc-12 cells, *Biochem. Biophys. Res. Commun.,* 167, 265, 1990.

59. **Noda, M., Ko, M., Oquia, A., Liu, D., Amano, T., Takano, D., and Ikawa, Y.,** Sarcoma viruses carrying *ras* oncogenes induce differentiation associated properties in a neuronal cell line, *Nature,* 318, 73, 1985.

60. **Mohr, C., Just, I., Hall, A., and Aktories, K.,** Morphological alteration of Xenopus oocytes induced by valine-14 p21*rho* depend on isoprenylation and are inhibited by Clostridium botulinum C3 ADP-ribosyltransferase, *FEBS Lett.,* 275, 168, 1990.

61. **Birchmeier, C., Brock, D., and Wigler, M.,** *Ras* protein can induce meiosis in Xenopus oocytes, *Cell,* 43, 615, 1988.

62. **Sekine, A., Motohatsu, F., and Narumiya, S.,** Asparagine residue in the *rho* gene product is the modification site for botulinum ADP-ribosyltransferase, *J. Biol. Chem.,* 264, 8602, 1989.

63. **Paterson, H. F., Self, A. J., Garrett, M. D., and Just, I.,** Microinjection of recombinant p21*rho* induces rapid changes in cell morphology, *J. Cell. Biol.,* 111, 1001, 1990.

64. **Avraham, H. and Weinberg, R. A.,** Characterization and expression of the human *rhoH12* gene product, *Mol. Cell. Biol.,* 9, 2058, 1989.

65. **Avraham, H.,** *rho* gene product, *Biochem. Biophys. Res. Commun.,* 168, 114, 1990.

66. **Perona, R., Esteve, P., Jiménez, B., Ballestero, R. P., and Lacal, J. C.,** Tumorigenic activity of rhogenes from *Aplysia californica, Oncogene,* in press, 1993.

67. **Gillies, R. J., Didier, N., and Denton, M.,** Determination of cell number in monolayer cultures, *Anal. Biochem.,* 159, 109, 1986.

68. **Vincent, S., Jeanteur, P., and Fort, P.,** Growth-regulated expression of rho G, a new member of the ras homolog gene family, *Mol. Cell. Biol.,* 12, 3136, 1992.

69. **Reiber, L., Dorsevil, O., Stancou, O., Bertogli, O., and Gacon, G. A.,** Hemopoyetic specific gene encoding a small GTP binding protein is overexpressed during T-cell activation, *Biochem. Biophys. Res. Commun.,* 175, 451, 1991.

70. **Hirata, K., Kikuchi, A., Sasaki, T., Kuroda, S., Kaibuchi, K., Matsuura, Y., Seki, H., Saida, K., and Takai, Y.,** Involvement of *rho* p21 in the GTP-enhanced calcium ion sensitivity of smooth muscle contraction, *J. Biol. Chem.,* 267, 8719, 1992.

71. **Adamson, P., Paterson, H., and Hall, A.,** Intracellular localization of the P21 rho proteins, *J. Cell Biol.,* 119, 617, 1992.

72. **Settelman, J., Narashiman, U., Foster, L., and Weinberg, R.,** Molecular cloning of cDNA encoding the GAP-associated protein p190, implications for a signaling pathway from ras to the nucleus, *Cell,* 69, 539, 1992.

73. **Ellis, C., Moran, M., McCormick, F., and Pawson, T.,** Phosphorilation of GAP and GAP-associated proteins by transforming and mitogenic tyrosine kinases, *Nature,* 343, 377, 1990.

74. **Moran, M., Polakis, P., McCormick, F., Pawson, T., and Ellis, C.,** Protein-tyrosine kinases regulate the phosphorilation, protein interaction, subcellular distribution and activity of p21 ras GTPase activating protein, *Mol. Cell. Biol.,* 11, 1804, 1991.

75. **Settelman, J., Albright, C., Foster, L., and Weinberg, R.,** Association between GTPase activators for rho and ras families, *Nature,* 359, 153, 1992.

76. **Ridley, A. and Hall, A.,** The small GTP binding protein rho regulates the assembly of focal adhesion and actin stress fibers in response to growth factors, *Cell,* 70, 389, 1992.

77. **Ridley, A., Paterson, H. F., Johnston, C. L., Dickman, D., and Hall, A.,** The small GTP binding protein *rac* regulates growth factors-induced membrane ruffling, *Cell,* 70, 401, 1992.

Chapter 14

rac GENE PRODUCTS

Shuh Narumiya, Yasuo Nemoto, and Narito Morii

TABLE OF CONTENTS

I. PERSPECTIVES AND SUMMARY

The *rac* gene family consisting of *rac1* and *rac2* is the latest member of the *ras* superfamily of GTP-binding proteins. Products of the *rac* family, like those of the *rho* genes, undergo ADP-ribosylation by botulinum C3 ADP-ribosyltransferase (C3 exoenzyme). Hence the name *rac* is derived (*ras*-related C3 botulinum toxin substrate). Among the *ras* superfamily, the primary structure of *rac* is most related to those of CDC42 and *rho*. To date, the *rac* family has been studied only in human tissues and cells of human origin, and its phylogenic distribution has not been reported. Human *rac* genes, particularly *rac2*, are expressed extensively in cells of myeloid origin, suggesting that they may not be a housekeeping gene but play specific roles in special types of cells. A *rac* gene product was purified partially only from human platelets and its biochemical properties are not known in detail. Its cellular functions have also not been clarified yet. Botulinum C3 exoenzyme has been used to analyze functions of the ADP-ribosylation substrates, *rho* and *rac* proteins, in the cell. This ADP-ribosylation occurs at an asparagine residue (Asn⁴¹) in the putative effector domain of a *rho* protein. The *rac* proteins have an asparagine residue at the corresponding position and are presumed to be ADP-ribosylated there. C3 exoenzyme incubated with or microinjected into cultured cells causes ADP-ribosylation of cellular substrates and evokes several phenotypic changes in the cells. These studies have suggested that *rho* and/or *rac* proteins are involved in cytoskeletal organization and may have some interaction with a membrane receptor. However, none of these studies was carried out in cells with identified ADP-ribosylation substrates, which makes it difficult to attribute specific roles to each species of *rho* and *rac* proteins. Future studies with C3 exoenzyme will be carried out in defined systems and clarify physiological functions for each member of the ADP-ribosylation substrates. Phylogeny of *rac* genes will be revealed and their actions will also be analyzed in more simple systems like yeast.

II. MOLECULAR STRUCTURES AND PROPERTIES

A. PRIMARY STRUCTURES

The *rac* family was first identified by Didsbury et al.[1] in a cDNA library of differentiated HL-60 cells by low stringency hybridization screening with oligonucleotides corresponding to a partial amino acid sequence of G25K (G$_p$, later identified as a human homolog of cdc42). Two members are known in this family and are named *rac1* and *rac2*. Their nucleotide and deduced amino acid sequences are shown in Figure 1. Each *rac* encodes a protein of 192 amino acids. *rac1* and *rac2* are 77% homologous in nucleotide sequence of the coding regions and 92% homologous in deduced amino acid sequences. The nucleotide differences between the two are randomly distributed in the sequences. This suggests that the two are not caused by alternative splicing of a single gene but derived from two separate genes. Their genomic structures, however, have not

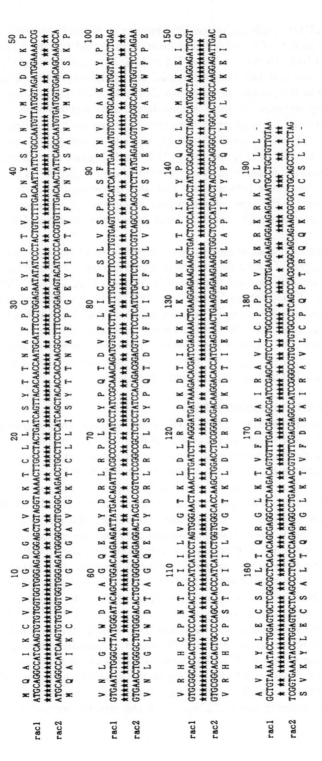

FIGURE 1. Nucleotide and deduced amino acid sequences of *rac1* and *rac2*. Identical nucleotides are indicated by asterisks. (Modified from Disbury et al., *J. Biol. Chem.*, 264, 16380, 1989.)

been clarified yet. Deduced amino acid sequences of *rac1* and *rac2* contain consensus sequences of postulated guanine nucleotide-binding domains in amino acid residues 10 to 16, 57 to 62, and 116 to 120. They also have the Cys-X-X-X-COOH sequence at their carboxyl termini and a polybasic motif just upstream of them. The former, as seen in *ras*-p21s and other related GTP-binding proteins, is presumably required for their isoprenylation and, together with the latter structure, involved in membrane attachment. *rac1* and *rac2* are different from other small molecular weight GTP-binding proteins in amino acid residues 22 to 27, 49 to 54, 121 to 153, and 170 to 188. These regions may, therefore, be involved in specific recognition by *rac* proteins of their regulators and effectors. The greatest divergence between *rac1* and *rac2* is observed at residues 180 to 190. This region is the hypervariable region among various *ras*-related proteins and presumed to confer different functional properties to each member of these proteins.

Amino acid sequence homology of *rac* proteins with other *ras*-related small molecular weight GTP-binding proteins is shown in Figure 2. The sequences of *rac* proteins are 70% identical to those of G25Ks (G$_p$, CDC42Hs),[2,3] 58% identical to those of *rho* proteins,[4-6] and only 30% identical to those of *ras*-p21s. G25K is a mammalian homolog of the yeast cdc42 gene product which is involved in normal bud orientation in *Saccharomyces cerevisiae*. Although G25K can substitute yeast cdc42, neither *rac1* nor *rac2* can complement the cdc42-1 mutation in *S. cerevisiae*,[2] suggesting that in spite of their close structural similarity, the *rac* and cdc42 families have different functions.

B. BIOCHEMICAL PROPERTIES

Polakis et al.[7] purified several GTP-binding proteins from membranes of human blood platelets and identified one of them as a *rac* protein. It was solubilized from the membranes with sodium cholate and purified by successive column chromatographs on DEAE-Sephacel, Ultrogel AcA 34, hydroxyapatite, and DEAE-fractogel, and finally, purified by preparative SDS-polyacrylamide gel electrophoresis. Partial amino acid sequences of the eluted protein were determined and identified it as *rac1* protein. According to this study, *rac1* protein is one of the major small molecular weight GTP-binding proteins in platelet membranes, obtained in half an amount of the most abundant protein, G25K.[8] Unlike most small molecular weight GTP-binding proteins, *rac1* and G25K did not bind [α-^{32}]GTP significantly after SDS-polyacrylamide gel electrophoresis and transfer to nitrocellulose membranes. Except this work, there are no other reports on identification of *rac* proteins in any tissue or cell. However, there have been reports on small molecular weight GTP-binding proteins in neutrophils which may be related to *rac* proteins. Bokoch et al.[9] purified three species of small molecular weight GTP-binding proteins from membranes of human neutrophils, G22K, G24K, and G26K. One of them, G22K, undergoes ADP-ribosylation by botulinum C3 enzyme in the presence of a cytosolic protein factor. These authors reported that G22k was not a *rho* protein, because it migrated

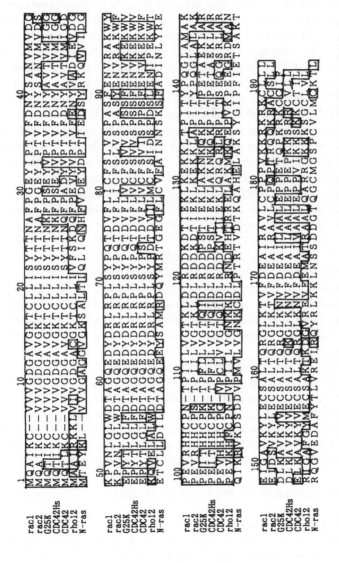

FIGURE 2. Comparison of the *rac* protein sequences with those of the G25K, *CDC42Hs*, *CDC42*, *rho12* (*rhoA*), and N-*ras* gene products. Identical amino acids are boxed.

differently from a recombinant *rho* protein on SDS-polyacrylamide gel electrophoresis, was not recognized by anti-*rho* polyclonal antibody, and required a cytosolic factor for ADP-ribosylation.[10] In addition, ADP-ribosylation of G22K, unlike *rho,* was inhibited by prior incubation of this protein with GTPγS. Although the initial study showed that the purified G22k contained sequences of *rap1* protein, *rap1* protein itself does not serve as an ADP-ribosylation substrate. On the basis of these results and that *rac* proteins are abundantly expressed in neutrophils, Didsbury et al.[1] presumed that purified G22k is a mixture of *rap1* and *rac* proteins, although ultimate identification has not been done. Small molecular weight GTP-binding proteins in neutrophils were also examined in another report. Philips et al.[11] examined subcellular distribution of substrates for botulinum C3 ADP-ribosyltransferase in human neutrophils. They reported that C3 enzyme ADP-ribosylated proteins of 16, 20, and 24 kDa in these cells. They reported that all of the three proteins are associated with specific granule membranes, the latter two were also found in plasma membranes and none with azurophilic granule membrane and cytosol. Identification of these substrate proteins was not carried out, though the authors inferred from the mobility on the electrophoresis that the 20-kDa protein may be a *rac* protein.

As discussed above, the carboxy terminal of *rac* proteins indicate that these proteins also undergo isoprenylation. Didsbury et al.[12] examined possible post-translational modification of *rac* proteins by expressing *rac* cDNA clones in cultured COS cells and incubating the cells with [³H] mevalonic acid. They found that *rac* proteins expressed in the cells were present both in the cytosol and membranes, and that only those associated with the membranes were radiolabeled with [³H] mevalonic acid, while no labeling was found in the cytosolic *rac* proteins. From these results they suggested that *rac* proteins, like *ras*-p21s, are posttranslationally modified with a isoprenoid group and translocated to cellular membranes.

The activity of *ras*-related small molecular weight GTP-binding proteins is regulated by several factors including GTPase activating proteins, GDP-GTP exchange factors (GDP dissociation stimulators, GDSs) and GDP dissociation inhibitors (GDIs). These factors are usually specific for each group of GTP-binding proteins. At present any factors specific for *rac* proteins are not known. On the other hand, GAP, GDI, and GDS specific for *rho* proteins have already been reported.[13-15] Among them, *rho*-GAP and *rho*-GDI were purified to homogeneity.[16,17] Although *rac* proteins have extensive homology to *rho* proteins, it remains to be examined whether these factors could act also on *rac* proteins.

C. TISSUE DISTRIBUTION

Distribution of *rac* mRNAs was examined by Northern blot analysis in several human tissues and cell lines of human origin.[1] The *rac1* mRNA found as a doublet of 2.4 and 1.1 kb is rich in placenta, and detected in brain and liver. It is abundant in human neutrophils and also in differentiated HL-60 cells with neutrophilic phenotype. On the other hand, it was only detectable in human mononuclear cells and cultured U-937 cells with monocytic appearance, hardly detectable in SK-7 cells (a T cell hybridoma cell line), and absent in a Jurkat T cell line. The ratio of the 2.4- and 1.1-kb species is different among the cells and tissues, and the 1.1-kb species was apparently increased with neutrophilic differentiation of HL-60 cells. The *rac2* mRNA detected as a single 1.45-kb species is less ubiquitously distributed with little or no mRNA detected in brain, liver, or Jurkat cells. It is, on the other hand, abundant in cells of myeloid origin, in particular mature neutrophils. Contrary to *rac1*, it is also present in significant amounts in mononuclear cells and in increased amounts with differentiation of U-937 cells. Although *rac1* and *rac2* are 92% homologous in their amino acid sequence and both present abundantly in neutrophils, the above differences suggest that their functions are not identical. In addition, the absence of both mRNAs in some cell lines suggest that *rac* genes may not be housekeeping genes and play specific roles in specialized cells.

III. BOTULINUM C3 ADP-RIBOSYLTRANSFERASE AND *rho* AND *rac* GENE PRODUCTS

A. BOTULINUM C3 ADP-RIBOSYLTRANSFERASE

An enzyme which specifically ADP-ribosylates 21- to 22-kDa cellular proteins was discovered in 1987 as an ADP-ribosyltransferase in C1 and D botulinum toxin preparations by Ohashi et al.[18,19] and as a distinct exoenzyme named C3 by Aktories et al.[20] Relation of the ADP-ribosyltransferase in the toxin preparations to C3 ADP-ribosyltransferase was confusing at first, but the studies followed have established that the ADP-ribosyltransferase activity could be separated from the neurotoxin actions and is due to C3 molecules in the toxin preparations.[21-23] The gene for C3 ADP-ribosyltransferase was isolated by two groups. Popoff et al.[24] cloned the gene from bacteriophage DNA isolated from South African strain *C. botulinum* type D. This clone encodes a partial gene covering most of the coding regions and 3' noncoding region, but lacks 5' noncoding region and a region coding the amino terminal part. In addition, expression of this clone was not reported. Nemoto et al.[25] recently isolated the whole structure gene from total DNA from *C. botulinum* type C strain 003-9. The gene encodes a protein of 244 amino acids with a relative molecular weight of 27,362 which begins with a putative signal peptide of 40 amino acids. Expression of this gene in *Escherichia coli* produced the active enzyme about 60% of which was secreted into culture medium. Identity of the enzyme was confirmed both biochemically and immunochemically. Interestingly, the amino acid sequence of this protein differs by 40% from that determined by Popoff et al. This may reflect different sources of DNA or a difference in botulinum strains used in the two studies and suggests that there may be several types of C3 exoenzyme. Genes for several botulinum neurotoxins have recently been cloned and their amino acid sequences deduced. Neither of the C3 exoenzyme sequences show homology to those of botulinum neurotoxins, confirming that the ADP-ribosyltransferase is not related to the neurotoxins.

B. SUBSTRATE PROTEINS AND ADP-RIBOSYLATION SITE

Botulinum C3 ADP-ribosyltransferase modifies a group of proteins of 20 to 25 kDa present both in the cytosol and membranes.[26] The *rho* gene products were substrate proteins first identified for this reaction. Three members, *rhoA*, *B*, and *C*, are now know in this gene family in mammals, and all the members are ADP-ribosylated by C3 enzyme. Morii et al. purified one of the substrates from bovine adrenal cytosol which was later identified as *rhoA* product.[27-29] Kikuchi et al.[30] purified an ADP-ribosylation substrate in bovine brain membranes. This was later identified as *rhoB* protein.[31] Following these studies, Chardin et al.[32] showed that recombinant human *rhoC* protein expressed in *E. coli* can also serve as substrates for this ADP-ribosylation. Thus, including two members of *rac* genes, there are now five known species of *ras*-related small molecular weight GTP-binding proteins as substrates for C3 enzyme-catalyzed ADP-ribosylation reaction (Figure 3A). It is not known if there are still other substrates for this reaction. Wang et al.[33] detected in calf and murine thymocytes two species of proteins ADP-ribosylated by C3 enzyme. While one of the substrates was identified as *rhoA* protein, the other protein named G21K has not been identified. This is a weak substrate for ADP-ribosylation and, from its partial amino acid sequences,[34] appears to be different from *rho* and *rac* proteins.

It is known that cholera toxin requires another protein factor, ADP-ribosylation factor, to ADP-ribosylate Gsα efficiently. A similar requirement was reported for some of the C3 enzyme-catalyzed ADP-ribosylation of G22k (presumably a mixture of *rap1* and *rac* proteins). Ohtsuka et al.[35] also found that ADP-ribosylation of 22- to 25-kDa substrate proteins in bovine brain cytosol by C3 enzyme requires an additional cytosolic factor. They separated this factor from the C3 enzyme-substrates and partially purified. This factor is protein in nature and its action is specific for C3 enzyme-catalyzed ADP-ribosylation; no activation was found for cholera toxin-catalyzed reaction. Contrary to these studies, other studies using purified and recombinant *rho* proteins (see for example, References 27 and 30) did not show such a requirement for C3 enzyme-catalyzed ADP-ribosylation. It is not known whether such a difference is due to different species or forms (with or without posttranslational modification) of ADP-ribosylation substrates, or to different species of C3 exoenzyme used in these studies. One such possibility is that ADP-ribosylation of *rac* proteins and not *rho* proteins requires such a factor. However, neither Quilliam et al.[10] nor Ohtsuka et al.[35] identified their substrate proteins and it is not known whether their finding is related to *rac* proteins.

The amino acid site for C3 enzyme-catalyzed ADP-ribosylation was analyzed by protease digestion of [^{32}P]ADP-ribosylated *rhoA* protein followed by amino acid sequencing of the radioactive peptides, and identified as an asparagine residue located at the 41st position from its amino terminal (Figure 3B).[36] This conclusion was supported by recent experiments with mutant *rho* proteins. Paterson et al.[37] introduced an Asn to Ile mutation at codon 41 of *rhoA* protein and found that this mutant protein does not undergo ADP-ribosylation of C3 enzyme. This reaction is, therefore, an *N*-glycosidation reaction between ribose

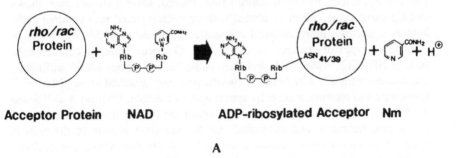

Botulinum C3 Enzyme

Acceptor Protein   NAD   ADP-ribosylated Acceptor   Nm

A

ADP-ribosylation site

rho (33–42)	Val	Tyr	Val	Pro	Thr	Val	Phe	Glu	Asn	Tyr
rac (31–40)	Glu	Tyr	Ile	Pro	Thr	Val	Phe	Asp	Asn	Tyr
H-ras (31–40)	Glu	Tyr	Asp	Pro	Thr	Ile	Glu	Asp	Ser	Tyr

B

FIGURE 3. (A) ADP-ribosylation of *rho* and *rac* proteins by botulinum C3 ADP-ribosyltransferase. (B) Comparison of the effector domains of the *rho*, *rac*, and *ras* gene products and the ADP-ribosylation site.

of ADP-ribose and β-amino group of asparagine. Both *rac* proteins have an asparagine residue at the corresponding position, Asn[39] (the middle line in Figure 3B), and are presumed to be ADP-ribosylated there. On the other hand, G25K (G_p or CDC42Hs) is highly homologous to *rac* genes and also has an asparagine residue at the corresponding position, but this protein appears not to be ADP-ribosylated as shown by the study using purified G_p[27] and by the recent study by Ohoka et al.[38] using a purified protein which appears to be CDC42Hs.

C. BIOCHEMICAL AND BIOLOGICAL EFFECTS OF THE ADP-RIBOSYLATION

Experiments using purified *rhoA* and *rhoB* proteins showed that the ADP-ribosylation by C3 enzyme does not affect either GTPase- or GTP-binding activity of the substrate proteins.[30,36] Nonetheless, as described below, this ADP-ribosylation causes marked phenotypic changes in cells, suggesting that it elicits its effect by a different mechanism. As shown in Figure 3B, this ADP-ribosylation occurs in the putative effector domain, a domain believed to be the site for interaction with a putative effector molecule downstream in the transduction pathway. From this finding, it was proposed that this ADP-ribosylation interferes with coupling of the substrate GTP-binding proteins with their target molecules, and thereby, affects its signal transduction. Indeed, several studies have shown that C3 exoenzyme, when incubated with or microinjected into cultured cells, induces significant morphological changes concomitant with this ADP-ribosylation. Two types of changes were observed. One was found in cultured cells such as NIH 3T3, Swiss 3T3, and Vero cells; the cell bodies round profoundly and become refractile with beaded dendritic processes attached to the plate.[32,37,39] Binucleate and multinucleate cells appear and accumulate. Chardin et al.[32] found in Vero cells that actin microfilaments disappeared with this rounding by C3 exoenzyme treatment and suggested that the substrate protein in the cells is involved in microfilament organization and its ADP-ribosylation causes disassembly of the filaments. This view is supported by a work by Paterson et al.[37] They made an active form of *rho* protein by *in vitro* mutagenesis of codon 14 from Gly to Val and found that this mutant *rho* protein, microinjected into Swiss 3T3 cells, caused marked phenotypic change quite different from that observed with C3 enzyme treatment, and the ADP-ribosylation of the mutant protein abolished this effect. They further found that this mutant protein caused actin microfilament organization when injected into quiescent Swiss 3T3 cells. These results suggest that *rho* protein may be involved in microfilament assembly and ADP-ribosylated protein may exert antagonistic action in this process. The effect of C3 exoenzyme was also examined on maturation of *Xenopus laevis* oocyte. Rubin et al.[39] microinjected the enzyme into an oocyte and found that it induces migration of the germinal vesicle by itself and enhances the maturation caused by progesterone. This effect may relate to that of ADP-ribosylation on microfilament assembly in cultured mammalian cells. The other type of morphological change was found in cells of neuronal origin such as PC12 rat pheochromocytoma cells and GOTO human neuroblastoma cells.[39,40] In these cells, C3 treatment induces flattened cell bodies with extended long neurites, an appearance resembling that induced by nerve growth factor, which occurred concomitant with ADP-ribosylation of cellular protein. At present it is not known whether this type of morphological change relates to a change in microfilament assembly described above. In addition to causing these morphological changes in cells, some reports suggest that *rho* and/or *rac* proteins may couple to a membrane receptor and play some role in transmembrane signal transduction. Wieland et al.[41] found that bovine rod outer segment contains several small molecular weight GTP-binding proteins, one of which undergoes ADP-ribosylation by botulinum C3 enzyme. Interestingly, this ADP-ribosylation was attenuated when the membranes were exposed to light prior to the reaction. This attenuation occurs in a manner very similar to that found in pertussis toxin-catalyzed ADP-ribosylation of transducin in the same membranes. Because rhodopsin molecules in the membranes are excited by light illumination, these authors suggested that the C3 substrate in ROS membranes interacts with photoexcited rhodopsin and mediate some signal. Another suggestion on the role of the ADP-ribosylation substrate(s) in signal transduction was given by Wang et al.[33] They incubated electropermeabilized thymocytes or their membranes with C3 enzyme and found that the enzyme activates phosphoinositide breakdown in these systems in a manner dependent on the enzyme and NAD. However, increase in inositol phosphates formation by the enzyme was relatively small, and the identity and role of the ADP-ribosylation substrate in this system is not known, leaving this work difficult to evaluate.

NOTE ADDED IN PROOF

Substantial advances have been made recently in the *rho* and *rac* protein research, some of which should be mentioned here. First, *rac* proteins have been shown to be very poor substrates for C3 exoenzyme.[42,43] ADP-ribosylation of *rac1* protein was found at least 100-fold less than that of *rho* proteins under the same reaction conditions, and significant ribosylation of it was observed only when it was denatured with 0.01% SDS. Moreover, when homogenates of blood platelets were subjected to the ADP-ribosylation, only one ADP-ribosylated substrate was found that was identified as *rhoA* protein, suggesting that the *rac* protein was hardly ADP-ribosylated in cells.[44] Thus, biological effects by C3 exoenzyme described in this chapter could reasonably be attributed to ADP-ribosylation of *rho* proteins and not that of *rac* proteins. Furthermore, proteins regulating GTP binding and GTPase activities of *rac* proteins have been reported. The GAP, GDI, and GDS proteins originally found for *rho* proteins were shown to act also on *rac* proteins and CDC42Hs.[45-48] Moreover, partial amino acid sequences of *rho*-GAP were shown to have significant homology to the product of *bcr* (breakpoint cluster region) gene, and the *bcr* product itself as well as *n*-chimerin, which has a *bcr* domain, was shown to have GAP activity for *rac* proteins but not for *rho* proteins.[45] Recently, p190, a protein co-precipitated with *ras*-GAP on mitogenic stimulation, was cloned and shown to have some *bcr* homology and to have GAP activity for *rho*, *rac*, and CDC42Hs proteins.[49] Third, *rac* proteins, *rac1* and *rac2*, were identified as a component in NADPH oxidase system in phagocytes (neutrophils and mononuclear cells) and shown to be involved in superoxide formation.[50,51] These studies, together with other reports,[52] suggest that a *rac* protein stimulates translocation of two other cytosoluble components of the oxidase, p47- and p67-phox, to cell membrane and makes the active oxidase complex with cytochrome b_{558} there. In addition to this, microinjection experiments with native and mutant *rac* proteins suggest that it receives a signal from growth factor receptor activation and induces actin filament organization in cell periphery to evoke membrane ruffling and pinocytosis.[53] On the other hand, *rho* proteins, particularly *rhoA*, have been shown to be involved in stimulus-evoked stress fiber formation in fibroblasts,[43] and a recent study has shown that *rhoA* regulates progression of cells from G1 to S through regulation of focal adhesion.[54] *rhoA* protein has been identified as a major *rho* protein in blood platelets and lymphocytes and has been shown to work downstream of c-kinase and activate integrin function in the process of adhesion and aggregation of these cells.[44,55,56]

ACKNOWLEDGMENTS

The authors thank Ms. Y. Akiyama for her secretarial work. This work was supported in part by Grants-in-aid for Scientific Research from the Ministry of Education, Science and Culture of Japan.

REFERENCES

1. **Didsbury, J., Weber, R. F., Bokoch, G. M., Evans, T., and Snyderman, R.,** *rac*, a novel *ras*-related family of proteins that are botulinum toxin substrates, *J. Biol. Chem.*, 264, 16378, 1989.
2. **Shinjo, K., Koland, J. G., Hart, M. J., Narasimhan, V., Johnson, D. J., Evans, T., and Cerione, R. A.,** Molecular cloning of the gene for the human placental GTP-binding protein G_p (G25K): identification of this GTP-binding protein as the human homolog of the yeast cell-division-cycle protein CDC42, *Proc. Natl. Acad. Sci. U.S.A.*, 87, 9853, 1990.
3. **Munemitsu, S., Innis, M. A., Clark, R., McCormick, F., Ullrich, A., and Polakis, P.,** Molecular cloning and expression of a G25K cDNA, the human homolog of the yeast cell cycle gene *CDC42*, *Mol. Cell. Biol.*, 10, 5977, 1990.
4. **Madaule, P. and Axel, R.,** A novel *ras*-related gene family, *Cell*, 41, 31, 1985.
5. **Yeramian, P., Chardin, P., Madaule, P., and Tavitian, A.,** Nucleotide sequence of human rho cDNA clone 12, *Nucleic Acid Res.*, 15, 1869, 1987.
6. **Chardin, P., Madaule, P., and Tavitian, A.,** Coding sequence of human *rho* cDNAs clone 6 and clone 9, *Nucleic Acid Res.*, 16, 2717, 1988.
7. **Polakis, P. G., Weber, R. F., Nevins, B., Didsbury, J. R., Evans, T., and Snyderman, R.,** Identification of the *ral* and *rac*1 gene products, low molecular mass GTP-binding proteins from human platelets, *J. Biol. Chem.*, 264, 16383, 1989.
8. **Polakis, P. G., Snyderman, R., and Evans, T.,** Characterization of G25K, a GTP-binding protein containing a novel putative nucleotide binding domain, *Biochem. Biophys. Res. Commun.*, 160, 25, 1989.
9. **Bokoch, G. M., Parkos, C. A., and Mumby, S. M.,** Purification and characterization of the 22,000-dalton GTP-binding protein substrate for ADP-ribosylation by botulinum toxin, G_{22k}, *J. Biol. Chem.*, 263, 16744, 1988.
10. **Quilliam, L. A., Lacal, J.-C., and Bokoch, G. M.,** Identification of *rho* as a substrate for botulinum toxin C3-catalyzed ADP-ribosylation, *FEBS Lett.*, 247, 221, 1989.
11. **Philips, M. R., Abramson, S. B., Kolasinski, S. L., Haines, K. A., Weissmann, G., and Rosenfeld, M. G.,** Low molecular weight GTP-binding proteins in human neutrophil granule membranes, *J. Biol. Chem.*, 266, 1289, 1991.
12. **Didsbury, J. R., Ushing, R. J., and Snyderman, R.,** Isoprenylation of the low molecular mass GTP-binding proteins *RAC* 1 and *RAC* 2: possible role in membrane localization, *Biochem. Biophys. Res. Commun.*, 171, 804, 1990.
13. **Garret, M. D., Self, A. J., van Oers, C., and Hall, A.,** Identification of distinct cytoplasmic targets for ras/R-ras and rho regulatory proteins, *J. Biol. Chem.*, 264, 10, 1989.
14. **Ohga, N., Kikuchi, A., Ueda, T., Yamamoto, J., and Takai, Y.,** Rabbit intestine contains a protein that inhibits the dissociation of GDP from and the subsequent binding of GTP to *rho*B p20, a *ras* p21-like GTP-binding protein, *Biochem. Biophys. Res. Commun.*, 163, 1523, 1989.

15. **Isomura, M., Kaibuchi, K., Yamamoto, T., Kawamura, S., Katayama, M., and Takai, Y.**, Partial purification and characterization of GDP dissociation stimulator (GDS) for the *rho* proteins from bovine brain cytosol, *Biochem. Biophys. Res. Commun.*, 169, 652, 1990.

16. **Morii, N., Kawano, K., Sekine, A., Yamada, T., and Narumiya, S.**, Purification of GTPase-activating protein specific for the rho gene products, *J. Biol. Chem.*, 266, 7646, 1991.

17. **Ueda, K., Kikuchi, A., Ohga, N., Yamamoto, J., and Takai, Y.**, Purification and characterization from bovine brain cytosol of a novel regulatory protein inhibiting the dissociation of GDP from the subsequent binding of GTP to rhoB p20, a ras p21-like GTP-binding protein, *J. Biol. Chem.*, 265, 9373, 1990.

18. **Ohashi, Y. and Narumiya, S.**, ADP-ribosylation of a Mr 21,000 membrane protein by type D botulinum toxin, *J. Biol. Chem.*, 262, 1430, 1987.

19. **Ohashi, Y., Kamiya, T., Fujiwara, M., and Narumiya, S.**, ADP-ribosylation by type C1 and D botulinum neurotoxins: stimulation by guanine nucleotides and inhibition by guanidino-containing compounds, *Biochem. Biophys. Res. Commun.*, 142, 1032, 1987.

20. **Aktories, K., Weller, U., and Chhatwal, G. S.**, *Clostridium botulinum* type C produces a novel ADP-ribosyltransferase distinct from botulinum C2 toxin, *FEBS Lett.*, 212, 109, 1987.

21. **Rösener, S., Chhatwal, G. S., and Aktories, K.**, Botulinum ADP-ribosyltransferase C3 but not botulinum neurotoxins C1 and D ADP-ribosylates low molecular weight mass GTP-binding proteins, *FEBS Lett.*, 224, 38, 1987.

22. **Adam-Vizi, V., Rösener, S., Aktories, K., and Knight, D. E.**, Botulinum toxin-induced ADP-ribosylation and inhibition of exocytosis are unrelated events, *FEBS Lett.*, 238, 277, 1988.

23. **Morii, N., Ohashi, Y., Nemoto, Y., Fujiwara, M., Ohnishi, Y., Nishiki, T., Kamata, Y., Kozaki, S., Narumiya, S., and Sakaguchi, G.**, Immunochemical identification of the ADP-ribosyl transferase in botulinum C1 neurotoxin as C3 exoenzyme-like molecule, *J. Biochem. (Tokyo)*, 107, 769, 1990.

24. **Popoff, M., Boquet, P., Gill, D. M., and Eklund, M. W.**, DNA sequence of exoenzyme C3, an ADP-ribosyltransferase encoded by *Clostridium botulinum* C and D phages, *Nucleic Acid Res.*, 18, 1291, 1990.

25. **Nemoto, Y., Namba, T., Kozaki, S., and Narumiya, S.**, *Clostridium botulinum* C3 ADP-ribosyltransferase gene: cloning, sequencing and expression of a functional protein in *Escherichia coli*, *J. Biol. Chem.*, 266, 19312, 1991.

26. **Narumiya, S., Morii, N., Ohno, K., Ohashi, Y., and Fujiwara, M.**, Subcellular distribution and isoelectric heterogeneity of the substrate for ADP-ribosyltransferase from *Clostridium botulinum*, *Biochem. Biophys. Res. Chem.*, 150, 1122, 1988.

27. **Morii, N., Sekine, A., Ohashi, Y., Nakao, K., Imura, H., Fujiwara, M., and Narumiya, S.**, Purification and properties of the cytosolic substrate for botulinum ADP-ribosyltransferase. Identification of an Mr 22,000 guanine nucleotide-binding protein, *J. Biol. Chem.*, 263, 12420, 1988.

28. **Narumiya, S., Sekine, A., and Fujiwara, M.**, Substrate for botulinum ADP-ribosyltransferase, Gb, has an amino acid sequence homologous to a putative *rho* gene product, *J. Biol. Chem.*, 263, 17255, 1988.

29. **Ogorochi, T., Nemoto, Y., Nakajima, M., Nakamura, E., Fujiwara, M., and Narumiya, S.**, cDNA cloning of Gb, the substrate for botulinum ADP-ribosyltransferase from bovine adrenal gland, and its identification as a *rho* gene product, *Biochem. Biophys. Res. Commun.*, 163, 1175, 1989.

30. **Kikuchi, A., Yamamoto K., Fujita, T., and Takai, Y.**, ADP-ribosylation of the bovine brain *rho* protein by botulinum toxin type C1, *J. Biol. Chem.*, 263, 16303, 1988.

31. **Hoshijima, M., Kondo, J., Kikuchi, A., Yamamoto, K., and Takai, Y.**, Purification and characterization from bovine brain membranes of a GTP-binding protein with a M_r of 21,000 ADP-ribosylated by an ADP-ribosyltransferase contaminated in botulinum toxin type C1-Identification as the *rhoA* gene product, *Mol. Brain Res.*, 7, 9, 1990.

32. **Chardin, P., Boquet, P., Madaule, P., Popoff, M. R., Rubin, E. J., and Gill, D. M.,** The mammalian G protein *rhoC* is ADP-ribosylated by *Clostridium botulinum* exoenzyme C3 and affects actin microfilaments in Vero cells, *EMBO J.*, 8, 1087, 1989.
33. **Wang, P., Nishihata, J., Makishima, F., Moriishi, K., Syuto, B., Toyoshima, S., and Osawa, T.,** Low-molecular-weight GTP-binding proteins serve as ADP-ribosylation substrate for ADP-ribosyltransferase from *Clostridium botulinum* and their relation to phosphoinositides metabolism in thymocytes, *J. Biochem. (Tokyo)*, 108, 879, 1990.
34. **Wang, P., Nishihata, J., Takabori, E., Yamamoto, K., Toyoshima, S., and Osawa, T.,** Purification and partial amino acid sequence of a phospholipase C-associated GTP-binding protein from calf thymocytes, *J. Biochem. (Tokyo)*, 105, 461, 1989.
35. **Ohtsuka, T., Nagata, K., Iiri, T., Nozawa, Y., Ueno, K., Ui, M., and Katada, T.,** Activator protein supporting the botulinum ADP-ribosyltransferase reaction, *J. Biol. Chem.*, 264, 15000, 1989.
36. **Sekine, A., Fujiwara, M., and Narumiya, S.,** Asparagine residue in the *rho* gene product is the modification site for botulinum ADP-ribosyltransferase, *J. Biol. Chem.*, 264, 8602, 1989.
37. **Paterson, H. F., Self, A. J., Garrett, M. D., Just, I., Aktories, K., and Hall, A.,** Microinjection of recombinant p21[rho] induces rapid changes in cell morphology, *J. Cell Biol.*, 111, 1001, 1990.
38. **Ohoka, Y., Imai, S., Kozaka, T., Maehama, T., Takahashi, K., Kaziro, Y., Ui, M., and Katada, T.,** Purification and characterization of a new GTP-binding protein of Mr 24,000 in bovine brain membranes, *J. Biochem. (Tokyo)*, 109, 428, 1991.
39. **Rubin, E. J., Gill, D. M., Boquet, P., and Popoff, M. R.,** Functional modification of a 21-kilodalton G protein when ADP-ribosylated by exoenzyme C3 of *Clostridium botulinum*, *Mol. Cell. Biol.*, 8, 418, 1988.
40. **Nishiki, T., Narumiya, S., Morii, N., Yamamoto, M., Fujiwara, M., Kamata, Y., Sakaguchi, G., and Kozaki, S.,** ADP-Ribosylation of the *rho/rac* proteins induces growth inhibition, neurite outgrowth and acetylcholine esterase in cultured PC-12 cells, *Biochem. Biophys. Res. Commun.*, 167, 265, 1990.
41. **Wieland, T., Ulibarri, I., Aktories, K., Gierschik, P., and Jakobs, K. H.,** Interaction of small G proteins with photoexcited rhodopsin, *FEBS Lett.*, 263, 195, 1990.
42. **Just et al.,** *J. Biol. Chem.*, 267, 10274, 1992.
43. **Ridley and Hall,** *Cell*, 70, 389, 1992.
44. **Nemoto et al.,** *J. Biol. Chem.*, 267, 20916, 1992.
45. **Diekmann et al.,** *Nature*, 351, 400, 1991.
46. **Mizuno et al.,** *J. Biol. Chem.*, 267, 10215, 1992.
47. **Hart et al.,** *J. Biol. Chem.*, 266, 20840, 1991.
48. **Leonard et al.,** *J. Biol. Chem.*, 267, 22860, 1992.
49. **Settleman et al.,** *Nature*, 359, 153, 1992.
50. **Abo et al.,** *Nature*, 353, 668, 1991.
51. **Knaus et al.,** *Science*, 254, 1512, 1992.
52. **Park and Babior,** *J. Biol. Chem.*, 267, 19901, 1992.
53. **Ridley et al.,** *Cell*, 70, 401, 1992.
54. **Yamamoto et al.,** *Oncogene*, in press, 1993.
55. **Morii et al.,** *J. Biol. Chem.*, 267, 20921, 1992.
56. **Tominaga et al.,** *J. Cell Biol.*, in press, 1993.

Chapter 15

CDC42: A MEMBER OF THE *ras* SUPERFAMILY INVOLVED IN THE CONTROL OF CELLULAR POLARITY DURING THE *Saccharomyces cerevisiae* CELL CYCLE

Douglas I. Johnson

TABLE OF CONTENTS

0-8493-5214-2/93/$0.00 + $.50
© 1993 by CRC Press, Inc.

I. INTRODUCTION

A. GTP-BINDING PROTEINS IN *Saccharomyces cerevisiae*

GTP-binding proteins play vital roles in the control of a variety of cellular functions in both higher and lower eukaryotes, but the molecular mechanisms involved in this control have yet to be fully elucidated.[1,2] At least 15 GTP-binding proteins have been found to date in the yeast *S. cerevisiae* and their cellular functions are quite different, including the control of cellular morphogenesis (see below),[3-5] cell growth,[6,7] secretion and intraorganellar transport,[8-13] pheromone response,[14] and chromosome stability.[15] The evidence that functional overlap exists between some yeast and mammalian low molecular weight GTP-binding proteins, especially between CDC42Sc and its human homolog CDC42Hs/G25K (see below), suggests that common cellular roles for these gene products may exist in higher and lower eukaryotes. The understanding of cellular processes in yeast will, therefore, contribute to the understanding of normal and abnormal cellular processes in higher eukaryotic cells.

B. CELLULAR MORPHOGENESIS DURING THE *S. cerevisiae* CELL CYCLE

Cellular morphogenesis is the process by which the complex, three-dimensional organization of the cell is developed and maintained. This process includes the generation of cell polarity, the determination of cell shape, the intracellular movement of organelles, and the directed secretion and deposition of new cellular constituents.[16] The study of cellular morphogenesis in the yeast *S. cerevisiae* is especially attractive since (1) much of the cellular machinery involved in these processes is similar to that utilized in the developmental processes of higher eukaryotes, (2) many temperature-sensitive lethal *(ts)* mutations affecting these processes have been isolated,[17-19] and (3) both classical and molecular genetic techniques can be easily used in yeast to study these processes.[20]

Sequential morphogenetic events that occur during the *S. cerevisiae* cell cycle include (1) selection of a nonrandom bud site at one of the two cell poles; (2) reorientation of the actin-based microfilament network toward the site of bud emergence along with formation of a ring of 10-nm filaments at the bud site; (3) formation of a chitin ring ("bud scar") at the bud site; (4) localized growth of the new cell wall at the bud site resulting in the appearance and selective growth of the bud; (5) selective and polar growth at the tip of the bud followed by uniform bud cell-wall growth, leading to its normal ellipsoidal shape; and (6) cytokinesis and cell separation. Cellular polarity in the budding yeast is manifested in two ways: selection of a nonrandom budding site at one of its two cell poles and directional growth toward and within the growing bud.

Cytoskeletal elements, such as microtubules, actin-based microfilaments, and intermediate filaments, along with other phenomena, such as transcellular ion currents,[21] have been implicated in the generation of cellular polarity in higher eukaryotes. The actin-based microfilament network in *S. cerevisiae* is

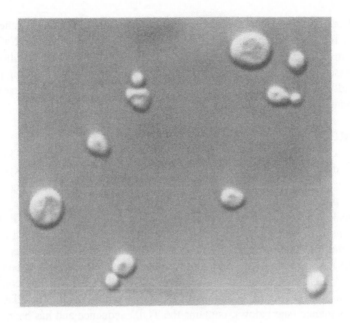

FIGURE 1. Morphological phenotype of a *cdc42* mutant. Hoffman Modulation Contrast photomicrograph of *cdc42-1ts* mutant cells grown at the permissive temperature of 23°C (small budded and unbudded cells) or at the restrictive temperature of 36°C for 6 h (large unbudded cells).

homologous to microfilaments in higher eukaryotes[22,23] and it plays a critical role in establishing and maintaining yeast morphology and cellular polarity.[24-29] The existence of ion channels in yeast[30] raises the possibility that transcellular ion currents and ion fluxes (especially calcium currents) play a role in the generation of cellular polarity in yeast.

II. THE *CDC42* GENE AND ITS PRODUCT

A. MORPHOLOGY OF *cdc42* MUTANTS

The phenotypes associated with the *cdc42-1 ts* mutation indicate that the *CDC42* gene product is involved in the control of cell polarity during the *S. cerevisiae* cell cycle.[3,17] These mutants grow and bud normally at permissive temperatures; at restrictive temperatures, the nuclear cycle continues but bud formation is blocked, resulting in cell-cycle arrest. Cell mass and volume continue to increase, however, which results in greatly enlarged, unbudded cells (Figure 1), some of which are multinucleate. Budding and nuclear division are, therefore, uncoupled in this mutant. The intracellular actin network in these cells is disorganized[17,24] and chitin and other cell surface materials are deposited randomly or uniformly throughout the enlarging cell walls, in contrast to their normal highly localized patterns of deposition. Mutations in the yeast actin gene,[25] as well as in several actin-binding proteins,[26-29] lead to similar morphological defects, suggesting a possible interaction between the *CDC42* gene product and the actin-based microfilament network. Deletion of the *CDC42* gene results in cell death, indicating that *CDC42* is essential for cell viability. Overexpression of the *CDC42* gene product alters the normal polar budding pattern to a random budding pattern, suggesting that it is involved in the initial selection and organization of the budding site.[3]

B. STRUCTURE OF THE *CDC42* GENE PRODUCT

The *CDC42* gene has been isolated and sequenced.[3] It resides 58 cM centromere-distal from the *RDN1* locus on the right arm of chromosome XII.[31] The CDC42 sequence is similar to the ras superfamily of low molecular weight GTP-binding proteins from *S. cerevisiae* and higher eukaryotes,[2] especially in those domains that have been implicated in the binding and hydrolysis of GTP (Figure 2, regions I, II, and III) and in carboxyl terminal isoprenylation, which leads to membrane localization (Figure 2, region IV). The CDC42Sc sequence differs from the RAS protein in one region of the guanine-nucleotide-binding consensus sequence; CDC42Sc contains the sequence TQID[118] rather than the consensus sequence NKXD (Figure 2, region III; X is any amino acid). The CDC42Hs/G25K sequence (see below), contains the TQID sequence and has been shown to bind and hydrolyze GTP[34,35] (see Chapter 16).

C. HUMAN HOMOLOGS OF *CDC42*

The CDC42 sequence shows the greatest degree of amino acid identity to the predicted products of two related human cDNAs, *CDC42Hs*[32] and *G25K*,[33] which encode isoforms of the low molecular weight GTP-binding protein previously referred to as G_p or G25K[34-37] (see Chapter 16). Their predicted amino acid sequences are 80% identical (88% related) to the *S. cerevisiae CDC42* gene product, which we have redesignated *CDC42Sc,* and 95% identical to each other (Figure 2). These human cDNAs, when expressed in *S. cerevisiae* under the control of a yeast promoter on a multicopy plasmid, can functionally complement the *cdc42-1[ts]* mutation.[32,33] Therefore, both are functional homologs of the *CDC42Sc* gene product. In addition, we have recently isolated the *CDC42* homolog from the fission yeast *Schizosaccharomyces pombe (CDC42Sp)* by functional complementation of the *cdc42-1[ts]* mutation using an *S. pombe* cDNA library (P. Miller and D. I. J., unpublished results). The existence of the CDC42Sp homolog and its high degree of amino acid identity to both CDC42Sc and CDC42Hs ($\approx$85% identical) reinforces the central role of the CDC42 protein in controlling eukaryotic cellular polarity.

The *CDC42Hs* and *G25K* gene products are identical in both nucleotide and predicted amino acid sequences up to amino acid residue 163 but are divergent from residues 163 to 191, suggesting that these isoforms are differential splicing products of a single gene. Biochemical analyses of the G_p/G25K protein[34-37] may have been done on either one of these gene products or a mixture of the *CDC42Hs* and *G25K* gene products; therefore, they will be referred to as the CDC42Hs/

```
                        ____I____                                    34
CDC42Sc   M Q T L K C V V V G D G A V G K T C L L I S Y T T N Q F P A D Y V P
CDC42Hs   - - - * - - - - - - - - - - - - - - - - - - - - - K - - S * - - -
G25K      - - - * - - - - - - - - - - - - - - - - - - - - - K - - S * - - -
RAC1      - - A * - - - - - - - - - - - - - - - - - - - - - A - - * * - * -
H-RAS     - T E Y - L - - - - A - G - - - * A - T - Q L I Q - H - V D * - D -


                                          ____II____                 68
CDC42Sc   T V F D N Y A V T V M I G D E P Y T L G L F D T A G Q E D Y D R L R
CDC42Hs   - - - - - - - - - - - - - G - - - - - - - - - - - - - - - - - - -
G25K      - - - - - - - - - - - - - G - - - - - - - - - - - - - - - - - - -
RAC1      - - - - - - S * N - - * D G K - V N - - - W - - - - - - - - - - -
H-RAS     - * E - S - R K Q - V - D G - T C L - D * L - - - - - - * - S A M -


                                                                     101
CDC42Sc   P L S Y P S T D V   F L V C F S V I S P P S F E N V K E K W F P E V
CDC42Hs   - - - - - Q - - -   - - - - - - - - * - - S - - - - - - - - V - - I
G25K      - - - - - Q - - -   - - - - - - - - * - - S - - - - - - - - V - - I
RAC1      - - - - - Q - - -   - - * - - - * * - - A - - - - - * A - - Y - - -
H-RAS     D Q - M R - G E G - - C V - A * N N T K - - - D * H Q Y R E Q I K


                            ____III____                              133
CDC42Sc   H H H C P G   V P C L V V G T Q I D L R D D K V I I E K L   Q R Q R
CDC42Hs   T - - - - K   T - F - * - - - - - - - - - - P S T - - - -   A * N *
G25K      T - - - - K   T - F - * - - - - - - - - - - P S T - - - -   A * N *
RAC1      R - - - - N   T - I * * - - - K * - - - - - D T - - - -     K E K *
H-RAS     R V K D S D D - - M * * - - N K C - - A A R T   * E S *     - A A D


                                                                     167
CDC42Sc   L R P I T S E Q G S R L A R E L K A V K Y V E C S A L T Q R G L K N
CDC42Hs   Q * - - - P - T A E * - - - * - - - - - - - - - - - * - - - - - - -
G25K      Q * - - - P - T A E * - - - * - - - - - - - - - - - - - - - - - - -
RAC1      - T - - - Y P - - L A M - * - * G - - - - * - - - - - - - - - - - T
H-RAS     -                       - - S Y G *   P - * - T - - K - R Q - * E D


                                                           ____IV____191
CDC42Sc   V F D E A I V A A L E P P               V I K K S K K C T I L
CDC42Hs   - - - - - - * - - - - -                 E P - - - * * - V * -
G25K      - - - - - - * - - - - -                 E T Q P K * - - C - F
RAC1      - - - - - - R - * - C - - P             - K - * K * - - L * -
H-RAS     * - Y T * * R E * R Q H K L R K L N P P D E S G P G C M - C - - V * S
```

FIGURE 2. Predicted amino acid sequence comparisons. Comparisons are between CDC42Sc,[3] CDC42Hs,[32] G25K,[33] rac 1,[39] and human H-*ras*.[54] The one letter amino acid code is used. Numbering corresponds to the CDC42Sc amino acid sequence. Dashes and asterisks indicate amino acids identical and similar to CDC42Sc, respectively. (Amino acids considered similar are aspartate and glutamate; asparagine and glutamine; isoleucine, leucine, valine, and alanine; serine and threonine; and lysine and arginine). Regions I, II, and III correspond to the putative GTP-binding and hydrolysis domains and region IV corresponds to the putative isoprenylation site.

G25K protein. The CDC42Hs/G25K protein is identical to the CDC42Sc protein in the GTP-binding domains and similar in the carboxyl terminal isoprenylation site. It has been shown to bind and hydrolyze GTP[34,35] (see Chapter 16) and to contain an all-*trans* geranylgeranyl cysteinyl methyl ester at its carboxyl terminus.[37,38]

FIGURE 3. Western blot analysis of CDC42Sc proteins. Primary antibodies are directed against the p1 peptide of the CDC42Hs/G25K protein.[36] Secondary antibodies are horseradish peroxidase-conjugated goat anti-rabbit antibodies. Lanes 1 and 2 are *E. coli* whole-cell extracts from the strain BL21(DE3) carrying the T7 promoter expression plasmid pET3b[56] containing the *CDC42Sc* gene in the incorrect (lane 1) or correct (lane 2) orientation for T7-promoted expression. Lanes 3–6 are extracts from the *cdc42-1ᵗˢ* strain DJMD22-3B carrying the *CDC42Sc* gene under the control of its own promoter on the high copy number plasmid YEp103 (lane 3); YEp103 alone (lane 4); DJMD22-3B alone (lane 5); and DJMD22-3B carrying the *CDC42Sc* gene under the control of the galactose-inducible *GAL10* promoter (lane 6) grown on 2% galactose.

Another human gene product, *rac 1*,[39] is 74% identical to Cdc42Sc (Figure 2). This gene cannot complement the *cdc42-1ᵗˢ* mutation when expressed either under the control of an inducible promoter on a multicopy plasmid[32] or under the control of the strong, constitutive *ADH1* promoter (J. O'Brien and D. I. J., unpublished results). Interestingly, the only region of rac 1 that is significantly different from CDC42Sc and CDC42Hs/G25K is residues 41 to 52. The analogous region in RAS proteins overlaps with the "effector" region, which is believed to be the site of interaction between RAS and its GTPase-activating protein (GAP).[2] The inability of rac 1 to complement the *cdc42-1ᵗˢ* mutation may, therefore, be due to an inability to interact with a CDC42-specific GAP.

D. IMMUNOLOGICAL CHARACTERIZATION OF CDC42Sc

Cross reactivity is seen between the yeast and human CDC42 proteins in immunological analyses. Antibodies (kindly provided by T. Evans) directed against the p1 peptide of the CDC42Hs/G25K protein[36] (NVFDEAILAALEPPEPK; underlined amino acids are different from CDC42Sc) crossreact with the Cdc42Sc protein when it is overproduced in either *S. cerevisiae* (Figure 3, lane 6) or *E. coli* (Figure 3, lane 2). CDC42Sc-specific antibodies react with purified Cdc42Hs protein, CDC42Sc protein overproduced in *S. cerevisiae*, CDC42Sc protein overproduced in *E. coli*, and the Cdc42Sp protein (data not shown). These antibodies were raised against the 17 amino acid synthetic peptide NVFDEAI-VAALEPPVIK, which corresponds to residues 165 to 181 of CDC42Sc and is analogous to the p1 peptide of CDC42Hs/G25K.[57]

The multiple protein bands recognized by the CDC42Hs/G25K-specific antibodies when the CDC42Sc protein is overproduced in *S. cerevisiae* (Figure 3, lane 6) may be different modified forms of the protein. This possibility is particularly interesting considering that the CDC42Hs/G25K protein is phosphorylated *in vitro* by the EGF receptor tyrosine kinase[35] (see Chapter 16). The CDC42Sc protein overproduced in *E. coli* comigrates with the lower yeast protein bands and the multiple bands are not seen in *E. coli* extracts, suggesting that the potential modifications are yeast-specific and not simply proteolysis. We see very weak reaction of the CDC42Hs/G25K-specific antibodies with protein extracts from plasmid-free *S. cerevisiae* cells (Figure 3, lane 5) and from cells carrying a high copy number plasmid containing *CDC42Sc* under the control of its own promoter (Figure 3, lane 3). This raises the possibility that the expression of *CDC42Sc* is transcriptionally regulated.

E. MUTATIONAL ANALYSIS OF THE *CDC42Sc* GENE

1. Analogies to *ras* Mutations

In order to further elucidate the role of the CDC42Sc protein in controlling cell polarity, we have generated new mutations in *CDC42Sc* using site-directed mutagenesis (Figure 4).[57] Mutations at positions 12 (Gly to Val), 61 (Gln to Leu), and 118 (Asp to Ala) are analogous to *ras* mutations that lead to a transforming phenotype (see Chapter 1).[1] The mutation at position 188 (Cys to Ser) is analogous to a *ras* mutation that results in the loss of carboxyl terminal modifications, such as isoprenylation and carboxymethylation, and leads to mislocalization of the mutant protein and loss of function. By analogy to mutations in other GTP-binding proteins, mutations at positions 12 and 61 are predicted to result in a decrease in the intrinsic GTPase activity of Cdc42Sc, thereby, locking the mutant protein in an "activated" GTP-bound state. A mutation at position 118 (position 119 in *ras*) is also predicted to lead to an "activated" GTP-bound protein due to an increased GDP dissociation rate of the protein that, in combination with a high GTP:GDP ratio of the cell, results in a higher probability of the mutant protein being in a GTP-bound state.[2,40]

2. The *CDC42*[Val12] and *CDC42*[Leu61] Mutations Lead to a Dominant-Lethal Activated Phenotype

The *CDC42*[Val12] and *CDC42*[Leu61] mutations result in dominant lethality to both the wild-type and *cdc42-1*[ts] alleles (Figure 4). This dominant-lethal phenotype suggests that these mutant proteins are irreversibly associated with cellular factors necessary for the budding process, thereby, preventing these factors from interacting with endogenous functional CDC42Sc proteins. Such cellular factors, which could include a Cdc42Sc-specific GAP, may act after bud emergence (downstream effectors) since cells containing these mutant genes under the control of a galactose-inducible promoter die with an abnormal- or multi-budded phenotype.[57] This phenotype suggests that these constitutively active CDC42Sc proteins cannot regulate proper bud emergence (i.e., more than one bud is formed on the same mother cell) and proper bud morphogenesis. We also observed aberrant actin-containing structures in these mutant cells, suggesting that normal

A.

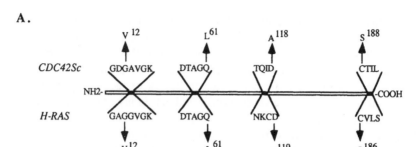

B.

Allele	Mutation	Phenotype	Functional protein
CDC42-3	Gly to Val[12]	Dominant lethal	NA
CDC42-4	Gln to Leu[61]	Dominant lethal	NA
CDC42-5	Asp to Ala[118]	Dose-dependent dominant lethal	NA
cdc42-6	Cys to Ser[188]	Recessive	−
cdc42-7	Gly to Val[12] Cys to Ser[188]	Recessive	−
cdc42-8	Gln to Leu[61] Cys to Ser[188]	Recessive	−
cdc42-9	Asp to Ala[118] Cys to Ser[188]	Recessive	−

FIGURE 4. (A) Comparison of mutations in *CDC42Sc* to mutations in H-*ras* that alter either the GTP-binding/hydrolysis of p21[H-ras] (Val[12], Leu[61], Ala[119]) or isoprenylation (Ser[186]). (B) Summary of phenotypes of *CDC42Sc* mutant alleles. Phenotype refers to the effect of these mutations in the presence of the wild-type *CDC42Sc* allele. The functionality of mutant proteins was determined by assaying the growth of *cdc42-1[ts]* cells containing mutant genes on plasmids on selective media at 36°C. A "+" indicates uniform growth at 36°C, therefore, a functional protein; a "−" indicates no growth at 36°C and a nonfunctional protein; "NA" indicates that the criterion is not applicable.

actin assembly is either unstable or improperly localized. It is unclear whether the abnormal cellular morphologies are due to unproductive interactions of cellular factors with the mutant CDC42Sc proteins or simply due to the disruption of the actin microfilaments.

3. The *CDC42[Ala118]* Mutation Leads to an Inactivated Protein Phenotype

If analogous *ras* and *CDC42Sc* mutations result in analogous biochemical defects, then the *CDC42[Ala118]* mutant would be predicted to show the same phenotype as the *CDC42[Val12]* and *CDC42[Leu61]* mutants. The *CDC42[Ala118]* mutant phenotype, however, resembles the *cdc42-1[ts17]* and null[3] mutant phenotypes (i.e., loss of function) with respect to arrested cell morphology (Figure 1; large, round, unbudded cells) and delocalized actin staining.[57] This effect suggests that the CDC42^{Ala118} protein is in an inactivated state rather than an activated state as predicted by analogy to similar *ras* mutants. This different effect suggests that aspects of the guanine-nucleotide binding and/or hydrolytic properties of the CDC42Sc protein differ from those of the RAS protein. Analysis of the GTP-binding capability of this mutant protein may elucidate this unexpected phenotype.

The dose-dependent lethality of *CDC42*^{Ala118} (Figure 4) indicates that it is not simply a null allele. This dose-dependent lethality is a function of the expression levels of Cdc42^{Ala118} protein and the lower levels of endogenous CDC42Sc protein in *cdc42-1^{ts}* cells vs. *CDC42Sc* cells.[57] This dominant-negative phenotype suggests that the CDC42^{Ala118} protein may irreversibly bind to cellular factors necessary for CDC42Sc activation, thereby, preventing activation of endogenous functional CDC42Sc proteins and initiation of bud emergence. These factors, which could include a CDC42Sc-specific guanine nucleotide release protein (GNRP),[1] may act before bud emergence (upstream effectors) since cells overexpressing the *CDC42*^{Ala118} allele cannot form buds.

4. Role of Potential Membrane Localization in CDC42 Function

The *cdc42*^{Ser188} mutation is a recessive, nonfunctional allele (Figure 4) and is predicted to affect the carboxyl terminal modification of the CDC42Sc protein such that it would not be isoprenylated and hence, would not be membrane-localized. There are three lines of evidence that suggest that proper membrane localization of the CDC42Sc protein is critical to its function. First, plasmids containing the *cdc42*^{Ser188} mutation cannot complement the *cdc42-1^{ts}* mutation, even when the CDC42^{Ser188} mutant protein is overproduced under the control of an inducible yeast promoter.[57] The mobility rate of the CDC42^{Ser188} mutant protein on polyacrylamide gels is slower than the wild-type protein, suggesting that it is unprocessed. Second, the *cdc42*^{Ser188} mutation can suppress the lethality associated with the *CDC42*^{Val12}, *CDC42*^{Leu61}, and *CDC42*^{Ala118} mutations in double mutant constructs, suggesting that proper localization of the mutant proteins is necessary for them to exert their lethal effects.[57] Finally, overproduction of the CDC42^{Ser188} protein does not lead to abnormal positioning of budding sites, which is in contrast to the random budding pattern observed when the wild-type CDC42Sc protein is overproduced.[3]

The putative membrane attachment of the CDC42Sc protein could be important in the initial selection of the bud site on the cell surface (i.e., budding-site marker) and/or in directing the localization of secretory vesicles containing cell-wall digestive enzymes necessary for bud emergence to the bud site, depending on to which cellular membrane the CDC42Sc protein is localized. Determination of the location of the CDC42Sc protein within the cell at various stages of the cell cycle using immunofluorescence localization and cell-fractionation studies will be crucial to deciphering its cellular function.

TABLE 1
Cell Polarity Genes in *S. cerevisiae*

Gene	Structure/function	Ref.
CDC42	≈21 kDa *ras* superfamily GTP-binding protein	3, 17
CDC43/CAL1	≈42 kDa component of protein geranylgeranyltransferase	17, 45, 58
CDC24/CLS4	≈84 kDa putative Ca^{+2}-binding protein	41–43, 47–50
RSR1/BUD1	≈30 kDa *ras* superfamily GTP-binding protein	5
BUD5	≈61 kDa putative GDP/GTP exchanger	51, 59
MSB1	≈130 kDa multicopy suppressor of *cdc24* and *cdc42*	5
MSB2	Unknown structure; multicopy suppressor of *cdc24*	5
BEM1	Unknown structure; mutant requires *MSB1*	51, 55
BEM2	Unknown structure; mutant requires *MSB1*	55
BUD2	Unknown structure; mutation leads to random budding pattern	53
BUD3	Unknown structure; mutation leads to bipolar budding pattern	53
BUD4	Unknown structure; mutation leads to bipolar budding pattern	53

III. OTHER GENE PRODUCTS INVOLVED IN CONTROLLING CELL POLARITY

It is clear that the CDC42Sc protein does not act alone in controlling cell polarity (Table 1; Figure 5). Both *cdc43*[ts17] and *cdc24*[ts41-43] mutants have identical phenotypes to the *cdc42-1*[ts] mutant. The phenotype of a *cdc42-1*[ts] *cdc43*[ts] double mutant is more severe (i.e., lethal) at the normally permissive temperature than either single mutant.[17] This "synthetic lethality" suggests that the CDC42Sc and CDC43 proteins interact with one another. In addition, the ability of *CDC42Sc* when present on plasmids to suppress a *cdc24*[ts] mutation[5] suggests that these two gene products may interact. Alternatively, the CDC42 protein may act downstream of the CDC24 protein.

Another mutant originally identified as a calcium-dependent mutant *(call)*,[44] has been shown to be an allele of *cdc43*, suggesting a role for calcium in *CDC43* function. The *CDC43* and *CAL1* gene products are identical[45,58] (hereafter referred to as *CDC43*) and are similar to the *DPR1/RAM1* gene product, which is implicated in the protein farnesyltransferase activity that modifies the carboxyl termini of *S. cerevisiae* RAS proteins.[46] There exists a *CDC43*-dependent protein geranylgeranyl transferase activity in *S. cerevisiae* and the CDC42Sc and CDC42Hs proteins are modified by this activity with the addition of a geranylgeranyl isoprene to presumably the Cys^{188} residue.[38] The CDC43 product, therefore, is either the protein geranylgeranyltransferase or necessary for transferase activity and is likely to act as an upstream effector of CDC42Sc function (Figure 5). Both genetic and biochemical data support the idea that the function of the Cdc43 protein is to modify the carboxyl terminus of the CDC42Sc protein, possibly in response to an intracellular calcium signal, and that this modification is necessary for the anchoring of the Cdc42Sc protein into a cellular membrane.

Calcium may also play a role in the function of another protein, CDC24, involved in controlling cell polarity. A mutant *(cls4)* originally identified among

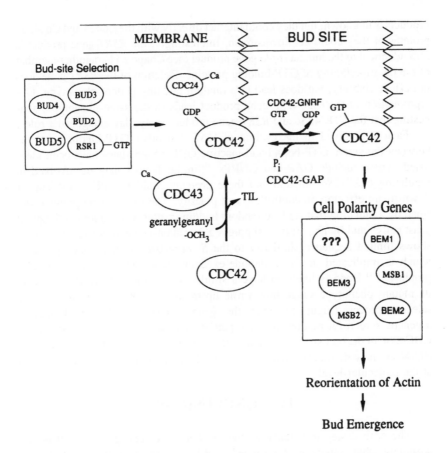

FIGURE 5. A model for CDC42Sc-mediated control of cell polarity. It is unknown to which cellular membrane the CDC42Sc protein is localized. See text for full details.

a collection of calcium-sensitive mutants[47,48] was shown to an allele of *cdc24*,[47,50] suggesting that the *CDC24* gene product interacts with calcium. The *CDC24* gene has been isolated[47,49] and sequenced[50] and its predicted gene product contains two putative calcium-binding domains. Other *cdc24^ts* mutants show abnormal positioning of budding sites when grown at permissive temperatures, suggesting that the *CDC24* gene product is involved in the initial selection and organization of the budding site.[17,43] The *G25K* cDNA, when expressed under the control of a yeast promoter on a plasmid, can suppress a *cdc24^ts* mutation,[33] as can the *CDC42Sc* gene when present on a multicopy plasmid.[5] This suggests that the CDC42Hs/G25K protein can functionally replace the Cdc42Sc protein in its interaction with the CDC24 protein.

Three other genes, *RSR1*, *MSB1*, and *MSB2*, were isolated by virtue of their ability to suppress a *cdc24^ts* mutation when present on a multicopy plasmid.[5] The *MSB1* gene can also suppress a *cdc42^ts* mutation. The function of these gene products is unknown, but they could interact directly with the Cdc24 and Cdc42Sc proteins or they could act downstream. Interestingly, the *RSR1* gene product is 56% identical to the human *rap1a* gene product (see Chapter 11), another member of the ras superfamily of GTP-binding proteins. Deletion of the *RSR1* gene does not affect viability, but does lead to a random budding pattern phenotype. Overproduction of the *CDC42Sc* gene product leads to the same phenotype, suggesting that the CDC42Sc and *RSR1* gene products may play antagonistic roles.

Factors involved in the cycling of RAS and other GTP-binding proteins between the active GTP-bound and inactive GDP-bound states have been identified. These include GAP and GNRPs.[1] Similar factors may be involved in regulating *CDC42Sc* function and these proteins could be either upstream or downstream effectors in controlling cell polarity. The *BUD5* gene product, defined by a mutation that leads to random budding patterns in haploid and diploid yeast cells[51] and by its ability to suppress a dominant-negative *RAS2* mutation,[59] shows limited sequence similarity to the *S. cerevisiae CDC25* gene product, which is implicated in the nucleotide-exchange reaction with the yeast RAS proteins.[52,60] Bud5 is likely to have a similar interaction with RSR1 based on its mutant phenotype suggesting a role upstream of CDC42Sc in the bud-site selection process (Figure 5). Four other genes *(BUD1-BUD4)* were defined in a genetic screen for random budding patterns and are likely to act upstream of CDC42Sc in the bud-site selection process.[53] *BUD1* is allelic to *RSR1* and *BUD2-BUD4* are new genes involved in bud-site selection and have not been analyzed at the molecular level.

IV. CONCLUSIONS

The critical step in elucidating the mechanism of cell polarity will be determining what signals are transduced, and by means of what downstream effectors (Figure 5). The molecular nature of this process in *S. cerevisiae* remains unclear but it seems likely to involve a response to an intracellular signal, possibly a calcium signal. As described above, many components involved in controlling cell polarity in *S. cerevisiae* have been identified and both calcium-binding and GTP-binding proteins have been implicated in this process. Much remains to be elucidated concerning the order of action and mechanism of these components.

The high degree of sequence similarity between CDC42Sc, CDC42Sp, and CDC42Hs/G25K suggests that functional homologs of many of these additional components may exist in higher eukaryotes and should exhibit significant sequence similarities to their yeast counterparts. Determination of the roles of these gene products in controlling cell polarity in *S. cerevisiae* may help elucidate control mechanisms for morphogenetic events of higher eukaryotes.

ACKNOWLEDGMENTS

I would like to thank Janet Kurjan, Carlene Raper, Michael Ziman, and Jeanne O'Brien for helpful discussions and critical reading of this manuscript.

Research in the author's laboratory is supported by the American Cancer Society Grant MV-469, a grant from the Gustavus and Louise Pfeiffer Research Foundation, and the National Science Foundation-VT EPSCoR Grant R11-8610679.

REFERENCES

1. **Bourne, H. R., Sanders, D. A., and McCormick, F.,** The GTPase superfamily: a conserved switch for diverse cell functions, *Nature,* 348, 125, 1990.
2. **Hall, A.,** The cellular functions of small GTP-binding proteins, *Science,* 249, 635, 1990.
3. **Johnson, D. I. and Pringle, J. R.,** Molecular characterization of *CDC42,* a *Saccharomyces cerevisiae* gene involved in the development of cell polarity, *J. Cell Biol.,* 111, 143, 1990.
4. **Madaule, P., Axel, R., and Myers, A. M.,** Characterization of two members of the rho gene family from the yeast *Saccharomyces cerevisiae, Proc. Natl. Acad. Sci. U.S.A.,* 84, 779, 1987.
5. **Bender, A. and Pringle, J. R.,** Multicopy suppression of the *cdc24* budding defect in yeast by *CDC42* and three newly identified genes including the *ras*-related gene *RSR1, Proc. Natl. Acad. Sci. U.S.A.,* 86, 9976, 1990.
6. **Tatchell, K.,** 1986. *RAS* genes and growth control in *Saccharomyces cerevisiae, J. Bacteriol.,* 166, 364, 1986.
7. **Gibbs, J. B. and Marshall, M. S.,** The *ras* oncogene — an important regulatory element in lower eukaryotic organisms, *Microbiol. Rev.,* 53, 171, 1989.
8. **Salminen, A. and Novick, P. J.,** A *ras*-like protein is required for a post-Golgi event in yeast secretion, *Cell,* 49, 527, 1987.
9. **Goud, B., Salminen, A., Walworth, N. C., and Novick, P. J.,** A GTP-binding protein required for secretion rapidly associates with secretory vesicles and the plasma membrane in yeast, *Cell,* 53, 753, 1988.
10. **Botstein, D., Segev, N., Stearns, T., Hoyt, M. A., Holden, J., and Kahn, R. A.,** Diverse biological functions of small GTP-binding proteins in yeast, in *Cold Spring Harbor Symposium on Quantitative Biology, Vol. LIII, Molecular Biology of Signal Transduction,* Cold Spring Harbor Laboratories, Cold Spring Harbor, New York, 1988, 629.
11. **Baker, D., Wuestehube, L., Schekman, R., Botstein, D., and Segev, N.,** GTP-binding Ypt1 protein and Ca^{+2} function independently in a cell-free protein transport reaction, *Proc. Natl. Acad. Sci. U.S.A.,* 87, 355, 1990.
12. **Stearns, T., Kahn, R. A., Botstein, D., and Hoyt, M. A.,** ADP ribosylation factor is an essential protein in Saccharomyces cerevisiae and is encoded by two genes, *Mol. Cell. Biol.,* 10, 6690, 1990.
13. **Stearns, T., Willingham, M., Botstein, D., and Kahn, R.,** ADP-ribosylation factor is functionally and physically associated with the Golgi-complex, *Proc. Natl. Acad. Sci. U.S.A.,* 87, 1238, 1990.
14. **Kurjan, J. and Whiteway, M.,** The *Saccharomyces cerevisiae* pheromone response pathway, in *Seminars in Developmental Biology: Developmental Systems in Fungi,* Raper, C. A. and Johnson, D. I., Eds., W. B. Saunders, Philadelphia, PA, 1990, 151.
15. **Stearns, T., Hoyt, M. A., and Botstein, D.,** Yeast mutants sensitive to antimicrotubule drugs define three genes that affect microtubule function, *Genetics,* 124, 251, 1990.
16. **McIntosh, J. R.,** *Spatial Organization of Eukaryotic Cells (Modern Cell Biology, Vol. 2),* Alan R. Liss, New York, 1983.
17. **Adams, A. E. M., Johnson, D. I., Sloat, B. F., and Pringle, J. R.,** *CDC42* and *CDC43,* two additional genes involved in budding and the establishment of cell polarity in the yeast *Saccharomyces cerevisiae, J. Cell Biol.,* 111, 131, 1990.

18. **Pringle, J. R. and Hartwell, L. H.,** The *Saccharomyces cerevisiae* cell cycle, in *The Molecular Biology of the Yeast Saccharomyces Life Cycle and Inheritance,* Strathern, J. N., Jones, E. W., and Broach, J. R., Eds., Cold Spring Harbor Laboratory, Cold Spring Harbor, New York, 1981, 97.

19. **Wheals, A. E.,** Biology of the cell cycle in yeasts, in *The Yeasts, Vol. 1,* Rose, A. H. and Harrison, J. S., Eds., Academic Press, London, 1987, 284.

20. **Botstein, D. and Fink, G. R.,** Yeast: an experimental organism for modern biology, *Science,* 240, 1439, 1988.

21. **Nuccitelli, R.,** Transcellular ion currents: signals and effectors of cell polarity, in *Modern Cell Biology, Vol. 2, Spatial Organization of Eukaryotic Cells,* McIntosh, J. R., Ed., Alan R. Liss, New York, 1983, 451.

22. **Ng, R. and Abelson, J.,** Isolation and sequence of the gene for actin in *Saccharomyces cerevisiae, Proc. Natl. Acad. Sci. U.S.A.,* 77, 3912, 1980.

23. **Thomas, J. H., Novick, P., and Botstein, D.,** Genetics of the yeast cytoskeleton, in *Molecular Biology of the Cytoskeleton,* Borisy, G. G., Cleveland, D. W., and Murphy, D. G., Eds., Cold Spring Harbor Laboratory, Cold Spring Harbor, New York, 1984, 153.

24. **Adams, A. E. M. and Pringle, J. R.,** Relationship of actin and tubulin distribution to bud growth in wild-type and morphogenetic-mutant *Saccharomyces cerevisiae, J. Cell Biol.,* 98, 934, 1984.

25. **Novick, P. and Botstein, D.,** Phenotypic analysis of temperature-sensitive yeast actin mutants, *Cell,* 40, 405, 1985.

26. **Drubin, D. G., Miller, K. G., and Botstein, D.,** Yeast actin-binding proteins: evidence for a role in morphogenesis, *J. Cell Biol.,* 107, 2551, 1988.

27. **Haarer, B. K., Lillie, S. H., Adams, A. E. M., Magdolen, V., Bandlow, W., and Brown, S. S.,** Purification of profilin from *Saccharomyces cerevisiae* and analysis of profilin-deficient cells, *J. Cell Biol.,* 110, 105, 1990.

28. **Liu, H. and Bretscher, A.,** Disruption of the single tropomyosin gene in yeast results in the disappearance of actin cables from the cytoskeleton, *Cell,* 57, 233, 1989.

29. **Amatruda, J. F., Cannon, J. F., Tatchell, K., Hug, C., and Cooper, J. A.,** Disruption of the actin cytoskeleton in yeast capping protein mutants, *Nature,* 344, 352, 1990.

30. **Gustin, M. C., Martinac, B., Saimi, Y., Culbertson, M. R., and Kung, C.,** Ion channels in yeast, *Science,* 233, 1195, 1986.

31. **Johnson, D. I., Jacobs, C. W., Pringle, J. R., Robinson, L. C., Carle, G. F., and Olson, M. V.,** Mapping of the *Saccharomyces cerevisiae CDC3, CDC25,* and *CDC42* genes to chromosome XII by chromosome blotting and tetrad analysis, *Yeast,* 3, 243, 1987.

32. **Shinjo, K., Koland, J. G., Hart, M. J., Narasimhan, V., Johnson, D. I., Evan, T., and Cerione, R. A.,** Molecular cloning of the gene for the human placental GTP-binding protein, G_p (G25K): identification of this GTP-binding protein as the human homolog of the yeast cell-division cycle protein, CDC42, *Proc. Natl. Acad. Sci. U.S.A.,* 87, 9853, 1990.

33. **Munemitsu, S., Innis, M. A., Clark, R., McCormick, F., Ullrich, A., and Polakis, P.,** Molecular cloning and expression of a G25K cDNA, the human homolog of the yeast cell cycle *CDC42, Mol. Cell. Biol.,* 10, 5977, 1990.

34. **Evans, T., Brown, M. L., Fraser, E. D., and Northup, J. K.,** Purification of the major GTP-binding proteins from human placental membranes, *J. Biol. Chem.,* 261, 7052, 1986.

35. **Hart, M. J., Polakis, P. G., Evans, T., and Cerione, R. A.,** The identification and characterization of an epidermal growth factor-stimulated phosphorylation of a specific low molecular weight GTP-binding protein in a reconstituted phospholipid vesicle system, *J. Biol. Chem.,* 265, 5990, 1990.

36. **Polakis, P. G., Snyderman, R., and Evans, T.,** Characterization of G25K, a GTP-binding protein containing a novel putative nucleotide binding domain, *Biochem. Biophys. Res. Commun.,* 160, 25, 1989.

37. **Yamane, H. K., Farnsworth, C. C., Xie, H., Evans, T., Howald, W. N., Gelb, M. H., Glomset, J. A., Clarke, S., and Kung, B. K.-K.,** Membrane-binding domain of the small G protein G25K contains an *S*-(all-*trans*-geranylgeranyl)cysteine methyl ester at its carboxyl terminus, *Proc. Natl. Acad. Sci. U.S.A.,* 88, 286, 1991.

38. **Finegold, A. A., Johnson, D. I., Farnsworth, C. C., Gelb, M. H., Judd, S. R., Glomset, J. A., and Tamanoi, F.**, Protein geranylgeranyltransferase of *Saccharomyces cerevisiae* is specific for Cys-Xaa-Xaa-Leu motif proteins and requires the *CDC43* gene product but not the *DPR1* gene product, *Proc. Natl. Acad. Sci. U.S.A.*, 88, 4448, 1991.

39. **Polakis, P. G., Weber, R. F., Nevins, B., Didsbury, J. R., Evans, T., and Snyderman, R.**, Identification of the *ral* and *rac1* gene products, low molecular mass GTP-binding proteins from human platelets, *J. Biol. Chem.*, 264, 16383, 1989.

40. **Casey, P. J. and Gilman, A. G.**, G protein involvement in receptor-effector coupling, *J. Biol. Chem.*, 263, 2577, 1988.

41. **Hartwell, L. H., Culotti, J., Pringle, J. R., and Reid, B. J.**, Genetic control of the cell division cycle in yeast, *Science*, 183, 46, 1974.

42. **Sloat, B. F. and Pringle, J. R.**, A mutant of yeast defective in cellular morphogenesis, *Science*, 200, 1171, 1978.

43. **Sloat, B. F., Adams, A., and Pringle, J. R.**, Roles of the *CDC24* gene product in cellular morphogenesis during the *Saccharomyces cerevisiae* cell cycle, *J. Cell Biol.*, 89, 395, 1981.

44. **Ohya, Y., Ohsumi, Y., and Anraku, Y.**, Genetic study of the role of calcium ions in the cell division cycle of *Saccharomyces cerevisiae*: a calcium-dependent mutant and its trifluoperazine-dependent pseudorevertants, *Mol. Gen. Genet.*, 193, 389, 1984.

45. **Johnson, D. I., O'Brien, J. M., and Jacobs, C. W.**, Isolation and sequence analysis of *CDC43*, a gene involved in the control of cell polarity in *Saccharomyces cerevisiae*, *Gene*, 90, 93, 1990.

46. **Goodman, L. E., Perou, C. M., Fujiyama, A., and Tamanoi, F.**, Structure and expression of yeast *DPR1*, a gene essential for the processing and intracellular localization of *ras* proteins, *Yeast*, 4, 271, 1988.

47. **Ohya, Y., Miyamoto, S., Ohsumi, Y., and Anraku, Y.**, Calcium-sensitive *cls4* mutant of *Saccharomyces cerevisiae* with a defect in bud formation, *J. Bacteriol.*, 165, 28, 1986a.

48. **Ohya, Y., Ohsumi, Y., and Anraku, Y.**, Isolation and characterization of Ca^{2+}-sensitive mutants of *Saccharomyces cerevisiae*, *J. Gen. Microbiol.*, 132, 979, 1986b.

49. **Coleman, K. G., Steensma, H. Y., Kaback, D. B., and Pringle, J. R.**, Molecular cloning of chromosome I DNA from *Saccharomyces cerevisiae*: isolation and characterization of the *CDC24* gene and adjacent regions of the chromosome, *Mol. Cell. Biol.*, 6, 4516, 1986.

50. **Miyamoto, S., Ohya, Y., Ohsumi, Y., and Anraku, Y.**, Nucleotide sequence of the *CLS4* (*CDC24*) gene of *Saccharomyces cerevisiae*, *Gene*, 54, 125, 1987.

51. **Chant, J., Corrado, K., Pringle, J. R., and Herskowitz, I.**, Yeast *BUD5*, encoding a putative GDP-GTP exchange factor, is necessary for bud site selection and interacts with bud formation gene *BEM1*, *Cell*, 65, 1213, 1991.

52. **Broek, D., Toda, T., Michaeli, T., Levin, L., Birchmeier, C., Zoller, M., Powers, S., and Wigler, M.**, The *S. cerevisiae CDC25* gene product regulates the *RAS*/adenylate cyclase pathway, *Cell*, 48, 789, 1987.

53. **Chant, J. and Herskowitz, I.**, Genetic control of bud site selection in yeast by a set of gene products that comprise a morphogenetic pathway, *Cell*, 65, 1203, 1991.

54. **Capon, D. J., Chen, E. Y., Levinson, A. D., Seeburg, P. H., and Goeddel, D. V.**, Complete nucleotide sequence of the T24 human bladder carcinoma oncogene and its normal homologue, *Nature*, 302, 33, 1983.

55. **Bender, A. and Pringle, J. R.**, Use of a screen for synthetic lethality and multicopy suppressee mutants to identify two new genes involved in morphogenesis in *Saccharomyces cerevisiae*, *Mol. Cell. Biol.*, 11, 1295, 1991.

56. **Studier, F. W. and Moffatt, B. A.**, Use of bacteriophage T7 RNA polymerase to direct selective high-level expression of cloned genes, *J. Mol. Biol.*, 189, 113, 1986.

57. **Ziman, M., O'Brien, J. M., Ouellette, L. A., Church, W. R., and Johnson, D. I.**, Mutational analysis of *CDC42Sc*, a *Saccharomyces cerevisiae* gene that encodes a putative GTP-binding protein involved in the control of cell polarity, *Mol. Cell. Biol.*, 11, 3537, 1991.

58. **Ohya, Y., Goebl, M., Goodman, L. E., Petersen-Bjorn, S., Friesen, J. D., Tamanoi, F., and Anraku, Y.,** Yeast *CAL1* is a structural and functional homologue to the *DPR1 (RAM)* gene involved in *ras* processing, *J. Biol. Chem.,* 266, 12356, 1991.
59. **Powers, S., Gonzales, E., Christensen, T., Cubert, J., and Broek, D.,** Functional cloning of *BUD5,* a *CDC25*-related gene from *S. cerevisiae* that can suppress a dominant-negative *RAS2* mutant, *Cell,* 65, 1225, 1991.
60. **Jones, S., Vignais, M. L., and Broach, J. R.,** The *CDC25* protein of *Saccharomyces cerevisiae* promotes exchange of guanine nucleotides bound to Ras, *Mol. Cell. Biol.,* 11, 2647, 1991.

Chapter 16

CDC42Hs: THE HUMAN HOMOLOG OF A YEAST CELL-DIVISION CYCLE PROTEIN

Matthew J. Hart, David Leonard, Katsuhiro Shinjo, Tony Evans, and Richard A. Cerione

TABLE OF CONTENTS

0-8493-5214-2/93/$0.00 + $.50
© 1993 by CRC Press, Inc.

I. INTRODUCTION

One of the first RAS-related proteins to be identified and purified was a 22-kDa protein, designated G_p (also later called G25K). Initial isolation was based on the ability of the protein to bind radiolabeled guanine nucleotides and utilized a detergent extract of a human placental particulate fraction as a source.[1] Subsequently, the same or very similar protein was identified and purified from detergent extracts of human platelet[2] and bovine brain membranes.[3] Primary amino acid sequence obtained from peptide fragments derived from both the human placental and platelet preparations demonstrated the existence of highly conserved regions implicated in GTP-binding/GTP hydrolysis and allowed for the generation of highly specific antipeptide sera.[4] The availability of primary sequence also allowed for the cloning of the cDNA's encoding this RAS-related protein.[5,6] Here we will focus on recent studies suggesting a link between this protein and signaling pathways activated by receptor tyrosine kinases and also the information currently available on the protein-protein interactions that serve to regulate their GDP-binding/GTPase cycle.

II. RECONSTITUTION STUDIES WITH THE EGF RECEPTOR/TYROSINE KINASE

A. IDENTIFICATION OF A 22-kDa GTP-BINDING PROTEIN AS A PHOSPHOSUBSTRATE FOR THE EGF RECEPTOR IN RECONSTITUTED PHOSPHOLIPID VESICLE SYSTEMS

A possible involvement of the G_p (G25K) protein in growth factor action was suggested from reconstitution experiments performed with the epidermal growth factor (EGF) receptor.[7] These studies were prompted by the question of whether GTP-binding proteins could serve as phosphosubstrates for the EGF receptor tyrosine kinase. Based on these studies, it appeared that none of the well-characterized heterotrimeric GTP-binding proteins served as effective phosphosubstrates, with only the α-subunits of the bovine brain G_i (mainly G_{i1} and G_{i2}) and G_o proteins showing a measurable, EGF-stimulated phosphorylation ($\ll 1$ mol $^{32}P_i$ incorporated per mol of G protein). However, it was during the course of these studies that a low molecular mass protein ($M_r \sim 22,000$), which was sometimes present in relatively low amounts in the bovine brain G_i and G_o preparations, was shown to be an excellent phosphosubstrate for the EGF receptor in the reconstituted phospholipid vesicle systems. This phosphorylation was found to be specifically and effectively attenuated (by 60 to 95%) by the addition of guanine nucleotides, over a nanomolar to micromolar concentration range. This attenuation suggested that the phosphosubstrate may be a high affinity guanine nucleotide-binding protein; given the apparent molecular weight of this protein, it then seemed likely that it may be a member of the superfamily of *ras*-related GTP-binding proteins (which have molecular weight values ranging from 20 to 28,000).

The 22-kDa phosphosubstrate was purified following its solubilization from bovine brain membranes using a series of chromatographic steps which included DEAE-Sepharose, Ultrogel AcA34, phenyl-Sepharose, hydroxyapatite, and FPLC/ Mono-Q chromatography. The guanine nucleotide-dependent, EGF-stimulated phosphorylation reaction was used as an assay for the 22-kDa protein during these purification steps. When the highly purified 22-kDa protein was coinserted into phosphatidylcholine vesicles with the purified, human placental EGF receptor, a highly effective, EGF-stimulated phosphorylation of the 22-kDa protein was observed (Figure 1A) with stoichiometries of $^{32}P_i$ incorporation approaching 2 mol$^{32}P_i$ per mol [^{35}S]GTPγS binding activity. Phosphoamino acid analyses showed that the phosphorylation of the 22-kDa protein in these reconstituted systems occurred exclusively on tyrosine residues, as was also the case for the EGF receptor autophosphorylation reaction (Figure 1B).

B. CHARACTERIZATION OF THE BOVINE BRAIN 22-kDa PHOSPHOSUBSTRATE

The levels of EGF-stimulated phosphorylation of the 22-kDa protein are the highest that have been observed with any tyrosine kinase and GTP-binding protein. It also compares well with lipocortin,[8] which is typically phosphorylated at levels below 1 mol $^{32}P_i$ per mol of protein and with the phospholipase C-γ,[9] which serves as a phosphosubstrate for the EGF receptor in intact cells, and which is phosphorylated in vitro with a stoichiometry of ~0.5 mol $^{32}P_i$ per mol enzyme. The EGF-stimulated phosphorylation of the brain 22-kDa protein could not be mimicked by the insulin receptor tyrosine kinase nor by the src tyrosine kinase. However, recombinant forms of the neu tyrosine kinase, another member of the EGF receptor tyrosine kinase family, elicits an effective phosphorylation of the brain 22-kDa protein.

The stoichiometric phosphorylation of the 22-kDa GTP-binding protein was observed under conditions where the amount of the 22-kDa protein in the phospholipid vesicles exceeded the amount of EGF receptor by as much as six-fold, suggesting that the receptor can act catalytically to elicit the phosphorylation of several GTP-binding proteins. This in turn implies that once the phosphorylation event has occurred, the receptor dissociates from the 22-kDa protein.

The EGF-stimulated phosphorylation of the 22-kDa protein was absolutely dependent on the insertion of this protein into phospholipid vesicles. At the present time, it is not known whether this reflects a specific role for phospholipids in influencing the orientation of the 22-kDa protein relative to the tyrosine kinase domain of the EGF receptor or if such results simply reflect the inhibitory actions of detergents, either in masking potential phosphorylation sites on the GTP-binding protein or actually preventing the coupling of the GTP-binding protein to the EGF receptor. The fact that the addition of recombinant forms of the tyrosine kinase domain of the EGF receptor to phospholipid vesicles containing the 22-kDa protein yields low extents of phosphorylation of the 22-kDa protein suggests that the proper orientation of the EGF holoreceptor and the GTP-binding

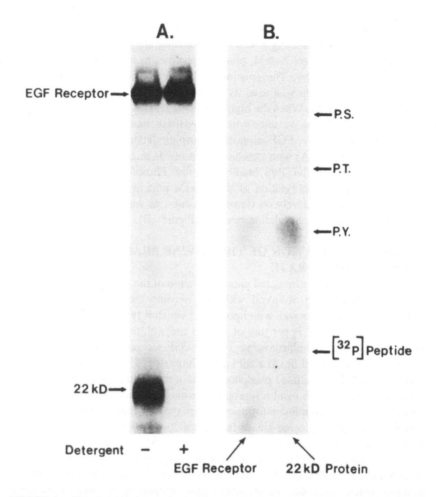

FIGURE 1. The phosphorylation of a 22-kDa protein from bovine brain by the EGF receptor. (A) The human placental EGF receptor and a bovine brain 22-kDa GTP-binding protein were purified and reconstituted into phosphatidylcholine vesicles as described in Reference 7. Aliquots from the reconstituted lipid vesicles were added to phosphorylation incubations in the absence (lane 1) and the presence (lane 2) of 0.4% CHAPS for 15 minutes at room temperature. (B) Demonstration that the EGF-stimulated phosphorylation of the brain 22-kDa protein occurs on tyrosine residues. The purified human placental EGF receptor and the bovine brain 22-kDa protein were added to a phosphorylation incubation for 50 min at room temperature. The EGF receptor and the 22-kDa GTP-binding protein were excised from a polyacrylamide gel, hydrolyzed, and subjected to amino acid analyses as outlined in Reference 7. (From Hart, M. J. et al., *J. Biol. Chem.*, 265, 5990, 1990. With permission.)

protein in the lipid vesicles may be an important factor in the phosphorylation reaction.

The inhibition of the EGF-stimulated phosphorylation of the 22-kDa protein by guanine nucleotides occurred over an essentially identical concentration range as that observed in direct binding studies, confirming the suggestion that the guanine nucleotide-occupied state of the 22-kDa protein was very poorly phosphorylated. Since both GTP and GTPγS inhibited the EGF-stimulated phosphorylation of the brain 22-kDa protein, it seemed possible that the activation of the GTP-binding protein might uncouple it from the EGF receptor. This then would be analogous to the situation with hormone receptors and heterotrimeric G proteins where the activation of the G protein leads to the release of the transducer from the receptor. However, nanomolar concentrations of GDP were also found to inhibit the EGF-stimulated phosphorylation, thereby, suggesting that the inhibition of the phosphorylation is a general effect arising from guanine nucleotide occupancy.

C. IDENTIFICATION OF THE BRAIN 22-kDa PHOSPHOSUBSTRATE AS A FORM OF THE HUMAN PLATELET/HUMAN PLACENTAL G_p (G25K) PROTEIN

Highly purified preparations of the 22-kDa phosphosubstrate from bovine brain membranes were found to crossreact with two different peptide-specific antibodies that were raised against sequences derived from human placental and human platelet Gp (G25K) proteins.[4] One of these peptide-specific antibodies appeared to specifically immunoprecipitate the phosphorylated brain 22-kDa protein following its solubilization from phospholipid vesicles. These peptide antibodies have been shown to be highly specific for the G_p (G25K) proteins and have not been found to react with any other identified member of the superfamily of *ras*-related GTP-binding proteins. In addition to this crossreactivity, the reconstitution of the purified human platelet G_p (G25K) protein with the purified EGF receptor resulted in an EGF-stimulated phosphorylation of the platelet GTP-binding protein with all of the same characteristics as observed with the brain 22-kDa GTP-binding protein. Thus, taken together, these results strongly suggested that the brain 22-kDa phosphosubstrate and the human placental and human platelet GTP-binding proteins, originally designated G_p or G25K, were one and the same (or at least isoforms).

III. IDENTIFICATION OF THE G_p (G25K) PROTEIN AS THE MAMMALIAN HOMOLOG OF THE YEAST CELL-DIVISION CYCLE PROTEIN CDC42

A. MOLECULAR CLONING

We felt that it would be important to obtain a cDNA for the G_p (G25K) protein in order to use gene transfer approaches as a means to further examine the relationship between the EGF receptor tyrosine kinase and the G_p (G25K) protein, as well as to obtain insight into the physiological role of this GTP-binding protein. The molecular cloning of two cDNAs for the G_p (G25K) protein was accomplished from λgt11 cDNA libraries from human placenta[6] and human fetal brain,[5] using two oligonucleotide probes that were prepared from the available peptide sequences for the human placenta/platelet G_p (G25K) protein.[4] The two cDNAs predicted amino acid sequences that were 95% identical and differed primarily in 8 of the carboxyl terminal 25 amino acids (Figure 2, lines 1 and 2). The predicted amino acid sequences were ~30% identical to the sequences for the *ras* proteins, ~50% identical to the sequences of the *rho* proteins, and ~70% identical to the *rac1* and *rac2* GTP-binding proteins. The two forms of the GTP-binding proteins contain three amino acids which when individually mutated in *ras*, result in oncogenic transformation, i.e., the glycine at position 12, the alanine at position 59, and the glutamine at position 61. These GTP-binding proteins also contain a CAAX box and recently it has been shown that the carboxyl terminal cysteine residue undergoes geranylgeranylation as well as carboxymethylation (see below). A potentially important difference between these GTP-binding proteins and most other members of the family (including *ras* and the heterotrimeric G proteins) is that they contain the sequence TQID (residues 115 to 118) instead of the consensus sequence NKXD. This suggests that the G_p (G25K) proteins may have different intrinsic guanine nucleotide-binding or GTPase activities compared to most other members of the family; interestingly, some differences have been observed in GTPase experiments (see below).

Certainly, one of the most interesting outcomes of the cloning of the cDNAs for the G_p (G25K) proteins was the finding that their predicted amino acid sequences were 80% identical to the yeast cell-division cycle protein, CDC42, which is essential for yeast cell growth and for the establishment of the proper budding sites[10] (also, see Chapter 15). The homology extends throughout most of the reading frames of the human and yeast proteins and includes the TQID sequence. In addition, both human cDNAs (i.e., the human placental and fetal brain clones) will completely complement the temperature-sensitive mutation in *Saccharomyces cerevisiae,* whereas, none of the other structurally related GTP-binding proteins (e.g., *rac1* and *rac2*) show complementation. This may be a reflection of the high degree of sequence identity between the yeast and human proteins between residues 30 to 42, i.e., the putative effector region for the *ras*-related GTP-binding proteins.[11] Thus far, we have found the GTP-binding protein that is coded for by the human placental cDNA to be ubiquitous, whereas, the message which codes for the fetal brain GTP-binding protein apparently is present in low abundance and is not detectable by PCR in most sources (including human placenta) that we have examined with the exception of bovine retina. In order to avoid any confusion regarding either the molecular size of these GTP-binding proteins (as might be implied by the name G25K) or their physiological function (i.e., the name G_p implies a role in phosphatidylinositol lipid turnover), we now refer to these proteins, collectively, as CDC42 (and refer to the human proteins as CDC42Hs).

B. cDNA TRANSFECTION STUDIES

As a first step toward determining whether the CDC42Hs protein has an effect on mammalian cell growth, we have used gene transfer approaches to

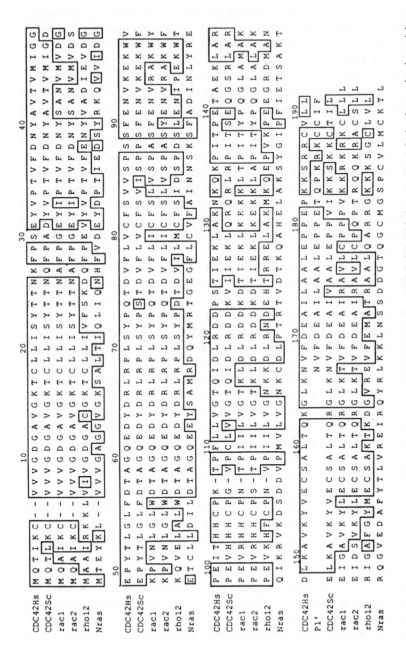

FIGURE 2. Comparison of the amino acid sequence of the human placental CDC42Hs with the sequences of other low molecular weight GTP-binding proteins. The sequence G_p1' represents the carboxyl terminal of the human fetal brain GTP-binding protein which was originally designated G25K.[5] Residues identical to those of the CDC42Hs protein are boxed. (From Shinjo, K. et al., *Proc. Natl. Acad. Sci. U.S.A.*, 87, 9853, 1990.)

introduce the cDNAs for the wild-type CDC42Hs protein (with a glycine residue at position 12, designated CDC42Hs^{gly12}), and a GTPase-defective form of CDC42Hs (which contains a valine residue at position 12, designated CDC42Hs^{val12}) into various fibroblasts (Shinjo et al, in preparation). The results of these studies indicated that neither the expression of the wild-type CDC42Hs nor the CDC42Hs^{val12} species in NIH 3T3 cells, nor the expression of these CDC42Hs proteins in NR6 fibroblasts which lack the EGF receptor tyrosine kinase, resulted in a stimulation of the rate of growth of these cells or in an enhanced tendency toward focus formation. In the NR6 cells, a slight inhibitory effect on the rate of cell growth accompanied the expression of the GTPase-defective CDC42Hs^{val12}. In addition, the expression of the CDC42Hs^{val12} protein in NR6 cells, and in some cases, the overexpression of the wild-type CDC42Hs protein was accompanied by a multinucleate phenotype. This phenotype may be related to a similar (multinucleate) phenotype observed in *S. cerevisiae* by Johnson and colleagues upon expression of various GTPase-defective forms of CDC42Sc[12] (also see Chapter 15) and to multinucleate phenotypes observed in NIH 3T3 cells upon the overexpression of the *dbl* oncogene product[13] (see Section IV.B, below).

Thus, taken together, the results obtained upon the expression of the different CDC42Hs species in NIH 3T3 or NR6 fibroblasts did not mimic the actions of the *ras* oncogene products. However, some interesting results were obtained upon the overexpression of the different CDC42Hs cDNAs into NR6 fibroblasts which also expressed the EGF receptor (20,000 to 50,000 receptors per cell) or the EGF receptor/erbB-2 chimera (200,000 to 500,000 receptors per cell). Specifically, in these cases the expression of the CDC42Hs^{val12} species, or the overexpression of the CDC42Hs^{gly12} (wild-type) species, was accompanied by inhibitions of the rate of growth of these fibroblasts. The most dramatic results were obtained upon the expression of the CDC42Hs^{val12} species in NR6 fibroblasts containing the EGF receptor/*erb B-2* chimera (which is comprised of the extracellular domain of the EGF receptor and the membrane-spanning helix and tyrosine kinase domain of *erb B-2*). In these cases, there was a striking EGF-dependent inhibition of the rate of cell growth (relative to the growth of NR6-EGFR/*erb B-2* cells which were transfected with the control pZIP-neo vector, i.e., which lacks the cDNA for CDC42Hs^{val12}). These EGF-dependent inhibitory effects were accompanied by striking changes in cell morphology; specifically, the fibroblasts rounded up, began to lift off from the plates, and actually formed clumps of cells.

Overall, the results of these gene transfer experiments implied some type of interaction between the signaling pathways of mitogenic tyrosine kinases like the EGF receptor and *erb B-2*, and that of the CDC42Hs protein; however, the ultimate outcome of these interactions was a negative effect on mitogenesis. Although the detailed molecular mechanisms underlying these negative-growth signals still remain to be delineated, the results from the transfection studies appear to be consistent with the need for the CDC42Hs protein to go through a complete GTP-binding/GTPase cycle in order to trigger (or at least contribute to) a normal growth signal. Thus, it not only is important for CDC42Hs to bind GTP (which is the primary activation step for the *ras* proteins) but also to ultimately hydrolyze the GTP to GDP. This hydrolytic event may be necessary to release an effector protein from the GTP-bound CDC42Hs species, thus the effector can then contribute to a normal (positive) growth signal. Given the apparent importance of the proper regulation of the GTP-binding/GTPase cycle of CDC42Hs, we set out to identify mammalian proteins that participate in this regulation. These regulatory factors, one of which appears to be influenced by the state of isoprenylation of the CDC42Hs protein, will be described in more detail below.

IV. REGULATION OF THE CDC42 PROTEINS

A. ISOPRENYLATION

The CDC42Hs protein contains the signature -Cys-Xaa-Xaa-Xaa sequence at its carboxyl terminus for isoprenylation. Yamane and Fung first demonstrated that the CDC42 protein purified from bovine brain membranes is carboxyme-thylated.[14] When the brain CDC42 protein was subjected to limited trypsin treatment, an ~1 kDa carboxymethylated protein fragment was removed. Reconstitution of the cleaved protein fragment with stripped bovine brain membranes resulted in an association of the carboxymethylated fragment, but not the CDC42 protein which lacks this fragment, with the membranes. These results then indicated that the site of carboxymethylation and the region responsible for membrane anchoring exists at the carboxyl terminus.

Yamane and his colleagues[15] subsequently demonstrated that the brain CDC42 protein is modified by geranylgeranylation, similar to the case for the *S. cerevisiae* GTP-binding protein[16] (also see Chapter 15). Specifically, the prenyl moiety is covalently attached to the cysteine sulfhydryl group via a thioether linkage; the three carboxyl terminal residues are then proteolytically removed and the terminal geranylgeranylated cysteine is carboxymethylated. While these modifications are clearly felt to be critical for the proper membrane-association of the mammalian (and presumably the human) CDC42 protein, it also seems likely that these modifications influence the interactions of other regulatory proteins with CDC42 (see below).

B. CDC42Hs-REGULATORY PROTEINS
1. CDC42Hs-GAP

Since the initial discovery of a distinct protein that stimulates the GTPase activity of *ras*,[17,18] a great amount of effort has been directed toward the biochemical characterization of various members of the family of GTPase-activating proteins (GAPs). Since neither the *ras*- nor *rap*-GAPs[19] were capable of stimulating the GTPase activity of CDC42Hs, we set out to identify a distinct GAP for this GTP-binding protein. After screening a number of tissue extracts,

CDC42Hs-GAP activity was found both in the cytosolic and membrane fractions of human platelets.[20] The cytosolic GAP was not sufficiently stable to allow further purification. However, the membrane-associated GAP was further purified following its solubilization with Triton® X-100. This involved a series of steps which included S-Sepharose, hydroxyapatite, Mono-Q, and Mono-S chromatographies. At the Mono-Q step, the peak fractions containing CDC42Hs-GAP activity appeared to correspond to a protein with an apparent molecular mass of ~25 kDa. This was further indicated by gel filtration chromatography, using CDC42Hs-GAP preparations that were purified ~4000-fold.

The Mono-Q-purified GAP specifically stimulates the initial rate of GTP hydrolysis by the wild-type form of CDC42Hs which is overexpressed in and purified from *Escherichia coli* (Figure 3A). However, the GAP has no ability to stimulate the GTPase activity of a mutant CDC42Hs protein in which a valine residue has been substituted for a glycine at position 12. These results are similar to those obtained with the *ras*-GTP-binding proteins where the GTPase activities of only the wild-type *ras* protein, and not val12-mutants of *ras*, are stimulated by the *ras*-GAP.[21,22] The CDC42Hs-GAP appears to stimulate the GTPase activities of the wild-type, *E. coli*, recombinant CDC42Hs, the wild-type, recombinant CDC42Hs isolated from *S. frugiperda* cells (following baculovirus infection), and the native CDC42Hs purified from platelets to essential identical extents (i.e., typically a 5- to 10-fold stimulation of the initial rate of GTP hydrolysis); (Figure 3B). These results then imply that the isoprenylation (geranylgeranylation) of CDC42Hs is not prerequisite for the functional coupling of this GTP-binding protein to its GAP.

The ability of the platelet CDC42Hs-GAP to stimulate the GTPase activity of other *ras*-related GTP-binding proteins has been investigated. The CDC42Hs-GAP shows no stimulatory activity for the *ras* or *rap1A* GTP-binding proteins. However, the CDC42Hs-GAP is able to stimulate the GTPase activities of the *rac 1* and *rac 2* proteins, essentially to the same extent as CDC42Hs (i.e., a 5- to 10-fold stimulation of the initial rate of GTP hydrolysis). The CDC42Hs-GAP also shows a measurable stimulation of the GTPase activity for the mammalian *rho*A protein (~2-fold under conditions where it stimulates the GTPase activity of CDC42Hs by greater than 5-fold). Overall, these results suggest that the human platelet CDC42Hs-GAP is capable of varying degrees of crossreactivity with other members of the *rho* subgroup of the ras superfamily, but is likely to be ineffective with members outside of this subgroup.

Both the apparent molecular masses of the platelet-membrane-associated CDC42Hs-GAP (~25 kDa) and the *rho*-GAPs (~29 kDa) purified from the cytosol of spleen[23] and bovine adrenal glands,[24] coupled with the crossreactivity observed with these GAPs and the *rho* and CDC42Hs proteins, suggested that these two GAPs were likely to have highly similar structures. In fact, comparisons of the existing amino terminal sequences for these GAPs have reinforced this notion, i.e., their amino terminal sequences appear to be nearly identical. Hall and his colleagues (cf. Diekman et al.[25]) found that some other interesting proteins

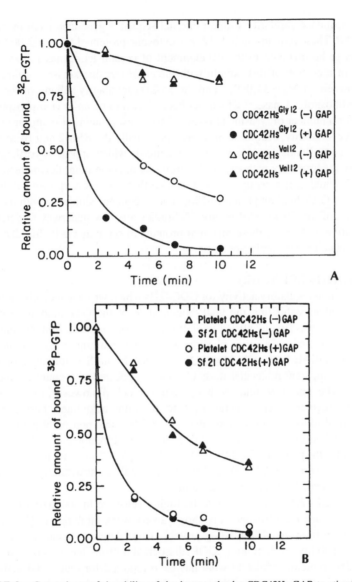

FIGURE 3. Comparisons of the ability of the human platelet CDC42Hs-GAP to stimulate the GTPase activities of the *E. coli* expressed CDC42Hs, the *S. frugiperda* expressed CDC42Hs, and the human platelet CDC42Hs. (A) The CDC42Hs protein with a glycine residue at position 12 (○,●), and a mutated protein with a valine residue at position 12 (△,▲), were expressed and purified from *E. coli* as described in Hart et al.[20] The different CDC42Hs proteins (2 μg) were preloaded with [γ^{32}P]GTP and then added to assay incubations containing the CDC42Hs-GAP (●,▲), or control buffer (○,△). The GAP activity was assayed for the indicated time periods at room temperature and was measured as the loss of the bound radioactivity (i.e., ^{32}P$_i$ release due to GTP hydrolysis) from CDC42Hs. (B) The CDC42Hs GTP-binding protein was purified from human platelet membranes (△,○) or expressed and purified from *S. frugiperda* cells (▲,●) as described by Hart et al.[20] The different CDC42Hs proteins (2 μg) were preloaded with [γ^{32}P]GTP and added to assay incubations containing the CDC42Hs GAP (○,●) or control buffer (△,▲) as described in A. (From Hart, M. J. et al., *J. Biol. Chem.*, 266, 20840, 1991. With permission.)

shared sequence similarity with an internal sequence generated from the spleen *rho*-GAP. These proteins included *bcr*, a cytosolic protein of M_r ~160,000 which appears to be involved in the development of certain leukemias following the fusion of a portion of this protein with the c-*abl* tyrosine kinase,[26] chimerin, a brain protein of M_r ~34,000,[27] and the 85-kDa regulatory subunit of the phosphatidylinositol-3 kinase,[28] which has been shown to be an *in vivo* phosphosubstrate for a number of tyrosine kinases.[29] Two yeast proteins also appear to share sequence similarity with these proteins, specifically the BEM2 and BEM3 gene products (A. Bender, personal communication), which appear to participate in the bud-site assembly pathway in *S. cerevisiae* at some point downstream from CDC42. Finally, it recently has been shown that the carboxyl terminal domain of the *ras*-GAP-binding protein, p190, shares sequence similarity with the other members of this "*rho*-GAP family."[30] Studies are now underway to determine the relative abilities of these different proteins to serve as GAPs for CDC42Hs or for related GTP-binding proteins.

2. CDC42Hs-GDI Activity

A second regulatory activity for CDC42Hs which we have recently identified is involved in stabilizing the GDP-bound state of the GTP-binding protein.[31] This activity was first identified in dialyzed cytosolic extracts from bovine brain and then purified by a series of steps which included ammonium sulfate fractionation, DEAE-Sephacel, Mono-Q, and then Mono-S chromatographies. A protein of apparent molecular mass of 28 kDa appeared to account for the GDI activity. The amino terminus of this protein is blocked; however, four cyanogen bromide-fragments were generated (10 to 20 amino acids) and these appear to be identical to four regions within the carboxyl terminal portion of the *rho*-GDI, first identified by Ueda et al.[32] and then cloned by Fukumoto et al.[33] Thus, like the case for the platelet CDC42Hs-GAP and the spleen *rho*-GAP, this may reflect a situation where the CDC42Hs and *rho* proteins utilize a structurally very similar, if not identical, GDI regulatory protein.

The CDC42Hs-GDI appears to be highly selective in stabilizing the GDP-bound state of CDC42Hs and only shows a very weak inhibition of [^{35}S]GTPγS dissociation from the GTP-binding protein. This GDI also appears to have little or no capability of inhibiting GDP dissociation from the *E. coli*-expressed CDC42Hs protein, similar to what has been reported for other GDI activities. These results then suggest that the isoprenylation of the carboxyl terminal cysteine of the *ras*-related GTP-binding proteins is essential for proper coupling to the GDI protein. It is interesting that we find that the GDI will effectively counteract the ability of another regulatory protein (i.e., the exchange factor or GDS protein) to catalyze GDP-GTP exchange on CDC42Hs. This does not appear to be the outcome of a competition between the GDI and GDS proteins for CDC42Hs, i.e., apparently these two regulatory proteins bind to distinct sites on the GTP-binding protein. Thus, it is likely that some type of regulatory event (e.g., phosphorylation) is necessary to ultimately promote the release of the GDI from CDC42Hs and enable the GDS to functionally couple to the GTP-binding protein.

In addition to stabilizing the GDP-bound state of CDC42Hs (and opposing the actions of the GDS protein), the GDI also elicits the dissociation of CDC42Hs from membranes. The mechanism by which the GDI triggers the release of the GTP-binding protein from membrane fractions is not known. This effect by the purified GDI is titratable and appears to occur equally well under conditions where CDC42Hs is in a GTPγS-bound state, as well as when CDC42Hs is in a GDP-bound state. These results then suggest that the GDI may be capable of coupling to both the GDP- and GTP-bound states of CDC42Hs. Therefore, it will be of interest to determine whether the GDI is capable of influencing other actions of the GTP-bound CDC42Hs protein, for example, the interactions between this protein and downstream effectors (or GAPs).

3. CDC42Hs-GDS Activity

A number of attempts to identify the presence of a CDC42Hs-GDS activity in various tissue extracts have not been successful; this probably reflects the fact that these extracts also contain the GDI activity, which effectively opposes CDC42Hs-GDS action. We obtained a clue to the identity of a possible GDS activity from the *S. cerevisiae* bud-site assembly pathway. Specifically, this involved the sequence similarity between the CDC24 gene product[34] (mutations of which are overcome by the expression of CDC42Sc in *S. cerevisiae*) and the *dbl* oncogene product.[13,35–37] The region of homology between CDC24 and both proto-*dbl*, and the *dbl* oncogene encompasses 238 amino acids; there is ~30% identity between CDC24 and proto-*dbl* (or *dbl*) in this region and ~70% similarity when counting conservative amino acid changes.[13] The *dbl* oncogene was first identified by Eva, Aaronson and their colleagues following the transfection of NIH 3T3 cells with the DNA from B-cell lymphomas.[35] Proto-*dbl* encodes an ~115 kDa cytosolic protein; oncogenic activation occurs as an outcome of a recombination event in which the 5′ terminal half of proto-*dbl* is replaced with some unidentified human sequences, yielding an ~66 kDa oncogenic protein. Both proto-*dbl* and *dbl* appear to be associated with the cytoskeletal matrix.[38] This, then, has led to the suggestion that the *dbl* oncogene product (which is not structurally similar to any membrane-associated or nuclear oncogene products) may be representative of a new class of oncogenic proteins that are cytoskeletal-based. Given the sequence similarity between *dbl* and CDC24, it was of interest to examine the effects of *dbl* on the GTP-binding/GTPase cycle of CDC42Hs.

Using *S. frugiperda* lysates expressing the *dbl* oncogene product, we first showed that *dbl* was not capable of stimulating the GTPase activity of the human platelet CDC42Hs protein. However, the insect cell expressed *dbl* did show a marked effect on guanine nucleotide-binding to CDC42Hs, specifically, it strongly catalyzed GDP dissociation.[39] In the absence of *dbl*, the half-time for GDP dissociation from CDC42Hs was >20 min (at Mg^{2+} > 1 mM), however, in the presence of lysates expressing *dbl* (or after immunoprecipitation of *dbl* from these lysates), the half-time for GDP dissociation was ~2 min and complete dissociation occurred within 10 min. The effects of *dbl* appeared to be highly

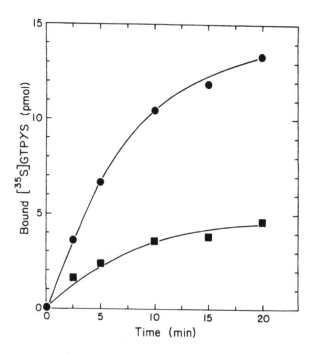

FIGURE 4. Effects of *S. frugiperda* lysates containing the 66-kDa oncogenic *dbl* on [^{35}S]GTPγS binding to the platelet CDC42Hs. The purified platelet CDC42Hs was preloaded with GDP and then ~0.3 µg of the GTP-bound CDC42Hs was added to an incubation containing 10 mM MgCl$_2$, 5 µM [^{35}S]GTPγS and 4 µl of either the control lysates from *S. frugiperda* cells (■) or lysates expressing oncogenic *dbl* (●). The binding of [^{35}S]GTPγS was measured at the indicated times. (From Hart, M. J. et al., *Nature*, 354, 311, 1991. With permission.)

specific for CDC42Hs, i.e., under conditions where *dbl* effectively catalyzed the dissociation of GDP from the human platelet CDC42Hs protein dbl showed only a weak ability to catalyze GDP dissociation from the human platelet *rac 1* protein. *dbl* was not active with either the c-H-*ras* or *rap 1a* proteins. It also exhibited a high degree of specificity with regard to the guanine nucleotide that was dissociated, specifically, *dbl* elicited only a slight dissociation of [^{35}S]GTPγS from CDC42Hs and no detectable dissociation of [γ^{32}P]GTP. Thus, it has all the appropriate characteristics for a GDS protein and in fact, it will clearly stimulate the binding of radiolabeled GTP analogs to CDC42Hs proteins that have been preloaded with (cold) GDP (Figure 4). Like the GAP activity, *dbl* can functionally couple to the human platelet CDC42Hs, the insect cell expressed CDC42Hs, and the *E. coli* expressed CDC42Hs. This suggests that the GDS protein does not require isoprenylation of CDC42Hs in order for functional coupling to occur.

An important question concerns the tissue distribution of *dbl*. It appears to be present mainly in brain, adrenal glands, and gonads,[36] whereas, CDC42Hs appears to be ubiquitously distributed. It, therefore, is of interest to determine whether other *dbl*-related molecules are capable of regulating the guanine nucleotide exchange activity of the CDC42Hs protein. There are two particularly interesting candidates. One is the *bcr* protein, which like *dbl*, contains a region of 200 amino acids that are homologous to a stretch of amino acids of CDC24.[13] The other is the recently identified *vav* oncogene product[40] which contains the CDC24/*dbl* homology region together with some other potentially interesting regulatory domains.[41] Studies are now in progress to determine whether either the *bcr* or *vav* protein is capable of stimulating nucleotide exchange in CDC42Hs or other related proteins.

REFERENCES

1. Evans, T., Brown, M. L., Fraser, E. D., and Northup, J. K., *J. Biol. Chem.*, 261, 7052, 1986.
2. Polakis, P. G., Weber, R. F., Nevins, B., Didsbury, J. R., Evans, T., and Snyderman, R., *J. Biol. Chem.*, 264, 16383, 1989.
3. Waldo, G. L., Evans, T., Fraser, E. D., Northup, J. K., Martin, M. W., and Harden, T. K., *Biochem. J.*, 246, 431, 1987.
4. Polakis, P. G., Snyderman, R., and Evans, T., *Biochem. Biophys. Res. Commun.*, 160, 25, 1989.
5. Munimetsu, S., Innis, M. A., Clark, R., McCormick, F., Ulrich, A., and Polakis, P., *Mol. Cell. Biol.*, 10, 5977, 1990.
6. Shinjo, K., Koland, J. G., Hart, M. J., Narasimhan, V., Johnson, D. I., Evans, T., and Cerione, R. A., *Proc. Natl. Acad. Sci. U.S.A.*, 87, 9853, 1990.
7. Hart, M. J., Polakis, P. G., Evans, T., and Cerione, R. A., *J. Biol. Chem.*, 265, 5990, 1990.
8. Fava, R. A. and Cohen, S., *J. Biol. Chem.*, 259, 2636, 1984.
9. Wahl, M. I., Nishibe, S., Suh, P.-G., Rhee, S. G., and Carpenter, G., *Proc. Natl. Acad. Sci. U.S.A.*, 86, 1568, 1989.
10. Johnson, D. I. and Pringle, J. R., *J. Cell Biol.*, 111, 143, 1990.
11. Sigal, I. S., Gibbs, J. B., D'Alonzo, J. S., and Scolnick, E. M., *Proc. Natl. Acad. Sci. U.S.A.*, 83, 4725, 1986.
12. Ziman, M., O'Brien, J. M., and Johnson, D. I., *Mol. Cell Biol.*, 11, 3537, 1991.
13. Ron, D., Zannini, M., Lewis, M., Wickner, R. B., Hunt, L. T., Graziani, G., Tronick, S. R., Aaronson, S. A., and Eva, A., *New Biol.*, 3, 372, 1991.
14. Yamane, H. K. and Fung, B. K.-K., *J. Biol. Chem.*, 264, 20100, 1989.
15. Yamane, H. K., Farnsworth, C. C., Xie, H., Evans, T., Howald, W. N., Gelb, M. H., Glomset, J. A., Clarke, S., and Fung, B. K.-K., *Proc. Natl. Acad. Sci. U.S.A.*, 88, 286, 1991.
16. Finegold, A. A., Johnson, D. I., Farnsworth, C. C., Gelb, M. H., Judd, S. R., Glomset, J. A., and Tamanoi, F., *Proc. Natl. Acad. Sci. U.S.A.*, 88, 4448, 1991.
17. McCormick, F., *Cell*, 56, 5, 1989.
18. Hall, A., *Cell*, 61, 921, 1990.
19. Polakis, P. G., Rubinfeld, B., Evans, T., and McCormick, F., *Proc. Natl. Acad. Sci. U.S.A.*, 88, 239, 1991.
20. Hart, M. J., Shinjo, K., Hall, A., Evans, T., and Cerione, R. A., *J. Biol. Chem.*, 266, 20840, 1991.
21. Trahey, M. and McCormick, F., *Science*, 238, 542, 1987.

22. Gibbs, J. B., Schaber, M. D., Allard, W. J., Sigal, I. S., and Scolnick, E. M., *Proc. Natl. Acad. Sci. U.S.A.*, 81, 5704, 1988.

23. Garrett, M. D., Self, A. J., van Oers, C., and Hall, A., *J. Biol. Chem.*, 264, 10, 1989.

24. Morii, N., Kawano, K., Sekine, A., Yamada, T., and Narumiya, S., *J. Biol. Chem.*, 266, 7646, 1991.

25. Diekmann, D., Brill, S., Garrett, M. D., Totty, N., Hsuan, J., Monfries, C., Hall, C., Lim, L., and Hall, A., *Nature*, 351, 400, 1991.

26. Heisterkamp, N., Stam, K., Groffen, J., DeKline, A., and Grasveld, G., *Nature*, 315, 758, 1985.

27. Hall, C., Monfries, C., Smith, P., Lim, H. H., Kozma, R., Ahmed, S., Vanniasingham, V., Leung, T., and Lim, L., *J. Mol. Biol.*, 211, 11, 1990.

28. Otsu, M., Hiles, I., Gout, I., Fry, M. J., Ruiz-Larrea, F., Panayotou, G., Thompson, A., Dhand, R., Hsuan, J., Totty, N., Smith, A. D., Morgan, S. J., Courtneidge, S. A., Parker, P. J., and Waterfield, M. D., *Cell*, 65, 91, 1991.

29. Cantley, L. C., Auger, K. R., Carpenter, C., Duckworth, B., Graziani, A., Kapeller, R., and Soltoff, S., *Cell*, 64, 281, 1991.

30. Settleman, J., Narasimhan, V., Foster, L. C., and Weinberg, R. A., *Cell*, 69, 539, 1992.

31. Leonard, D., Hart, M. J., Eva, A., Henzel, W., Evans, T., and Cerione, R. A., *J. Biol. Chem.*, 267, 22860, 1992.

32. Ueda, T., Kikuchi, A., Ohga, N., Yamamoto, J., and Takai, Y., *J. Biol. Chem.*, 265, 9373, 1990.

33. Fukumoto, Y., Kaibuchi, K., Hori, Y., Fujioka, H., Araki, S., Ueda, T., Kikuchi, A., and Takai, Y., *Oncogene*, 5, 1321, 1990.

34. Sloat, B. F., Adams, A., and Pringle, J. R., *J. Cell Biol.*, 89, 395, 1981.

35. Eva, A. and Aaronson, S. A., *Nature*, 316, 273, 1985.

36. Ron, D., Tronick, S. R., Aaronson, S. A., and Evan, S. A., *EMBO J.*, 7, 2465, 1988.

37. Eva, A., Vecchio, G., Rao, C. D., Tronick, S. R., and Aaronson, S. A., *Proc. Natl. Acad. Sci. U.S.A.*, 85, 2061, 1988.

38. Graziani, G., Ron, D., Eva, A., and Srivastava, S. K., *Oncogene*, 4, 823, 1989.

39. Hart, M. J., Eva, A., Evans, T., Aaronson, S. A., and Cerione, R. A., *Nature*, 354, 311, 1991.

40. Katzav, S., Martin-Zanca, D., and Barbacid, M., *EMBO J.*, 8, 2283, 1989.

41. Adams, J. M., Houston, H., Allen, J., Lints, T., and Harvey, R., *Oncogene*, 7, 611, 1992.

Chapter 17

IDENTIFICATION OF NOVEL *ras* FAMILY GENES IN A HUMAN TERATOCARCINOMA CELL LINE BY OLIGONUCLEOTIDE SCREENING

George T. Drivas, Mark G. Rush, and Peter D'Eustachio

TABLE OF CONTENTS

0-8493-5214-2/93/$0.00 + $.50

© 1993 by CRC Press, Inc.

I. INTRODUCTION

Guanine nucleotide-binding and hydrolyzing proteins have been the subject of intense investigation in recent years. We now recognize their fundamental importance in normal and aberrant cellular physiology, and the diversity of functions that they have assumed in eukaryotic cells. These regulatory proteins all share a common mechanism based on the nature of the bound nucleotide and the rates of nucleotide hydrolysis and exchange. When bound to GTP these proteins interact with one set of cellular components, yet interact with another set in their GDP-bound forms. In most cases, the release of GDP is slow and is catalyzed by other protein factors. Certain classes of GTPases with slow intrinsic rates of hydrolysis also interact with catalyzing cofactors which greatly accelerate their return to a conformationally "inactive" form. Eukaryotic cells have repeatedly employed this switching mechanism to regulate such diverse cellular functions as protein synthesis, vesicle targeting, signal transduction, cellular proliferation, and differentiation.[1-4]

In the last decade we have witnessed a tremendous increase in the number of identified guanine nucleotide binding and hydrolyzing proteins. Perhaps the most rapidly expanding class is that of the ras family. This large gene family codes for various small GTPases (180 to 220 amino acids) with common amino acid sequences and overall structures that suggest evolution from a prototypical ancestral gene. Comparison of these protein sequences and examination by parsimony analysis both indicate that the evolutionary history of this gene family is complex, likely involving the accumulation of nucleotide substitutions, deletions, and insertions, as well as both ancient and more recent gene duplications. However, one striking feature of the *ras* gene family is the extraordinary conservation of some of these coding sequences between species separated by over a billion years of evolution, suggesting that these proteins play important roles in basic cellular processes.

All ras family proteins share four highly conserved domains that have been shown, through both mutagenic[5] and X-ray crystallographic studies,[6-8] to be involved in the binding and hydrolysis of guanine nucleotides. Some ras family proteins contain two additional domains that mediate interactions with other cellular proteins and proper subcellular localization. On the basis of amino acid sequence homology, this gene family has been divided into four major groups termed: true *ras*, *ras*-like (about 50% homology to true *ras*), *rho* (about 35% homology to true *ras*), and YPT/RAB (about 30% homology to true *ras*). Members of the true *ras* and *ras*-like subfamilies are believed to function in signal transduction and the regulation of cell growth,[9,10] and most YPT/RAB members are thought to participate in vesicle trafficking.[11] A common function for the *rho* subfamily has not been established, but it may relate to the maintenance of cell structure.[12,13] Certain ras family proteins may play regulatory roles in other basic cellular processes (see Section III.D.)

II. SCREENING STRATEGIES FOR IDENTIFYING NEW ras FAMILY GENES

The first members of the *ypt/rab* and *rho* subfamilies to be isolated were discovered fortuitously,[14,15] whereas, several members of the *ras*-like group were identified by low stringency screening with full-length true *ras* group probes.[16,17] The conserved sequence motifs of ras family proteins have allowed the design of mixed oligonucleotide probes reactive with a broad range of ras family genes (see for example References 18 and 19). In all, the various strategies and subsequent searches for homologous sequences have been extremely fruitful, such that nearly 100 genes and interspecies homologs have now been identified. In our own attempts to define the functions of members of this gene family, we sought to isolate and characterize *ras*-related genes that are expressed in human cells. We have utilized two approaches, both based on the use of synthetic oligonucleotides, to identify novel *ras*-related genes.

In the first approach, we used a mixed oligonucleotide probe designed by Touchot et al.,[19] corresponding to the conserved ras family domain Asp-Thr-Ala-Gly-Gln-Glu (see Figure 1), a sequence that is invariant in nearly all ras family members with the exception of some *ras*-like group proteins. This resulted in the identification of four novel *ras*-related genes in a cDNA library prepared from the N-Tera2 human teratocarcinoma cell line.[20] One, TC21, is part of the *ras*-like subfamily; two, TC25 and TC10, are part of the RHO subfamily; and one, TC4, shows a distant relationship to members of the YPT/RAB group. (The relationships of the TC genes to other ras family genes is described more precisely in Section III.A.)

Detection of multigene family members by oligonucleotide probing generally requires both exceptional nucleotide conservation and the use of mixed probes. Detection of multigene family members by the polymerase chain reaction (PCR), on the other hand, allows for more flexibility in primer-target homology. In order to determine the applicability of the PCR procedure for the detection of ras family members, oligonucleotide primers complementary to the four conserved GTP-binding regions (Figure 1), designed on the basis of consensus sequences obtained from all four *ras* subfamilies, were used to amplify sequences from two human cDNA libraries. The results of these studies are shown in Figure 2. This approach was successful in identifying members of each of the four groups of the ras family. Forty of the fifty-five amplification products examined represented previously isolated ras family genes, and three others represented a novel *ras*-related gene which we termed YL8[21] (see also Section III.A).

The results obtained from these two oligonucleotide-based screening approaches indicate that the PCR complements oligonucleotide probing as a strategy for identifying members of the *ras* gene family. Of the five ras family members detected by PCR, none was detected by conventional library screening using the downstream member of primer pair 1 as a probe. Conversely, none of the four novel members detected by conventional screening using this probe (TC21,

```
                                    25                                    50
                                     |                                     |
              #   ##    ##                 #                        ######
H-RAS                 MTEYKLVVVGAGGVGKSALTIQLIQNHFVDEYDPTIEDSY-RKQVVIDGETCLLDILDTAGQEEYSAMR
TC21   MAAAAGGRLRQEK.R.....G......TC.L.SYTT.A.PG....I..VF.N.-SAN.MV..KPVN.GLW.......FG...
TC25          .QAI.C......D.A...TC.L.SYTT-A.PG....I..VF.N.-SAN.MV..KPVN.GLW.......D.DRL.
TC10   MAHCPGALMI.C....D.A-..TC.LMSYANDA.PE..V..VF.H.-AVS.TVG.KQY..GLY.......D.DRL.
TC4    MAAQGEPQVQF...L..D..T..TTFVKRHLTGE.EKK.VA.LGVEVHPLVFHTNRGPIKFNVW.......KFGGL.
YL8    MGTRDDEYDYHF.VDLI.DS....N.I.SRFTR.E.NL.SKS..GVEFATRSIQVDGKTIKAQ.W.......R.R.IT
                        >>>>>>>                                          <<<<<<
                            1                                              2

                   75                           100                          125
                    |                             |                            |
       #               #                 ##  #     #                   #
H-RAS  D-QYMRTGEGFLCVFAINNTKSFEDIHQYREQIKRVKDSDDVPMVLVGNKCDLAAR---------TVESRQAQDL
TC21   E-.........L.SVTDRG...E.YKFOR.L....R.EF..I.I..A..DHQ---------RQ.TQEG.Q.
TC25   PLS.PQ.DV-..IC.SLVSPA..NVRAKWPEV.HHCP-NT.II...T.L..RDDKDTIEKLKEKKLTPITYP.GLAM
TC10   PLS.PM.DV-..IC.SLVSPA..QNVKEEWPELKEYAP-N..FL.I.TQI..RDDPKTLARLNDMKEKPICVE.G.K.
TC4    .-G.YIQAQCAIM.DVTSRVTYKNVPNWHRDLV.CE--NI.I..C..V.IKDR---------KVKAKSIV
YL8    S-A.Y.GAV.A.I.YD.AKHLTY.NVERWLKELRDHA..-NIVIM....S..RHL---------RAVPTDEARA
       #                                                               >>>>>>>
                                                                           3

                     150                           175
                      |                             |
          ##   #
H-RAS  AR-SYGIPYIETSAKTRQGVEDAFYTLVREIRQHKLRKLNPPDESGPG-------------CMSCKCVLS
TC21   ..-QLKVT.M.A...I.MN.DQ..HE...V..KFQEQECP.SP.PTRKEKDKG-------------H..IF
TC25   .KEIGAVK.L.C..L.QR.LKTV.DEAI.AVLCPPPV.KRKR-------------.L.L
TC10   .KEIGACC.V.C..L.QK.LKTV.DEAIIA.LTP.KHTVKKRIG.R-------------CINC.LIT
TC4    FHRKKNLQ.YDI...SNYNF.KP.LW.A.KLIGDPNLEFVAMPALA.PEVVMDPALAAQYEHDLEVAQTALPDEDDDL
YL8    FAEKN.LSF.....LDSTN..A..Q.ILT..YRIVSQ.QMSDRRENDMSPSNNVVPIHVPPTTENKPKVQ---C.QNI
                  <<<<<<
                    4
```

FIGURE 1. Homologies of teratocarcinoma (TC and YL8) clones to human H-*ras*, and orientation of PCR primers with respect to RAS protein sequences. The amino acid sequences were aligned with the sequence of the human H-*ras* protein, with gaps (–) inserted to maximize homology. Other symbols: ., identities between a teratocarcinoma sequence and the H-*ras* sequence; #, residues identical among all six sequences. The four conserved domains of the ras family are underlined and numbered, and arrowhead underlining indicates the four regions used for PCR primer design. The reference numbering is that of the H-*ras* protein, and standard single letter amino acid abbreviations are used.

TC25, TC10, TC4) were detected by the PCR. These results are best understood by comparing primer and probe target site homologies.[21] From our analysis of these findings, we conclude that: (1) our data concerning PCR support the generally held view, noted previously[22,23] that poor overall primer-target homology (50 to 75%) can *sometimes* be compensated for by good 3' primer-target homology, and that relatively good overall primer-target homology can *sometimes* be obviated by poor 3' primer-target matching and (2) the PCR is much more flexible in terms of primer-target homology than is conventional screening. While only a 5% variation in probe-target homology will often differentiate positive and negative signals in conventional screening, a 25% mismatch in primer-target homology is often compatible with successful amplification. In this particular application, and quite likely in general, it appears that the PCR and oligo probing will detect distinct sets of genes when applied to analysis of gene families.

III. CHARACTERIZATION OF NOVEL TERATOCARCINOMA cDNA CLONES

A. SEQUENCE ANALYSIS

The open reading frames of the teratocarcinoma (TC) clones encode proteins of 192 to 216 amino acids with predicted molecular weights of 21,430 to 24,375 Da. Alignment of the coding regions with that of the human H-*ras* protein (Figure 1), reveals that the four conserved GTP-binding domains are present in all five TC coding sequences. Residues outside these domains, specifically Phe28 which interacts with the guanine ring, and Thr35, a residue important for coordination of the Mg^{2+} ion and interaction with one of the phosphate groups, are also strictly conserved. Other residues, such as Tyr32, Arg68, Tyr71, Pro110, and Phe156, which are conserved in most ras family genes, are present in four of the five coding sequences. (In YL8, there are substitutions at three of these five positions.) These features suggest that the proteins encoded by the TC genes will have biochemical properties similar to those of proteins encoded by ras family genes, i.e., guanine nucleotide binding and GTPase activities. Overexpression of the five TC cDNAs in monkey COS cells (accounting for 2 to 5% of the total protein in transfected cell extracts) and subsequent analysis by SDS-PAGE, confirms that all five cDNAs encode proteins in the range of 21,000 to 25,000 Da. Of the five, only YL8 shows appreciable GTP-binding activity in a GTP blot overlay assay.[24] TC21, TC10, and TC25 terminate with a CAAX motif (where C is cysteine, A is an aliphatic residue, and X is any amino acid), a feature common to many members of the ras family (see chapter on processing of *ras* proteins). Both TC21 and TC10 contain an additional cysteine residue just upstream of this domain, suggesting that they may be posttranslationally modified by both isoprenylation and palmitoylation. In the case of TC25, the region preceding this domain does not contain a proximal cysteine residue, but rather a stretch of five basic amino acids which may serve to increase the avidity of this protein for a membrane compartment.

PRIMERS:

upstream:

1)
```
                     G   G
5' GT-GGTGITGGTGGTGT-GG-AAG 3'
               T        T
```

2)
```
             T    G  G  G  G
5' T-GTGCT-GT-GG-AACAA- 3'
             G    T  T  T    A
```

downstream:
```
           G   G   G
5' TCTTCTTG-CC-GC-GTGTC 3'
           T   T   T
```
```
           T  C     T  GT
5' C-GGT-TTGGC-GAG--TTC 3'
           G        A    CA
```

AMPLIFIED SEQUENCES:

1) N-ras
TC cDNA (7)
```
5' AGCCGCACTGACAATCCAGCTAATCCAGAACCACTTTGTAGATGAATATGATCCCCATAGAGGATTCTTACAGAAAACAAGTGGTTATAGATGGTGAAACCTGTTTGTTGGACATACTG 3'
   S  A  L  T  I  Q  L  I  Q  N  H  F  V  D  E  Y  D  P  T  I  E  D  S  Y  R  K  Q  V  V  I  D  G  E  T  C  L  L  D  I  L
```

rhoA
TC cDNA (8)
```
5' ACATGCTTGCTCATAGTCTTCAGCAAGGACCAGTCCCAGAGGTGTATGTGCCCACAGTGTTTGAGAACTATGTGGCAGATATCGAGGTGGATGGAAAGCAGGTAGTGTTGGCTTTGTGTG 3'
   T  C  L  L  I  V  F  S  K  D  Q  F  P  P  E  V  Y  V  V  P  T  V  F  E  N  Y  V  V  A  D  I  E  V  D  G  K  Q  V  E  L  A  L  M
```

YL8
TC cDNA (3)
```
5' AGTAATCTCCTGTCTGATTTACTGAAATGAGTTTAATCTGGAAGCAAGAGACCACCATTGGAGTAGAGTTTGCAACAAGAAGCATCCAGGTTGATGGGAAAAACAATAAAGGCACAGATATGG 3'
   S  N  L  L  S  R  F  F  T  R  N  E  F  P  N  L  E  S  K  S  T  I  G  V  E  F  A  T  R  S  I  Q  V  D  G  K  T  I  K  A  Q  I  M
```

2) H-ras
PTL cDNA (1)
```
5' TGTGACCTGGCTGCACGCACTGTGGAATCTCGGCAGGCTCAGGACCTGCCCGAGCCATCCCCTACATC 3'
   C  D  L  A  A  R  T  V  E  S  R  Q  A  Q  D  L  A  R  S  Y  G  I  P  Y  I
```

N-ras
PTL cDNA (7)
TC cDNA (8)
```
5' TGTGATTTGCCAACAAGGACAGTTGATACAAAAGAAGCCCACGAACTGGCCAAGAGTTACGGGATTCCATTCATT 3'
   C  D  L  P  T  R  T  V  D  T  K  Q  A  H  E  L  A  K  S  Y  G  I  P  F  F  I
```

ralA
PTL cDNA (5)
TC cDNA (4)
```
5' TCAGATTTAGAAGATAAAGACAGGTTCTGTAGAAGAGGCAAAAAACAGAGCTGAGCAGTGAATGTTAACTACGTG 3'
   S  D  L  E  D  K  R  Q  V  S  V  E  E  A  K  N  R  A  E  Q  W  N  V  N  Y  V  V
```

FIGURE 2. Sequences of ras family members present in PCR-amplified fragments. Two different primer pairs, 1 (corresponding to the first two conserved domains of RAS proteins) and 2 (corresponding to the third and fourth conserved domains), were used to amplify two cDNA libraries, TC and PTL (peripheral T lymphocyte) and the resulting fragments were subcloned and sequenced. The nature of the primers are shown under Primers, and the nature of the subcloned fragments under Amplified Sequences, with those derived from primer pair 1 (N-ras, rhoA and YL8) in the top group, and those derived from primer pair 2 (H-ras, N-ras, and ralA) in the bottom. The template DNAs, TC cDNA and/or PTL cDNA are also indicated, along with the number of independent subclones of each type sequenced (numbers in parentheses). The internal nucleotide and translated amino acid sequences (not including primers) of each type are shown. Amino acid sequences are indicated by the single letter code. (From Drivas, G. T. et al., *Oncogene*, 6, 3, 1991. With permission.)

The coding sequences of the five novel genes are compared to those of representative members of each of the *ras* subfamilies in Figure 3. TC21 shares 55% homology with the human true *ras* proteins, and 65% identity with the *ras*-like R-*ras* protein. The "effector" domain of RAS proteins (amino acids 32 to 40 of H-*ras*), required for both the transforming activity of RAS proteins and their interaction with GAP, is strictly conserved in TC21. These features suggest that TC21, like R-*ras*,[25] may also respond to the action of the RAS-GAP protein.

TC25 and TC10 exhibit 60 to 70% homology with proteins of the *rho* subfamily. The coding sequence of TC25 is identical to the *rac 1* gene, isolated from an undifferentiated HL-60 cell cDNA library.[26] Certain members of this group have been shown to serve as substrates for the exoenzyme C3 of *Clostridium botulinum*.[12,26] The site of ADP ribosylation is an asparagine residue within the putative effector region of these proteins,[27] which is found in all members of this group, except TC10.

Comparison of YL8 with other RAS family proteins clearly indicates that it is a member of the YPT/RAB group, showing greatest overall homology (70%) to the *Schizosaccharomyces pombe* YPT3 protein.[28] YL8, thus, appears to be the human homolog of this essential fission yeast gene. In contrast to the other three *ras* subfamilies, members of the YPT/RAB group do not terminate with CAAX motifs, but rather with one or two cysteine residues, and this type of terminal sequence presumably is important in targeting these proteins to intracellular compartments. In this respect, YL8 and the recently reported identical canine sequence, *rab11*[29] are unusual, terminating with the sequence -Cys-Cys-Gln-Asn-Ile, which is neither a CAAX motif nor a typical YPT/RAB sequence terminus. While human YL8 may perform the same function as its fission yeast counterpart, it is also possible that sequence differences may have altered their target specificities and/or cellular locations.

TC4 shares only 25 to 30% homology with other ras family genes (Figure 3), as well as with other types of small GTP-binding proteins believed to be involved in the regulation of vesicular transport, such as ADP-ribosylation factor (ARF) and Sar1[30,31] (not shown). The homology of TC4 to ras family sequences is restricted almost entirely to those regions important for guanine nucleotide binding and GTP hydrolysis. (It should be noted that the TC4 amino acid sequence presented here differs at its carboxyl terminal end from that reported previously.[20] The correctness of the present sequence has been verified by extensive resequencing of the human cDNA clone as well as by comparison to the sequence of a mouse TC4 cDNA.) We have chosen to classify TC4 as a ras family member because it encodes all four *ras* homology domains and shares some limited overall sequence homology with certain members of the YPT/RAB group.[20] The carboxyl terminal region of TC4 gives no indication that it undergoes posttranslational modifications, and thus, TC4 may not associate with a cellular membrane compartment. Recently, we have become aware of a coding sequence of *S. pombe* that shares about 80% amino acid identity with the human TC4 protein.[32]

	HRAS	RRAS	TC21	RAP1A	HRALA	RHOA	TC25	TC10	G25	HRAB1	HRAB4	YL8	YPT3	TC4
HRAS	--													
RRAS	55	--												
TC21	55	65	--											
RAP1A	55	49	52	--										
HRALA	48	43	45	51	--									
RHOA	34	33	33	34	36	--								
TC25	36	36	33	33	34	59	--							
TC10	32	29	29	31	32	51	63	--						
G25	38	31	36	28	34	53	69	65	--					
HRAB1	35	31	28	33	35	28	27	29	28	--				
HRAB4	31	36	29	32	29	31	26	21	24	42	--			
YL8	36	32	27	31	31	31	31	25	29	48	44	--		
YPT3	35	32	32	33	34	28	33	30	27	49	41	70	--	
TC4	23	26	26	28	25	30	28	22	27	29	28	31	30	--

FIGURE 3. Amino acid sequence comparisons among ras family members. Numbers shown indicate percent identity between two sequences, as determined by the GAP program alignment.

Figure 4 shows the results of a parsimony analysis of the TC protein sequences and 10 other representative RAS family proteins. In this analysis, done with the PROTPARS program,[33] the tree topology is not determined by percent amino acid identity alone. The program insists that any change of amino acids be consistent with the genetic code, and considers only nucleotide substitutions that result in amino acid substitutions. Thus, if two proteins differ at an amino acid position, the program recognizes that this difference may have required more than one change at the nucleotide level. The branch lengths of this tree are not proportional to evolutionary time and have been drawn symmetrically. The tree presented here should be considered a grouping with a strong bias toward phylogenetic relationships, rather than a true evolutionary tree. In essence, this type of analysis of the ras family[34] is consistent with that expected on the basis of amino acid homology, classifying the 15 sequences into the four recognized subfamilies. It is worth noting that parsimony analysis also recognizes a relationship between TC4 and YPT/RAB group proteins, though the function of TC4 in eukaryotic cells may be quite different than that proposed for many members of the YPT/RAB subfamily (see Note Added in Proof at end of chapter).

B. GENOMIC ANALYSIS

A number of ras family genes (e.g., *ras, rho,* and *ypt1*) are known to be well-conserved among different eukaryotes. Sequences highly homologous to human YL8 and TC4 are present in species as evolutionarily distant as fission yeast. In order to assess the genomic organization and extent of evolutionary conservation of all the TC genes, each was used to probe Southern blots of *EcoRI* digested genomic DNA from various mammalian and one avian species. Representative blots for TC4 and TC21 are shown in Figure 5. At high stringency, sequences reactive with all five human probes were found in all species tested. In the case of TC4, the complex hybridizing pattern seen suggests that this gene may belong to a group of closely related sequences.

To further characterize these clones genetically, we looked for restriction-fragment length variants associated with each (TC4, TC10, TC25, TC21 — YL8 was not done) among inbred strains of mice. The inheritance patterns of these variants allowed the definition of eleven distinct loci, one each for TC10 and TC21, two for TC25, and seven for TC4, some of which are likely to represent murine pseudogenes. Seven of the eleven loci could be assigned to linkage groups when compared to patterns obtained for previously mapped loci in recombinant inbred mouse strains.[35] Interestingly, two possible clusters of ras family and/or previously mapped G protein genes were identified, one on distal chromosome 9, and one on proximal chromosome 17.

C. TRANSCRIPTIONAL ANALYSIS

Comparison of the transcriptional patterns of these genes (except YL8) in undifferentiated N-Tera2 cells (the cellular source of these cDNAs) and N-Tera2 cells induced to differentiate into neuronal derivatives by treatment with retinoic

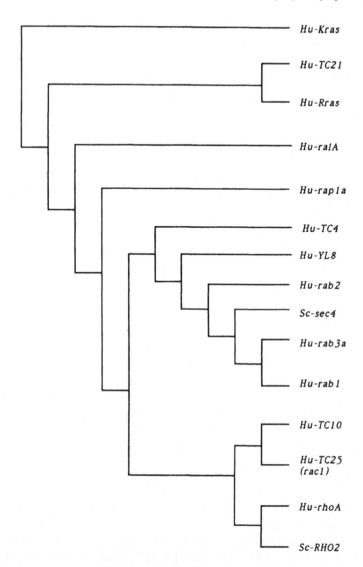

FIGURE 4. Evolutionary grouping of teratocarcinoma and selected *ras* family amino acid sequences. The prefix (Hu-) indicates a human sequence, and the prefix (Sc-) a sequence from *Saccharomyces cerevisiae*. The Hu-K-*ras* sequence is used as the outlying related protein. Hu-*ralA* and Hu-*rap*1a are considered part of the *ras*-like subfamily, and Sc-sec4 is a member of the YPT/ RAB subfamily. Since the parsimony program used to generate this tree considers the minimum number of nucleotide changes consistent with a given amino acid change, the percent amino acid identity alone does not determine the tree topology. (Adapted from Drivas, G. T. et al., *Biochem. Biophys. Res. Commun.*, 176, 1130, 1991.)

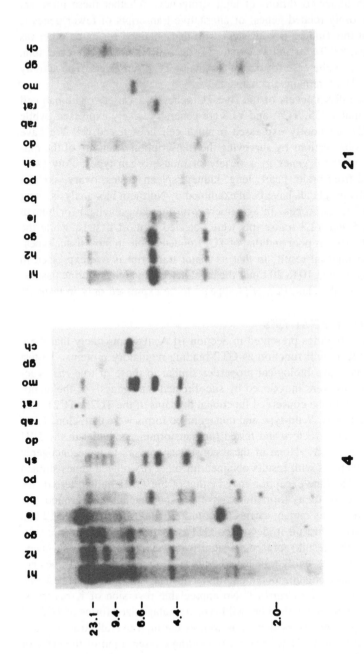

FIGURE 5. Southern blot analysis of various mammalian and chicken DNAs, probed with clones TC4 and TC21. Samples (10 μg) were digested with *EcoRI* and analyzed by Southern blotting with pUC19 subclones of TC4 and TC21. A marker lane containing *Hind*III-digested lambda DNA fragments was electrophoresed in a lane adjacent to h1, and the migration distances of selected size fragments (in kb) are indicated at the left. The lanes contain the following DNAs: N-Tera2 cell (h1); human placenta (h2); gorilla (go); lemur (le); bovine (bo); porcine (po); dog (do); sheep (sh); rabbit (rab); rat (rat); mouse (mo); guinea pig (gp); and chicken (ch). Filters were washed at 65°C in 0.015 *M* NaCl/1.5 m*M* trisodium citrate (0.1 × SSC). (From Drivas, G. T. et al., *Mol. Cell. Biol.*, 10, 1793, 1991. With permission.)

acid, is shown in Figure 6. In all cases, Northern blot analysis reveals multiple transcripts even under conditions of high stringency. Whether these represent transcripts of closely related genes, or alternative transcripts of fewer genes is not known, but this finding is consistent with that obtained for many other ras family members.[36-38] TC10 and TC21 steady-state mRNA levels decrease upon differentiation, as is the case for one of the major TC4 transcripts. The activity and pattern for TC25 remains unchanged.

Analysis of mRNA levels of the five TC genes in a variety of human cell lines indicates that TC25, TC4, and YL8 are generally widely expressed, while TC10 and TC21 are poorly expressed in most cell lines tested.[20,21] We have extended these observations by surveying the transcriptional activity of the murine homologs of these genes in a variety of mouse organ types. Poly $(A)^+$ RNAs prepared from brain, heart, lung, kidney, spleen, testes, ovary, skeletal muscle, and salivary glands have been examined by Northern blot analysis using the human cDNAs as probes. In agreement with the data obtained for human cell lines, TC25 and YL8 transcripts were detected in most tissues, while the expression of TC10 was poor and that of TC21 undetectable in most tissue types. TC4 yielded an unusual result, in that its major transcript is overexpressed in mouse testes, present at 10 to 20 times the level detected in other murine tissues. In addition, a less abundant transcript appears to be expressed only in testes.[24]

D. FUNCTIONAL ANALYSIS

Based on the findings presented in Section III.A, it seems likely that these five cDNA products will function as GTP-binding regulatory proteins. To test whether they may have biological properties similar to those of true *ras*, "activating" mutations were introduced by site-directed mutagenesis at the analogous positions within the conserved functional domains of the TC25, TC21, and TC4 coding sequences. Wild-type and mutagenized forms were then cloned into eukaryotic expression vectors and tested for transforming potential in the NIH 3T3 focus-forming assay. None of these constructs appears to have oncogenic potential, in agreement with results obtained for other *ras*-related genes studied in this way.[39,40] The lone exception may be that of the TC21 mutant tested, for which a low level of focus formation above the spontaneous background rate has been observed. This mutant carries substitutions analogous to the Val12 and Leu61 of true *ras* that result in decreased GTPase activity and oncogenic activation. Whether this result represents a significant finding is now being investigated in 3T3 cell lines stably transfected with TC21 constructs. TC21 does not appear to have antioncogenic potential similar to that found for a related *ras*-like group member, K*rev-1/rap1a*.[10] No appreciable reversion of K-*ras*-transformed DT cells was seen with either wild-type or mutagenized forms of TC21.[24] In view of this preliminary evidence, as well as the high overall structural and sequence similarity of TC21 to true *ras* (including conservation of the effector and membrane localization domains), and its restricted patterns of expression, it remains plausible that TC21, like true *ras*, plays a role in cellular proliferation.

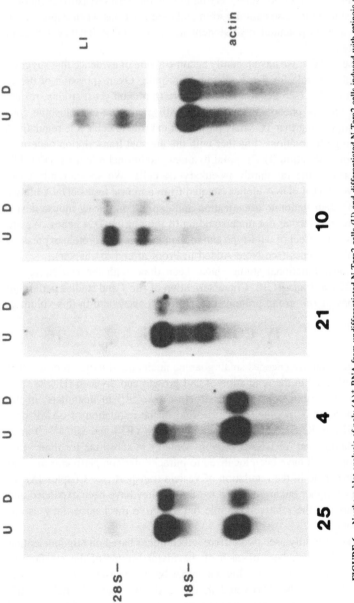

FIGURE 6. Northern blot analysis of poly (A)$^+$ RNA from undifferentiated N-Tera2 cells (U) and differentiated N-Tera2 cells induced with retinoic acid (D), probed with four TC clones.[20] Samples (2.5 μg) of poly (A)$^+$ RNA were electrophoresed in a denaturing formaldehyde-agarose gel, blotted, and hybridized to pUC19 subclones of TC25, TC4, TC21, or TC10 as indicated below the panels. L1 and actin control experiments are shown in the two panels at the right. Human L1 repetitive DNA is abundantly transcribed in undifferentiated N-Tera cells but disappears upon differentiation, and so this probing was done to demonstrate that the cell population had been properly induced. The actin panel shows the relative amounts of RNA in each lane. The migration distances of 28S and 18S rRNAs are shown at the left. The same filter was used for each experiment, and was stripped between probings.

YL8 as a member of the YPT/RAB group is likely to function in the regulation of vesicle transport in eukaryotic cells. The exceptional amino acid conservation (70%) between YL8 and *S. pombe* YPT3, indicates that both proteins may perform the same function in all eukaryotes. Experiments are underway to determine its subcellular localization, the nature of the posttranslational modification that results in its association with membrane fractions when expressed in COS cells,[24] and its potential involvement in *in vitro* GTP-dependent protein transport systems.

In the case of TC4, we have recently become aware of evidence that suggests a unique regulatory role for this RAS family protein. Overexpression of the *S. pombe* TC4 homolog (see Section III.A) in certain mutant yeast strains, results in the suppression of phenotypes associated with premature chromosome condensation,[32] suggesting that TC4 may play a critical role in cell cycle regulation. These intriguing observations, together with the unusual transcription pattern of TC4 in the mouse (Section III.C), point to a new, additional role for small GTP-binding proteins of the ras family in eukaryotic cells. We are currently characterizing several TC4 cDNA clones isolated from a mouse testes cDNA library and examining their genomic organization and expression during mouse development, in order to further our understanding of the murine TC4 genes. We are also examining the effect of wild-type and mutant human TC4 protein expression in a variety of cell types (see Note Added in Proof at end of chapter).

No additional functional studies have been done with the *rho* subfamily members TC25/*rac 1* and TC10. Characterization of *rac 1* and studies pertaining to the functions of *rho* group proteins can be found elsewhere in this volume.

IV. CONCLUSIONS

The *ras* gene family encodes small guanine nucleotide-binding proteins that have been implicated in the regulation of cell growth and division (H-, K-, and N-*ras*[2,5]) and the maintenance of cell structure (*rho*[12,13]) in mammals, in the control of mating in *S. pombe* (*ras1*[41,42]), and in the regulation of cAMP production (RAS1 and RAS2[43,44]) and vesicle targeting (YPT1 and SEC4[45-47]) in *S. cerevisiae*. Mammalian YPT/RAB proteins which can substitute for their yeast counterparts or which have been localized to intracellular compartments are also thought to participate in the regulation of vesicle transport (see Chapter 18). In the evolution of higher eukaryotes, the *ras* family may have been expanded and selected because of the relatively simple but effective mechanism they use to transmit signals.

We have successfully used two different techniques based on oligonucleotide screening to isolate members of each of the major groups of the ras family. There is no reason to suspect that this search has been exhausted, and many new members are likely to be discovered in the next few years. In light of the information that has been amassed concerning the conservation and functions of RAS family proteins, and some of the new data beginning to emerge about *ras-*

related genes such as TC4, it is clear that the elucidation of their regulatory roles will provide us with new insights into many eukaryotic cellular processes.

NOTE ADDED IN PROOF

The RAS-related protein encoded by the human teratocarcinoma cDNA TC4 is predominantly localized in the nuclei of rodent and primate cells.[48,19] It has been renamed Ran/TC4 (Ras-related nuclear protein originally identified in a teratocarcinoma cDNA library).

Consistent with our identification of multiple mouse genomic DNA sequences reactive with a Ran/TC4 probe,[35] we have identified at least two transcribed Ran genes in mice. One of these, Ran/M1, is 95% identical in its predicted amino acid sequence to human Ran/TC4; the other, Ran/M2, is completely identical (unpublished data).

Ran/TC4 interacts with a second protein, RCC1 (regulator of chromosome condensation-1).[48,50] RCC1 is a chromatin-associated DNA binding protein, mutations in which result in premature initiation of mitosis in both fission yeast[50] and mammals.[51] Specifically, in the absence of wild-type RCC1 function, the cell cycle checkpoint that prevents the onset of mitosis until the completion of DNA synthesis fails to operate. Such cells enter mitosis even in the presence of DNA synthesis inhibitors due to the premature activation of mitosis promoting factor (MPF).

A central role for Ran proteins in this aspect of cell cycle regulation was suggested by the finding that a fission yeast RCC1 defect can be suppressed by overexpression of the yeast homolog of Ran/TC4 (SPI1, 80% amino acid identity to Ran/TC4),[50] and by the observation that purified human RCC1 protein binds to and catalyzes guanine nucleotide exchange on purified Ran/TC4.[48,52]

These and other observations have led to the proposal that Ran/TC4 is a GTP binding and hydrolyzing protein whose switching between GTP- and GDP-bound forms is coupled to cell cycle progression.[48,49,52,53] In the model shown in Figure 7, START represents the commitment to DNA synthesis and FINISH the end of DNA synthesis. Key features of this model are the RCC1-stimulated generation of Ran/TC4·GTP at START, the role of Ran/TC4·GTP as an inhibitor of mitosis (by preventing the activation of MPF), and the conversion of Ran/TC4·GTP to Ran/TC4·GDP at FINISH by specific GAP. We and others have recently identified a Ran/TC4-specific GAP activity (unpublished data).

To test this model directly, we have transiently expressed a variety of Ran/TC4 mutant proteins in monkey (COS) and human (293/Tag) cells. Expression of a GTPase-defective Ran/TC4 double mutant protein (analogous to a RAS gly12→val12, gln61→leu61 GTPase defective mutant) inhibited both extra-chromosomal and chromosomal DNA synthesis in COS cells, while expression of wild type Ran/TC4 had no effect.[49] Expression of the GTPase defective mutant in 293/Tag cells also inhibited DNA synthesis (M. Ren and E. Coutavas, unpublished data). Due to the very high transfection efficiency of the 293/Tag

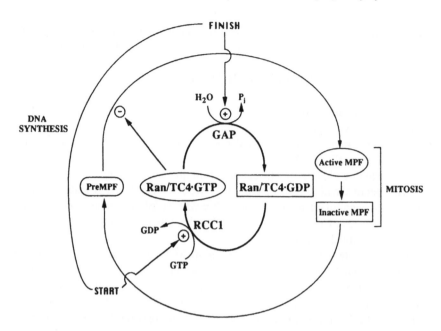

FIGURE 7. Coupling of a Ran/TC4 GTPase cycle to the cell cycle. The inner circle represents the Ran/TC4 GTPase cycle and the outer circle represents the MPF (mitosis promoting factor) cycle. MPF is a complex of a mitotic cyclin and a cyclin-dependent protein kinase. In most eukaryotic cells an inactive precursor form, preMPF, is generated during late G1 or S phase and is activated at the onset of mitosis. In this model, Ran/TC4 is present predominantly in its GTP-bound form during S phase and acts to inhibit the conversion of preMPF to MPF. Stimulation of the guanine nucleotide exchange activity of RCC1 throughout S phase links the Ran/TC4 GTPase cycle to the cell cycle.

cells, we have been able to use flow cytometric analysis of transfected 293/Tag cell populations to show that DNA synthesis inhibition is due predominantly to arrest in the G2 phase of the cell cycle (M. Ren and E. Coutavas, unpublished data). This result is consistent with the model shown in Figure 7, as accumulation of Ran/TC4 in its GTP-bound form (caused here by expression of a mutant protein that binds GTP efficiently but fails to hydrolyze it) should inhibit conversion of preMPF to MPF, arresting cells at the G2/M boundary.

Further studies are underway *in vivo* and in cell-free systems to define other functional domains of the Ran/TC4 protein and to characterize the effector proteins with which it interacts.

REFERENCES

1. **Freissmuth, M., Casey, P. J., and Gilman, A. G.,** G proteins control diverse pathways of transmembrane signalling, *FASEB J.,* 3, 2125, 1989.
2. **Santos, E. and Nebreda, A. R.,** Structural and functional properties of *ras* proteins, *FASEB J.,* 3, 2151, 1989.
3. **Bourne, H. R., Sanders, D. A., and McCormick, F.,** The GTPase superfamily: conserved structure and molecular mechanism, *Nature,* 349, 117, 1991.
4. **Bourne, H. R., Sanders, D. A., and McCormick, F.,** The GTPase superfamily: a conserved switch for diverse cell functions, *Nature,* 348, 125, 1990.
5. **Barbacid, M.,** *ras* genes, *Annu. Rev. Biochem.,* 56, 779, 1987.
6. **De Vos, A. M., Tong, L., Milburn, M. V., Matias, P. M., Jancarik, J., Miura, K., Ohtsuka, E., Noguchi, S., Nishimura, S., and Kim, S.-H.,** Three dimensional structure of an oncogene protein: catalytic domain of human c-Ha-ras p21, *Science,* 239, 888, 1988.
7. **Pai, E. F., Kabsch, W., Krengel, U., Holmes, K. C., John, J., and Wittinghofer, A.,** Structure of the guanine-nucleotide-binding domain of the Ha-*ras* oncogene product p21 in the triphosphate conformation, *Nature,* 341, 209, 1989.
8. **Krengel, U., Schlichting, I., Scherer, A., Schumann, R., Frech, M., John, J., Kabsch, W., Pai, E. F., and Wittinghofer, A.,** Three dimensional structures of H-*ras* mutants: molecular basis for their inability to function as signal switch molecules, *Cell,* 62, 539, 1990.
9. **Downward, J.,** The ras superfamily of small GTP-binding proteins, *Trends Biochem. Sci.,* 15, 469, 1990.
10. **Kitiyama, H., Sugimoto, Y., Matsuzaki, T., Ikawa, Y., and Noda, M.,** A *ras* related gene with transformation suppressor activity, *Cell,* 56, 77, 1989.
11. **Balch, W. E.,** Small GTP-binding proteins in vesicular transport, *Trends Biochem. Sci.,* 15, 473, 1990.
12. **Chardin, P., Boquet, P., Madaule, P., Popoff, M. R., Rubin, E. J., and Gill, D. M.,** The mammalian G protein rhoC is ADP-ribosylated by *Clostridium botulinum* exoenzyme C3 and affects actin microfilaments in Vero cells, *EMBO J.,* 8, 1087, 1989.
13. **Paterson, H. F., Self, A. J., Garrett, M. D., Just, I., Aktories, K., and Hall, A.,** Microinjection of recombinant p21*rho* induces rapid changes in cell morphology, *J. Cell Biol.,* 111, 1001, 1990.
14. **Gallwitz, D., Donath, C., and Sander, C.,** A yeast gene encoding a protein homologous to the human *c-has/bas* protooncogene product, *Nature,* 306, 704, 1983.
15. **Madaule, P. and Axel, R.,** A novel *ras*-related gene family, *Cell,* 41, 31, 1985.
16. **Lowe, D. G., Capon, D. J., Delwart, E., Sakaguchi, A. Y., Naylor, S. L., and Goeddel, D. V.,** Structure of the human and murine R-*ras* genes, novel genes closely related to the *ras* protooncogenes, *Cell,* 48, 137, 1987.
17. **Pizon, V., Chardin, P., Lerosey, I., Olofsson, B., and Tavitian, A.,** Human cDNAs rap1 and rap2 homologous to the *Drosophila* gene Dras3 encode proteins closely related to *ras* in the "effector" region, *Oncogene,* 3, 201, 1988.
18. **Chardin, P. and Tavitian, A.,** The *ral* gene: a new *ras*-related gene isolated by the use of a synthetic probe, *EMBO J.,* 5, 2203, 1986.
19. **Touchot, N., Chardin, P., and Tavitian, A.,** Four additional members of the *ras* gene superfamily isolated by oligonucleotide strategy: molecular cloning of YPT-related cDNAs from a rat brain library, *Proc. Natl. Acad. Sci. U.S.A.,* 84, 8210, 1987.
20. **Drivas, G. T., Shih, A., Coutavas, E., Rush, M. G., and D'Eustachio, P.,** Characterization of four novel *ras*-like genes expressed in a human teratocarcinoma cell line, *Mol. Cell. Biol.,* 10, 1793, 1990.
21. **Drivas, G. T., Shih, A., Coutavas, E. E., D'Eustachio, P., and Rush, M. G.,** Identification and characterization of a human homolog of the *Schizosaccharomyces pombe ras*-like gene YPT-3, *Oncogene,* 6, 3, 1991.
22. **Sommer, R. and Tautz, D.,** Minimal homology requirements for PCR primers, *Nucleic Acids Res.,* 17, 6749, 1989.

23. **Kwok, S., Kellogg, D. E., McKinney, N., Spasic, D., Goda, L., Levenson, C., and Sninsky, J. J.,** Effects of primer-template mismatches on the polymerase chain reaction: human immunodeficiency virus type 1 model studies, *Nucleic Acids Res.,* 18, 999, 1990.

24. **Drivas, G. and Palmieri, S.,** unpublished data.

25. **Garrett, M. D., Self, A. J., van Oers, C., and Hall, A.,** Identification of distinct cytoplasmic targets for ras/R-ras and rho regulatory proteins, *J. Biol. Chem.,* 264, 10, 1989.

26. **Didsbury, J., Weber, R. F., Bokoch, G. M., Evans, T., and Snyderman, R.,** *rac,* a novel *ras*-related family of proteins that are botulinum toxin substrates, *J. Biol. Chem.,* 264, 16378, 1989.

27. **Sekine, A., Fujiwara, M., and Narumiya, S.,** Asparagine residue in the *rho* gene product is the modification site for botulinum ADP-ribosyltransferase, *J. Biol. Chem.,* 264, 8602, 1989.

28. **Miyake, S. and Yamamoto, M.,** Identification of *ras*-related YPT family genes in *Schizosaccharomyces pombe, EMBO J.,* 9, 1417, 1990.

29. **Chavrier, P., Vingron, M., Sander, C., Simons, K., and Zerial, M.,** Molecular cloning of YPT1/SEC4-related cDNAs from an epithelial cell line, *Mol. Cell. Biol.,* 10, 6578, 1990.

30. **Kahn, R. A., Kern, F. G., Clark, J., Gelmann, E. P., and Rulka, C.,** Human ADP-ribosylation factors, a functionally conserved family of GTP-binding proteins, *J. Biol. Chem.,* 2606, 1991.

31. **Nakano, A. and Muramatsu, M.,** A novel GTP-binding protein, Sarlp, is involved in transport from the endoplasmic reticulum to the golgi apparatus, *J. Cell Biol.,* 109, 2677, 1989.

32. **Beach, D.,** personal communication, 1991. **Sazer, S. and Nurse, P.,** personal communication, 1991.

33. **Felsenstein, J.,** PHYLIP Manual, Version 3.3, University Herbarium, University of California, Berkeley, CA, 1990.

34. **Drivas, G. T., Palmieri, S., D'Eustachio, P., and Rush, M. G.,** Evolutionary grouping of the RAS-protein family, *Biochem. Biophys. Res. Commun.,* 176, 1130, 1991.

35. **Drivas, G., Massey, R., Chang, H.-Y., Rush, M. G., and D'Eustachio, P.,** *Ras*-like genes and gene families in the mouse, *Mamm. Genome,* 1, 112, 1991.

36. **Zahraoui, A., Touchot, N., Chardin, P., and Tavitian, A.,** The human *rab* genes encode a family of GTP-binding proteins related to yeast YPT1 and SEC4 products involved in secretion, *J. Biol. Chem.,* 264, 12394, 1989.

37. **Bucci, C., Frunzio, R., Chiaroitti, L., Brown, A. L., Rechler, M. M., and Bruni, C. B.,** A new member of the *ras* gene superfamily identified in a rat liver cell line, *Nucleic Acids Res.,* 16, 9979, 1988.

38. **Olofsson, B., Chardin, P., Touchot, N., Zahraoui, A., and Tavitian, A.,** Expression of the *ras*-related ralA, *rho*12, and *rab* genes in adult mouse tissues, *Oncogene,* 3, 231, 1988.

39. **Lowe, D. G. and Goedell, D. V.,** Heterologous expression and characterization of the human R-*ras* gene product, *Mol. Cell. Biol.,* 7, 2845, 1987.

40. **Avraham, H. and Weinberg, R. A.,** Characterization and expression of the human *rhoH12* gene product, *Mol. Cell. Biol.,* 9, 2058, 1989.

41. **Fukui, Y., Kozasa, T., Kaziro, Y., Takeda, T., and Yamamoto, M.,** Role of a *ras* homolog in the life cycle of *Schizosaccharomyces pombe, Cell,* 44, 329, 1986.

42. **Fukui, Y., Kaziro, Y., and Yamamoto, M.,** Mating pheromone-like diffusible factor released by *Schizosaccharomyces pombe, EMBO J.,* 5, 1991, 1986.

43. **Toda, T., Uno, I., Ishikawa, T., Powers, S., Kataoka, T., Broek, D., Cameron, S., Broach, J., Matsumoto, K., and Wigler, M.,** In yeast, RAS proteins are controlling elements of adenylate cyclase, *Cell,* 40, 27, 1985.

44. **Broek, D., Samily, N., Fasano, O., Fujiyama, A., Tamanoi, F., Northup, J., and Wigler, M.,** Differential activation of yeast adenylate cyclase by wild type and mutant RAS proteins, *Cell,* 41, 763, 1985.

45. **Segev, N., Mulholland, J., and Botstein, D.,** The yeast GTP-binding YPT1 protein and a mammalian counterpart are associated with the secretory machinery, *Cell,* 52, 915, 1988.

46. **Bacon, R. A., Salminen, A., Ruohala, H., Novick, P., and Ferro-Novick, S.,** The GTP-binding protein YPT1 is required for transport in vitro: the golgi apparatus is defective in ypt1 mutants, *J. Cell Biol.,* 109, 1015, 1989.
47. **Salminen, A. and Novick, P. J.,** A *ras*-like protein is required for a post-Golgi event in yeast secretion, *Cell,* 49, 527, 1987.
48. **Bischoff, F. R. and Ponstingl, H.,** Mitotic regulator protein RCC1 is complexed with a nuclear ras-related polypeptide, *Proc. Natl. Acad. Sci. U.S.A.,* 88, 10830, 1991.
49. **Ren, M., Drivas, G., D'Eustachio, P., and Rush, M. G.,** Ran/TC4: a small nuclear GTP-binding protein that regulates DNA synthesis, *J. Cell. Biol.,* in press, 1993.
50. **Matsumoto, T. and Beach, D.,** Premature initiation of mitosis in yeast lacking RCC1 or an interacting GTPase, *Cell,* 66, 347, 1991.
51. **Seki, T., Yamashita, K., Nishitani, H., Takagi, T., Russell, P., and Nishimoto, T.,** Chromosome condensation caused by loss of RCC1 function requires the cdc25 protein that is located in the cytoplasm, *Mol. Biol. Cell,* 3, 1373, 1992.
52. **Bischoff, F. R. and Ponstingl, H.,** Catalysis of guanine nucleotide exchange on Ran by the mitotic regulator RCC1, *Nature,* 354, 80, 1991.
53. **Roberge, M.,** Checkpoint controls that couple mitosis to completion of DNA replication, *Trends Cell Biol.,* 2, 177, 1992.

370

Chapter 18

THE *rab* GENE FAMILY

Armand Tavitian and Ahmed Zahraoui

TABLE OF CONTENTS

0-8493-5214-2/93/$0.00 + $.50
© 1993 by CRC Press, Inc.

I. INTRODUCTION

The acronym ''rab'' was first used in 1987 when four additional members of the *ras* gene family were isolated from a rat brain cDNA library by means of synthetic oligonucleotide probes.[1] It was observed that these mammalian *rab* genes constituted a distinct branch of the ras superfamily, more closely related to two genes discovered in *Saccharomyces cerevisiae:* the *YPT1* gene found serendipitously in 1983 by Gallwitz et al.[2] and the newly characterized *SEC4* gene involved in yeast secretion.[3]

The *rab* genes have been the focus of intense research for the last four years and, indeed, this branch has become the largest. Additional members have been characterized by several groups: the YPT cDNA of mouse capable of complementing that of yeast confirmed that *rab1 (rab1A)* was the homolog of yeast *YPT1* gene.[4] G proteins purified from bovine brain permitted to design oligonucleotide that revealed additional members related to *rab3* (denoted *smg* 25A, B, and C).[5] The search for human counterparts of the rat *rab* cDNAs in human cDNA libraries revealed the same four (*rab1, 2, 3, 4*) and additional *rab3B, 5,* and 6.[6] BRL-*ras,* isolated from a rat liver cell line cDNA library,[7] is referred as *rab7.* Another clone, very close to *rab1*A was isolated from a rat brain cDNA library and denoted *rab1B.*[8] sas1 and sas2 were isolated from *Dictyostelium discoideum.*[9] More recently additional *rab8, 9, 10,* and 11 to 20 were characterized in canine cell.[10-11] The fission yeast *Schizosaccharomyces pombe* has three genes ypt1, ypt2 (that may be functionally equivalent to *YPT1* and *SEC4*) and *ryh,* for which the human counterpart is *rab6.*[12-14]

At present, there are approximately 26 members, characterized in mammals, that pertain to the *rab* branch of the family. This figure may represent only a fraction of the total number of the existing rab proteins, even though different approaches have often led to the rediscovery of previously known members. One may foretell that there are, perhaps, as many as 40 different genes pertaining to the *rab* family.

The chromosome mapping of six human *rab* genes has been reported. It confirms the wide dispersion of the *rab* genes throughout the genome.[15-17]

II. STRUCTURAL FEATURES

The small G proteins of the *rab* branch have molecular weights of 22 to 26 kDa. They are very well-conserved throughout the philogeny, especially between mammalian species. For instance, the percentage of identity between the human rab1A and the YPT protein is 75%. The rat and the human rab1A proteins are identical. There is only one conservative change between the human and rat rab3A proteins. The rab proteins display typical features of *ras*-like GTP-binding proteins (Figure 1). The four domains of the RAS proteins that were shown to be involved in the binding of GTP/GDP and which correspond to some 7 to 8 amino acid residues around positions 15, 60, 115, and 145 are highly conserved. Some variations are observed, however, in region 10 to 17 where the glycines in position 12 and sometimes 13 of ras are replaced by other conservative (or even nonconservative for rab6) amino acids.

Three additional sequences, corresponding to residues 35 to 40 (known as the effector region in *ras*), 51 to 56, and 63 to 85 (flanking the DTAGQE sequence), are highly conserved among all the rab proteins but diverge from the corresponding sequences of p21*ras*. In particular, region 35 to 40 points to different effectors from the *ras*. There are probably different effectors for the various rab proteins; it is remarkable that the YPT2 protein, whose sequence in the effector domain is identical to the corresponding SEC4 sequence, can complement the SEC4 mutations and cannot complement the YPT1 mutations.[12]

III. UBIQUITOUS AND NONUBIQUITOUS EXPRESSION OF THE rab PROTEINS; rab PROTEINS AND THE NERVOUS SYSTEM

The growing family of the small G proteins has prompted several groups to ask the question whether all the members were expressed in all cell types.

It was soon observed that some G proteins were more specifically expressed in some cell types and not in others and the most remarkable observation was that of *rab*3A being found expressed in the brain, almost exclusively.[18] This fact has prompted groups of researchers to study the developmental and the regional expression of the rab genes in the brain of different animal species.

In the mouse brain rab1A and rab2 mRNA were expressed late, at days E15 to 17 during embryogenesis. At that time, the vast majority of neurons do not divide but start to differentiate. This suggest a participation of rab1A and rab2 in neuronal motivation.[19]

Expression of rab3A is exclusively neuronal. It starts at day 11 (E11) of embryogenesis and persists to be highly expressed until adulthood. It was also observed that expression of rab3 was not evenly distributed in the brain, but apparently more abundant in the mid-brain.

In 1985, Bar-Sagi and Feramisco showed that the ras p21 could induce neuronal differentiation in the pheochromocytoma cell line PC12, in a manner similar to that exerted by the nerve growth factor (NGF).[20] Microinjection of the activated p21-*ras* provoked neuronal differentiation and cell division arrest. These observations were corroborated by several other experiments using viral infection by Harvey or Kirsten sarcoma viruses and introduction of activated N-*ras*.[21] Other contributions followed, in this line of research, showing that injection of anti-p21-*ras* antibody in the PC12 cell line does inhibit cellular differentiation induced by NGF but not that induced by cyclic AMP.[22-24] Later, Borasio et al.[25] demonstrated that primary cultures of neurons, microinjected with p21-*ras,* could survive and differentiate in the absence of growth factors. It should be noted that in all these experiments, the activated RAS proteins were used, since they exerted much more marked effects than the wild-type p21-*ras*. Similarly, incubating p21-*ras* with the nonhydrolyzable analog of GTP, GTPγS, also activates p21-*ras* and renders it capable of inducing PC12 differentiation.[26]

```
                                    21                     40                     60
HrablA  MSS----------MNPEYD  YLFKLLLIGDSGVGKSCLLL  RFADDTYTESYISTIGVDFK  IRTIELDGKTIKLQIWDTAG
Hrab1B  M-----------NPEYD    YLFKLLLIGDSGVGKSCLLL  RFADDTYTESYISTIGVDFK  IRTIELDGKTIKLQIWDTAG
Hrab2   MA-----------YA      YLFKYIIIGDTGVGKSCLLL  QFTDKRFQPVHDLTIGVEFG  ARMITIDGKQIKLQIWDTAG
Hrab3A  MASATDSRYGQKESSDQNFD  YMFKILIIGNSSVGKTSFLF  RYADDSFTPAFVSTVGIDFK  VKTIYRNDKRIKLQIWDTAG
Hrab3B  MASVTDGKHGVKDASDQNFD  YMFKLLIIGNSSVGKTSFLL  RYADDTFTPAFVSTVGIDFK  VKTVYRBEKRVKLQIWDTAG
Hrab4   MS----------ETYD     FLFKFLVIGNAGTGKSCLLH  QFIEKKFKDDSNHTIGVEFG  SKIINVGGKYVKLQIWDTAG
Hrab5   MASRGATRPGPNTGNKI--   CQFKLVLLGESAVGKSSLVL  RFVKGQFHEFQESTIGAAFL  TQFVCLDDTTVKFEIWDTAG
Hrab6   MSTGGDFGNPL----------  RKFKLVFLGEQSVGKTSLIT  RFMYDSFDNTYQATIGIDFL  SKTMYLEDRTVRLQLWDTAG
Hrab8   MAK----------TYD     YLFKLLLIGDSGVGKTCVLF  RFSEDAFNSTFISTIGIDFK  IRTIELDGKRIKLQIWDTAG
Crab10  MAKK----------TYD    LLFKLLLIGDSGVGKTCVLF  RFSDDAFNTTFISTIGIDFK  IKTVELQGKKIKLQIWDTAG
Rrab7   M----------          LL-KVIILGDSGVGKTSLMN  QIVNKKFSNQYKATIGADFL  TKEVMVDDRLVTMQIWDTAG
CRab9   ----------           ----------            ---NKFDTQLFHIGVEFL    NKDLEVDGHFVTMQIWDTAG
Hrab11  MGTRDD----------EYD  YLFKVVLIGDSGVGKSNLLS  RFTRNEFNLESKSTIGVEFA  TRSIQVDGKTIKAQIWDTAG
YPT1    MNS----------EYD     YLFKLLLIGNSGVGKSCLLL  RFSDDTYTNDYISTIGVQFK  IKTVELDGKTVKLQIWDTAG
SEC4    MSG--LRTVSASSGNGKSYD  SIMKILLIGDSGVGKSCLLV  RFVEDKFNPSFITTIGIDFK  IKTVDINGKKVKLQLWDTAG
K-ras   M----------          TEYKLVVVGAGGVGKSALTI  QLIQNHFVDEYDPTIEDSY-  RKQVIDGETCLLDILDTAG
```

FIGURE 1. Alignment of 13 rab protein sequences with yeast YPT1p and SEC4p, and the human K-*ras* (exon 4B) protein. Sets of identical or conservative residues are indicated by italic and bold-face type. The reference numbering is that of K-*ras*. H is for human; C for canine; and R for rat.

	80	100	117	136
Hrab1A	QERFRTITSSYYRGAHGIIV	VYDVTDQESFNNVKQWLQEI	DRYA-----SENVNKLLVGNK	CDLTTKKVVDYTTAKEFADS
Hrab1B	QERFRTVTSSYYRGAHGIIV	VYDVTDQESYANVKQWLQEI	DRYA-----SENVNKLLVGNK	SDLTTKKVVDNTTAKEFADS
Hrab2	QESFRSITRSYYRGAAGALL	VYDITRRDTFNHLTTWLEDA	RQHS-----NSNMVLMLIGNK	SDLESRREVKKEEGEAFARE
Hrab3A	QERYRTITTAYYRGAMGFIL	MYDITNEESFNAVQDWSTQI	KTYS-----WDNAQVLLVGNK	CDMEDERVVSSERGRQLADH
Hrab3B	QERYRTITTAYYRGAMGFIL	MYDITNEESFNAVQDWATQI	KTYS-----WDNAQVILVGNK	CDMEEERVVPTEKGQLLAEQ
Hrab4	QERFRSVTRSYYRGAAGALL	VYDITSRETNALTNWLTDA	RMLA-----SQNIVILCGNK	KDLDADREVTFLEASRFAQE
Hrab5	QEGYHSLAPMYYRGAQAAIV	VYDITNEESFARAKNWVKEL	QRQA-----SPNIVIALSGNK	ADLANKRAVDFQEAQSYADD
Hrab6	QERFRSLIPSYIRDSTVAVV	VYDITNVNSFQQTTKWIDDV	RTER-----GSDVIIMLVGNK	TDLADKRQVSIEEGERKAKE
Hrab8	QERFRTITTAYYRGAMGIML	VYDITNEKSFDNIRNWIRNI	EEHA-----SADVEKMILGNK	CDVNDKRQVSKERGEKLALD
Crab10	QERFETITTSYYRGAMGIML	VYDITNGKSFENISKWLRNI	DEHA-----NEDVERMLLGNK	CDMDDKRVVPKGKGEQIARE
Rrab7	QERFQSLGVAFYRGADCCVL	VFDVTAPNTFKTLDSWRDEF	LIQASPRDPENFPFVVLGNK	IDLEN-RQVATKRAQAWCYS
Crab9	QERFRSLRTPFYRGSDCCLL	TFSVDDSQSFQNLSNWKKEF	IYYADVKEPESFPFVILGNK	IDISE-RQVSTEEAQAWCRD
Hrab11	QERYRAITSAYYRGAVGALL	VYDIAKHLTENVERWLKEL	RDHA-----DSNIVIMLVGNK	SDLRHLRAVPTDEARAFAEK
YPT1	QERFRTITSSYYRGSHGIII	VYDVTDQESFNGVKMWLQEI	DRYA-----TSTVLKLLVGNK	CDLKDKRVVEYDVAKEFADA
SEC4	QERFRTITTAYYTGAMGIIL	VYDVTDERTFTNIKQWFKTV	NEHA-----NDEAQLLLVGNK	SDM-ETRVVTADQGEALAKE
K-ras	QEEYSAMRDQYMRTGEGFLC	VFAINNTKSFEDIHHYREQI	KRVK-----DSEDVPMVLVGNK	CDLPS-RTVDTKQAQDLARS

FIGURE 1 (continued).

	155	173	183	187
Hrab1A	LGIP-FLETSAKNATNVEQSF	MTMAAEIKKRMGPGATAGGA	EKSNVKIQSTPVK-------	-QS----GGGCC
Hrab1B	LGVP-FLETSAKNATNVEQAF	MTMAAEIKKRMGPGAASGG-	ERPNLKIDSTPVK-------	-SA----SGGCC
Hrab2	HGLM-FMETSAKTASNVEEAF	INTAKEIYEKIQEGVFDINN	EANGIKIGPQHAATNATHAG	NQGGQQAGGCC
Hrab3A	LGFE-FFEASAKDNINVKQTF	ERLVDVICEKMSESLDTADP	AVTGAKQGPQLSDQ------	-QVPP-HQDCAC
Hrab3B	LGFD-FFEASAKENISVRQAF	ERLVDAICDKMSDSLDT-DP	SMLGSSKNTRLSDTPPLL--	-Q-----QNCSC
Hrab4	NEIM-FLETSALTGEDVEEAF	VQCARKILNKIESGELDPER	MGSGIQYGDAALRQLRSPRR	TQA--PNAQECGC
Hrab5	NSLL-FMETSAKTSMNVNEIF	MAIAKKLPKNEPQNPGANSA	RGGGVDLTEPTQPTRN----	-Q-------CCSN
Hrab6	LNVM-FIETSAKAGYNVKQLF	RRVAAALPGMESTQDRSRED	MIDIKLEKPQE---------	-QP--VSEGGCSC
Hrab8	YGIK-FMETSAKANINVENAF	FTLARDIKAKMDKKLEGNSP	QGSNQGVKITPDQ-------	-QK-RSSFFRCVLL
Crab10	HGIR-FFETSAKVNINIEKAF	LTLAEDILRKTPVKEPNSEN	VDISSGGV-----------	--T--GWKSKCC
Rrab7	KNNIPYFETSAKEAINVEQAF	QTIARNALKQETEVELYNEF	PEPIKLDKNERA--------	--K--ASAESCSC
Crab9	NGDYPYFETSAKDATNVAAAF	EEAVRRVLATEDRSDHLIQT	DTVSLHR------------	-KP--KPSSSCC
Hrab11	NGLS-FIETSALDSTNVEAAF	QTILTEIYRIVSQKQMSDRR	ENDMSPSNNVVPIHVPPTTE	NKP--KVQCCQNI
YPT1	NKMP-FLETSALDSTNVEDAF	LTMARQIKQSMSQQNLNETT	QKKEDKGNVNLKG------	-QSLTNTGGGCC
SEC4	LGIP-FIESSAKNDDNVNEIF	FTLAKLIQEKIDSNKLYGVG	NGQEGNISINSGSG-----	-NS---SKSNCC
K-ras	YGIP-FIETSAKTRQGVDDAF	YTLVREIRKHKEKMSKD--G	KKKKKKSKTK--------	--------CVIM

FIGURE 1 (continued).

The possible effects of five rab proteins on neuronal differentiation (rab1A, 2, 3A, 4, and 6) were tested on embryonic neurons from the mid-brain of rat. Whereas rab3A, 4, and 6 did not exert any evident morphological changes *in vitro*, rab1A and rab2 provoked distinct and opposite effects. The introduction of rab2 was spectacular, quite specific, and obtained with the wild-type protein. As soon as 3 h after the introduction of rab2p, one could observe an increase of adhesion to the substrate of neuronal cells with an apparent increase of the cell body surface, together with an outgrowth of neurites.[27] In contrast, rab1A seemed to increase the proportion of unipolar neurons and diminish cell adhesion to the substrate (A. Prochiantz, personal communication).

IV. BIOCHEMICAL PROPERTIES

As for all the proteins of the *ras* superfamily, the rab proteins have the ability to bind GTP and GDP and they exert biochemical properties similar to those of the members of the *ras* and *rho* families. The intrinsic GTPase activity appears to vary between members: it is quite low for rab5 and rab6 and quite similar to the activity of the p21-*ras* for rab3A and rab2.[6] For YPT1, rab1A, and rab1B, the association to the guanidylic phosphates depends on Mg^{2+} and the GTPase activity of these proteins is modified accordingly with the induced mutation described for RAS.[28-29] It might well be, as suggested, that intrinsic GTP hydrolysis is affected by specific deformations of the phosphate site induced by variation at positions 12 or 13 of *ras*.

V. REGULATORY PROTEINS

A. IS THERE A GAP FOR rab?
Several groups have looked for the existence of a GAP that would specifically activate the intrinsic GTPase of a given rab protein. Such activities have been reported in cellular extracts from yeast for the SEC4p (P. Novick, personal communication) for YPT1p,[30] and for the mammalian rab3Ap.[31] Recently, Macara's group[31] has brought evidence for the presence of a rab3A-specific GAP in rat brain. This activity was also detected in tissues where the rab3A protein could not be detected. The GAP factor in rat brain has a molecular size of 400 kDa. This is, in fact, the first GAP activity described for a member of the rab branch of the family. It should be noted that the rab3A GAP has not yet been purified and could stimulate the rab3A GTPAse only five-fold. This was in contrast with the other GAP activities detected in brain cytosol that stimulate the GTPases of the RAS or the rap proteins much more efficiently than they stimulated rab protein GTPases.

B. THE GDP DISSOCIATION INHIBITOR (GDI) AND THE GDP DISSOCIATION STIMULATOR (GDS)
Takai's group[32] has defined proteins that regulate the dissociation of GDP from the small G proteins. For rab, these regulators have been studied on *smg*

p25A (i.e., the rab3A protein). It is well known that the intracellular pool of GTP is about 10 times higher than that of GDP. Nevertheless, the small G proteins extracted from the cell are in the GDP-bound form. This not only suggests that the G proteins rapidly interact with target proteins such as GAP or GAP-complexes, but also, that proteins are inhibiting the dissociation of GDP from the G protein. In fact, such a protein was purified to near homogeneity from bovine brain cytosol. It inhibits the dissociation of GDP from (and thereby, the binding of GTP to) the rab3A protein (*smg* p25A).[32] It was shown that this factor (GDI) had no effect on ras, rho, or rap proteins, but could interact with another small G protein in human platelet membranes with an apparent molecular weight of 24 kDa. It seems, therefore, that the rab3A-GDI does recognize close member(s) in platelets. Conversely, a GDI activity can be detected in human platelet cytosol which is recognized by a polyclonal antibody against p25A-GDI (rab3A-GDI). This GDI shares some weak amino acid homology with the yeast *S. cerevisiae* CDC25-encoded protein, which is thought to serve as the GTP-GDP exchange factor for the yeast RAS2 protein.[33] It was also clearly shown that this GDI was present in many tissues where rab3A is undetectable, suggesting, therefore, that GDI does interact with other small G proteins. One might presume that these proteins pertain to the rab branch. In agreement with this, it has been shown recently that rab3A-GDI also interacts with SEC4.[34] The GDI seems to both inhibit the attachment of rab3Ap GDP-bound form and favor its release from membranes.[35]

Kinetic analysis of the guanine nucleotide binding to the bovine brain Rab3A protein, revealed that there is also a protein stimulating the dissociation of GDP from the G protein and promoting, therefore, the binding of GTP. More details on these regulatory proteins are found in Chapter 27.

VI. POSTTRANSLATIONAL MODIFICATIONS

All the proteins of the ras superfamily (ras, rab, rho, . . .) except arf share one or two cysteine residues at, or near, their C terminal extremity, which are necessary for their membrane attachment and function. The rab proteins can be classified into at least three groups according to their carboxyl terminal sequences (see Figure 2).

In the case of *ras* and some *ras*-related proteins which end with a Cys-AAX motif, the C terminal cystein is first farnesylated (p21-*ras*) or geranylgeranylated (rap1A, rap1B, smg p21A, B, and *rho*A). Then the AAX sequence is removed and the exposed cysteine α-carboxyl group is methylated. In some RAS proteins (N-*ras*, H-*ras*) membrane anchoring is enhanced by palmitoylation of an upstream cysteine from the farnesylated residue.[36] Less is known about the post-translational processing of rab proteins. Nevertheless, it has been reported recently that rab1B, 2, 5, and 3A proteins are geranylgeranylated and that rab3A is also carboxymethylated.[37-38]

```
Group 1 : XXXCC

        Rab1A    TPVKQSGGGCC
        Rab1B    TPVKSASGGCC
        Rab2     QGGQQAGGGCC
        YPT1     QSLTNTGGGCC

        Rab9     HRKPKPSSSCC
        Rab10    GGVTGWKSKCC
        SEC4     GSGNSSKSNCC


Group 2 : XXXCAC    A: aliphatic residue

        Rab3A    QQVPPHQDCAC
        Rab3B    TPPLLQQNCSC
        Rab3C    TPPPPQPNCGC
        Rab4     TQAPNAQECGC
        Rab6     EQPVSEGGCSC
        Rab7     RAKASAESCSC


Group 3 : XXXCAAX
        Rab8     PKRSSFFRCCVLL

        Rab11    TENKPKVQCCQNI
        Rab5A    EPTQPTRNQCCSN
```

FIGURE 2. C terminal sequences of mammalian *rab*, yeast YPT1, and SEC4 proteins.

VII. CELLULAR FUNCTION OF rab PROTEINS FAMILY

A. EFFECT OF GTPγS ON VESICULAR TRANSPORT

Evidence for the role of GTP hydrolysis in vesicular traffic comes from studies in cell-free systems. These systems reconstitute vesicular traffic between different cellular compartments. They measure vectorial transport and/or fusion of vesicles with their acceptor membrane. Cell-free assays have shown that several steps of exocytic and endocytic membrane traffic in mammalian cells are sensitive to GTPγS. Along the secretory pathway, the slowly hydrolyzable GTP analog, GTPγS, irreversibly and dramatically inhibits transport between the endoplasmic reticulum (ER) and *cis*-Golgi compartment, and between successive cisternae of the Golgi stack.[39-40] It also inhibits the recycling of the mannose 6-phosphate receptor through the *trans*-Golgi network (TGN).[41] Recently, it has been shown that GTPγS blocks the transport of membrane proteins between TGN and plasma membrane (PM).[42-43] In addition, GTPγS inhibits the formation of constitutive secretory vesicles and immature secretory granules.[44]

Along the endocytic pathway, GTP hydrolysis is needed for the fusion events between plasma membrane-derived vesicles and early endosomes as well as for the fusion between endosomes.[45]

All these results indirectly show that vesicle trafficking is controlled by guanine nucleotide-binding proteins:

B. YEAST SEC4p AND YPT1p

There is strong genetic evidence that SEC4 and YPT1: two small GTP-binding proteins highly related to mammalian rab proteins, are required for intracellular transport in *S. cerevisiae*. Thermosensitive mutations of the SEC4 gene impair the fusion of post-Golgi vesicles with the PM, suggesting that its product is required for the targeting and/or the fusion of vesicles with PM. In agreement with this, the SEC4p (p for protein) is found associated with the cytoplasmic face of vesicles in transit to the cell surface and the inner face of the PM.[46] Further genetic studies have identified at least 10 genes to be essential for post-Golgi secretory events. One of these, SEC15 localized on secretory vesicles, may be a SEC4 target. This suggests that SEC4 controls the formation of a complex between a docking protein, SEC15, and a PM target.[47]

Conditional mutations of the *YPT1* gene cause an early block in secretion with accumulation of abnormal Golgi structures.[48-49] These defects can be suppressed by addition of calcium, suggesting that the function of YPT1p is primarily to regulate cytoplasmic Ca^{2+} concentration.[50] However, using an assay that reconstitutes protein transport *in vitro* through the Golgi apparatus, Bacon et al.[51] arrived at a different conclusion and suggested that YPT1p controls vesicle trafficking between ER and *cis*-Golgi. It is now known, that both Ca^{2+} and GTP are required at distinct steps between ER and Golgi. In *in vitro* assays, it has been shown that the antibodies to YPT1p block transport from ER to Golgi and that Ca^{2+} is essential at a later stage, probably for the fusion of vesicles with *cis*-Golgi membrane.[52]

C. MAMMALIAN rab PROTEINS

Three lines of evidence show that rab proteins are involved in the regulation of several steps of exocytic and endocytic pathways.

1. Rab proteins share much higher similarity with YPT1p and SEC4p, particularly in the region known as the putative effector domain, than with other *ras*-like GTP-binding proteins. The *rab*1A gene can complement *YPT1* deletions in *S. cerevisiae*.[4] This suggests that rabps, YPT1p, and SEC4p can interact with related effectors and, thus, exert related functions. In agreement with this, it has been shown that a peptide with two substitutions (rab3AL) corresponding to the effector domain of rab3Ap blocks ER to Golgi and intra-Golgi transport *in vitro*.[53]

2. rab proteins share highly variable C terminal domains that specify their localization in appropriate secretory compartments.[54] Many rab proteins have been localized, they have been found in distinct subcompartments along the exocytic and endocytic pathways (Figure 3). rab1Ap is associated with ER and/or Golgi structures[49] (B. Goud personal communication). rab2p is localized in the intermediate compartment between ER and Golgi. rab6p is associated with *cis*, medial, and especially with *trans*-Golgi and rab3Ap is present on synaptic vesicles and chromaffin granules. Three other rab proteins are found on endocytic organelles: rab4p and rab5p are associated with PM and early endosomes and rab7p with late endosomes.[55-59]

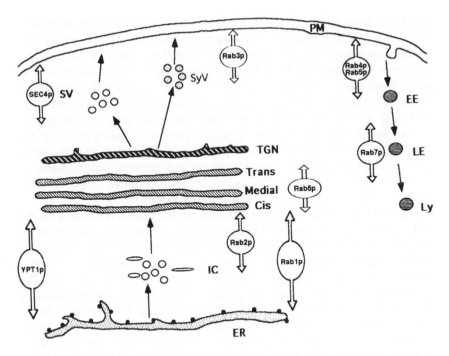

FIGURE 3. Subcellular localization of rab/YPT1/SEC4 protein family. A scheme of different exocytic and endocytic compartments is shown. The figure displays the distribution of the rab/YPT1/ SEC4 proteins along the secretory and endocytic pathways. ER, endoplasmic reticulum; IC, intermediate compartment; *cis*, medial, and *trans*-Golgi; TGN, trans-Golgi network; SV, secretory vesicles; SyV, synaptic vesicles; PM, plasma membrane; EE, early endosome; LE, late endosome; and LY, lysozome.

3.   Recently, direct lines of evidence have shown that rab proteins control vesicle transport. Using an assay which reconstitutes transport between the ER and the *cis*-Golgi compartment, Plutner et al.[60] have shown that rab1Bp is required in an initial step in export of protein from the ER. It is also required for protein transport between *cis*- and medial Golgi cisternae. rab5p is needed for the fusion of early endosomes, as was demonstrated by the use of specific antibodies to rab5p in a cell-free assay. Conversely, fusion of endosomes is stimulated by an excess of wild-type rab5p.[61] It was shown that rab3Ap dissociates from synaptic vesicle fraction after calcium-dependent exocytosis in synaptosomes, and that this dissociation is partially reversible during recovery after stimulation.[62] These results show an association-dissociation cycle of rab3Ap during exocytosis.

VIII. A MODEL FOR THE MECHANISM OF rab ACTION IN VESICULAR TRANSPORT

In mammalian cells, proteins destined for organelles or for secretion undergo a series of enzymatic processing steps in ER and through the Golgi compartments.

Movement between these compartments occurs by vesicular transport, and mechanisms for targeting vesicles to the appropriate locations is essential. Every vesicle that buds off a donor membrane organelle (ER, Golgi, PM, endosomes, etc.) must target specifically to the exact membrane acceptor compartment. It seems likely that rab proteins have specific roles in each step occuring during vesicular trafficking: budding, targeting, and fusion with acceptor membrane. Such roles are supported by the observations cited in the preceding paragraph.

At present, the exact function of the rab proteins in vesicle traffic is still suggestive. Several models have been proposed[63-64] in which GTP would serve as a "conformational switch" for the G protein involved in the vectorial transport of secretory vesicles between different membrane compartments. Figure 4 shows a possible role of a rab protein in vesicular targeting.

The rab-GDP protein would recognize a specific protein R (receptor) at the surface of the secretory vesicle. This recognition stabilizes the association of the rabp with the membrane and a GDI protein binds to the rabp. The complex, so-formed with the associated vesicle, detaches from donor membrane and migrates to the acceptor membrane where it is recognized by a specific target (T). The interaction with the T provokes the release of GDP, the subsequent binding of GTP and the release of GDI. In this hypothesis, a localized nucleotide exchange protein (GDS) would catalyze the insertion of the rab-vesicle complex to the acceptor membrane. The rabp presumably cooperates, with other membrane-associated proteins, to promote the attachment of the vesicle to the acceptor membrane. The complex with rab^{GTP} is then recognized by a GAP-like protein which stimulates the rab-GTPase activity. The hydrolysis of GTP induces a conformational change that allows the release of rab^{GDP}, leaving the complex R-T-Vesicle in a position to initiate the process of membrane fusion.

The model, proposed here, takes into account the following observations: GTPγS has an inhibitory effect on the acceptor membrane, but has no effect upon the activity of the donor. Moreover, it appears that GTPγS inhibits transport after the attachment of secretory vesicles, but prior to their fusion with the acceptor membrane.[40] This suggests that GTP/GDP exchange, followed by GTP hydrolysis, occurs on the acceptor compartment.

In the case of a role for rabp in vesicular budding, nucleotide exchange occurs on the budding vesicle of the donor membrane. The hydrolysis of the GTP triggers the vesicle detachment.

Finally, in addition to constitutive transport, mammalian cells also show more differentiated functions, such as regulated secretion in secretory cells, axonal traffic in neurons, and polarized transport in epithelial cells. The small GTP-binding proteins are, most probably, also involved in these specialized processes.

NOTE ADDED IN PROOF

Prenylation of rab1A and rab3A proteins is catalysed by a Rab geranylgeranyl transferase. This enzyme is composed of two components, A and B. B component

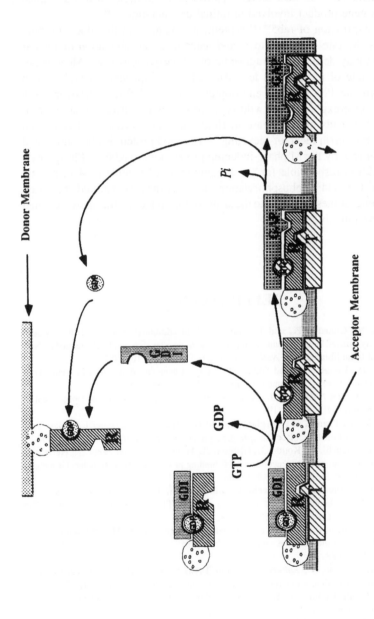

FIGURE 4. A model for the role of rab proteins in vesicular transport. The figure represents one transport cycle of a vesicle from a donor compartment to an acceptor membrane. The rabp (circle) under its inactive form (rab^{GDP}) binds to a receptor (R) on the surface of a budding vesicle. The so-formed complex detaches from donor compartment through interaction with GDI. The insertion of this complex into the acceptor membrane is coupled with the GTP/GDP exchange. The hydrolysis of GTP (which can be blocked by GTPγS) ends the cycle, provokes the dissociation of Rab^{GDP}, and triggers the fusion of the transport vesicle with the acceptor membrane. Then, the complex dissociates and R is recycled to the donor compartment.

consists of two 60- and 38-kDa proteins that may be analogous to the α and β subunits of ras farnesyl transferase. The amino acid sequence of six peptides from component A, a 95-kDa protein, shows high similarity with the human choroidermia gene product involved in retinal degeneration.[65,66]

In cells, expression of rab5[Ile133] protein, a mutant form that does not bind GTP, inhibits the rate of endocytosis and leads to an accumulation of tubular structures that may derive from fragmentation of early endosomes. Meanwhile, however, the rate of recycling is less affected. Rab5 appears to regulate both transport from the PM to early endosomes and lateral fusion between early endosomes.[67] Overexpression of wild type rab4 does not affect initial rates of endocytosis, but leads to a decrease in fluid phase endocytosis and causes an alteration in transferrin receptor recycling.[68] The rab4 protein is phosphorylated *in vitro* and *in vivo* by the M phase-inducing protein kinase p34[cdc2]. Phosphorylation of ser196 is responsible for the accumulation of soluble rab4 in mitotic cells. It seems that p34[cdc2] phosphorylation may alter the association/dissociation of rab4 from endosomes and leads to the arrest of the endocytic transport mediated by rab4 during mitosis.[69,70]

REFERENCES

1. **Touchot, N., Chardin, P., and Tavitian, A.,** Four additional members of the *ras* gene superfamily isolated by an oligonucleotide strategy: molecular cloning of YPT-related cDNAs from a rat brain library, *Proc. Natl. Acad. Sci. U.S.A.*, 84, 8210, 1987.
2. **Gallwitz, D., Donath, C., and Sander, C.,** A yeast gene encoding a protein homologous to the human c-*has/bas* proto-oncogene product, *Nature*, 306, 704, 1983.
3. **Salminen, A. and Novick, P. J.,** A ras like protein is required for a post-Golgi event in yeast secretion, *Cell*, 49, 527, 1987.
4. **Haubruck, H., Prange, R., Vorgias, C., and Gallwitz, D.,** The ras-related mouse ypt1 protein can functionally replace the *ypt1* gene product in yeast, *EMBO J.*, 8, 1427, 1989.
5. **Matsui, Y., Kikuchi, A., Kondo, J., Hishida, T., Teranishi, Y., and Takai, Y.,** Nucleotide and deduced amino acid sequences of a GTP-binding protein family with molecular weights of 25000 from bovine brain, *J. Biol. Chem.*, 263, 11071, 1988.
6. **Zahraoui, A., Touchot, N., Chardin, P., and Tavitian, A.,** The human *rab* genes encode a family of GTP-binding proteins related to yeast YPT1 and Sec4 products involved in secretion, *J. Biol. Chem.*, 264, 12394, 1989.
7. **Bucci, C., Frunzio, R., Chiarotti, L., Brown, A. L., Rechier, M. M., and Bruni, C. B.,** A new member of the *ras* gene superfamily identified in a rat liver cell line, *Nucleic Acids Res.*, 16, 997, 1988.
8. **Vielh, E., Touchot, N., Zaharoui, A., and Tavitian, A.,** Nucleotide sequence of a rat cDNA: rab1B encoding a rab1A-YPT-related protein, *Nucleic Acids Res.*, 17, 1770, 1989.
9. **Saxe, S. A. and Kimmel, A. R.,** Genes encoding novel GTP binding proteins in Dictyostelium, *Dev. Genet.*, 9, 259, 1988.
10. **Chavrier, P., Vingron, M., Sander, C., Simons, K., and Zerial, M.,** Molecular cloning of YPT1/SEC4-related cDNAs from an epithelial cell line, *Mol. Cell. Biol.*, 10, 6578, 1990.

11. **Chavrier, P., Simons, K., and Zerial, M.,** The complexity of the rab and rho GTP-binding protein subfamilies revealed by a PCR cloning approach, *Gene,* 112, 261, 1992.
12. **Miyake, S. and Yamamoto, M.,** Identification of *ras*-related *ypt* family genes in *Schizosaccharomyces pombe, EMBO J.,* 9, 1417, 1990.
13. **Haubruck, H., Engelke, U., Mertins, P., and Gallwitz, D.,** Structural and functional analysis of ypt2 an essential *ras*-related gene in the fission yeast *Schizosacharomyces pombe* encoding a sec4 protein homologue, *EMBO J.,* 9, 1957, 1990.
14. **Hengst, L., Lehmeier, T., and Gallwitz, D.,** The *ryh*1 gene in the fission yeast *Schizosaccharomyces pombe* encoding a GTP-binding protein related to *ras, rho* and *ypt*. Structure, expression and identification of its human homologue, *EMBO J.,* 9, 1949, 1990.
15. **Rousseau-Merck, M. F., Zahraoui, A., Bernheim, A., Touchot, N., Miglierina, R., Tavitian, A., and Berger, R.,** Chromosome mapping of the human *ras*-related *rab*3A gene to 19p13.2, *Genomics,* 5, 694, 1989.
16. **Rousseau-Merck, M. F., Zahraoui, A., Touchot, N., Tavitian, A., and Berger, R.,** Chromosome assignment of four *ras*-related *rab* genes, *Hum. Genet.,* 86, 350, 1991.
17. **Nimmo, E. R., Sanders, P. G., Padua, R. A., Hughes, D., Williamson, R., and Johnson, K. J.,** The *mel* gene: a new member of the RAB/YPT class of *ras*-related genes, *Oncogene,* 6, 1347, 1991.
18. **Olofsson, B., Chardin, P., Touchot, N., Zahraoui, A., and Tavitian, A.,** Expression of the *ras* related *ral*A, *rho*12 and *rab* genes in adult mouse tissues, *Oncogene,* 3, 231, 1988.
19. **Ayala, J., Olofsson, B., Touchot, N., Zahraoui, A., Tavitian, A., and Prochiantz, A.,** Developmental and regional expression of three new members of the *ras*-gene family in the mouse brain, *J. Neurosci. Res.,* 22, 384, 1989.
20. **Bar-Sagi, D. and Feramisco, J. R.,** Microinjection of the *ras* oncogene protein into PC12 cells induces morphological differentiation, *Cell,* 42, 841, 1985.
21. **Noda, M., Ko, M., Ogura, A., Liu, D., Amano, T., Takano, T., and Ikawa, Y.,** Sarcoma viruses carrying *ras* oncogenes induce differentiation-associated properties in a neuronal cell line, *Nature,* 318, 73, 1985.
22. **Guerrero, I., Wong, H., Pellicer, A., and Burstein, D. E.,** Activated N-*ras* gene induces neuronal differenciation of PC12 rat pheochromocytoma cells, *J. Cell Physiol.,* 129, 71, 1986.
23. **Hagag, N., Halegoua, S., and Viola, M.,** Inhibition of growth factor-induced differentiation of PC12 cells by microinjection of antibody to rasp21, *Nature,* 319, 680, 1986.
24. **Mattson, M. P., Taylor-Hunter, A., and Kater, S. B.,** Neurite outgrowth in individual neurons of a neuronal population is differentially regulated by calcium and cyclic AMP, *J. Neurosci.,* 8, 1704, 1988.
25. **Borasio, G. D., Wittinghofer, J. J., Barde, Y. A., Sendter, M., and Heumann, R.,** Ras p21 protein promotes survival and fiber outgrowth of cultured embryonic neurons, *Neuron,* 2, 1087, 1989.
26. **Satoh, T., Nakamura, S., and Kaziro, Y.,** Induction of neurite formation in PC12 cells by microinjection of proto-oncogenic Ha-ras protein preincubated with guanosine-5'-*O*-(3-thiotriphosphate), *Mol. Cell. Biol.,* 7, 4553, 1987.
27. **Ayala, J., Touchot, N., Zahraoui, A., Tavitian, A., and Prochiantz, A.,** The product of *rab*2, a small GTP binding protein, increases neuronal adhesion, and neurite growth *in vitro, Neuron,* 4, 797, 1990.
28. **Touchot, N., Zahraoui, A., Wiehl, E., and Tavitian, A.,** Biochemical properties of the YPT-related rab1B protein. Comparison with rab1A, *FEBS Lett.,* 256, 79, 1989.
29. **Wagner, P., Molenaar, C. M. T., Rauh, A. J. G., Brökel, R., Schmitt, H. D., and Gallwitz, D.,** Biochemical properties of the ras-related YPT protein in yeast: a mutational analysis, *EMBO J.,* 6, 2373, 1987.
30. **Becker, J., Tan, T. J., Trepte, H. H., and Gallwitz, D.,** Mutational analysis of the putative effector domain of the GTP-binding Ypt1 protein in yeast suggests specific regulation by a novel GAP activity, *EMBO J.,* 10, 785, 1991.

31. **Burstein, E. S., Linko-Stentz, K., Lu, Z., and Macara, I. G.,** Regulation of the GTPase activity of the *ras*-like protein p25^rab3A, *J. Biol. Chem.*, 266, 2689, 1990.

32. **Sasaki, T., Kikuchi, A., Araki, S., Hata, Y., Isomura, M., Kuroda, S., and Takai, Y.,** Purification and characterization from brain cytosol of a protein that inhibits the dissociation of GDP from and the subsequent binding of GTP to smgp25A, a ras p21-like GTP-binding protein, *J. Biol. Chem.*, 265, 2333, 1990.

33. **Matsui, Y., Kikuchi, A., Araki, S., Hata, Y., Kondo, Y., Teranishi, Y., and Takai, Y.,** Molecular cloning and characterization of a novel type of regulatory protein (GDI) for smgp25A, a *ras* p21-like GTP-binding protein, *Mol. Cell Biol.*, 10, 4116, 1990.

34. **Sasaki, T., Kaibuchi, K., Kabcenell, A. K., Novick, P. J., and Takai, Y.,** A mammalian inhibitory GDP/GTP exchange protein (GDP dissociation inhibotor) for smg p25A is active on the yeast SEC4 protein, *Mol. Cell. Biol.*, 2909, 1991.

35. **Araki, S., Kikuchi, A., Hata, Y., Isoaura, M., and Takai, Y.,** Regulation of reversible binding of smg25A, a Ras p21-like GTP-binding protein, to synaptic plasma membranes and vesicles by its specific regulatory protein, GDP dissociation inhibitor, *J. Biol. Chem.*, 265, 13007, 1990.

36. **Hancock, J. F., Magee, A. I., Childs, J. E., and Marshall, C. J.,** All ras proteins are polyisoprenylated but only some are palmitoylated, *Cell*, 57, 1167, 1989.

37. **Farnsworth, C. C., Kawata, M., Yoshida, Y., Takai, Y., Gelb, M. H., and Glomset, J. A.,** C terminus of the small GTP-binding protein smgp25A contains two geranylgeranylated cysteine residues and a methyl ester, *Proc. Natl. Acad. Sci. U.S.A.*, 88, 6196, 1991.

38. **Kinsella, B. T. and Maltese, W. A.,** *rab* GTP-binding proteins implicated in vesicular transport are isoprenylated *in vitro* at cysteines within a novel carboxyl-terminal motif, *J. Biol. Chem.*, 266, 8540, 1991.

39. **Beckers, C. J. M. and Balch, W. E.,** Calcium and GTP: Essential components in vesicular trafficking between the endoplasmic reticulum and Golgi apparatus, *J. Cell. Biol.*, 108, 1245, 1989.

40. **Melançon, P., Glick, B. S., Malhotra, V., Weidman, P. J., Serafini, T., Gleason, M. L., Orci, L., and Rothman, J. E.,** Involvement of GTP-binding "G" proteins in transport through the Golgi stack, *Cell*, 51, 1053, 1987.

41. **Goda, Y. and Pfeffer, R.,** Selective recycling of the mannose 6-phosphate/IGF-II receptor to the trans-golgi network *in vitro, Cell*, 55, 309, 1988.

42. **Gravotta, D., Adesnik, M., and Sabatini, D. D.,** Transport of inluenza Ha from the trans-golgi network to the apical surface of MDCK cells permeabilized in their basolateral plasma membranes: energy dependence and involvement of GTP-binding proteins, *J. Cell Biol.*, 111, 2893, 1990.

43. **Miller, S. F. and Moore, H. P. H.,** Reconstitution of constitutive secretion using semi-intact cells: regulation by GTP but not calcium, *J. Cell Biol.*, 112, 39, 1991.

44. **Mayorga, L. S., Diaz, R., and Stahl, P. D.,** Regulatory role for GTP binding proteins in endocytosis, *Science*, 244, 1475, 1989.

46. **Goud, B., Salminen, A., Walworth, N. C., and Novick, P. F.,** A GTP binding protein required for secretion rapidly associates with secretory vesicles and the plasma membrane in yeast, *Cell*, 53, 753, 1988.

47. **Salminen, A. and Novick, P.,** The Sec15 protein responds to the function of the GTP binding protein, Sec4, to control vesicular traffic in yeast, *J. Cell. Biol.*, 109, 1023, 1989.

48. **Segev, N. and Botstein, D.,** The *ras*-like yeast *YPT1* gene is itself essential for growth, sporulation and starvation response, *Mol. Cell. Biol.*, 7, 2367, 1987.

49. **Segev, N., Mulholland, J., and Botstein, D.,** The yeast GTP-binding YPT1 protein and a mammalian counterpart are associated with the secretion machinery, *Cell*, 52, 915, 1988.

50. **Schmitt, H. D., Puzicha, M., and Gallwitz, D.,** Study of a temperature-sensitive mutant of the *ras*-related *YPT1* gene product in yeast suggests a role in the regulation of intracellular calcium, *Cell*, 53, 635, 1988.

51. **Bacon, R. A., Salminen, A., Ruohola, H., Novick, P., and Ferro-Novick, S.,** The GTP-binding protein YPT1 is required for transport *in vitro:* the Golgi apparatus is defective in ypt1 mutants, *J. Cell. Biol.*, 109, 1015, 1989.

52. **Baker, D., Wuestehube, L., Schekman, R., Botstein, D., and Segev, N.,** GTP-binding ypt1 protein and Ca^{2+} function independently in a cell-free protein transport reaction, *Proc. Natl. Acad. Sci. U.S.A.,* 87, 355, 1990.

53. **Plutner, H., Schwaninger, R., Pind, S., and Balch, W. E.,** Synthetic peptides of the rab effector domain inhibit vesicular transport through the secretory pathway, *EMBO J.,* 9, 2375, 1990.

54. **Chavrier, P., Gorvel, J. P., Stelzer, E., Simons, K., Gruenberg, J., and Zerial, M.,** Hypervariable C-terminal domain of rab proteins acts as a targeting signal, *Nature,* 353, 769, 1991.

55. **Chavrier, P., Parton, R. G., Hauri, H. H., Simons, K., and Zerial, M.,** Localization of low molecular weight GTP-binding proteins to exocytic and endocytic compartments, *Cell,* 62, 317, 1990.

56. **Goud, B., Zahraoui, A., Tavitian, A., and Saraste, J.,** Small GTP-binding protein associated with Golgi cisternae, *Nature,* 345, 553, 1990.

57. **Mollard, G. F. V., Mignery, G. A., Baumert, M., Perin, M. S., Hanson, T. J., Burger, P. M., Jahn, R., and Südhof, T. C.,** rab3 is samll GTP-binding protein exclusively localized to synaptic vesicles, *Proc. Natl. Acad. Sci. U.S.A.,* 87, 1988, 1990.

58. **Darchen, F., Zahraoui, A., Hammel, F., Monteils, M., Tavitian, A., and Scherman, D.,** Association of the GTP-binding protein Rab3A with bovine adrenal chromaffin granules, *Proc. Natl. Acad. Sci. U.S.A.,* 87, 5692, 1990.

59. **Van der Sluijs, P., Hull, M., Zahraoui, A., Tavitian, A., Goud, B., and Mellman, I.,** The small GTP-binding protein rab4 is associated with early endosomes, *Proc. Natl. Acad. Sci. U.S.A.,* 86, 6313, 1991.

60. **Plutner, H., Cox, A. D., Pind, S., Khosravifer, R., Bourne, J. R., Schwaninger, R., Der, C. J., and Balch, W. E.,** Rab1b regulates vesicular transport between the endoplasmic reticulum and successive golgi compartments, *J. Cell Biol.,* 115, 1, 1991.

61. **Gorvel, J. P., Chavrier, P., Zerial, M., and Gruenberg, J.,** Rab5 controls early endosome fusion *in vitro, Cell,* 64, 915, 1991.

62. **Fischer von Mollard, G., Südhof, T. C., and Jahn, R.,** A small GTP-binding protein dissociates from synaptic vesicles during exocytosis, *Nature,* 349, 79, 1991.

63. **Bourne, H. R.,** Do GTPases direct membrane traffic in secretion?, *Cell,* 53, 669, 1988.

64. **Chardin, P.,** The ras family proteins: a conserved functional mechanism?, *Cancer Cells,* 3, 117, 1991.

65. **Seabra, M. C., Brown, M. S., Slaughter, C. A., Südhof, T. C., and Goldstein, J. L.,** Purification of component A of Rab geranylgeranyl transferase: possible identity with the choroidermia gene product, *Cell,* 70, 1049, 1992.

66. **Seabra, M. C., Brown, M. S., and Goldstein, J. L.,** Retinal degeneration in choroidermia: deficiency of Rab geranylgeranyl transferase, *Science,* 259, 377, 1993.

67. **Bucci, C., Parton, R. G., Mather, I. H., Stunnenberg, H., Simons, K., Hoflack, B., and Zerial, M.,** The small GTPase rab5 functions as a regulatory factor in the early endocytic pathway, *Cell,* 70, 715, 1992.

68. **Van der Sluijs, P., Hull, M., Webster, P., Mâle, P., Goud, B., and Mellman, I.,** The small GTP-binding protein rab4 controls an early sorting event on endocytic pathway, *Cell,* 70, 729, 1992.

69. **Bailly, E., McCaffrey, M., Touchot, N., Zahraoui, A., Goud, B., and Bornens, M.,** Phosphorylation of two small GTP-binding proteins of the Rab family by p34^{cdc2}, *Nature,* 350, 715, 1991.

70. **Van der Sluijs, P., Hull, M., Huber, L. A., Mâle, P., Goud, B., and Mellman, I.,** Reversible phosphorylation-dephosphorylation determines the localisation of rab4 during the cell cycle, *EMBO J.,* 11, 4379, 1992.

Chapter 19

THE YPT PROTEIN FAMILY IN YEAST

Warren A. Kibbe, Ludger Hengst, and Dieter Gallwitz

TABLE OF CONTENTS

0-8493-5214-2/93/$0.00 + $.50

367

I. INTRODUCTION

GTP-binding proteins of the Ypt family are members of the ras superfamily of proteins. The first example of a YPT gene, *YPT1*, was cloned and sequenced as part of the actin-β-tubulin gene cluster in the yeast *Saccharomyces cerevisiae* and the homology of the Ypt1 protein (Ypt1p) with ras proteins was immediately noted.[1] The subsequent identification by cDNA cloning of mammalian proteins that are strikingly similar in primary structure to the yeast Ypt1p[2,3] suggested the existence of a larger family of evolutionarily conserved proteins distinct from *ras* gene products. In fact, multiple members of the Ypt family have been found in the evolutionarily distant yeasts *S. cerevisiae* and *Schizosaccharomyces pombe*.[1,4-7] The existence in mammals of 20 or more Ypt-related proteins, designated Rab,[2,3,8-10] signifies the importance of this still-growing family.

The yeasts *S. cerevisiae* and *S. pombe* are particularly amenable to genetic manipulations and biochemical characterization, and much of the work on Ypt/Rab proteins has been carried out in these organisms. For example, the essential gene *YPT1* was analyzed using gene disruptions replaced with *GAL10-YPT1* promoter fusions and the phenotypes of these and other *ypt1* mutants were intensely studied.[11-15] Unfortunately, the varied phenotypes of the mutants made the assignment of a primary role for Ypt1p unclear. It was not until *SEC4*, a gene previously identified as part of the secretory pathway,[16] was identified as a Ypt family member that the assignment of a function for Ypt proteins became clearer.[4] The *sec4* mutants have been shown to accumulate Golgi-derived vesicles that would normally fuse with the plasma membrane.[4,17,18] Sec4p is associated with post-Golgi vesicles and the plasma membrane,[19] whereas Ypt1p has been localized to 50 nm-vesicles derived from the endoplasmic reticulum (ER)[20,21] and to Golgi membranes.[13] It is now clear that Sec4p and Ypt1p fulfill essential functions at defined steps of the secretory pathway, are involved in membrane targeting and/or membrane fusion, and that the cycle of these proteins between a GDP- and a GTP-bound form is central to understanding their biological and biochemical function.

From the work performed in yeast, Bourne[22] first suggested a general mechanism for Ypt protein involvement in vesicle transport akin to the proofreading function of EF-Tu, a member of a distantly related class of GTPases. EF-Tu is coupled to the translation process where the energy of the GTP hydrolysis is used to drive protein synthesis forward by catalyzing macromolecular interactions. In line with recent findings that show a striking conservation of the structure, function, and molecular components of the secretory machinery between yeast and mammals[23] Rab proteins in mammalian cells seem to perform similar functions in the unidirectional vesicle transport as their yeast counterparts.[7,24,25] Considering known and hypothesized proteins interacting with Ypt/Rab proteins, their functional cycle is shown schematically in Figure 1.

This chapter presents a short review of the different members of the Ypt protein family in the yeasts *S. cerevisiae* and *S. pombe*, their biochemical prop-

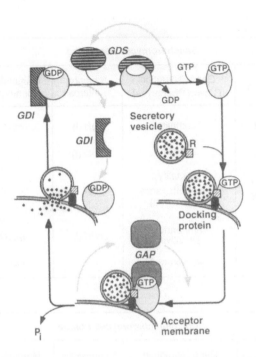

FIGURE 1. Putative functional cycle of Ypt proteins. This figure graphically depicts the association of the Ypt protein cycle with the fusion of vesicles at the acceptor membrane. The Ypt protein bound with GDP is associated with a GDI (GDP dissociation inhibitor) in the upper left hand corner. The cycle proceeds when the GDI is exchanged for a GDS (GDP dissociation stimulator, or guanine nucleotide exchange factor), and GDP is released. Ypt protein rapidly incorporates GTP, and the GTP-bound form of the Ypt protein associates with a hypothesized receptor protein found on the surface of the secretory vesicle and recognizes the proper docking protein on the surface of the acceptor membrane. After the recognition/docking event occurs, GAP (GTPase-activating protein) stimulates the hydrolysis of GTP to GDP and subsequently, Yptp is released from the vesicle/acceptor membrane complex and is again competent for the next cycle.

erties, and their mode of action that is emerging from genetic and biochemical analyses.

II. YPT GENES IN *S. cerevisiae*

Six genes encoding Ypt proteins have been identified thus far in baker's yeast (Figure 2). A link between Ypt proteins and intracellular transport processes became evident from observations showing that mutants of the most intensively studied genes, *YPT1* and *SEC4*, accumulate membrane-enclosed structures of the secretion machinery, i.e., ER and Golgi-derived vesicular intermediates,[4,13,14] and are defective in the passage of proteins through different compartments of the secretory pathway. As outlined below, preliminary studies show that other, recently identified Ypt proteins in *S. cerevisiae* are likely to also act as regulators in the endocytic pathway. Clearly, the maintenance of the organelles involved

A

Saccharomyces cerevisiae			
Gene	Gene Disruption (Growth Phenotype)	S.pombe Counterpart	Mammalian Counterpart
YPT1	Lethality	*ypt1* (!)	*rab1* (!)
SEC4	Lethality	*ypt2* (!)	?
YPT3A *YPT3B*	Lethality (when both genes are disrupted)	*ypt3*	*rab11*
YPT6	Temperature sensitivity	*ryh1* (!)	*rab6* (!)
YPT7	Like wild-type	?	*rab7*

B

Schizosaccharomyces pombe			
Gene	Gene Disruption (Growth Phenotype)	S.cerevisiae Counterpart	Mammalian Counterpart
ypt1	Lethality	*YPT1*	*rab1*
ypt2	Lethality	*SEC4*	?
ypt3	Lethality	*YPT3A,B*	*rab11*
ypt5	?	?	*rab5*
ryh1	Temperature sensitivity	*RYH1*	*rab6*(!)[1]

FIGURE 2. Members of Ypt family in the yeasts *S. cerevisiae* (A) and *S. pombe* (B). Exclamation mark in parentheses signifies that the gene can complement the functional loss of corresponding yeast gene.

in protein transport is a dynamic process that requires a constant balance of incoming and outgoing membranes (vesicles) and there must be membrane-associated and/or membrane-spanning factors that are responsible for establishing the identity of each organelle uniquely. Likewise, the targeting of vesicles must be carefully controlled, so that incoming vesicles recognize the proper acceptor membrane, successfully dock, and fuse. Finally, there must be processes that maintain an equilibrium between incoming and outgoing vesicles, most probably a recycling mechanism that scavenges received membrane material (from the

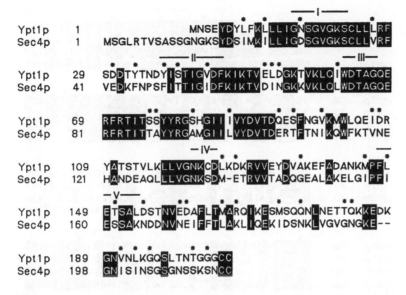

FIGURE 3. *S. cerevisiae* Ypt1p and Sec4p sequence comparisons. Ypt1p and Sec4p sequences are shown, with identical sequences indicated on black background. Conservative variations are indicated with a star over the sequence. The G-1 through G-5 conserved regions are denoted by roman numerals over the center of the conserved region. The C termini have two consecutive cysteine residues found in many Ypt family proteins.

vesicles) and sends it back to a donor organelle. It is these membrane transport processes where small guanine nucleotide-binding proteins of the Ypt/Rab family seem to play a key regulatory role in all eukaryotic cells.

A. *YPT1*

The *YPT1* gene in *S. cerevisiae* is located on chromosome VI in a gene cluster containing the single copy of actin *(ACT1)* and β-tubulin *(TUB2)*. The genes are oriented with *YPT1* in the middle of the cluster and the direction of transcription toward *ACT1*, with *TUB2* and *YPT1* sharing a short promoter region.[1,26] The gene encodes a protein of 206 amino acids (Figure 3) with all the structural domains typical for guanine nucleotide-binding proteins of the ras superfamily. The extent of sequence identity to Ras proteins of various species is limited to about 35%.[1] In contrast to Ras and Rho proteins, Ypt1p terminates with two consecutive cysteine residues, at least one of which is absolutely required for membrane association and functional integrity of the protein.[27]

YPT1 gene disruption leads to lethality.[11,12] The vital role of Ypt1p in cell growth and proliferation has also been shown with *GAL10-YPT1* promoter fusions, allowing the switching off of Ypt1p synthesis,[11] and with conditional *ypt1* mutants.[12-14] Of the many phenotypic alterations observed, specific cytoskeletal lesions in *ypt1* mutants[11,12] and the partial rescue of mutants with high extracellular calcium[14] were misleading in assigning a function to Ypt1p. Finally, the

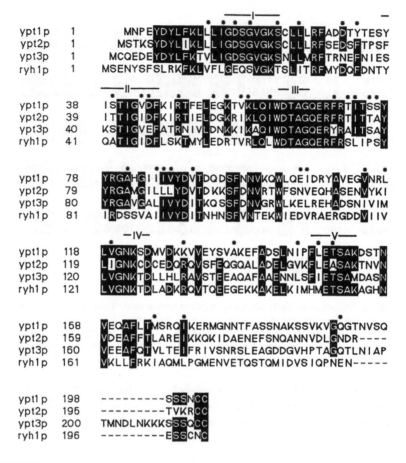

FIGURE 4. *S. pombe* Yptp family sequence comparisons. The *S. pombe* ypt1p (homolog of *S. cerevisiae* Ypt1p and mammalian Rab1p), ypt2p (homolog of *S. cerevisiae* Sec4p), ypt3p (homolog of *S. cerevisiae* Ypt3Ap and Ypt3Bp and mammalian Rab11p), and ryh1p (homolog of *S. cerevisiae* Ypt6p and mammalian Rab6p) sequences are shown, with sequences identical in three or more proteins indicated by black background with white letters. Positions with conservative variations are indicated with a star over the sequence. The G-1 through G-5 conserved regions are denoted by roman numerals over the center of the conserved region.

discovery of the essential role in secretion of Sec4p, a Ypt1p-related protein,[4] proved helpful for defining the function of Ypt1p. Mutants of *ypt1* accumulate immature proteins *in vivo* that are ER core-glycosylated, indicating a block in transport at an early step of the secretory pathway.[13-15] The requirement of a functional Ypt1 protein for ER-to-Golgi transport was also documented in studies employing *in vitro* transport systems.[15,29] This transport is via a vesicle intermediate[20,21,30] and is dependent on Ca^{2+} and ATP. Accumulation of transport vesicles has also been observed in a conditional-lethal *ypt1* mutant *in vivo*.[28] The evidence supports the involvement of Ypt1p in vesicle targeting to and/or fusion with the acceptor membrane, presumably the *cis*-Golgi compartment.

Organism	Protein	Effector region			C-terminus
S. c.	Yptlp	35 ND	YIS TIGVDF	KI	202 GGGCC
S. p.	yptlp	35 ES	YIS TIGVDF	KI	199 SSNCC
H. s.	rablp	38 ES	YIS TIGVDF	KI	201 GGGCC
S. c.	Sec4p	47 PS	FIT TIGIDF	KI	211 KSNCC
S. p.	ypt2p	36 PS	FIT TIGIDF	KI	196 VKRCC
S. c.	Ypt3Ap	40 MD	SKS TIGVEF	AT	219 GNNCC
S. c.	Ypt3Bp	40 IE	SKS TIGVEF	AT	218 SSNCC
S. p.	ypt3p	37 IE	SKS TIGVEF	AT	210 SSQCC
C. f.	rabllp	38 LE	SKS TIGVEF	AT	212 CCQNI
S. c.	Ypt6p	37 DH	YQA TIGIDF	LS	212 SACQC
S. p.	ryhlp	38 NT	YQA TIGIDF	LS	197 SSCNC
H. s.	rab6p	40 NT	YQA TIGIDF	LS	205 SGCSC
S. c.	Ypt7p	35 QQ	YKA TIGADF	LT	204 NSCSC
C. f.	rab7p	35 NQ	YKA TIGADF	LT	203 ESCSC

FIGURE 5. Comparison of effector region and C-termini of Ypt family proteins. The effector region and C-terminal sequences of Yptlp, Sec4p, Ypt3p, Ypt6p, and Ypt7p are shown for ascomycetes and their mammalian counterparts. Note that interspecies counterparts have identical effector regions. The C-termini generally contain two cysteine residues, and at least one cysteine is required for *in vitro* function in all Ras-related GTPases. Numbers in front of amino acid sequences indicate the position of the first residue of that region in the respective protein. S.c., *S. cerevisiae;* S.p., *S. pombe;* H.S., *Homo sapiens;* c.f., *Canus familiaris.*

S. cerevisiae Yptlp and its mammalian homolog, Rablp/Yptlp, share 71% of identical sequence.[2,3] This remarkable conservation applies not only to the primary sequence but also to the function of the protein. A deletion of the essential *S. cerevisiae YPT1* gene can be suppressed by the expression of its mammalian homolog,[24] and more importantly, Rablp has been demonstrated to be indispensable for ER-to-Golgi protein transport *in vitro*.[31,32] In several cases, it has been clearly demonstrated that Ypt/Rab proteins are capable of cross-species complementation as long as their effector regions (corresponding to amino acids 32 to 40 of mammalian Ras proteins[33,34]) have identical sequence (Figure 2; Figure 5). A mutational analysis of the Yptlp effector domain revealed that this region is critical for the interaction with a specific GTPase-activating protein (GAP), yptGAP.[28,35] The generation of Yptlp-depleted mutants has allowed the identification of suppressor genes that can overcome the otherwise lethal loss of Yptlp function.[36] This class of suppressors, the *SLY* genes (for suppressors of the loss of *YPT1* function) were isolated by replacing the *YPT1* gene with a *GAL10-YPT1* promoter fusion in a haploid strain (allowing the shut-off of *YPT1* expression in glucose media). Four *SLY* genes have been identified by this method: *SLY1, SLY2, SLY12,* and *SLY41*.[36] Like Yptlp, the *SLY* gene-encoded proteins are involved in ER-to-Golgi protein transport,[37] but their mode of action is presently unclear.

B. *SEC4*

The *SEC4* gene was first identified genetically in a search for genes involved in the secretory pathway.[16] Secretion in yeast has been extensively characterized genetically with roughly 30 genes in the secretory pathway currently identified.[16,36,38] At least 10 of these genes function in protein transport from the *trans*-Golgi compartment to the plasma membrane. A temperature-sensitive mutant of one of these genes, *SEC4*, accumulates secretory vesicles at the nonpermissive temperature.[17]

The *SEC4* gene was isolated from a centromere-based genomic yeast library as an extragenic suppressor of *sec15-1*, another mutant in the Golgi-to-plasma membrane transport step. After this gene was cloned, sequenced, and analyzed, *SEC4* was predicted to encode a 215 amino acid, 23.5 kDa protein containing the canonical GTP-binding domains and sharing a 47% homology to Ypt1p (Figure 3), and a 32% homology with the human H-*ras* protein.[4] Thus, *SEC4* gave the first evidence that a ras superfamily member was directly involved in protein transport. Subsequent studies by Novick and co-workers have revealed that Sec4p is synthesized as a soluble protein and rapidly associates with secretory vesicles and the plasma membrane.[18,19] Membrane binding of Sec4p requires, like Ypt1p,[27] at least one of its C terminal cysteine residues.[19] Recent studies show that both Sec4p and Ypt1p require the *BET2* gene for membrane attachment. Bet2p has a significant homology with known prenyltransferases.[39]

It has been described that Sec4p binds GTP with a higher relative affinity than GDP.[40] A mammalian protein, smg p25A GDI, originally characterized as inhibitor of the GDP-GTP exchange reaction of the Rab3A,[41] has been shown to also act on *S. cerevisiae* Sec4p,[42] suggesting the existence in yeast of a protein with similar properties (Figure 1).

C. OTHER YPT GENES

Recently, four additional genes coding for new members of the Ypt protein family of *S. cerevisiae* were characterized in our laboratory. Two of them, *YPT6* and *YPT3A*, were isolated by their respective sequence homology to the *ryh1* gene[7] and the *ypt3* gene[5] of the fission yeast *S. pombe* (L. Hengst, M. Benli and D. Gallwitz, unpublished results). *YPT3B* and *YPT7* were isolated using polymerase chain reaction (PCR) products generated from Ypt family conserved sequences (H. Wichmann and D. Gallwitz, unpublished results). The coding regions of these four genes contain no introns, as was also the case with *YPT1* and *SEC4*.

The newly discovered YPT genes code for GTP-binding proteins of 208 to 223 amino acids length. The deduced amino acid sequences of the putative effector regions of the Ypt proteins are identical to their mammalian counterparts (rab family). The relationships between the mammalian and yeast proteins were assigned by comparison of the effector region, the conserved "cysteine motif" at the C-terminus, and by overall amino acid homology. *S. cerevisiae* Ypt3Ap and Ypt3Bp have been putatively assigned as Rab11p[9] homologs, Ypt6p to mammalian Rab6p,[8] and Ypt7p to Rab7p[9] (Figure 2, Figure 5). In addition, three of the newly isolated genes have known counterparts in the fission yeast *S. pombe:* Ypt3Ap and Ypt3Bp are closely related to ypt3p; the deduced primary structure of Ypt6p exhibits over 60% sequence identity to ryh1p of fission yeast, including the identical effector region. The functional homology of the *YPT6* gene and its counterparts from mammalian cells or fission yeast is demonstrated by the specific complementation of defects of the *ypt6* null mutant by expression of the human rab6 cDNA or the ryh1 cDNA of fission yeast (L. Hengst and D. Gallwitz, unpublished results).

Surprisingly, two of the newly discovered Ypt proteins are not essential for growth. Disrupting the *YPT7, YPT3A,* or *YPT3B* genes has no detectable effect on cell growth, whereas, a deletion of essential parts of the *YPT6* gene causes a temperature-sensitive growth phenotype. However, deleting both *YPT3* genes in a haploid strain is lethal, strongly suggesting that *YPT3A* and *YPT3B* are functionally related and demonstrating that Ypt3p function is essential for cell viability (M. Benli, H. Wichmann, and D. Gallwitz, unpublished results).

Preliminary evidence suggests that Ypt7p and Ypt6p may be regulatory components of the endocytic pathway, as null mutants of either gene are characterized by defective maturation of vacuolar enzymes and by a fragmentation of the vacuolar compartment.

III. YPT GENES IN *S. pombe*

The fission yeast *S. pombe* is only distantly related to the baker's yeast *S. cerevisiae,* but equally suitable to genetic analyses. Considering certain cellular features, like gene structure, organelle morphology (distinct Golgi compartments!), or cell cycle events, the fission yeast is thought to be the better model organism for studies related to higher eukaryotes.[43] In addition, Ras proteins of the two ascomycetes appear to be integrated into different regulatory pathways, i.e., sexual differentiation in *S. pombe*[44] and cAMP production in *S. cerevisiae.*[45] For these reasons, three groups set out to isolate and study members of the Ypt gene family in fission yeast.[5-7,46,47] Five genes have thus far been discovered whose protein products share extensive structural homology with Ypt proteins of *S. cerevisiae* and the corresponding Rab proteins of mammalian species (Figure 2). Three of the genes, *ypt2, ypt3,* and *ryh1,* have been localized to different parts of chromosome I, and *ypt1* is closely linked to the topoisomerase I gene *(top1)* on chromosome II.[48]

A. *ypt1*
The gene was independently isolated by Fawell et al.[46] using the *S. cerevisiae* *SEC4* gene as hybridization probe and by Miyake and Yamamoto[5] in a screen for additional *ras* genes by using oligonucleotides corresponding to protein sequences conserved in Ras proteins. Because of the high degree of sequence identity of its protein product with the *S. cerevisiae* Ypt1p and mammalian Rab1/ Ypt1 proteins (Figures 4 and 5), the gene was named *ypt1.*[5] The protein-coding

region of *ypt1* is interrupted by four short introns. None of the intron positions matches those found in other *S. pombe ypt* genes, i.e., *ypt3* or *ryh1*. Coincidentally, the third intron separating the isoleucine-61 and tryptophan-62 codons has its counterpart in the mouse *Ypt1/Rab1* gene at the identical position.[49]

No functional analysis of ypt1p has thus far been performed. Like *YPT1* in *S. cerevisiae*, *ypt1* in *S. pombe* has an essential function, since gene disruption is lethal. It has also been shown that the *S. pombe ypt1* gene, but not the *ypt3* gene, can functionally replace the essential *YPT1* gene in baker's yeast, suggesting similar functions of their protein products in both ascomycetes.[5]

B. *ypt2*

The *ypt2* gene was detected and cloned by crosshybridization to a *S. cerevisiae YPT1* gene probe.[6] In contrast to the other *S. pombe ypt* genes, the protein-coding region of *ypt2* is continuous. The deduced amino acid sequence (Figure 4) exhibits the highest degree of identity to *S. cerevisiae* Sec4p (64.7%)[4] and the *Dictyostelium discoideum sas1* gene product (76.6%).[50] Most importantly, the effector region of these proteins is identical, and it has been shown that a conditional-lethal *S. cerevisiae sec4* mutant can be complemented by high expression of the *S. pombe ypt2* gene.[6] Like *ypt1* and *ypt3*, the *ypt2* gene serves an essential function.[6]

C. *ypt3*

The *ypt3* gene was cloned with the help of synthetic oligonucleotides.[5] It contains two introns interrupting codons 142 and 170. Gene disruption is lethal. Mammalian Rab11p[9] and two recently discovered *S. cerevisiae* Ypt3 proteins (M. Benli, H. Wichmann, and D. Gallwitz, unpublished results) have the highest homology with the 214 amino acid long *S. pombe* ypt3p. These proteins not only share between 60 and 70% identical amino acid residues, they also have an identical effector region (Figure 5). The question of whether these proteins are functionally interchangeable or not has not yet been addressed experimentally.

D. *ryh1*

A fragment of the protein-coding region of *S. cerevisiae YPT1* served to identify and clone the *ryh1* gene.[7] The coding part of *ryh1* is discontinuous, three short introns are located between codons 41 and 42 and within codons 95 and 135. The gene encodes a protein of 201 amino acids that contains all structural domains typical for GTP-binding proteins of the ras superfamily (Figure 4). As some of the ryh1p sequences adjacent to the conserved nucleotide-binding regions differ from other yeast Ypt proteins, the gene was provisionally given its name to stress its relatedness to Ras, Rho, and Ypt (Ras/Rho/Ypt homolog).[7] The ryh1 protein exhibits a high degree of structural homology (more than 70% of identical amino acid residues) to mammalian Rab6p,[8] with the effector regions and C termini (CysXCys) being identical (Figure 5).

Null mutants of *ryh1* are viable but temperature-sensitive. The temperature-sensitive (ts) phenotype can be complemented by expressing human rab6 cDNA in the yeast mutant cells, suggesting similar functions of the proteins in yeast and mammals.[7]

E. OTHERS

Two other genes encoding ypt1-related proteins have been isolated (J. Armstrong, personal communication). The protein product of one of them, designated *ypt5*, is identical with mammalian Rab5p in more than 70% of its amino acid residues. A structural analog for the other protein has not been identified among the Ypt/Rab proteins of other species.

IV. COMMON STRUCTURAL AND BIOCHEMICAL FEATURES

Sequence divergence as well as placement and conservation of short peptide domains has allowed the classification of the ras superfamily of GTP-binding proteins into the Ras, Rho, Ypt/Rab, and Arf families.[51-55] From crystal structures of the bacterial elongation factor EF-Tu[56,57] and the human H-*ras* p21,[34,58] it can be readily observed that the tertiary structure of the nucleotide-binding pocket of these two distantly related proteins is remarkably similar, even though the amino acid sequence similarity of the domains making up the binding pocket is poorly conserved. This illustrates the highly constrained nature of the GTP-binding pocket and indicates that the tertiary structure of closer related GTP-binding proteins, i.e., the Ras and Ypt families, will be quite similar. Unfortunately, no crystals of Ypt1p or other Ypt proteins have been obtained thus far.

Five regions, G-1 through G-5, have been identified that are conserved in all ras superfamily proteins and these regions are directly involved in guanine nucleotide-binding.[54,55] Another characteristic feature of Ypt and other Ras-related proteins (except the Arf family members) is the occurrence of one or more cysteine residues at or near the C-terminus. These cysteines are subject to covalent modifications and are required for membrane association of the proteins.

Ypt proteins bind guanine nucleotides specifically and they have a low intrinsic GTPase activity. With nearly all yeast Ypt proteins, their capacity to bind GTP and GDP has been demonstrated by the so-called GTP-blot analysis where the electrophoretically separated proteins are blotted onto nitrocellulose or nylon membranes and, after renaturation, are used for binding of ^{32}P-labeled nucleotides.[6,7,11,18,59] Guanine nucleotide-binding to purified Ypt proteins in solution has only been performed with Ypt1p[60] and Sec4p.[40] The effects of amino acid substitutions on the biochemical and functional properties of Ypt1p have been particularly helpful to define important sequence segments in Ypt/Rab proteins in general.[11,14,28,59]

A. CONSERVED REGIONS FOR NUCLEOTIDE BINDING

The five sequence segments (G-1 to G-5) shown from X-ray studies of human H-*ras* p21[34,55] to interact with the bound nucleotide are conserved in all yeast Ypt proteins (Figures 3 and 4).

Substitution of lysine with methionine in G-1 (GXXXXGK$^S/_T$) renders Ypt1p nonfunctional and leads to inviable cells. Nucleotide binding of the mutant protein on filters was no longer detectable.[59] This result supports the hypothesis that the Ras and Ypt proteins' tertiary structures are highly conserved, since it is known that the corresponding lysine in Ras protein contacts the β- and γ-phosphates.[61] Substitution of alanine with threonine in G-3 (DTAG) led to autophosphorylation of Ypt1p[59] in exactly the same way as the identical substitution in oncogenic viral Ras proteins.[62] Surprisingly, this mutation in Ypt1p did not result in an impairment of yeast cell viability.[59]

In contrast, substitution of asparagine with isoleucine in G-4 (NKXD) had a dominant-lethal effect[11] and significantly impaired the capacity of the Ypt1p to bind GTP.[59] It should be noted that the same substitution in Sec4p[19] and mammalian Rab5p[63] has comparable deleterious consequences. Likewise, this mutation significantly affects the guanine nucleotide-binding properties of human Ras proteins.[64] The conserved asparagine residue in G-4 participates in binding of the guanine base to the H-*ras* p21,[34,61] thus the effects on nucleotide binding following asparagine > isoleucine substitutions in Ypt/Rab proteins once again indicates very similar tertiary structure for all of these proteins.

The intrinsic GTPase activity of Ypt proteins, 0.006 min^{-1} for Ypt1p[59] and 0.0012 min^{-1} for Sec4p[40] (measured at 30°C), is even lower than that determined for H-*ras* p21 (0.028 min^{-1}, at 37°C).[60] Enhancement of the intrinsic GTPase activity of Ypt1p by replacing serine 17 with glycine within the conserved G-1 segment did not impair the growth properties of mutant cells.[59] Interestingly, mutating the corresponding amino acid in Ras proteins, glycine-12, leads to a decrease in GTPase activity and results in a transforming potential for the mutant proteins.[65]

For determining the rate constants for association and dissociation of Ypt1p and guanine nucleotides, nucleotide- and Mg^{2+}-free protein was used as starting material. At 30°C, the calculated GDP binding affinity constant of Ypt1p was $2.2 \times 10^9 \ M^{-1}$,[60] which is comparable to the high binding constant determined for Ras p21·GDP.[67]

B. EFFECTOR REGION

One of the most interesting regions in proteins of the Ras superfamily is the G-2 domain, otherwise known as the effector region. This sequence segment is highly conserved between interspecies homologs, and the precise sequence found in this region is indicative of functional similarity. In Figure 2 this relationship can be seen. The effector domain is highly solvated and is one of the two regions (the other being G-3) of the GTP-binder that undergoes a drastic conformation change on hydrolysis of GTP to GDP, as shown by the high resolution crystal structure of Ras p21 in the GTP- and GDP-bound forms.[58,68] It would be expected, therefore, that molecules interacting with Yptp that differentiate between the GTP- and GDP-bound forms (such as GAP and GDI) would interact with one or both of these regions. In the case of H-*ras* p21, this region (spanning residues 32–40) has been shown by mutational analysis to interact with ras-GAP.[69,70]

The primary sequence of the putative effector regions of Ypt and Rab proteins is similar to that of Ras proteins, but individual members of the Ypt/Rab family are characterized by slight differences of that domain (see Figure 5).[6,8,9,10] Since, in all cases where yeast Ypt mutants have been complemented with Ypt/Rab proteins from other species, complementations were successful only with proteins having the identical effector region (Figure 2),[5-7,24] it is evident that the effector domain plays a pivotal role for the functional specificity of different Ypt/Rab proteins. In a mutational analysis of the effector domain of *S. cerevisiae* Ypt1p, several amino acid substitutions led to either lethality or conditional lethality.[28,71] Most importantly, mutations of the effector region resulting in an impairment of cell growth also had a profound negative effect on the activation of the intrinsic GTPase activity of the mutant proteins by ypt-GAPs partially purified from either yeast or mammalian sources.[28,35] This suggests, by analogy to the interaction of ras-GAP and H-*ras* p21, that ypt-GAP interacts with Ypt1p via the effector region of the GTP-binding protein and this interaction is crucial to the regulation of the GTPase activity of Ypt1p and intracellular function.

C. C-TERMINUS

The C-termini of members of the ras superfamily contain a critical cysteine residue necessary for membrane attachment. In the case of Ras proteins, the C-terminal sequence is the conserved motif CAAX, where C is cysteine, A is any aliphatic amino acid, and X is any amino acid. The modification of the CAAX motif during maturation is necessary for the proper attachment of Ras proteins to the plasma membrane, but is not in itself sufficient for membrane association.[72] In mammalian cells as well as in the yeast *S. cerevisiae,* the C-terminal cysteine residue is modified by covalent linkage with a farnesyl moiety.[73-75] Farnesylation is catalyzed by a transferase composed of two nonidentical subunits.[76,77]

The Ypt/Rab family allows more variation of the C-terminus than the Ras family, as is shown in Figure 5, but also requires a C-terminal cysteine. For yeast Ypt1p, the presence of at least one of the two C-terminal cysteine residues is required for the proper membrane association of the protein.[27] There is evidence that Ypt1p is palmitoylated, but it is unclear which amino acid residue of the protein is linked to the fatty acid. The deletion or substitution with serine of the C-terminal cysteines resulted in a mutant Ypt1 protein that was not localized to membrane fractions and had no apparent biological function, although the GTP binding and hydrolysis rates for the mutant protein were normal.[27] The cysteine residues of yeast Sec4p are likewise required for membrane attachment and biological function of the protein.[19]

Most of the mammalian Rab proteins and all of the yeast Ypt proteins currently known terminate with either two cysteines or with a CysXCys motif (Figure 5). There is increasing evidence that the members of the Ypt/Rab family are modified by geranylgeranylation and that individual transferases are composed of two subunits.[39,78-81] Interestingly, the *BET2* gene product initially identified to be required for ER-to-Golgi transport in yeast exhibits structural homology to Dpr1/Ram1p, one of the subunits of the *S. cerevisiae* farnesyl transferase. *BET2* is required for membrane attachment both of Ypt1p and Sec4p,[39] suggesting that it encodes a subunit of a Ypt protein modification enzyme, presumably a geranylgeranyl transferase.

V. OUTLOOK

Evidence is accumulating that most, if not all, of the small guanine nucleotide-binding proteins belonging to the Ypt/Rab family serve regulatory functions in intracellular vesicle transport. The recent development of *in vitro* transport systems[21,29,38] promises to be a rich source of information regarding the biochemical identities and mechanisms of inter- and intracompartmental protein transport, with the results of current and future genetic and biochemical analysis of Ypt proteins providing fundamental insights and molecular handles into the function and interaction of transport processes.

For a detailed understanding of how Ypt/Rab proteins act, key questions to be answered are: (1) when in the functional cycle is hydrolysis of protein-bound GTP required; (2) what proteins interact with the GTP binders during their cycle between the membrane-attached and the soluble form; and (3) whether, and how, interacting proteins, like GTPase activators and nucleotide exchange factors, themselves are regulated during cell growth and proliferation. One might expect that the attachment of different Ypt proteins to distinct cellular compartments (ER, Golgi cisternae, various transport vesicles, endosomes, etc.) requires specific receptors. This is suggested from a recent study with mammalian Rab proteins showing that a signal for specific membrane association lies within the variable C-terminal domain.[82] It also seems likely that some component for docking of transport vesicles to the proper cellular compartment exists. Furthermore, specific proteins required for nucleotide exchange, inhibitors (GDI)[41,42] or activators (GDS),[83,84] are likely to be essential components of the functional cycle of the Ypt/Rab protein (Figure 1). None of those proteins has yet been identified in yeast. There is, however, firm evidence for the existence of GTPase-activating proteins specific for different members of the Ypt family of proteins in yeast (Reference 35 and M. Strom and D. Gallwitz, unpublished results).

It can be anticipated that because of its easy handling and its genetic amenability, the unicellular yeast will continue to be a pacesetter in this field of research.

NOTE ADDED IN PROOF

We have recently shown that Ypt7p most likely acts in protein transport between early and late endosome-like compartments.[85] As Rab7p, the mammalian counterpart of the yeast Ypt7p, appears to be localized primarily on late endosomes, our findings suggest that both proteins might perform similar functions. By screening a yeast genomic DNA library for high expression of GTPase-activating proteins, we isolated a gene encoding a GAP with high specificity of Ypt6p. The *GYP6* gene product (for GAP of Ypt6 protein) is a protein of 458 amino acids that does not display a significant similarity to known GTPase-activating proteins with specificity for Ras or Rho family members.[86]

REFERENCES

1. **Gallwitz, D., Donath, C., and Sander, C.,** A yeast gene encoding a protein homologous to the human c-*has*/*bas* proto-oncogene product, *Nature,* 306, 704, 1983.
2. **Haubruck, H., Disela, C., Wagner, P., and Gallwitz, D.,** The *ras*-related *ypt* protein is a ubiquitous eukaryotic protein: isolation and sequence analysis of mouse cDNA clones highly homologous to the yeast *YPT1* gene, *EMBO J.,* 6, 4049, 1987
3. **Touchot, N., Chardin, P., and Tavitian, A.,** Four additional members of the ras gene superfamily isolated by an oligonucleotide strategy: molecular cloning of YPT-related cDNAs from a rat brain library, *Proc. Natl. Acad. Sci. U.S.A.,* 84, 8210, 1987.
4. **Salminen, A. and Novick, P.,** A *ras*-like protein is required for a post-Golgi event in yeast secretion, *Cell,* 49, 527, 1987.
5. **Miyake, S. and Yamamoto, M.,** Identification of *ras*-related, *YPT* family genes in *Schizosaccharomyces pombe, EMBO J.,* 9, 1417, 1990.
6. **Haubruck, H., Engelke, U., Mertins, P., and Gallwitz, D.,** Structural and functional analysis of *ypt2,* an essential *ras*-related gene in the fission yeast *Schizosaccharomyces pombe* encoding a Sec4 protein homologue, *EMBO J.,* 9, 1957, 1990.
7. **Hengst, L., Lehmeier, T., and Gallwitz, D.,** The *ryh1* gene in the fission yeast *Schizosaccharomyces pombe* encoding a GTP-binding protein related to ras, rho and ypt: structure, expression and identification of its human homologue, *EMBO J.,* 9, 1949, 1990.
8. **Zahraoui, A., Touchot, N., Chardin, P., and Tavitian, A.,** The human *Rab* genes encode a family of GTP-binding proteins related to yeast *YPT1* and *SEC4* products involved in secretion, *J. Biol. Chem.,* 264, 12394, 1989.
9. **Chavrier, P., Parton, R. G., Hauri, H. P., Simons, K., and Zerial, M.,** Localization of low molecular weight GTP binding proteins to exocytic and endocytic compartments, *Cell,* 62, 317, 1990.
10. **Chavrier, P., Simons, K., and Zerial, M.,** The complexity of the rab and rho GTP binding protein subfamilies revealed by a PCR cloning approach, *Gene,* 112, 261, 1992.
11. **Schmitt, H. D., Wagner, P., Pfaff, E., and Gallwitz, D.,** The *ras*-related *YPT1* gene product in yeast: a GTP-binding protein that might be involved in microtubule organization, *Cell,* 47, 401, 1986.
12. **Segev, N. and Botstein, D.,** The ras-like *YPT1* gene is itself essential for growth, sporulation and starvation response, *Mol. Cell. Biol.,* 7, 2367, 1987.
13. **Segev, N., Mulholland, J., and Botstein, D.,** The yeast GTP-binding YPT1 protein and a mammalian counterpart are associated with the secretion machinery, *Cell,* 52, 915, 1988.
14. **Schmitt, H. D., Puzicha, M., and Gallwitz, D.,** Study of a temperature-sensitive mutant of the *ras*-related *YPT1* gene product in yeast suggests a role in the regulation of intracellular calcium, *Cell,* 53, 635, 1988.
15. **Bacon, R. A., Salminen, A., Ruohola, H., Novick, P., and Ferro-Novick, S.,** The GTP-binding protein ypt1 is required for transport in vitro: the Golgi apparatus is defective in ypt1 mutants, *J. Cell. Biol.,* 109, 1015, 1989.

16. **Novick, P., Field, C., and Schekman, R.,** Identification of 23 complementation groups required for post-translational events in the yeast secretory pathway, *Cell,* 21, 205, 1980.

17. **Novick, P., Ferro, S., and Schekman, R.,** Order of events in the yeast secretory pathway, *Cell,* 25, 461, 1981.

18. **Goud, B., Salminen, A., Walworth, N. C., and Novick, P. J.,** A GTP-binding protein required for secretion rapidly associates with secretory vesicles and the plasma membrane in yeast, *Cell,* 53, 753, 1988.

19. **Walworth, N. C., Goud, B., Kabcenell, A. K., and Novick, P. J.,** Mutational analysis of SEC4 suggests a cyclical mechanism for the regulation of vesicular traffic, *EMBO J.,* 8, 1685, 1989.

20. **Segev, N.,** Mediation of the attachment or fusion step in vesicular transport by the GTP-binding of Ypt1 protein, *Science,* 252, 1553, 1991.

21. **Rexach, M. F. and Schekman, R. W.,** Distinct biochemical requirements for the budding, targeting, and fusion of ER-derived transport vesicles, *J. Cell. Biol.,* 114, 219, 1991.

22. **Bourne, H.,** Do GTPases direct membrane traffic in secretion?, *Cell,* 53, 669, 1988.

23. **Rothman, J. E. and Orci, L.,** Molecular dissection of the secretory pathway, *Nature,* 355, 409, 1992.

24. **Haubruck, H., Prange, R., Vorgias, C., and Gallwitz, D.,** The *ras*-related mouse ypt1 protein can functionally replace the *YPT1* gene product in yeast, *EMBO J.,* 8, 1427, 1989.

25. **Balch, W. E.,** Biochemistry of interorganelle transport, *J. Biol. Chem.,* 264, 16965, 1989.

26. **Halfter, H., Müller, U., Winnaker, E. L., and Gallwitz, D.,** Isolation and DNA-binding characteristics of a protein involved in transcription activation of two divergently transcribed, essential genes, *EMBO J.,* 8, 3029, 1989.

27. **Molenaar, C. M. T., Prange, R., and Gallwitz, D.,** A carboxyl-terminal cysteine residue is required for palmitic acid binding and biological activity of the *ras*-related yeast *YPT1* protein, *EMBO J.,* 7, 971, 1988.

28. **Becker, J., Tan, T. J., Trepte, H.-H., and Gallwitz, D.,** Mutational analysis of the putative effector domain of the GTP-binding Ypt1 protein in yeast suggests specific regulation by a novel GAP activity, *EMBO J.,* 10, 785, 1991.

29. **Baker, D., Wuestehube, L., Schekman, R., Botstein, D., and Segev, N.,** GTP-binding Ypt1 protein and Ca^{2+} function independently in a cell-free protein transport reaction, *Proc. Natl. Acad. Sci. U.S.A.,* 87, 355, 1990.

30. **Groesch, M. E., Ruohola, H., Bacon, R., Rossi, G., and Ferro-Novick, S.,** Isolation of a functional vesicular intermediate that mediates ER to Golgi transport in yeast, *J. Cell. Biol.,* 111, 45, 1990.

31. **Beckers, C. J. M. and Balch, W. E.,** Calcium and GTP: essential components in vesicular trafficking between the endoplasmic reticulum and Golgi apparatus, *J. Cell. Biol.,* 108, 1245, 1989.

32. **Plutner, H., Cox, A. D., Pind, S., Khosravi-Far, R., Bourne, J. R., Schwaninger, R., Der, C. J., and Balch, W. E.,** Rab1b regulates vesicular transport between the endoplasmic reticulum and successive Golgi compartments, *J. Cell. Biol.,* 115, 31, 1991.

33. **Sigal, I., Gibbs, J., D'Alanzo, J., Temeles, G., Wolanski, B., Socher, S., and Scolnick, E.,** Mutant *ras*-encoded proteins with altered nucleotide binding exert dominant biological effects, *Proc. Natl. Acad. Sci. U.S.A.,* 83, 952, 1986.

34. **Pai, E. F., Kabsch, W., Krengel, U., Holmes, K. C., John, J., and Wittinghofer, A.,** Structure of the guanine-nucleotide-binding domain of the Ha-ras oncogene product p21 in the triphosphate conformation, *Nature,* 341, 209, 1989.

35. **Tan, T. J., Vollmer, P., and Gallwitz, D.,** Identification and partial purification of GTPase-activating proteins from yeast and mammalian cells that preferentially act on Ypt1/Rab1 proteins, *FEBS Lett.,* 291, 322, 1991.

36. **Dascher, C., Ossig, R., Gallwitz, D., and Schmitt, H. D.,** Identification and structure of four yeast genes *(SLY)* that are able to suppress the functional loss of *YPT1,* a member of the *ras* superfamily, *Mol. Cell. Biol.,* 11, 872, 1991.

37. **Ossig, R., Dascher, C., Trepte, H.-H., Schmitt, H. D., and Gallwitz, D.,** The yeast SLY gene products, suppressors of defects in the essential GTP-binding Ypt1 protein, may act in endoplasmic reticulum-to-Golgi transport, *Mol. Cell. Biol.,* 11, 2980, 1991.

38. **Newman, A. P. and Ferro-Novick, S.,** Characterization of new mutants in the early part of the yeast secretory pathway isolated by a [³H]mannose suicide selection, *J. Cell. Biol.,* 105, 1587, 1987.

39. **Rossi, G., Jiang, Y., Newman, A. P., and Ferro-Novick, S.,** Dependence of Ypt1 and Sec4 membrane attachment on Bet2, *Nature,* 351, 158, 1991.

40. **Kabcenell, A. K., Goud, B., Northrup, J. K., and Novick, P. J.,** Binding and hydrolysis of guanine nucleotides by Sec4p, a yeast protein involved in the regulation of vesicular traffic, *J. Biol. Chem.,* 265, 9366, 1990.

41. **Sasaki, T., Kikuchi, A., Araki, S., Hata, Y., Isomura, M., Kuroda, S., and Takai, Y.,** Purification and characterization from bovine brain cytosol of a protein that inhibits the dissociation of GDP from and the subsequent binding of GTP to *smg* p25A, a *ras*-like GTP-binding protein, *J. Biol. Chem.,* 265, 2333, 1990.

42. **Sasaki, T., Kaibuchi, K., Kabcenell, A. K., Novick, P. J., and Takai, Y.,** A mammalian inhibitory GDP/GTP exchange protein (GDI dissociation inhibitor) for *smg* p25A is active on the yeast *SEC4* protein, *Mol. Cell. Biol.,* 11, 2909, 1991.

43. **Russel, P. and Nurse, P.,** *Schizosaccharomyces pombe* and *Saccharomyces cerevisiae*: a look at yeasts divided, *Cell,* 45, 781, 1986.

44. **Fukui, Y., Kozasa, T., Kazira, Y., Takeda, T., and Yamamoto, M.,** Role of a *ras* homolog in the life cycle of *Schizosaccharomyces pombe, Cell,* 44, 329, 1986.

45. **Toda, T., Uno, I., Ishikawa, T., Powers, S., Kataoka, T., Broek, D., Cameron, S., Broach, J., Matsumoto, K., and Wigler, M.,** In yeast, *ras* proteins are controlling elements of adenylate cyclase, *Cell,* 40, 27, 1985.

46. **Fawell, E., Hook, S., and Armstrong, J.,** Nucleotide sequence of a gene encoding a *YPT1*-related protein from *Schizosaccharomyces pombe, Nucleic Acids Res.,* 17, 4373, 1989.

47. **Fawell, E., Hook, S., Sweet, D., and Armstrong, J.,** Novel *YPT1*-related genes from *Schizosaccharomyces pombe, Nucleic Acids Res.,* 18, 4264, 1990.

48. **Miyake, S., Tanaka, A., and Yamamoto, M.,** Mapping of four *ras* superfamily genes by physical and genetic means in *Schizosaccharomyces pombe, Curr. Genet.,* 20, 277, 1991.

49. **Wichmann, H., Disela, C., Haubruck, H., and Gallwitz, D.,** Nucleotide sequence of the mouse *ypt1* gene encoding a *ras*-related GTP-binding protein, *Nucleic Acids Res.,* 17, 6737, 1989.

50. **Saxe, S. A. and Kimmel, A. R.,** *SAS1* and *SAS2*, GTP-binding protein genes in *Dictyostelium discoideum* with sequence similarities to essential genes in *Saccharomyces cerevisiae, Mol. Cell. Biol.,* 10, 2367, 1990.

51. **Chardin, P., Touchot, N., Zahraoui, A., Pizon, V., Lerosey, I., Olofsson, B., and Tavitian, A.,** Structure of the human *ras* family, in *The Guanine-Nucleotide Binding Proteins,* Bosch, L., Kraal, B., and Parmeggiani, A., Eds., Plenum, New York, 1989, 153.

52. **Gallwitz, D., Haubruck, H., Molenaar, C., Prange, R., Puzicha, M., Schmitt, H. D., Vorgias, C., and Wagner, P.,** Structural and functional analysis of *ypt* proteins, a family of *ras*-related nucleotide-binding proteins in eukaryotic cells, in *The Guanine-Nucleotide Binding Proteins,* Bosch, L., Kraal, B., and Parmeggiani, A., Eds., Plenum, New York, 1989, 257.

53. **Hall, A.,** The cellular functions of small GTP-binding proteins, *Science,* 249, 635, 1990.

54. **Bourne, H. R., Sanders, D. A., and McCormick, F.,** The GTPase superfamily: conserved structure and molecular mechanism, *Nature,* 349, 117, 1991.

55. **Valencia, A., Chardin, P., Wittinghofer, A., and Sander, C.,** The *ras* protein family: evolutionary tree and role of conserved amino acids, *Biochemistry,* 30, 4637, 1991.

56. **La Cour, T. F. M., Nyborg, J., Thirup, S., and Clark, B. F. C.,** Structural details of the binding of guanosine diphosphate to elongation factor Tu from *E. coli* as studied by X-ray crystallography, *EMBO J.,* 4, 2385, 1985.

57. **Jurnak, F.,** Structure of the GDP domain of EF-Tu and location of the amino acids homologous to *ras* oncogene proteins, *Science,* 230, 32, 1985.
58. **Milburn, M. V., Tong, L., deVos, A. M., Brünger, A., Yamaizumi, Z., Nishimura, S., and Kim, S.-H.,** Molecular switch for signal transduction: structural differences between active and inactive forms of protooncogenic *ras* proteins, *Science,* 247, 939, 1990.
59. **Wagner, P., Molenaar, C. M. T., Rauh, A. J. G., Brökel, R., Schmitt, H. D., and Gallwitz, D.,** Biochemical properties of the *ras*-related *YPT* protein in yeast: a mutational analysis, *EMBO J.,* 6, 2373, 1987.
60. **Wagner, P., Hengst, L., and Gallwitz, D.,** Ypt proteins in yeast, *Methods Enzymol.,* 219, 369, 1992.
61. **Wittinghofer, A. and Pai, E. F.,** The structure of Ras protein: a model for a universal molecular switch, *Trends Biochem.,* 16, 382, 1991.
62. **Shih, T. Y., Stokes, P. E., Smythers, G. W., Dhar, R., and Oroszlan, S.,** Characterization of the phosphorylation sites and the surrounding amino acid sequences of the p21 transforming proteins coded for by the Harvey and Kirsten strains of murine sarcoma viruses, *J. Biol. Chem.,* 257, 11767, 1982.
63. **Gorvel, J.-P., Chavrier, P., Zerial, M., and Gruenberg, J.,** Rab5 controls early endosome fusion in vitro, *Cell,* 64, 915, 1991.
64. **Walter, M., Clark, S. G., and Levinson, A. D.,** The oncogenic activation of human p21^{ras} by a novel mechanism, *Science,* 233, 649, 1986.
65. **Seeburg, P. H., Colby, W. W., Capon, D. J., Goeddel, D. V., and Levinson, A. D.,** Biological properties of human c-Ha-*ras*1 genes mutated at codon 12, *Nature,* 312, 71, 1984.
66. **John, J., Sohmen, R., Feuerstein, J., Linke, R., Wittinghofer, A., and Goody, R. S.,** Kinetics of interaction of nucleotides with nucleotide-free H-ras p21, *Biochemistry,* 29, 6058, 1990.
67. **John, J., Frech, M., and Wittinghofer, A.,** Biochemical properties of Ha-ras encoded p21 mutants and mechanism of the autophosphorylation reaction, *J. Biol. Chem.,* 263, 11792, 1988.
68. **Schlichting, I., Almo, S. C., Rapp, G., Wilson, K., Petratos, K., Lentfer, A., Wittinghofer, A., Kabsch, W., Pai, E. F., Petsko, G. A., and Goody, R. S.,** Time-resolved X-ray crystallographic study of the conformational change in Ha-Ras p21 protein on GTP hydrolysis, *Nature,* 345, 309, 1990.
69. **Adari, H., Lowy, D. R., Willumsen, B. M., Der, C. J., and McCormick, F.,** Guanosine triphosphatase activating protein (GAP) interacts with the p21 *ras* effector binding domain, *Science,* 240, 518, 1988.
70. **Calés, C., Hancock, J. F., Marshall, C. J., and Hall, A.,** The cytoplasmic protein GAP is implicated as the target for regulation by the *ras* gene product, *Nature,* 332, 548, 1988.
71. **Gallwitz, D., Becker, J., Benli, M., Hengst, L., Mosrin-Huaman, C., Mundt, M., Tan, T. J., Vollmer, P., and Wichmann, H.,** The YPT-branch of the ras superfamily of GTP-binding proteins in yeast: functional importance of the putative effector region, in *The Superfamily of ras-Related Genes,* Spandidos, D. A., Ed., Plenum, New York, 1992, 121.
72. **Hancock, J. F., Cadwallader, K., and Marshall, C. J.,** Methylation and proteolysis are essential for efficient membrane binding of prenylated p21$^{K-ras(B)}$, *EMBO J.,* 10, 641, 1991.
73. **Wolda, S. L. and Glomset, J. A.,** Evidence for modification of lamin B by a product of mevalonic acid, *J. Biol. Chem.,* 263, 5997, 1988.
74. **Hancock, J. F., Magee, A. I., Childs, J. E., and Marshall, C. J.,** All *ras* proteins are polyisoprenylated but only some are palmitoylated, *Cell,* 57, 1167, 1989.
75. **Schafer, W. R., Trueblood, C. C., Yang, C.-Y., Mayer, M. P., Rosenberg, S., Poulter, C. D., Kim, S.-H., and Rine, J.,** Enzymatic coupling of cholesterol intermediates to a mating pheromone precursor and to the Ras protein, *Science,* 249, 1133, 1990.
76. **Goodman, L. E., Judd, S. R., Farnsworth, C. C., Powers, S., Gelb, M. H., Glomset, J. A., and Tamanoi, F.,** Mutants of *Saccharomyces cerevisiae* defective in the farnesylation of Ras proteins, *Proc. Natl. Acad. Sci. U.S.A.,* 87, 9665, 1990.

77. **Reiss, Y., Seabra, M. C., Armstrong, S. A., Slaughter, C. A., Goldstein, J. L., and Brown, M. S.,** Nonidentical subunits of p21$^{H\text{-}ras}$ farnesyltransferase, *J. Biol. Chem.,* 266, 10672, 1991.

78. **Finegold, A. A., Johnson, D. I., Farnsworth, C. C., Gelb, M. H., Judd, S. R., Glomset, J. A., and Tamanoi, F.,** Protein geranylgeranyltransferase of *Saccharomyces cerevisiae* is specific for Cys-Xaa-Xaa-Leu motif proteins and requires the *CDC43* gene product but not the *DPR1* gene product, *Proc. Natl. Acad. Sci. U.S.A.,* 88, 4448, 1991.

79. **Khosravi-Far, R., Lutz, R. J., Cox, A. D., Conroy, L., Bourne, J. R., Sinensky, M., Balch, W. E., Buss, J. E., and Der, C. J.,** Isoprenoid modification of rab proteins terminating in CC or CXC motifs, *Proc. Natl. Acad. Sci. U.S.A.,* 88, 6264, 1991.

80. **Moores, S. L., Schaber, M. D., Mosser, S. D., Rands, E., O'Hara, M. B., Garsky, V. M., Marshall, M. S., Pompliano, D. L., and Gibbs, J. B.,** Sequence dependence of isoprenylation, *J. Biol. Chem.,* 266, 14603, 1991.

81. **Kohl, N. E., et al.,** Structural homology among mammalian and *Saccharomyces cerevisiae* isoprenyl-protein transferases, *J. Biol. Chem.,* 266, 18884, 1991.

82. **Chavrier, P., Gorvel, J.-P., Stelzer, E., Simons, K., Gruenberg, J., and Zerial, M.,** Hypervariable C-terminal domain of rab proteins acts as a targeting signal, *Nature,* 353, 769, 1991.

83. **Huang, Y. K., Kung, H.-F., and Kamata, T.,** Purification of a factor capable of stimulating the guanine nucleotide exchange reaction of ras proteins and its effect on ras-related small molecular mass G proteins, *Proc. Natl. Acad. Sci. U.S.A.,* 87, 8008, 1990.

84. **Kaibuchi, K., Mizuno, T., Fujioka, H., Yamamoto, T., Kishi, K., Fukumoto, Y., Hori, Y., and Takai, Y.,** Molecular cloning of the cDNA for stimulatory GDP/GTP exchange protein for *smg* p21s (*ras*-like small GTP-binding proteins) and characterization of stimulatory GDP/GTP exchange protein, *Mol. Cell. Biol.,* 11, 2873, 1991.

85. **Wichmann, H., Hengst, L., and Gallwitz, D.,** Endocytosis in yeast: evidence for involvement of a small GTP-binding protein (Ypt7p), *Cell,* 71, 1131, 1992.

86. **Strom, M., Vollmer, P., Tan, T. J., and Gallwitz, D.,** Cloning of a gene encoding a GTPase-activating protein with specificity for a GTP-binding protein of the Ypt/Rab family, *Nature,* in press, 1993.

Chapter 20

THE *SAS* GENES: FUNCTIONALLY DISTINCT MEMBERS OF THE *YPT1/SEC4* FAMILY IN *Dictyostelium*

Alan R. Kimmel, Tracy Ruscetti, and James Cardelli

TABLE OF CONTENTS

0-8493-5214-2/93/$0.00 + $.50
© 1993 by CRC Press, Inc.

I. INTRODUCTION

Dictyostelium discoideum grows as a unicellular, ameboid organism. When cells exhaust their food supply, they initiate a developmental program which leads to the formation of a multicellular aggregate comprised of distinct, non-dividing cell types.[1,2] These cells, organized through morphogenetic movement and cell sorting, are nonterminally differentiated prestalk and prespore cells, the precursors to the terminally differentiated stalk and spore cells found in the fully mature fruiting body at the culmination of development.[1] Aggregation, differentiation, and cell sorting are directed by secreted (extracellular) cAMP which serves as both a chemoattractant and a hormone-like, primary signaling molecule.[2]

The prestalk and prespore cells first appear late in aggregation as multicellular mounds being to form.[1] Continued morphological and biochemical differentiation leads to the pseudoplasmodium, a migrating slug comprised of the precursor cells organized in a distinct spatial pattern. The prestalk cells are found predominantly in the anterior ~15% and at the very most posterior of the organism. The anterior prestalk tip is believed to act as the primary signaling and organizing center throughout development. Eventually slug migration arrests and a spatially organized mound of cells is reformed with an elongated tip. This structure develops into the fruiting body comprised of the mature spore and stalk cells.

The complex extracellular cAMP signaling system which organizes development is mediated through surface receptors linked to heterotrimeric G proteins.[2] *Dictyostelium* is an excellent system for study of GTP-binding regulatory proteins. Mutations in $G\alpha$ and *ras* genes interfere with cAMP signaling and development (see Chapter 8).[2] Other aspects of growth and/or development in *Dictyostelium* are also regulated by GTP-binding proteins. Such processes may include ion transport, intracellular signaling, secretion, and membrane trafficking. Intracellular vesicles are important structures throughout the *Dictyostelium* life cycle. For instance, endocytic and lysosomal vesicles are utilized during growth and vesicles localized specifically in prespore cells are required for spore coat formation.[3-5] Accordingly, we have begun studies to understand the role of GTP-binding proteins in the regulation of membrane trafficking in *Dictyostelium*.

We have identified SAS1 and SAS2, two genes of *Dictyostelium* which encode ~20kDa, GTP-binding proteins that are ~90% identical in sequence.[6] They are members of the ras superfamily but are most related to the YPT1/SEC4 family of proteins involved in trafficking of intracellular vesicles (see Chapter 19).[7-15] We have determined that SAS proteins are expressed during growth and development and that at least a fraction of the SAS protein may colocalize with structures of the endoplasmic reticulum.

We present evidence to suggest that the SAS proteins may function in a manner that is distinct from YPT1 or SEC4. However, they may share a common intracellular pathway and, similar to YPT1/SEC4, the SAS proteins may be essential throughout the *Dictyostelium* life cycle to regulate intracellular signals associated with membrane function.

```
SAS1          MTSPATNKSAAYDYLIKLLLIGDSGVGKSCLLLRFSEDSF
              :::::::::: ::::::.::::::::::::::::::::::. ::
SAS2          MTSPATNKPAAYDFLVKLLLIGDSGVGKSCLLLRFSDGSF

                      ********
SAS1          TPSFITTIGIDFKIRTIELEGKRIKLQIWDTAGQERFRTI
              :::::.::::::::::::::::::::::::::::::::::::
SAS2          TPSFIATIGIDFKIRTIELEGKRIKLQIWDTAGQERFRTI


SAS1          TTAYYRGAMGILLVYDVTDEKSFGNIRNWIRNIEQHATDS
              :::::::::::::::::::::::::: :::::::::::::.::
SAS2          TTAYYRGAMGILLVYDVTDEKSFGSIRNWIRNIEQHASDS


SAS1          VNKMLIGNKCDMAEKKVVDSSRGKSLADEYGIKFLETSAK
              :::::::::::::.:::::::::::::::::::::::::::
SAS2          VNKMLIGNKCDMTEKKVVDSSRGKSLADEYGIKFLETSAK

                          ◄---------------►
SAS1          NSINVEEAFISLAKDIKKRMIDTPNEQPQVVQPGTNLGAN
              ::.:::::::: ::::::::::::::::.         :   .
SAS2          NSVNVEEAFIGLAKDIKKRMIDTPNDP       DHTICITP


SAS1          NNKKKACC
              :::: .::
SAS2          NNKKNTCC
```

FIGURE 1. Amino acid sequence comparison between SAS1 and SAS2. Identities are indicated by double dots (:) and similarities based upon the grouping of Dayhoff et al.[16] by a single dot (.). The putative effector domain for the SAS proteins is noted (********). The 17-mer SAS1 peptide used to produce rabbit anti-SAS serum is indicated as (◄————►). (Adapted from Saxe, S. A. and Kimmel, A. R., *Mol. Cell. Biol.*, 10, 2367, 1990.)

II. RESULTS AND DISCUSSION

A. SEQUENCE COMPARISONS OF THE SAS1 AND SAS2 PROTEINS

The *SAS1* gene was first identified as a member of a specific class of developmentally regulated *Dictyostelium* genes.[6] Genomic DNA blot hybridization suggested the presence of two related genes in *Dictyostelium*. This was confirmed through library screening, polymerase chain reaction (PCR) analysis and blot hybridization. DNA sequence analysis of genomic and cDNA clones was used to predict and compare the amino acid sequence of the two proteins, SAS1 and SAS2 (Figure 1). *SAS1* and *SAS2* encode proteins of nearly identical size and sequence with differences primarily restricted to their carboxyl termini.

A FASTA comparison of the SAS1 amino acid sequence with those listed in the GenBank database indicated primary identity among the members of the ras super-family of GTPases (Figure 2).[16,17] Across this entire diverse grouping, identity was essentially restricted to domains that interact with guanine nucleotides and to functionally equivalent carboxyl termini, which are potential sites

Protein	% Total Identity	% N-Terminal Identity	C-Terminus Motif
SAS2	87	94	xxCC
YPT2	70	98	xxCC
RAB8	67	86	Cxxx
RAB10	60	77	xxCC
YPT1/RAB1	58	80	xxCC
SEC4	55	80	xxCC
RAB3	47	66	xCxC
RAB2	46	62	xxCC
YPT3	43	62	xxCC
RAB11	42	62	Cxxx
RAB4	41	59	xCxC
RAB5	37	44	CCxx
RAB6	33	43	xCxC
RHY1	35	43	xCxc
RAB7	32	43	xCxc
RAB9	32	43*	xxCC
RAS	30	30	Cxxx
RAL	29	34	CCxx
RAP	28	34	Cxxx
RHO	20	30	Cxxx
RAC	18	28	Cxxx
CDC42	17	28	Cxxx

FIGURE 2. Amino acid sequence relationship of SAS1 to group members within the ras super-family. The amino acid sequence of SAS1 was compared to a restricted class of ras family members using the FASTA alignment program.[17] Relatedness is approximate ($\pm$ 3%) varying with diversity of group members across broad species lines (i.e., yeast through mammals). Sequence identities are expressed as percent of total sequence or percent identity between amino acids ~10 and 100 of SAS1. The entire N terminus of RAB9 was unavailable for the N terminal comparison (*). Cysteines located at the carboxyl termini are also listed.

for lipid modification.[18-20] More extensive identity was revealed by comparison with the YPT1(RAB1)/SEC4 subgroup. Within this class the proteins are >55% identical in sequence, with identities of >80% observed within their N terminal regions.

The protein which shares the greatest identity with SAS1 is YPT2 of the fission yeast *Schizosaccharomyces pombe*.[21] Figure 3 indicates the identities and similarities of SAS1 and YPT2 throughout their entire amino acid sequences. Boxed regions indicate identities shared among all members of the YPT1/SEC4 group, whereas, only minimal identity is seen in comparison with the Dd *ras* protein (see Figures 2 and 3).

B. FUNCTIONAL ANALYSIS OF THE *SAS* GENES

YPT1 and *SEC4* are essential genes of the budding yeast *Saccharomyces cerevisiae* and do not complement one another.[8,10-13] SAS1 and SAS2 are more closely related to YPT1 and SEC4 than SEC4 and YPT1 are to each other.[6] Further, the similarities among SAS1, SAS2, YPT1, and SEC4 highlight the structural and functional aspects shared by the various members of the ras family subgroups. Many genes within a given subgroup of the ras superfamily can effectively substitute for one another and are, thus, considered functionally equivalent.[9,22-25] Therefore, we attempted to complement mutations of *ypt1* and *sec4* in *S. cerevisiae* by the heterologous expression of *Dictyostelium SAS1*.[6]

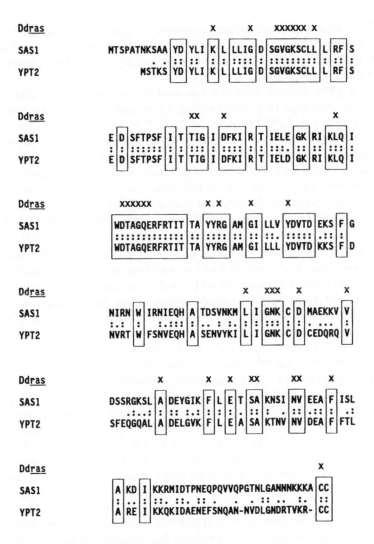

FIGURE 3. Amino acid sequence comparisons among SAS1, YPT2, SAS2, YPT1, SEC4, and Ddras. Amino acid identities between SAS1 and YPT2 are indicated by (:) and similarities by (.). Boxed regions are identical in sequence among SAS1, SAS2, YPT2, YPT1, and SEC4. Within these boxed regions RAB8 and RAB10 sequences possess only four and six conservative amino acids differences, respectively.

The wild-type *YPT1* gene on chromosome VI of *S. cerevisiae* was replaced with *YPT1* that had been fused to the inducible GAL10 promoter.[8] These cells (GYW7-1A) express *YPT1* and grow normally in the presence of galactose media. However, cells stop growing within ~10 h after transfer into glucose media which represses the GAL10 promotor and, thus, YPT1 expression.[8] *SAS1* was fused to the yeast ADH1 promoter, transformed into the GYW7-1A (GAL10-

YPT1) cells and growth monitored in glucose and galactose media. *SAS1* expression did not rescue the growth of the GAL10-YPT1 cells in glucose media.[6] In fact, these cells stopped their growth more quickly than did the parental cells after transfer into glucose.[6]

A slightly different approach was used for complementation studies with *SEC4*. *S. cerevisiae* (NY 456) carrying a temperature-sensitive *sec4* mutation grow normally at 25°C but poorly at 33°C.[10] SAS1 was fused to the GAL10 promotor and transformed into these temperature-sensitive cells.[6] Growth was monitored in parental and *SAS1*-transformed cells at permissive and nonpermissive temperatures, in both glucose and galactose media. No differences in growth rates between parental and transformed cells were seen in glucose media when *SAS1* is not expressed. In galactose media, the expression of SAS1 was not sufficient for normal growth of *sec4* cells at 33°C. Interestingly, at 25°C, the *SAS1* cells grew more poorly in galactose media than did the parental cells.[6]

Thus, in these studies, SAS1 is functionally distinct from YPT1 and SEC4 of *S. cerevisiae*. Expression of SAS1 in wild-type strains did not interfere with growth, whereas, expression of SAS1 in *S. cerevisiae* carrying mutations in *YPT1* or *SEC4* may actually impede growth. There may be sequences within SAS1 that interfere with the functions of YPT1 or SEC4. This may occur through competition for a common effector.[6]

Residues 32 through 42 of the RAS proteins comprise a putative effector region that is believed to interact with the GTPase-activating protein (GAP).[26] Studies using chimeric YPT1/RAS protein constructs indicate that this region is necessary but not sufficient for specific function.[24,27] The equivalent region of the proteins most related to SAS1 are aligned within this putative effector domain (Figure 4). The SAS1 region includes residues 44 through 52 and shares perfect sequence identity with SEC4. The amino acid differences between SAS1 and YPT1 are conservative changes based upon the Dayhoff et al. groupings.[16] Further, altering these sites toward the SAS1 sequence as single-site mutations is not predicted to significantly alter YPT1 function.[27] Thus, we suggest that SAS1, YPT1, and/or SEC4 may interact with a common effector but also with distinct intracellular components that preclude their functional equivalency. This conclusion may be supported by evidence that suggests that yeast-carrying mutated *yptl* or *sec4* genes grow more poorly if they also express *SAS1*.[6] Members of the ras superfamily which fall outside of the YPT1/SEC4 family may not be expected to interact with this potential common effector (see Figures 3 and 4).

Other data may not support a shared effector for SAS1, YPT1, and SEC4. The YPT2 protein of *S. pombe* is also essential for cell viability.[21] Its effector loop is identical in sequence to that of SAS1 and SEC4 but not of YPT1 (see Figure 4). In contrast to the functional experiments described for SAS1, YPT2 of *S. pombe* will complement *sec4* but not *yptl* in *S. cerevisiae*.[21] Similarly, mammalian RAB1 shares an identical effector loop with YPT1 and will function equivalently in *S. cerevisiae*.[24] A YPT1-dependent GAP activity has recently been identified.[27] It will be extremely interesting to determine the effector loop

```
SAS1        F I T T I G I D F

            : : * : : : : : :
SAS2        F I A T I G I D F

            : : : : : : : : :
YPT2        F I T T I G I D F

            : : * : : : : : :
RAB8        F I S T I G I D F

            : : * : : : : : :
RAB10       F I S T I G I D F

            * : * : : : * : :
YPT1/RAB1   Y I S T I G V D F

            : : : : : : : : :
SEC4        F I T T I G I D F

            : . * : . : : : :
RAB3        F V S T V G I D F

                  : : : * . :
RAB2        H D L T I G V E F
```

FIGURE 4. Sequence comparisons within putative effector domains of SAS-related proteins. The putative effector domain of SAS1 includes residues 44 through 52 (see Figure 1). Amino acid identities to SAS1 are indicated by (:) and similarities by (.). Differences at single sites which are suggested to have no effect on YPT1 activity are indicated by (*).

sequence specificity required for this activity and whether this ypt1-GAP is similarly capable of activating SAS1.

Many proteins of the ras superfamily are associated with specific membrane components and are involved in vesicular trafficking. Among the RAB proteins, RAB2 is located on compartments intermediate between the endoplasmic reticulum and the Golgi, RAB3 is on synaptic vesicles, and RAB7 is on late endosomes (see Chapter 18).[28-31] RAB5 is located on the cytoplasmic face of the plasma membrane and on early endosomes, and is implicated in endosome fusion.[32]

YPT1/RAB1 and SEC4 are targeted to separate membrane components and are required for trafficking of different intracellular vesicles. *SEC4* was isolated as a temperature-sensitive mutation in secretion and its protein is localized to the cytoplasmic face of plasma membranes and to secretory vesicles in transit to the cell surface.[10,12] The YPT1/RAB1 protein is associated with the Golgi apparatus and is essential for vesicular transport of proteins from the endoplasmic reticulum to the Golgi complex.[14-15]

It is interesting to consider the potential relationship of the SAS proteins to membrane targeting in *Dictyostelium*. Intracellular vesicles are prominent throughout the *Dictyostelium* life cycle. During growth and early development they are primarily associated with endocytosis and lysozomal action. By the time

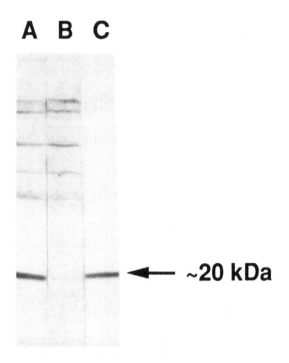

FIGURE 5.   Western blot analysis of anti-SAS peptide antibodies. Cell extracts from growing *Dictyostelium* were separated by gel electrophoresis and electroblotted onto nitrocellulose. Blots were incubated with total anti-SAS sera, affinity-purified, SAS-specific antibodies or nonspecific flow-through of a SAS peptide-coupled column. Crossreactivity was detected using goat-anti-rabbit antibodies conjugated with alkaline phosphatase and visualized using nitro blue tetrazolium/5-bromo-4-chloro-3-indolyl phosphate (NBT/BCIP). (A) Rabbit serum prior to affinity purification (anti-SAS sera). (B) Flow-through of SAS peptide-coupled column (nonspecific antibodies). (C) Antibodies eluted from SAS peptide-coupled column (SAS-specific antibodies). Arrow indicates SAS proteins of ~20 kDa.

of cytodifferentiation, specific vesicles are observed in prespore cells. To initiate a series of cytolocalization studies, we prepared antibodies that would specifically recognize the SAS proteins.

A cysteine-linked, peptide specific to SAS1 was synthesized and coupled to Keyhole Limpet Hemocyanin through the free sulfhydryl group of the terminal cysteine (see Figure 1). The coupled peptide was used to inject rabbits and serum was collected at 10 d following secondary and tertiary boosting. To prepare affinity-purified SAS antibodies, the peptide was coupled to a sulfhydryl-binding column and serum from one animal was chromatographed over the column. Nonspecific flow-through fractions were collected and bound antibodies were eluted with acidic 0.1 *M* glycine.

The specificity of the antibody preparation was tested by Western blot analysis. The results are shown in Figure 5. Only a single 20 kDa protein band is detected in cell extracts with the affinity-purified antibodies. Using these antibodies in blot analysis of subcellular fractions that had been separated on sucrose gradients, we observe SAS proteins distributed with structures of the endoplasmic reticulum. No detectable SAS protein cofractionated with Golgi membranes or with lysozomes. Thus, the SAS proteins appear to localize to a membrane compartment that is different from that targeted by YPT1 or SEC4. These data are consistent with a function of SAS that is distinct from that of YPT1/RAB1 and SEC4.

C. DEVELOPMENTAL EXPRESSION PATTERNS OF *SAS1* AND *SAS2*

To understand the function of the SAS proteins in *Dictyostelium* it is critical to determine their pattern of expression through development. Total *SAS* mRNA expression during development was analyzed by RNA blot hybridization (Figure 6). The relative expression patterns of *SAS1* to *SAS2* were determined by hybridization using gene-specific oligonucleotides and quantitative PCR amplification of mRNA.[6] *SAS* mRNAs are expressed during growth and throughout development. The level of total *SAS* expression increases by approximately three-fold until 15 h in development before declining. SAS1 mRNA comprises the bulk of SAS mRNA at all developmental times. SAS1 is never <85% of the total SAS mRNA at any developmental stage. SAS2 expression increases approximately five-fold during development, yet even by culmination (20 h), SAS2 mRNA comprises only ~15% of the total SAS mRNA population.

Dictyostelium is an excellent organism for targeted gene disruption experiments. It will be interesting to determine if the *SAS* genes are essential for normal growth and development of *Dictyostelium* and to determine if the SAS1 and SAS2 proteins are biologically equivalent. Maximal expression (15 h) of the *SAS* mRNAs is observed following cytodifferentiation. Other experiments in progress will determine the cell-type specific expression pattern and intracellular location of the SAS proteins throughout development. Finally, we have preliminary data to suggest the presence of genes in *Dictyostelium* that are related to but distinct from *SAS*. Their analysis will be critical to understanding the role of membrane targeting and trafficking during the growth and development of *Dictyostelium*.

NOTE ADDED IN PROOF

In a collaborative effort the laboratories of J. Cardelli and G. Weeks have isolated cDNAs for 23 additional members of the *ras* superfamily of GTPases from *Dictyostelium*. New cDNAs have been identified that encode: (1) 8 members of the *rho* family, including a *rac1* homolog; (2) 11 members of the *rab* family, including potential homologs to *rab1, rab2, rab4, rab7,* and *rab11*; (3) 3 members of the *ras* family; and (4) a potential *ran/TC4* homolog.

0 5 10 15 20

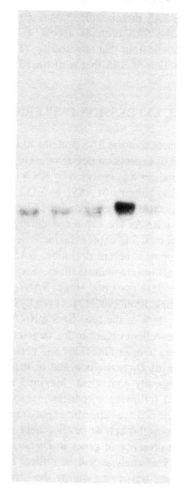

FIGURE 6. Expression of the SAS mRNAs during the development of *Dictyostelium*. Wild-type cells growing on bacteria were harvested on logarithmic phase, washed, plated on filter pads, and allowed to synchronously proceed through development. At 5 h intervals, cells were collected and poly (A)+ RNA-isolated. The RNAs were size-separated on denaturing gels and blotted for hybridization to *SAS1*. (Adapted from Saxe, S. A. and Kimmel, A. R., *Mol. Cell. Biol.*, 10, 2367, 1990.)

ACKNOWLEDGMENTS

We are extremely grateful to the collaborative efforts and discussions of our collegues. In particular we wish to thank Drs. D. Gallwitz, J. Louis, P. Novick, S. Saxe, and M. Wolcott.

REFERENCES

1. **Williams, J. G.**, Regulation of cellular differentiation during *Dictyostelium* morphogenesis, *Curr. Op. Gen. Dev.*, 1, 358, 1991.
2. **Kimmel, A. R. and Firtel, R. A.**, cAMP signal transduction pathways regulating development of *Dictyostelium discoideum*, *Curr. Op. Gen. Dev.*, 1, 383, 1991.
3. **Ebert, D. L., Freeze, H. H., Richardson, J. M., Dimond, R. L., and Cardelli, J. A.**, A *Dictyostelium discoideum* mutant that missorts (and oversecretes) lysosomal enzyme precursors is defective in endocytosis, *J. Cell Biol.*, 109, 1445, 1989.
4. **Cardelli, J. A., Schatzle, J., Bush, J. M., Richardson, J., Ebert, D. L., and Freeze, H. H.**, Biochemical and genetic analysis of the biosynthesis, sorting, and secretion of *Dictyostelium* lysosomal enzymes, *Dev. Gen.*, 11, 454, 1990.
5. **Cardelli, J.**, Regulation of lysosomal trafficking and function during growth and development of *Dictyostelium discoideum*, in *Endosomes and Lysosomses: A Dynamic Relationship*, Storrie, B. and Murphy, R., Eds., JAI Press, Greenwich, CT, 1992, in press.
6. **Saxe, S. A. and Kimmel, A. R.**, *SAS1* and *SAS2*, GTP-binding protein genes in *Dictyostelium discoideum* with sequence similarities to essential genes in *Saccharomyces cerevisiae*, *Mol. Cell. Biol.*, 10, 2367, 1990.
7. **Gallwitz, D., Donath, C., and Sander, C.**, A yeast gene encoding a protein homologous to the human c-*has/bas* proto-oncogene product, *Nature*, 306, 704, 1983.
8. **Schmitt, H. D., Wagner, P., Pfaff, E., and Gallwitz, D.**, The *ras*-related *YPT1* gene product in yeast: a GTP-binding protein that might be involved in microtubule organization, *Cell*, 47, 401, 1986.
9. **Haubruck, H., Disela, C., Wagner, P., and Gallwitz, D.**, The *ras*-related *ypt* protein is an ubiquitous eukaryotic protein: isolation and sequence analysis of mouse cDNA clones highly homologous to the yeast YPT1 gene, *EMBO J.*, 6, 4049, 1987.
10. **Salminen, A. and Novick, P. J.**, A *ras*-like protein is required for a post-Golgi event in yeast secretion, *Cell*, 49, 527, 1987.
11. **Segev, N. and Botstein, D.**, The *ras*-like yeast *YPT1* gene is itself essential for growth, sporulation, and starvation response, *Mol. Cell. Biol.*, 7, 2367, 1987.
12. **Goud, B., Salminen, A., Walworth, N. C., and Novick, P. J.**, A GTP-binding protein required for secretion rapidly associates with secretory vesicles and the plasma membrane in yeast, *Cell*, 53, 753, 1988.
13. **Schmitt, H. D., Puzicha, M., and Gallwitz, D.**, Study of a temperature-sensitive mutant of the *ras*-related YPT1 gene product in yeast suggests a role in the regulation of intracellular calcium, *Cell*, 53, 635, 1988.
14. **Segev, N., Mulholland, J., and Botstein, D.**, The yeast GTP-binding *YPT1* protein and a mammalian counterpart are associated with the secretion machinery, *Cell*, 52, 915, 1988.
15. **Plutner, H., Cox, A. D., Pind, S., Khosravi-Far, R., Bourne, J. R., Schwaninger, R., Der, C. J., and Balch, W. E.**, Rab1b regulates vesicular transport between the endoplasmic reticulum and successive Golgi components, *J. Cell Biol.*, 115, 31, 1991.
16. **Dayhoff, M. O., Schwartz, R. M., and Orcutt, B. C.**, A model of evolutionary change in proteins, in *Atlas of Protein Sequence and Structure*, Dayhoff, M. O., Ed., The National Biomedical Research Foundation, Washington, D.C., Vol. 5, Suppl. 3, 1978, 345.
17. **Pearson, W. R. and Lipman, D. J.**, Improved tools for biological sequence comparison, *Proc. Natl. Acad. Sci. U.S.A.*, 85, 2444, 1988.
18. **Halliday, K. R.**, Regional homology in GTP-binding proto-oncogene products and elongation factors, *J. Cyclic Nucl. Prot. Phosph. Res.*, 9, 435, 1983.
19. **Molenaar, C. M. T., Prane, R., and Gallwitz, D.**, A carboxyl-terminal cysteine residue is required for palmitic acid binding and biological activity of the *ras*-related yeast YPT1 protein, *EMBO J.*, 7, 971, 1988.
20. **Hancock, J. F., Magee, A. I., Childs, J. E., and Marshall, C. J.**, All *ras* proteins are polyisoprenylated but only some are palmitoylated, *Cell*, 57, 1167, 1989.

21. **Haubruck, H., Engelke, U., Mertins, and Gallwitz, D.**, Structural and functional analysis of *ypt2*, an essential *ras*-related gene in the fission yeast *Schizosaccharomyces pombe* encoding a Sec4 protein homologue, *EMBO J.*, 9, 1957, 1990.

22. **DeFeo-Jones, D., Scolnick, E. M., Koller, R., and Dhar, R.**, *ras*-related gene sequences identified and isolated from *Saccharomyces cerevisiae*, *Nature*, 306, 707, 1983.

23. **Kataoka, T., Powers, S., Cameron, S., Fasano, O., Goldfarb, M., Broach, J., and Wigler, M.**, Functional homology of mammalian and yeast *ras* genes, *Cell*, 40, 19, 1985.

24. **Haubruck, H., Prange, R., Vorgias, C., and Gallwitz, D.**, The *ras*-related mouse protein YPT1 can functionally replace the YPT1 gene product in yeast, *EMBO J.*, 8, 1427, 1989.

25. **Hengst, L., Lehmeier, T., and Gallwitz, D.**, The *ryh1* gene in the fission yeast *Schitzosaccharomyces pombe* encoding a GTP-binding protein related to ras, rho, and ypt: structure, expression and identification of its human homologue, *EMBO J.*, 9, 1949, 1990.

26. **Sigal, I. S., Gibbs, J. B., D'Alonzo, J. S., and Scolnick, E. M.**, Identification of effector residues and a neutralizing epitope of Ha-*ras*-encoded p21, *Proc. Natl. Acad. Sci. U.S.A.*, 83, 4725, 1986.

27. **Becker, J., Tan, T. J., Trept, H.-H., and Gallwitz, D.**, Mutational analysis of the putative effector domain of the GTP-binding Ypt1 protein in yeast suggests specific regulation by a novel GAP activity, *EMBO J.*, 10, 785, 1991.

28. **Touchot, N., Chardin, P., and Tavitian, A.**, Four additional members of the *ras* gene superfamily isolated by an oligonucleotide strategy: molecular cloning of YPT-related cDNAs from a rat brain library, *Proc. Natl. Acad. Sci. U.S.A.*, 84, 8210, 1987.

29. **Zahraoui, A., Touchot, N., Chardin, P., and Tavitian, A.**, Complete coding sequences of the *ras* related *rab* 3 and 4 cDNAs, *Nucleic Acids. Res.*, 16, 1204, 1988.

30. **Chavrier, P., Parton, R. G., Hauri, H. P., Simons, K., and Zerial, M.**, Localization of low molecular weight GTP binding proteins to exocytotic and endocytotic compartments, *Cell*, 62, 317, 1990.

31. **Matteoli, M., Takei, K., Cameron, R., Hurlbut, P., Joghnson, P. A., Sudhof, T. C., Jahn, R., and de Camilli, P.**, Association of Rab3A with synaptic vesicles at late stages of the secretory pathway, *J. Cell Biol.*, 115, 625, 1991.

32. **Gorvel, J.-P., Chavrier, P., Zerial, M., and Grueberg, J.**, rab5 controls early endosome fusion in vitro, *Cell*, 64, 915, 1991.

SECTION III
Functional Regulation of *ras* and *ras*-Related GTPases

Chapter 21

REGULATORS AND EFFECTORS OF *ras* PROTEINS

Frank McCormick

TABLE OF CONTENTS

0-8493-5214-2/93/$0.00 + $.50
© 1993 by CRC Press, Inc.

401

I. INTRODUCTION

ras-p21 proteins cycle between inactive and active states.[2,3] Conversion of the inactive, GDP-bound form to the active, GTP-bound form is mediated by proteins referred to as "exchange factors", "guanine-nucleotide releasing factors" or GDP dissociation stimulators (GDSs). Conversion of the active state to the inactive state is mediated by GTPase-activating proteins (GAPs). The level of active *ras*-p21 in cells is determined by the balance between exchange activity and GAP activity. Clearly, if these two activities were equal, *ras*-p21-GDP and *ras*-p21-GTP would exist in equal proportions. However, in most cells, the GDP form predominates[6,10,11,20,24] indicating that GAP activity exceeds exchange activity. This situation changes in activated cells, in which the GTP form accumulates. In this chapter, we will discuss ways in which these changes in GDP/GTP could be regulated, and the possible significance of this regulation. Elsewhere in this volume, the possible role of membrane localization and processing in *ras*-p21 will be discussed (Chapter 3).

II. CHANGES IN GDP/GTP BOUND TO *ras*-p21

The nucleotide state of *ras*-p21 in vertebrate cells was first determined by Trahey and McCormick (1987), who injected purified *ras*-p21 proteins into *Xenopus* oocytes labeled to high specific activity with 32-phosphate. Several hours after injection, immunoprecipitated proteins were denatured, and the bound nucleotide was identified by thin layer chromatography. Similar studies were performed using RAS proteins expressed in *Saccharomyces cerevisiae*:[11] cultures of growing cells were labeled with 32-phosphate, RAS proteins were immunoprecipitated, and the nucleotide with which the precipitated proteins were associated was determined. In both systems, wild-type *ras* proteins were found to be predominantly in their GDP-bound states. More recent improvements in this type of technique has made it possible to examine the nucleotide state of endogenous *ras*-p21 in mammalian cells, and thus, to facilitate estimation of the extent of *ras* activation in cells exposed to various stimuli. Increases in the proportion of *ras*-p21 in the GTP-state have been observed in cells stimulated by PDGF, EGF, IL-2, IL-3, GM-CSF,[19,20] insulin,[5] or tyrosine kinase oncogenes,[10] and in activated T cells.[6] During T-cell activation, for example, the level of *ras*-p21-GTP jumps from 5 to 80% in a period of only one minute.[6] Equally dramatic effects were observed during exposure of fibroblasts to insulin.[5] These effects must be due to increases in the ratio of exchange rate/GTPase activity.

In T cells, it has been possible to show that guanine nucleotide exchange rate, as measured by rate of binding of labeled GTP to *ras*-p21 in permeabilized cells, is comparable in resting or in activated cells, whereas, GAP activity decreases after activation. This drop in GAP activity is not sufficient to fully account for the dramatic shift in levels of *ras*-GTP: such a change demands a change in the ratio of exchange to GAP activity of about 100-fold.[15] Measurements of exchange activity in permeabilized cells and of GAP activity in extracts must, therefore, fail to accurately reflect the situation in cells. Nonetheless, the possibility that *ras*-p21 is regulated is through changes in GAP activity rather than exchange activity is a novel one that has interesting implications. Such a situation would differ radically from a mechanism by which signally heterotrimeric G proteins are regulated; these proteins remain in their stable, GDP-bound states until a signal provokes exchange of GDP for GTP. The signal input is clearly at the level of exchange and there is little evidence to suggest that the ensuing cycle is regulated at the level of GTP hydrolysis.

Changes in the GTP/GDP state of RAS proteins have been examined using strains of yeast that contain mutant RAS proteins, and in strains lacking CDC25 or IRA proteins. It is interesting that in exponentially growing yeast cells, most of the RAS proteins are in their inactive GDP state.[11] However, in these same strains, active RAS appears to be essential for viability, since loss of CDC25 (which converts RAS-GDP to RAS-GTP) is lethal.[9] These observations suggest RAS cycles constantly between inactive and active forms: although the GTP form is constantly being formed and plays an active role in supporting growth, it is converted to the GDP form rapidly, and the inactive form predominates at steady state. The high GDP/GTP exchange rate measured in T cells is consistent with the dynamic model deduced form yeast genetics, suggesting that RAS regulation in yeast is fundamentally similar to regulation in mammalian cells.

We do not know why RAS proteins cycle rapidly between two states. Analysis of RAS function in yeast reveals that rapid cycling is not necessary for RAS function, and that the level of RAS in the GTP state determines its activity rather than its rate of formation and degradation. This deduction is based on the observation that the toxic effects of inactivating CDC25 can be suppressed by deletion of IRA genes.[22] This means that strains of yeast that are defective in both exchange (CDC25) and GAP activity (IRA) are normal with respect to RAS function. In these strains, RAS proteins presumably depend on their intrinsic exchange and GTPase rates and thus, accumulate in the GTP states at levels sufficient to support growth. We expect, from comparison of intrinsic vs. catalyzed exchange and GTPase rates, that the rate of cycling between GDP- and GTP-bound states is greatly reduced relative to wild-type strains, yet this reduced rate of cycling does not appear to affect growth. We also expect that the ability of these strains to respond to signals through RAS pathways would be greatly reduced relative to wild-type strains, since these signals must depend on altered exchange of GTPase rates for their propagation.

III. REGULATION OF GDP/GTP EXCHANGE

The conversion to the active GTP form is a step that would appear to be a suitable opportunity to regulate *ras* activity. Certainly, this is a major step at which signaling heterotrimeric G proteins are regulated in their analogous cycle.

However, we have discussed above the fact that RAS proteins differ from G proteins in that they appear to cycle constantly between GDP and GTP states. We may, therefore, ask whether in fact, nucleotide exchange is indeed a regulated step in the RAS GDP/GTP cycle. Currently, we do not have a clear answer to this question since exchange cannot be measured accurately in cell-free systems or in living cells. In permeabilized T cells, exchange seems to be constitutively high and does not appear to be regulated during *ras* activation. It will be of great interest to determine whether this important result holds true in other cell systems and whether an experimental system can be established to confirm these results in a system that does not require permeabilization.

It has been possible to establish conditions in which exchange factors appear to be limiting. This has been done by introducing into cells (yeast or mammalian) mutant *ras* proteins such as Asn17 that are thought to interfere with *ras* activation.[7,21] The evidence that they perform in this way is based on the fact that, in yeast, their toxic properties can be overcome by overexpressing CDC25 and that they interfere much more effectively with normal *ras* proteins than with oncogenic *ras* proteins, which are thought to have a reduced requirement for exchange because of their reduced sensitivity to GAP.[17] The actual mechanism by which Asn17 *ras*-p21 inhibits exchange has not been demonstrated directly, but a working model has been proposed based on the known properties of the bacterial GDP/GTP exchange protein EF-Ts which catalyzes nucleotide exchange on the *ras*-like protein EF-Tu.[2] EF-Ts works by the following mechanism:

1.   EF-Ts binds to EF-Tu-GDP.
2.   The ternary complex has a reduced affinity for GDP, which now dissociates rapidly (without EF-Ts, the rate of dissociation would be very slow).
3.   EF-Tu complexed with EF-Ts now binds GTP; the affinity of the complexed form of EF-Tu for GTP is relatively low, but cells contain high concentrations of GTP (up to millimolar, much higher than GDP).
4.   EF-Ts dissociates from EF-Tu-GTP, to complete the exchange cycle.

If we assume that *ras*-p21 goes through a similar series of steps, we can understand the effect of Asn17 in the following way: Asn17 binds to the *ras* exchange factor (the EF-Ts equivalent) and releases GDP. The mutant protein has a reduced affinity for GTP, so that GTP can no longer bind and displace the exchange factor. As a result, the exchange factor forms a "dead-end" complex with the mutant *ras* protein, and cannot exert its effect on normal *ras*-p21 proteins in the same cell.

In cells expressing Asn17 *ras*-p21 proteins, it therefore, appears that *ras*-p21 activation is significantly reduced through titration of exchange factors, and that *ras* function is compromised in these cells. As a result, tyrosine kinase receptors and oncogenes fail to function efficiently. These experiments have demonstrated in a convincing way that tyrosine kinases require active *ras*-p21 for their function and show that when exchange factors are limiting, *ras* function is also limiting.

Genetic analysis of the *Drosophila sevenless* pathway has confirmed that tyrosine kinase signaling requires active *ras* function.[21a] Furthermore, in this system, it has been possible through manipulation of mutations in the *sevenless* pathway, to demonstrate that loss of one allele encoding an exchange factor for the *ras1* gene involved in *sevenless* signaling results in a measurable drop in *ras1* function. It is, therefore, clear that guanine nucleotide exchange is limiting in this system and that Ras1 signal output is proportional to exchange activity.

If inhibiting exchange activity reduces *ras*-p21 signaling, we might expect that increasing exchange activity would increase *ras*-p21 function. This has been demonstrated recently by Takai and co-workers, who have overexpressed RHO-GDS in mammalian cells, and found that this protein, which increases the rate of GDP dissociation from K-*ras* p21, causes oncogenic transformation.[13] Increased exchange can also be achieved by using mutant forms of *ras*-p21 that have high intrinsic dissociation rates for GDP. Such mutants have been described, and shown to be oncogenic.[27] It is, therefore, clear that the level of ras activation in cells can be altered by increasing or decreasing exchange factor activity: whether this is a mechanism by which *ras* proteins are normally regulated remains to be demonstrated directly. In addition, proteins may exist that regulate the rate of dissociation of GDP from *ras* proteins. Guanine nucleotide dissociation inhibitors (GDIs) for *ras*-related proteins have been identified[18,26] and cloned,[8,14] but to date no such GDI for *ras*-p21 has been identified.

IV. REGULATION OF GAP ACTIVITY

The issue of whether GAPs play a role in signaling from *ras* proteins has not yet been resolved:[12] they remain plausible candidates for *ras* effectors but direct proof is lacking. However, it is clear that GAPs are regulators of *ras* function in cells, acting by converting them to their inactive, GDP states. The predominance of *ras* proteins in GDP forms is evidence that GAPs are more potent than exchange factors under normal cellular conditions. Indeed, GAP was discovered as part of a search for a cellular factor that would selectively inhibit normal *ras*-p21 function without affecting oncogenic *ras,* and thus, account for the differences in biological potency between normal and oncogenic *ras* proteins.[24] Direct demonstration that GAP is able to regulate the GTP-state of *ras* proteins in cells has been achieved by Zhang and co-workers,[30] who have overexpressed full-length GAP or the catalytic domain in cells transformed by c-*ras*-p21: GAP overexpression inactivates normal *ras*-p21 function and reverts the transformed phenotype. Overexpression of GAP had no effect on transformation by v-*ras*, as expected from its known biochemical properties. In T cells, GAP activity is reduced following activation, but the reduction observed by measuring GAP activity in cell extracts is not sufficient to account for the increase in *ras*-p21 in the GTP state under these conditions. In other cells, changes in GAP activity following cell stimulation have been even less substantial. These results could mean that changes in GAP activity do not fully account for changes in *ras*-p21 GTP loading, thus, implying exchange as the main control point. Alternatively, inhibition of GAP activity may be substantial in activated cells, but these changes may escape detection using current techniques. For example, if GAPs were regulated by a second messenger molecule, it is possible that the effects of such a regulator would be missed when measuring GAP activity in cell extracts in which the inhibitor would be diluted or degraded. Evidence already exists to support this hypothesis. The activities of p120-GAP and NF-1-GAP are regulated by lipids and related compounds *in vitro*.[1,25] We must also consider the possibility that GAP is regulated by a covalent modification that is unstable, and is lost during preparation of cell extracts and measurement of GAP activity. Finally, GAP activity could be regulated by translocation from one cellular compartment to another.[16] This form of regulation would also escape detection using current *in vitro* assay systems. In conclusion, we cannot yet determine to what extent GAP activity is regulated in cells, but we can exclude the possibility that *ras*-p21 GTP levels are regulated exclusively by stable covalent GAP modification.

V. EFFECTORS OF *ras*-p21 ACTION

Effectors of *ras*-p21 action remain elusive. As discussed elsewhere, GAPs are currently the best known candidates for effectors of *ras* action: they interact with *ras*-p21 in a GTP-dependent manner, they appear to interact at a site on *ras*-p21 defined as the effector binding region, they interact with oncogenic *ras*-p21 proteins and in a cell-free system, and p120-GAP behaves in a *ras*-dependent fashion.[28,29] The idea that GAPs are effectors is attractive because it implies a tight coupling of signal transmission to downregulation: GAP binds to *ras*-p21 in the GTP-state, sends a pulse of signal, and quickly terminates the signal by hydrolyzing GTP to GDP. Oncogenic *ras* proteins, according to this model, bind to GAP and signal constitutively because the mutant *ras* protein cannot be induced to hydrolyze GTP. However, although this model is appealing, it is hard to prove. Furthermore, substantial evidence argues against the model.[12] Of these arguments, perhaps the most compelling is the discovery that in yeast, inactivation of GAP-like IRA genes causes the expected rise in RAS-GTP, but does not prevent signaling. Indeed, yeast defective in IRA genes behaves as if the RAS protein is in the fully active state.[22,23] It is, therefore, clear that IRA genes are not required for known RAS signaling pathways. We cannot, however, exclude the possibility that IRA proteins have effector functions that generate signals from RAS proteins that have thus far escaped detection. The precise effectors of RAS proteins in yeast have not yet been identified: RAS proteins certainly activate adenylyl cyclase in these organisms, but it is not yet clear whether this activation is mediated by direct binding of RAS proteins to the cyclase protein, or through binding to regulatory subunits.

VI. CONCLUSION

In summary, substantial progress has been made in recent years in determining how *ras*-p21 proteins are regulated: exchange factors, GAPs, and GDIs (for *ras* relatives) have been identified, and conditions in which *ras*-p21 proteins can be provoked to switch from inactive to active states have been described. Furthermore, a number of possibilities for regulating *ras*-p21 function through regulated association with the membrane have been presented through extensive analysis of *ras*-p21 processing events. In spite of this progress, we do not yet know how *ras*-p21 receives or transmits signals. We speculate that GAPs may be involved in signal transmission, but do not have conclusive evidence to support this speculation. The search for effectors of *ras*-p21 therefore continues, while the characterization of regulatory proteins provides a fertile area of investigation.

REFERENCES

1. **Bollag, G. and McCormick, F.**, Differential regulation of *ras*GAP and neurofibromatosis gene product activities, *Nature*, 351, 576, 1991.
2. **Bourne, H. R., Sanders, D. A., and McCormick, F.**, The GTPase superfamily: a conserved switch for diverse cell functions, *Nature*, 348, 125, 1990.
3. **Bourne, H. R., Sanders, D. A., and McCormick, F.**, The GTPase superfamily: conserved structure and molecular mechanism, *Nature*, 349, 117, 1991.
4. **Broach, J. R.**, *RAS* genes in *Saccharomyces cerevisiae*: signal transduction in search of a pathway, *Trends Genet.*, 7, 28, 1991.
5. **Burgering, B. M. T., Medema, R. H., Maasen, J. A., van de Wetering, M. L., van der Eb, A. J., McCormick, F., and Bos, J. L.**, Insulin stimulation of gene expression mediated by p21^{ras} activation, *EMBO J.*, 10, 1103, 1991.
6. **Downward, J., Graves, J. D., Warne, P. H., Rayter, S., and Cantrell, D. A.**, Stimulation of p21^{ras} upon T-cell activation, *Nature*, 346, 719, 1990.
7. **Feig, L. A. and Cooper, G. M.**, Inhibition of NIH 3T3 cell proliferation by a mutant *ras* protein with preferential affinity for GDP, *Mol. Cell. Biol.*, 8, 3235, 1988.
8. **Fukumoto, Y., Kaibuchi, K., Hori, Y., Fujioka, H., Araki, S., Ueda, T., Kikuchi, A., and Takai, Y.**, Molecular cloning and characterization of a novel type of regulatory protein (GDI) for *rho* proteins, *ras* p21-like small GTP-binding proteins, *Oncogene*, 5, 1321, 1990.
9. **Gibbs, J. B. and Marshall, M. S.**, The *ras* oncogene — an important regulatory element in lower eucaryotic organisms, *Microbiol. Rev.*, 53, 171, 1989.
10. **Gibbs, J. B., Marshall, M. S., Scolnick, E. M., Dixon, R. A., and Vogel, U. S.**, Modulation of guanine nucleotides bound to Ras in NIH3T3 cells by oncogenes, growth factors, and the GTPase activating protein (GAP), *J. Biol. Chem.*, 265, 20437, 1990.

11. **Gibbs, J. B., Schaber, M. D., Marshall, M. S., Scolnick, E. M., and Sigal, I. S.,** Identification of guanine nucleotides bound to *ras*-encoded proteins in growing yeast cells, *J. Biol. Chem.*, 262, 10426, 1987.

12. **Hall, A.,** *ras* and GAP — who's controlling whom?, *Cell,* 61, 921, 1990.

13. **Hiroyoshi, M., Kaibuchi, K., Kawamura, S., Hata, Y., and Takai, Y.,** Role of the C-terminal region of *smg* p21, a *ras* p21-like small GTP-binding protein, in membrane and *smg* p21 GDP/GTP exchange protein interactions, *J. Biol. Chem.*, 266, 2962, 1991.

14. **Matsui, Y., Kikuchi, A., Araki, S., Hata, Y., Kondo, J., Teranishi, Y., and Takai, Y.,** Molecular cloning and characterization of a novel type of regulatory protein (GDI) for *smg* p21A, a *ras* p21-like GTP-binding protein, *Mol. Cell. Biol.,* 10, 4116, 1990.

15. **McCormick, F., Adari, H., Trahey, M., Halenbeck, R., Koths, K., Martin, G. A., Crosier, W. J., Watt, K., Rubinfeld, B., and Wong, G.,** Interaction of *ras* p21 proteins with GTPase activating protein, *Cold Spring Harb. Symp. Quant. Biol.,* 2, 849, 1988.

16. **Molloy, C. J., Bottaro, D. P., Fleming, T. P., Marshall, M. S., Gibbs, J. B., and Aaronson, S. A.,** PDGF induction of tyrosine phosphorylation of GTPase activating protein, *Nature,* 342, 711, 1989.

17. **Powers, S., O'Neill, K., and Wigler, M.,** Dominant yeast and mammalian *RAS* mutants that interfere with the *CDC25*-dependent activation of wild-type *RAS* in *Saccharomyces cerevisiae, Mol. Cell. Biol.,* 9, 390, 1989.

18. **Sasaki, T., Kikuchi, A., Araki, S., Hata, Y., Isomura, M., Kuroda, S., and Takai, Y.,** Purification and characterization from bovine brain cytosol of a protein that inhibits dissociation of GDP from and the subsequent binding of GTP to *smg* p25A, a *ras* p21-like GTP-binding protein, *J. Biol. Chem.,* 265, 2333, 1990.

19. **Satoh, T., Endo, M., Nakafuku, M., Akiyama, T., Yamamoto, T., and Kaziro, Y.,** Accumulation of p21^{ras}-GTP in respond to stimulation with epidermal growth factor and oncogene products with tyrosine kinase activity, *Proc. Natl. Acad. Sci. U.S.A.,* 87, 7926, 1990a.

20. **Satoh, T., Endo, M., Nakafuku, M., Nakamura, S., and Kaziro, Y.,** Platelet-derived growth factor stimulates formation of active p21^{ras}-GTP complex in Swiss mouse 3T3 cells, *Proc. Natl. Acad. Sci. U.S.A.,* 87, 5993, 1990b.

21. **Sigal, I. S., Gibbs, J. B., D'Alonzo, J. S., Temeles, G. L., Wolanski, B. S., Socher, S. H., and Scolnick, E. M.,** Mutant *ras*-encoded proteins with altered nucleotide binding exert dominant biological effects, *Proc. Natl. Acad. Sci. U.S.A.,* 83, 952, 1986.

21a. **Simon, M. A., Bowtell, D. D. L., Dodson, G. S., Laverty, T. R., and Rubin, G. M.,** Ras1 and a putative guanine nucleotide exchange factor perform crucial steps in signalling by the sevenless protein tyrosine kinase, *Cell,* 67, 701–716, 1991.

22. **Tanaka, K., Matsumoto, K., and Toh-e, A.,** *IRA1,* an inhibitory regulator of the *RAS*-cyclic AMP pathway in *Saccharomyces cerevisiae, Mol. Cell. Biol.,* 9, 757, 1989.

23. **Tanaka, K., Nakafuku, M., Satoh, T., Marshall, M. S., Gibbs, J. B., Matsumoto, K., Kaziro, Y., and Toh-e, A.,** *S. cerevisiae* genes *IRA1* and *IRA2* encode proteins that may be functionally equivalent to mammalian *ras* GTPase activating protein, *Cell,* 60, 803, 1990.

24. **Trahey, M. and McCormick, F.,** A cytoplasmic protein stimulates normal N-*ras* p21 GTPase, but does not affect oncogenic mutants, *Science,* 238, 542, 1987.

25. **Tsai, M.-H., Hall, A., and Stacey, D. W.,** Inhibition by phospholipids of the interaction between R-*ras, rho,* and their GTPase-activating proteins, *Mol. Cell. Biol.,* 9, 5260, 1989.

26. **Ueda, T., Kikuchi, A., Ohga, N., Yamamoto, J., and Takai, Y.,** Purification and characterization from bovine brain cytosol of a novel regulatory protein inhibiting the dissociation of GDP from and the subsequent binding of GTP to *rho*B p20, a *ras* p21-like GTP-binding protein, *J. Biol. Chem.,* 265, 9373, 1990.

27. **Walter, M., Clark, S. G., and Levinson, A. D.,** The oncogenic activation of human p21^{ras} by a novel mechanism, *Science,* 233, 649, 1986.

28. **Yatani, A., Okabe, K., Polakis, P., Halenbeck, R., McCormick, F., and Brown, A. M.,** *ras* p21 and GAP inhibit coupling of muscarinic receptors to atrial K^+ channels, *Cell,* 61, 769, 1990.
29. **Yatani, A., Quilliam, L. A., Brown, A. M., and Bokoch, G. M.,** Rap1A antagonizes the ability of Ras and Ras-Gap to inhibit muscarinic K^+ channels, *J. Biol. Chem.,* 266, 22,222, 1991.
30. **Zhang, K., Papageorge, A. G., Martin, P., Vass, W. C., Olah, Z., Polakis, P. G., McCormick, F., and Lowy, D. R.,** Heterogeneous amino acids in ras and rap1A specify sensitivity to GAP proteins, *Science,* 254, 1630, 1991.

Chapter 22

PROPERTIES OF GAP PROTEINS

Carmela Calés

TABLE OF CONTENTS

I. INTRODUCTION

GAP was first described in 1987 as a protein present in cytosolic fractions of *Xenopus* oocytes and NIH 3T3 cells, able to stimulate the GTPase activity of mammalian normal p21*ras*.[1] This exciting finding explained at that time the differences found in the transforming ability of some oncogenic forms of *ras*.[1] Since then, much effort has been made to determine its precise function.

After a 120,000 Da protein was isolated with GAP activity specific for *ras*,[2-4] many groups have reported the existence of different proteins with GTPase-stimulating activity for distinct *ras*-like proteins.[5-16] This will probably help to elucidate the mechanism of action of these small G proteins. However, only the different GAPs for *ras* proteins (mammalian *ras*-GAP and *Saccharomyces cerevisiae* RAS-GAPs, *IRA1* and *IRA2*[15,16]) have been studied in detail.

Most likely, every *ras*-related protein has a specific GAP, and to date *rap1*,[5-8] *rho*,[9] *rab*,[10] *ypt1*,[11] *ral*,[12] and *rac*[13] appear to interact with one, and in some cases with more than one, specific GAP protein, although little is known about their function. Table 1 shows a summary of all the GAP proteins reported thus far. As the individual characteristics of these specific GAPs will be discussed in different chapters of this book, I will concentrate in reviewing all the data available on the mammalian *ras* p120-GAP protein. Excitingly, the product of the gene defective in human neurofibromatosis, NF-1[17-19] appears to be a new GAP for *ras*,[20] and the current investigations will also be described.

II. STRUCTURE AND EXPRESSION OF *ras* p120-GAP

Human p120-GAP cDNA clones were first obtained from a placenta cDNA library in Frank McCormick's laboratory.[2] The deduced amino acid sequence revealed a very high homology (96%) to the previously isolated bovine GAP.[3,4] The most common form of human GAP (type I GAP) is present in all tissues and cells thus far examined.[21] A schematic representation of its structure is shown in Figure 1. Different domains of the protein have been defined through both experimental and theoretical analysis. The carboxyl terminus of the molecule contains the catalytic domain, as shown by the fact that a truncated form, consisting in the last 344 amino acids, retains full GTPase-stimulating activity.[22] This is the most conserved sequence (98% homologous to bovine GAP[2,3], NF-1,[20] and *IRA1* and *IRA2*[15]). The rest of the molecule is most likely the target for regulation of its activity. This has been deduced through analysis of its sequence: three *src* homology (SH) regions (two SH2 and one SH3 regions) were identified, first in bovine, and then in human GAP.[4,24] These SH2 and SH3 domains are known to be regulatory sequences of different nonreceptor tyrosine kinases (*src, fps, abl,* etc.) and some cytosolic proteins like phospholipase C-gamma and *crk* oncogene.[23,73] These domains seem to be implicated in the interaction of GAP with tyrosine-phosphorylated molecules,[73] as it will be discussed below.

A potential membrane-association site could be provided by a stretch of 180 amino acids at the N terminus, particularly rich in hydrophobic residues.[2] Another

413

TABLE 1
GTPase-Activating Proteins for Different *ras*-Like Proteins

G protein specificity	Source of purification	Cellular localization	Nonreactive proteins	Size	Ref.
ras	Human placenta	Cytosol	*rap, rho, racl*	120 kDa	2
	Bovine brain	Cytosol	*ypt1*	125 kDa	3
	Recombinant NF-1	—	—	—	42–44
rap1/Krev-1	HL-60	Membrane	*ras, rho,* G25A, *racl*	88 kDa	8
	Human neutrophils[a]	Cytosol	—	250 and 300 kDa[b]	7
	Human platelets	Cytosol	*ras*	88 and 300 kDa[b]	6
	Bovine brain	Cytosol	*rho, ras, smg* 25A	—	5
	Bovine brain[a]	Membrane	—	—	65
	Recombinant (human placenta)		*rab3A, ral1b, ras, rap2*	88 kDa	65
ral1A	Rat brain and testes[a]	Cytosol	*ras*	—	12
rhoA, racl	Human spleen	Cytosol	*ras,* R-*ras*	27.5 kDa	72
rab3A	Rat brain	50% cytosol 50% membrane	—	400 kDa	10
racl	Recombinant *bcr*	—	*ras, rho*	—	13
	Recombinant chimerin	—			
ypt1, rab1p	Porcine liver[a]	Cytosol	*ras*	—	11
RAS (*S. cerevisiae*)	Recombinant	—	*ras*	[b]	16
ras1 (*S. pombe*)	c	—	—	—	14

[a] GAP activity detected in crude or partially purified fractions.

[b] Two distinct proteins detected.

[c] GAP detected by genetic analysis.

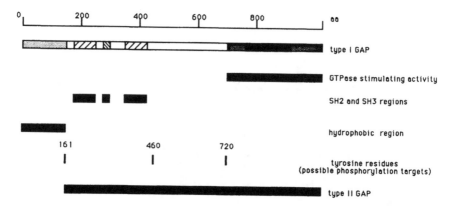

FIGURE 1. Structure of p120-GAP.

form of GAP referred to as type II has been detected only in certain human cell lines and in human placenta, where it represents 50% of total GAP.[21] This form is identical to the ubiquitous p120-GAP, except for the N terminal hydrophobic region (Figure 1). It is probably generated by alternative RNA splicing.[2] The existence of this ''hydrophilic'' GAP in certain cells could be interpreted, as the authors suggest, as a mechanism of regulating *ras* activity by altering GAP in some way, for instance, its cellular localization.[2,21]

GAP activity is mainly detected in the cytosolic fraction of most tissues and cell lines tested to date.[24] This cytosolic localization is somehow puzzling as one would expect it to be in the vicinity of *ras*. However, the observation that a significant amount of tyrosine-phosphorylated GAP is found in membrane fractions (presumably complexed with some growth factor receptors),[25] could not only explain GAP-*ras* interaction but also give some light on how GAP activity is regulated. Interestingly, the GAP activity due to the NF-1 product seems also to be mainly cytosolic,[26] in contrast with the *S. cerevisiae* protein *IRA2* which appears to be membrane-bound.[16]

III. BIOCHEMICAL PROPERTIES OF *ras* p120-GAP

GAP interacts with all three human *ras* proteins (H-, K-, and N-*ras*), and dramatically stimulates the intrinsically slow GTPase activity of normal *ras*.[1,3,27] It has been determined by competition experiments that GAP also interacts with oncogenically activated *ras* proteins,[4] but in this case, it is unable to increase the hydrolysis of the *ras*-bound GTP.[1,3,27] These oncogenic *ras* proteins bind to GAP more tightly than their normal counterparts.[3,28] In addition, its affinity is much higher for *ras*-GTP than for *ras*-GDP.[4,29] By using these competition experiments it has also been possible to show that proteins other than p21*ras* can interact with *ras* p120-GAP, and in some instances more efficiently than *ras* proteins do.[24] Yeast *RAS*,[15,30] R-*ras*,[9] and *rap1A*,[29,31] do indeed interact with GAP, whereas, other *ras*-related proteins, e.g., *rho* protein,[9] do not. This is in

TABLE 2
Biological Activity and GAP Sensitivity of *ras* Proteins Containing Point Mutations in the Effector Region

Residue	Mutation		GAP sensitivity	Biological activity	Ref.
	From	To			
33,34	D, P	H, S	−	+ (normal)	38
				− (oncogene)	38
35	T	A	−	−	27
		S	+	−	33
35, 36	T, I	A, L	−	−	36
36	I	L	+	+	33
		M	+	+	33
		A	−	−	33
38	D	A	−	−	36
		A	−	−	27
		E	−	−	27
		N	−	−	38
39	S	A	+	+	36
		C	+	+	33
40	Y	K	−	−	27
		F	+	+	27
		I	+	−	33
		S	+	−	33

Note: + , presence or − , absence of response (no quantitative meaning).

agreement with the fact that the interaction of GAP takes place through, at least, a region common to all those proteins.[32,33] This region is defined by *ras* 30 to 40 residues and was referred to as the effector region since point mutations lead to an inhibition of *ras* biological activity without altering its biochemical properties or its location in the plasma membrane.[34,35] Early reports showed that mutations in the effector region that lead to biological inactivation of *ras* were also insensitive to GAP-mediated GTPase stimulation.[27,36] Table 2 shows a summary of all effector mutations thus far analyzed. A significant correlation exists between transforming ability and stimulatory activity of GAP. Also, antibodies raised against this effector region are able to inhibit the GAP-*ras* interaction.[37] All these data strongly suggested that GAP may be the effector or be part of an effector complex of *ras* function. Some effector mutants could also be a valuable tool to determine some aspects of the interaction between *ras*-GAP and cellular or oncogenic *ras*.[38]

Although these effector mutant *ras* proteins do not interact with GAP,[4] recent studies show that at least one mutant (E38D) could still interact with it.[28] Other regions of *ras* likely to be involved in the interaction with GAP include the 50 to 60 amino acid region,[28,39] which is recognized by the neutralizing antibody

Y13-259.[40] This would explain the observation that this antibody impairs GAP "*in vitro*" activity.[1] On the other hand, the region of GAP involved in *ras* interaction is presumably the catalytic C terminus, since it is able to stimulate the GTPase activity as efficiently as the entire molecule.[22]

Experiments performed with NF-1-GRD (for NF-1 GAP-related domain), which is the region of the NF-1 gene highly homologous to p120-GAP and yeast *IRA1* and *IRA2*,[20] have demonstrated that this NF-1 "catalytic" region interacts with mammalian and yeast *ras* proteins, similarly to the classical p120-*ras*-GAP.[42-44] The main difference, in terms of its biochemical activity, resides in the affinity constant for *ras* (NF-1-GRD binds to the *ras* proteins up to 300 times more efficiently than p120-GAP[43]). This could be of physiological relevance at local high *ras* concentration.

IV. REGULATION OF *ras* p120-GAP ACTIVITY

Much effort has been made to define a physiological role for GAP, with little success up to now. A somehow confusing picture has emerged from the different approaches aimed to find a biological effect of GAP. A likely explanation for not finding an evident cellular function for GAP is that its activity could be regulated by external factors resulting in an alteration of its biochemical properties and/or its cellular localization, and consequently of its interaction with *ras*.

A. p120-GAP AND TYROSINE KINASES

A number of experiments have revealed that p120-GAP is phosphorylated in tyrosine residues upon activation of different tyrosine kinases.[25,45] Typically, in these experiments, GAP is immunoprecipitated with specific anti-GAP or antiphosphotyrosine antibodies and the immunoprecipitates are probed with antiphosphotyrosine or anti-GAP antibodies, respectively. Among those kinases are PDGFR,[25,46] EGFR,[45] CSF1-R,[47] but not InsulinR or bFGFR,[25] and different forms of oncogenically activated tyrosine kinases such as v-*src*,[45] v-*fps*,[45] v-*abl*,[45] and *erbB-2*.[48] In some cases it has been shown that p120-GAP physically interacts with the kinases.[46,47,49] The formation of stable complexes between the growth factor receptor and GAP seems to be dependent both on prior growth factor binding and on the receptor integrity, since mutations that impair kinase activity or autophosphorylation of PDGF or EGF receptors result in lack of GAP interaction.[50-52]

The phosphorylation of GAP on tyrosine is probably necessary for GAP interaction with other cellular components which are also phosphotyrosine proteins.[45,50,51] In particular, GAP has been found to be associated with two other proteins, p190 and p62, in cells transformed by oncogenes with tyrosine kinase activity (e.g., *src*[45,50,51]), or cells stimulated by EGF[45,50] or CSF-1.[47] The role of these phosphoproteins is unknown although certain additional data suggest that a delicate balance between free GAP and p190/p62-complexed GAP could exist in the cell for regulating GAP function. In particular, in v-*src*-transformed fibroblasts, GAP-p62 and GAP-p190 complexes have been found to have different cellular localization and even to affect the *in vitro* activity of GAP,[51] although it remains obscure whether this is of physiological relevance.

The interaction between GAP and all these proteins takes place through the regulatory SH regions.[52,54,55] For instance, SH2 and SH3 regions of GAP are essential for the interaction of GAP with the C terminal region of EGFR and p62.[55] Also, the SH2 regions of *fps* and *src* are necessary for the tyrosine phosphorylation of all three GAP, p62, and p190, and even for their association with each other.[50,51] The fact that EGFR phosphorylates GAP at a tyrosine residue adjacent to the SH2 domains[57] suggests that this modification could affect the conformation of the regulatory SH2 regions.

The effect of the tyrosine residues phosphorylation on GAP as well as its association with those proteins is not known, although one could expect to play an important role in the regulatory pathway of *ras* function. It is possible then that the phosphorylation of GAP leads to an alteration of its GTPase stimulating activity. In fact, the phosphorylation of GAP does not affect its *in vitro* properties, although the p62-GAP complex has been reported to have a decreased GTPase-stimulating activity.[51] Also, a significant amount of GAP phosphorylated on tyrosine is recovered in the particulate fraction of PDGF-stimulated cell lysates,[25,51] in agreement with the observation that the p62-GAP complex (which represents 8% of total GAP) is associated with the plasma membrane.[51]

The interaction of GAP with certain growth factor receptors would also help to understand the thus far unexplained relationship between *ras* and growth factor receptors with tyrosine kinase activity.[45] In this respect, it is worth noting that in *ras*-transformed cells, the GAP-PDGF receptor interaction is impaired.[46] Thus, GAP appears to be a good candidate to link the growth factor signaling pathway to *ras* activity.

B. p120-GAP AND PHOSPHOLIPIDS

A complementary aspect of GAP regulation has come from experiments aimed to determine the possible relationship of GAP with the phospholipid metabolism. It has been found that certain mitogenically related phospholipids specifically inhibit GAP activity, not only *ras*-GAP[57] but also *rho* and R-*ras* GAPs.[58] Interestingly, there is a correlation between production of these phospholipids (mainly derivatives of arachidonic acid) upon serum stimulation and their ability to inhibit *ras*GAP activity.[59] Although the existence of a *ras*GAP inhibitor protein has been reported.[60] Two different groups have demonstrated that GAP physically interacts with mitogenically active phospholipids through its catalytic domain, suggesting a direct effect on GAP.[61,62] In addition, the high local concentration of the lipids needed to be effective, could be reached within the membrane,[61] raising the possibility of GAP being trapped in particulate compartments and regulating *ras* activity in this way.

The lipid-mediated inhibition of the different proteins with GAP activity appears to be specific for every protein (e.g., *ras*-GAP vs. *rho*-GAP[58]). For instance, it has been shown that the *ras* GTPase stimulating activities of p120-GAP and NF1-GRD can be distinguished by their different sensitivity to phospholipid inhibition.[26,63] Furthermore, it has been reported that some arachidonic acid derivatives, such as different prostaglandins, stimulate GAP activity of p120-GAP, but not of NF1-GRD, while both are inhibited by arachidonate.[26] Also, NF1-GRD GAP-associated activity is inhibited at micromolar concentrations of arachidonate, phosphatidate, or PtIP2, whereas p120-GAP retains its full activity.[63] This is important not only from the point of view of a putative physiological function, but also for experimental purposes. In this respect, Bollag and McCormick[26] have been able to distinguish between two separate *ras*-GAP activities (p120-like and NF1-like) in different mammalian tissues and cell lines.

C. p120-GAP AND OTHER ras-LIKE PROTEINS

An alternative regulatory mechanism can be determined by the fact that GAP/*ras* interaction is inhibited by *rap1A* protein, due to its higher affinity for p120-GAP.[29,31] This is of great interest since *rap1A* is able to suppress *ras* transformation.[64] In addition, *ras* p120-GAP is unable to "deactivate" *rap1A*[29,65] and therefore, it is supposed to firmly bind to it, leaving *ras* in an activated form. It is important to note that a GAP for *rap1A* has already been cloned,[65] and this will probably lead to a better understanding of the *ras/rap1A* and their respective GAP's relationship.

V. BIOLOGICAL PROPERTIES OF ras p120-GAP

Two different and contradictory hypotheses have been raised.[32,33] either GAP is the target for *ras* activity, according to the mutational analysis of the *ras* effector region, or GAP acts as an attenuator, based on GAP biochemical activity. Different approaches have been attempted in order to assess a biological function for *ras* p120-GAP.

The overexpression of p120-GAP protein in normal and oncogenic *ras*-transformed cells has shown that high levels of GAP are able to suppress the transformation due to normal *ras* but not the one due to mutant *ras*.[66] This would be consistent with a role for GAP as a downregulator of normal *ras*. A similar conclusion can be drawn from the fact that in yeast, human GAP does the same work than *IRA1* and *IRA2*, that is to downregulate RAS activity by deactivating it.[30] However, some authors have argued that an overexpression of GAP in *ras*-transformed cells could alter a potentially important control mechanism, if a delicate stoichiometric balance between GAP and other molecules (growth factor receptors, p60 and p190) was to be needed. On the other hand, the yeast system could be missing some pieces of the puzzle, since the *ras* pathway appears to have introduced new elements during the evolution and experimentally both systems are not entirely interchangeable (e.g., neither *IRA1* nor *IRA2* interact with mammalian *ras*, although *ras*-GAP does, indeed, bind to yeast RAS[30]).

Recent reports also show that GAP is able to suppress *src*-driven transformation.[67,68] These results provide further evidence for *ras* being downstream of *src*, as it was strongly supported by the fact that microinjection of a *ras*-neutralizing antibody transiently impairs *src*-driven transformation.[40] These results also suggest a very close link between *ras*, GAP, and *src* signaling pathways. However, they do not help to elucidate the position of GAP with respect to *ras* in the signaling pathway, as it makes use of overexpression of GAP. In addition, the C terminal end of GAP is more effective than the full-length protein, suggesting that some regulatory mechanism could be missing in these experiments.

Thus far, the main evidence to support a physiological role of GAP as the effector molecule for *ras* comes from experiments using a patch-clamp system, where it was shown that GAP and *ras* cooperate in inhibiting the G protein-mediated coupling of a muscarinic signal and the atrial K^+ channel.[69] Whether this is a common mechanism of *ras*/GAP function or just a particular effect in this model is not known, although an important feature here is that GAP and *ras* are shown to work together in a biological system. In addition, experiments carried out in the *Xenopus* oocytes system suggest that GAP works downstream of *ras*.[70] In this system, a soluble form of oncogenic *ras* (with no biological activity, but still able to interact with GAP) inhibits the effect of a fully active *ras* on the germinal-vesicle breakdown.

In conclusion, little is known about what GAP is actually doing in a cell, possibly because of lack of pertinent experimental systems. It is possible that the use of more physiological situations could be of great help in the understanding of the role of GAP in *ras* function. In this respect, Downward and collaborators[71] have shown, in a T-cell system, that the endogenous GAP activity is diminished upon T-cell receptor activation apparently through the action of protein kinase C. Nevertheless, it has not been possible to determine a clear relationship between all the components of the system. This could be due to the fact that, as it is becoming more evident, more than one GAP could account for the effect we are able to detect, that is an alteration on *ras* GTPase activity, and possibly this is a very narrow vision of all the effects of the GAP/*ras* interaction. In fact, many researchers are now considering a dual role for GAP, as target and deactivator of *ras* activity. This would explain both the interaction of GAP through the *ras* effector region and all the reported biological effects of GAP.

What is becoming more evident is that we are dealing with a very complex system in which many actors are playing together at a given time and it appears, therefore, that none of the models that are established for the classical G proteins are entirely transposable to *ras*. The involvement of proteins, together with GAP, in *ras* pathway is becoming an evidence of the crosstalking between these mitogenic pathways and further investigation on GAP is likely to provide valuable information on how these different elements are integrated.

NOTE ADDED IN PROOF

Since this chapter was written, a number of relevant reports concerning GAP proteins have been published. Thus, much effort has been made in studying the function of the NF1 protein, neurofibromin. Two independent studies have shown that the levels of neurofibromin are dramatically reduced or even null in cells cultured from Schwannomas derived from NF1 tumors.[74,75] The authors also show that in those cell lines, p21*ras* is mainly in the activated, GTP-bound form. This, together with the fact that in some tumors neurofibromin has been found to carry mutations that inhibit its GAP activity, strongly suggest that NF1 could be a tumour suppressor gene and its product, a down regulator of p21*ras*.

Experiments carried out with p120GAP in different systems stress the importance of this *ras* GAP as a downstream modulator of p21*ras* activity. In one case[76] p120GAP is implicated in the mitogenic pathway that leads to *Xenopus* oocytes maturation through phosphatidylcholine hydrolyzing phospholipase C. Of particular interest is the report by Martin et al.,[77] which shows not only that the p120GAP N terminal domain is responsible for *ras*-dependent inhibition of muscarinic K+ channels in patch-clamped atrial cells, but also that p21*ras* is no longer required when the catalytic C terminal domain is deleted. Furthermore, the authors identify the SH regions as the domains of p120GAP responsible for this effect.

Both p190 and p62, the tyrosine-phosphoproteins that interact with cellular p120GAP, have been cloned. p62 shows a high regional homology to a putative hnRNP protein.[78] The recombinant protein shows some of the expected properties for a protein involved in RNA processing (DNA and mRNA binding, dimethylation on multiple arginine sites) and it binds tightly to GAP through the GAP SH2 region when p62 is tyrosine-phosphorylated. Its physiological role, however, is yet unknown.

The analysis of p190 primary structure[79] has revealed, in its N terminus, three moitifs indicative of a GTP-binding protein. Its central region encodes a transcriptional repressor of the glucocorticoid receptor, and in its C terminus a surprising homology (40%) to *rho* GAP, *bcr*, and n-chimerin has been discovered. Indeed, recombinant p190 has a specific GAP activity for members of the *rho* family,[80] suggesting that the association of p190 and p120GAP in growth factor-stimulated cells could couple *ras* and *rho* signalling pathways. These new discoveries show the existence of different GAP families for different *ras*-like proteins, and that it is likely to find common elements in the diverse signalling pathways of these small G-binding proteins.

REFERENCES

1. **Trahey, M. and McCormick, F.,** A cytoplasmic protein stimulates normal N-*ras* p21 GTPase, but does not affect oncogenic mutants, *Science,* 238, 542, 1987.
2. **Trahey, M., Wong, G., Halenbeck, R., Rubinfeld, I., Martin, G. A., Ladner, M., Long, C. M., Crosier, W. J., Watt, K., Koths, K., and McCormick, F.,** Molecular cloning of two types of GAP complementary DNA from human placenta, *Science,* 242, 1697, 1988.
3. **Gibbs, J. B., Schaber, M. D., Allard, W. J., Sigal, I. S., and Scolnick, E. M.,** Purification of *ras* GTPase activating protein from bovine brain, *Proc. Natl. Acad. Sci. U.S.A.,* 85, 5026, 1988.
4. **Vogel, U. S., Dixon, R. A. F., Schaber, M. D., Diehl, R. E., Marshall, M. S., Scolnick, E. M., Sigal, I. S., and Gibbs, J. B.,** Cloning of bovine GAP and its interaction with oncogenic *ras* p21, *Nature,* 335, 90, 1988.
5. **Kikuchi, A., Sasaki, T., Araki, S., Hata, Y., and Takai, Y.,** Purification and characterization from bovine brain cytosol of two GTPase-activating proteins specific for *smg* p21, a GTP-binding protein having the same effector domain as c-*ras* p21s, *J. Biol. Chem.,* 264, 9133, 1989.
6. **Ueda, T., Kikuchi, A., Ohga, N., Yamamoto, J., and Takai,Y.,** GTPase activating proteins for the *smg* p21 GTP-binding protein having the same effector domain as the *ras* proteins in human platelets, *Biochem. Biophys. Res. Commun.,* 159, 1411, 1989.
7. **Quilliam, L. A., Der, C. J., Clark, R., O'Rourke, E. C., Zhang, K., McCormick, F., and Bokoch, G. M.,** Biochemical characterization of baculovirus-expressed *rap*1A/K*rev*-1 and its regulation by GTPase-activating proteins, *Mol. Cell. Biol.,* 10, 2901, 1990.
8. **Polakis, P. G., Rubinfeld, B., Evans, T., and McCormick, F.,** Purification of a plasma membrane-associated GTPase-activating protein specific for *rap*1/K*rev*-1 from HL60 cells, *Proc. Natl. Acad. Sci. U.S.A.,* 88, 239, 1991.
9. **Garret, M. D., Self, A. J., VanOers, C., and Hall, A.,** Identification of distinct cytoplasmic targets for *ras*/R*ras* and *rho* regulatory proteins, *J. Biol. Chem.,* 264, 10, 1989.
10. **Burstein, E. S., Linko-Stentz, K., Lu, Z., and Macara, I. G.,** Regulation of the GTPase activity of the *ras*-like protein, p25^{rab3A}, *J. Biol. Chem.,* 266, 2689, 1991.
11. **Becker, J., Tan, T. J., Trepte, H-H., and Gallwitz, D.,** Mutational analysis of the putative effector domain of the GTP-binding *Ypt*1 protein in yeast suggests specific regulation by a novel GAP activity, *EMBO J.,* 10, 785, 1991.
12. **Emkey, R., Freedman, S., and Feig, L. A.,** Characterization of a GTPase-activating protein for the *ras*-related *ral* protein, *J. Biol. Chem.,* 266, 9703, 1991.
13. **Diekmann, D., Brill, S., Garret, M. D., Totty, N., Hsuan, J., Monfries, C., Hall, C., Lim, L., and Hall, A.,** *Bcr* encodes a GTPase-activating protein for p21^{rac}, *Nature,* 351, 400, 1991.
14. **Imai, Y., Mikaye, S., Hughes, D. A., and Yamamoto, M.,** Identification of a GTPase-activating protein homolog in *Schizosaccharomyces pombe, Mol. Cell. Biol.,* 11, 3088, 1991.
15. **Tanaka, K., Nakafuku, M., Satoh, T., Marshall, M. S., Gibbs, J. B., Matsumoto, K., Kaziro, Y., and Toh-e, A.,** *S. cerevisiae* genes *IRA*1 and *IRA*2 encode proteins that may be functionally equivalent to mammalian *ras* GTPase activating protein, *Cell,* 60, 803, 1990.
16. **Tanaka, K., Lin, B. K., Wood, D. R., and Tamanoi, F.,** *IRA*2, an upstream negative regulator of *RAS* in yeast, is a *RAS* GTPase activating protein, *Proc. Natl. Acad. Sci. U.S.A.,* 88, 468, 1991.
17. **Cawthon, R., Weiss, R., Xu, G., Viskochil, D., Culver, M., Stevens, J., Robertson, M., Dunn, D., Gesteland, R., O'Connell, P., and White, R.,** A major segment of the neurofibromatosis type 1 gene: cDNA sequence, genomic structure, and point mutations, *Cell,* 62, 193, 1990.

18. **Viskochil, D., Buchberg, A., Xu, G., Cawthon, R., Stevens, J., Wolff, R., Culver, M., Carey, J., Copeland, N., Jenkins, N., White, R., and O'Connell, P.**, Deletions and a translocation interrupt a cloned gene at the neurofibromatosis type 1 locus, *Cell*, 62, 187, 1990.

19. **Wallace, M., Marchuk, D., Andersen, L., Letcher, R., Odeh, H., Saulino, A., Fountain, J., Brereton, A., Nicholson, J., Mitchell, A., Brownstein, B., and Collins, F.**, Type 1 neurofibromatosis gene: identification of a large transcript disrupted in three NF1 patients, *Science*, 249, 181, 1990.

20. **Xu, G., O'Connell, P., Viskochil, D., Cawthon, R., Robertson, M., Culver, M., Dunn, D., Stevens, J., Gesteland, R., White, R., and Weiss, R.**, The neurofibromatosis type 1 gene encodes a protein related to GAP, *Cell*, 62, 599, 1990.

21. **Halenbeck, R., Crosier, W. J., Clark, R., McCormick, F., and Koths, K.**, Purification, characterization and Western blot analysis of human GTPase-activating protein from native and recombinant sources, *J. Biol. Chem.*, 265, 21,922, 1990.

22. **Marshall, M. S., Hill, W. S., Ng, A. S., Vogel, U. S., Schaber, M. D., Scolnick, E. M., Dixon, R. A. F., Sigal, I. S., and Gibbs, J. B.**, A C-terminal domain of GAP is sufficient to stimulate *ras* GTPase activity, *EMBO J.*, 8, 1105, 1989.

23. **Sadowski, I., Stone, J. C., and Pawson, T.**, A noncatalytic domain conserved among cytoplasmic protein-tyrosine kinases modifies the kinase function and transforming activity of Fujinami sarcoma virus p130$^{gag-fps}$, *Mol. Cell. Biol.*, 6, 4396, 1986.

24. **McCormick, F.**, The world according to GAP, *Oncogene*, 5, 1281, 1990.

25. **Molloy, C. J., Bottaro, D. P., Fleming, T. P., Marshall, M. S., Gibbs, J. B., and Aaronson, S.**, PDGF induction of tyrosine phosphorylation of GTPase activating protein, *Nature*, 342, 711, 1989.

26. **Bollag, G. and McCormick, F.**, Differential regulation of *ras*GAP and neurofibromatosis gene product activities, *Nature*, 351, 576, 1991.

27. **Calés, C., Hancock, J. F., Marshall, C. J., and Hall, A.**, The cytoplasmic protein GAP is implicated as the target for regulation by the *ras* gene product, *Nature*, 332, 548, 1988.

28. **Krengel, U., Schlichting, I., Scherer, A., Schumann, R., Frech, M., John, J., Kabsch, W., Pai, E. F., and Wittinghofer, A.**, Three-dimensional structures of H-*ras* p21 mutants: molecular basis for their inability to function as signal switch molecules, *Cell*, 62, 539, 1990.

29. **Frech, M., John, J., Pizon, V., Chardin, P., Tavitian, A., Clark, R., McCormick, F., and Wittinghofer, A.**, Inhibition of GTPase activating protein stimulation of *ras*-p21 GTPase by the K*rev*-1 gene product, *Science*, 249, 169, 1990.

30. **Ballester, R., Michaeli, T., Ferguson, K., Xu, H.-P., McCormick, F., and Wigler, M.**, Genetic analysis of mammalian GAP expressed in yeast, *Cell*, 59, 681, 1989.

31. **Hata, Y., Kikuchi, A., Sasaki, T., Schaber, M. D., Gibbs, J. B., and Takai, Y.**, Inhibition of the *ras*p21 GTPase-activating protein-stimulated GTPase activity of c-Ha-*ras* p21 by *smg* p21 having the same putative effector domain as *ras* p21s, *J. Biol. Chem.*, 265, 7104, 1990.

32. **Hall, A.**, The cellular functions of small GTP-binding proteins, *Science*, 249, 635, 1990.

33. **McCormick, F.**, *ras* GTPase activating protein: signal transmitter and signal terminator, *Cell*, 56, 5, 1989.

34. **Willumsen, B. M., Papageorge, A. G., Kung, H. F., Bekesi, K., Robins, T., Johnsen, M., Vass, W. C., and Lowy, D. R.**, Mutational analysis of a catalytic domain, *Mol. Cell. Biol.*, 6, 2646, 1986.

35. **Sigal, I. S., Gibbs, J. B., D'Alonzo, J. S., and Scolnick, E. M.**, Identification of effector residues and a neutralizing epitope of Ha-*ras*-encoded p21, *Proc. Natl. Acad. Sci. U.S.A.*, 83, 4725, 1986.

36. **Adari, H., Lowy, D. R., Willumsen, B. M., Der, C. J., and McCormick, F.**, Guanosine triphosphatase activating protein (GAP) interacts with the p21 *ras* effector binding domain, *Science*, 240, 518, 1988.

37. **Rey, I., Soubigou, P., Debussche, L., David, C., Morgat, A., Bost, P. E., Mayaux, J. F., and Tocqué, B.,** Antibodies to synthetic peptide from the residue 33 to 42 domain of c-Ha-*ras* p21 block reconstitution of the protein with different effectors, *Mol. Cell. Biol.,* 9, 3904, 1989.

38. **Farnsworth, C. L., Marshall, M. S., Gibbs, J. B., Stacey, D. W., and Feig, L. A.,** Preferential inhibition of the oncogenic form of *ras*[H] by mutations in the GAP binding/ "effector" domain, *Cell,* 64, 625, 1991.

39. **Srivastava, S. K., DiDonato, A., and Lacal, J. C.,** H-*ras* mutants lacking the epitope for the neutralizing monoclonal antibody Y13-259 show decreased biological activity and are deficient in GTPase activating protein interaction, *Mol. Cell. Biol.,* 9, 1779, 1989.

40. **Mulcahy, L. S., Smith, M. R., and Stacey, D. W.,** Requirement for *ras* proto-oncogene function during serum stimulated growth of NIH 3T3 cells, *Nature,* 313, 241, 1985.

41. **Buchberg, A. M., Cleveland, L. S., Jenkin, N. A., and Copeland, N. G.,** Sequence homology shared by neurofibromatosis type-1 gene and *IRA*-1 and *IRA*-2 negative regulators of the *RAS* cyclic AMP pathway, *Nature,* 347, 291, 1990.

42. **Ballester, R., Marchuk, D., Boguski, M., Saulino, A., Letcher, R., Wigler, M., and Collins, F.,** The NF1 locus encodes a protein functionally related to mammalian GAP and yeast *IRA* proteins, *Cell,* 63, 851, 1990.

43. **Martin, G. E., Viskochil, D., Bollag, G., McCabe, P. C., Crosler, W. J., Haubruck, H., Conroy, L., Clark, R., O'Connell, P., Cawthon, R. M., Innis, M. A., and McCormick, F.,** The GAP-related domain of the neurofibromatosis type 1 gene product interacts with *ras* p21, *Cell,* 63, 843, 1990.

44. **Xu, G., Lin, B., Tanaka, K., Dunn, D., Wood, D., Gesteland, R., White, R., Weiss, R., and Tamanoi, F.,** The catalytic domain of the neurofibromatosis type 1 gene product stimulates *ras* GTPase and complements *ira* mutants of *S. cerevisiae, Cell,* 63, 835, 1990.

45. **Ellis, C., Moran, M., McCormick, F., and Pawson, T.,** Phosphorylation of GAP and GAP-associated proteins by transforming and mitogenic tyrosine kinases, *Nature,* 343, 377, 1990.

46. **Kaplan, D. R., Morrison, D. K., Wong, G., McCormick, F., and Williams, L. T.,** PDGF beta-receptor stimulates tyrosine phosphorylation of GAP and association of GAP with a signaling complex, *Cell,* 61, 125, 1990.

47. **Reedijk, M., Liu, X., and Pawson, T.,** Interactions of phosphatidylinositol kinase, GTPase activating protein (GAP) and GAP-associated proteins with the colony-stimulating factor 1 receptor, *Mol. Cell. Biol.,* 10, 5601, 1990.

48. **Fazioli, F., Kim, U.-H., Rhee, S. G., Molloy, C. J., Segatto, O., and DiFiore, P. P.,** The *erb*B-2 mitogenic signalling pathway: tyrosine phosphorylation of phospholipase C-gamma and GTPase-activating protein does not correlate with *erb*B-2 mitogenic potency, *Mol. Cell. Biol.,* 11, 2040, 1991.

49. **Kazlauskas, A., Ellis, C., Pawson, T., and Cooper, J. A.,** Binding of GAP to activated PDGF receptors, *Science,* 247, 1578, 1990.

50. **Bouton, A. H., Kanner, S. B., Vines, R. R., Wang, H.-C. R., Gibbs, J. B., and Parsons, J. T.,** Transformation by pp60[src] or stimulation of cells with epidermal growth factor induces the stable association of tyrosine-phosphorylated cellular proteins with GTPase-activating protein, *Mol. Cell. Biol.,* 11, 945, 1991.

51. **Moran, M. F., Polakis, P., McCormick, F., Pawson, T., and Ellis, C.,** Protein-tyrosine kinases regulate the phosphorylation, subcellular distribution, and activity of p21[ras] GTPase-activating protein, *Mol. Cell. Biol.,* 11, 1804, 1991.

52. **Anderson, D., Koch, C. A., Grey, L., Ellis, C., Moran, M. F., and Pawson, T.,** Binding of SH2 domains on phospholipase C-gamma 1, GAP, and *src* to activated growth factor receptors, *Science,* 250, 979, 1990.

53. **Brott, B. K., Decker, S., Shafer, J., Gibbs, J. B., and Jove, R.,** GTPase-activating protein interactions with the viral and cellular *src* kinases, *Proc. Natl. Acad. Sci. U.S.A.,* 88, 755, 1991.

54. **Moran, M. F., Koch, C. A., Anderson, D., Ellis, C., England, L., Martin, G. S., and Pawson, T.,** *Scr* homology region 2 domains direct protein-protein interactions in signal transduction, *Proc. Natl. Acad. Sci. U.S.A.,* 87, 8622, 1990.

55. **Margolis, B., Li, N., Koch, A., Mohammadi, M., Hurwitz, D. R., Zilberstein, A., Ullrich, A., Pawson, T., and Schlessinger, J.,** The tyrosine phosphorylated carboxyterminus of the EGF receptor is a binding site for GAP and PLC-gamma, *EMBO J.,* 9, 4375, 1990.

56. **Liu, X. and Pawson, T.,** The epidermal growth factor receptor phosphorylates GTPase-activating protein (GAP) at tyr-460, adjacent to the SH2 domains, *Mol. Cell. Biol.,* 11, 2511, 1991.

57. **Tsai, M.-H., Yu, C.-L., Wei, F.-S., and Stacey, D. W.,** The effect of GTPase activating protein upon *ras* is inhibited by mitogenically responsive lipids, *Science,* 243, 522, 1989.

58. **Tsai, M.-H., Hall, A., and Stacey, D. W.,** Inhibition by phospholipids of the interaction between R-*ras, rho* and their GTPase-activating proteins, *Mol. Cell. Biol.,* 9, 5260, 1989.

59. **Yu, C.-L., Tsai, M.-H., and Stacey, D. W.,** Serum stimulation of NIH 3T3 cells induces the production of lipids able to inhibit GTPase-activating protein activity, *Mol. Cell. Biol.,* 10, 6683, 1990.

60. **Tsai, M.-H., Yu, C.-L., and Stacey, D. W.,** A cytoplasmic protein inhibits the GTPase activity of H-*ras* in a phospholipid-dependent manner, *Science,* 250, 982, 1990.

61. **Serth, J., Lautwein, A., Frech, M., Wittinghofer, A., and Pingoud, A.,** The inhibition of the GTPase activating protein H-*ras* interaction by acidic lipids is due to physical association of the C-terminal domain of the GTPase activating protein with micellar structures, *EMBO J.,* 10, 1325, 1991.

62. **Tsai, M.-H., Roudebush, M., Dobrowolski, S., Yu, C.-L., Gibbs, J. B., and Stacey, D. W.,** *Ras* GTPase-activating protein physically associates with mitogenically active phospholipids, *Mol. Cell. Biol.,* 11, 2785, 1991.

63. **Han, J.-W., McCormick, F., and Macara, I. G.,** Regulation of *ras*-GAP and the neurofibromatosis-1 gene product by eicosanoids, *Science,* 252, 576, 1991.

64. **Kitayama, H., Sugimoto, Y., Matsukaki, T., Ikawa, Y., and Noda, M.,** A *ras*-related gene with *ras* transformation suppression activity, *Cell,* 56, 77, 1989.

65. **Rubinfeld, B., Munemitsu, S., Clark, R., Conroy, L., Watt, K., Crosier, W. J., McCormick, F., and Polakis, P.,** Molecular cloning of a GTPase activating protein specific for the K*rev*-1 protein p21[rap]1, *Cell,* 65, 1033, 1991.

66. **Zhang, K., DeClue, J. E., Vass, W. C., Papageorge, A. G., McCormick, F., and Lowy, D. R.,** Suppression of c-*ras* transformation by GTPase-activating protein, *Nature,* 346, 754, 1990.

67. **Nori, M., Vogel, U. S., Gibbs, J. B., and Weber, M. J.,** Inhibition of v-*src*-induced transformation by a GTPase-activating protein, *Mol. Cell. Biol.,* 11, 2812, 1991.

68. **DeClue, J. E., Zhang, K., Redford, P., Vass, W. C., and Lowy, D. R.,** Suppression of *src* transformation by overexpression of full-length GTPase-activating protein (GAP) or of the GAP C terminus, *Mol. Cell. Biol.,* 11, 2819, 1991.

69. **Yatani, A., Okabe, K., Polakis, P., Halenbeck, R., McCormick, F., and Brown, A. M.,** *ras* p21 and GAP inhibit coupling of muscarinic receptors to atrial K$^+$ channels, *Cell,* 61, 769, 1990.

70. **Gibbs, J. B., Schaber, M. D., Schofield, T. L., Scolnick, E. M., and Sigal, I. S.,** *Xenopus* oocyte germinal-vesicle breakdown induced by (Val[12]) *ras* is inhibited by a cytosol-localized *ras* mutant, *Proc. Natl. Acad. Sci. U.S.A.,* 86, 6630, 1989.

71. **Downward, J., Graves, J. D., Warne, P. H., Rayter, S., and Cantrell, D. A.,** Stimulation of p21*ras* upon T-cell activation, *Nature,* 346, 719, 1990.

72. **Garrett, M. D., Major, G. N., Totty, N., and Hall, A.,** Purification and N-terminal sequence of the p21*rho* GTPase-activating protein, *rho* GAP, *Biochem. J.,* 276, 833, 1991.

73. **Koch, C. A., Anderson, D., Moran, M. F., Ellis, C., and Pawson, T.,** SH2 and SH3 domains: elements that control interactions of cytoplasmic signaling proteins, *Science,* 252, 668, 1991.

74. **DeClue et al.**, *Cell*, 69, 265, 1992.
75. **Basu et al.**, *Nature*, 356, 713, 1992.
76. **Domínguez et al.**, *EMBO J.*, 10, 3215, 1991.
77. **Martin et al.**, *Science*, 255, 192, 1992.
78. **Wong et al.**, *Cell*, 69, 551, 1992.
79. **Settleman et al.**, *Cell*, 69, 539, 1992.
80. **Settleman et al.**, *Nature*, 359, 153, 1992.

Chapter 23

THE *IRA* GENE FAMILY IN YEAST

Kazuma Tanaka, Akio Toh-e, and Kunihiro Matsumoto

TABLE OF CONTENTS

0-8493-5214-2/93/$0.00 + $.50
© 1993 by CRC Press, Inc.

I. INTRODUCTION

The *ras* proteins are members of the family of guanine nucleotide-binding proteins (GTP-binding proteins); they exhibit guanine nucleotide-binding activity and possess an intrinsic GTPase activity.[1-4] As in the case of heterotrimeric G proteins, these biochemical properties of *ras* proteins are central to their biological function: the GTP-bound form is an active conformation that can activate a target protein, a stimulation shut-off upon hydrolysis of GTP to GDP. The GTP-bound form is regenerated through the exchange of bound GDP for external GTP, a reaction probably promoted by an upstream signal (Figure 1). Thus, by analogy to G proteins, *ras* proteins are thought to function as signal transducing molecules involved in cell growth control. To understand the detailed mechanism of the *ras*-mediated signal transduction pathway, it is important to identify the upstream and downstream components interacting directly with *ras;* none, as yet, have been identified in mammalian systems.

The yeast *Saccharomyces cerevisiae* has two homologs of mammalian *ras* genes, *RAS1* and *RAS2*,[5,6] which are key regulatory molecules in a major signaling pathway linking growth respond to nutrients.[7,8] The yeast RAS proteins regulate the activity of adenylyl cyclase, which produces the second messenger, cyclic AMP (cAMP), responsible for the activation of cAMP-dependent protein kinase (A-kinase); (Figure 1).[9] The RAS2 protein has been shown to activate adenylyl cyclase when bound to GTP but not when bound to GDP.[10] Therefore, it appears that adenylyl cyclase is an effector of yeast RAS proteins. Human H-*ras* can substitute for yeast RAS in activating adenylyl cyclase,[11,12] although the ultimate target for mammalian *ras* may not be adenylyl cyclase in mammalian cells.[13,14] Moreover, a modified yeast *RAS1* that contains the activating mutation *RAS1*[Leu68] and a deletion of the unique 120 amino acids near the carboxyl terminal region can transform mouse NIH 3T3 cells in a gene transfer assay.[11] These structural and biochemical homologies between mammalian *ras* and yeast RAS proteins suggest that the identification of molecules interacting with RAS proteins in yeast is expected to facilitate the understanding of the function of mammalian *ras* protein. Genetic analysis identified regulatory genes, *CDC25, IRA1*, and *IRA2*, that modulate the activity of yeast RAS proteins (Figure 1). The *CDC25* gene product is proposed to be a positive regulator that exchanges the GDP bound to RAS proteins for GTP.[15-18] In contrast, the *IRA* gene products negatively regulate RAS activity. Recently, it has been demonstrated that mammalian cells possess a *ras* GTPase-activating protein (GAP) that stimulates the hydrolysis of GTP bound to *ras* proteins.[19,20] GAP has been shown to directly interact with *ras*, but it is still unknown whether GAP acts as a negative regulator of *ras* by inducing it to hydrolyze GTP, or if GAP acts as a positive effector of *ras*. Moreover, it has been found that neurofibromatosis type 1 (NF-1) protein shares homology with the catalytic region of GAP and a more extensive homology with yeast IRA proteins.[21] The first part of this review describes our current understanding of the function of IRA proteins in yeast. In the second part, we will

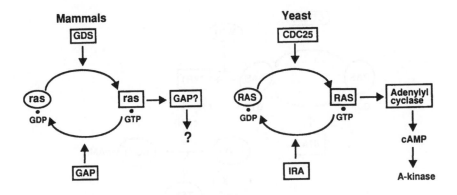

FIGURE 1. Regulation of *ras* proteins in mammals and yeast. GDS — GDP/GTP exchange protein.

discuss the structural and functional relationship among the mammalian GAP, NF-1 and yeast IRA proteins.

In this review, wild-type alleles and dominant mutations of yeast genes are denoted by capital italicized letters and defective mutations and gene disruptions by lower case italicized letters representing the gene followed by a superscript minus sign. The yeast proteins are denoted by capital Roman letters. Yeast in this review means budding yeast *Saccharomyces cerevisiae* and fission yeast is used to describe *Schizosaccharomyces pombe*.

II. IRA FUNCTION IN YEAST

A. IDENTIFICATION OF *IRA1* AND *IRA2* GENES

In *S. cerevisiae*, cAMP has a crucial role in cell cycle progression via activation of A-kinase.[22] A schematic representation of the cAMP pathway is shown in Figure 2. cAMP is produced by adenylyl cyclase which is encoded by *CYR1*[22,23] and yeast RAS proteins appear to be potent activators of this enzyme.[9] cAMP binds to the kinase regulatory subunit, encoded by the *BCY1* gene,[22,24] to release the catalytic subunits encoded by the *TPK1*, *TPK2*, and *TPK3* genes.[25] Free catalytic subunits then phosphorylate unknown target protein(s) whose phosphorylation is essential for cell-cycle progression. Mutations involved in the RAS-cAMP pathway can be divided into two groups. The first group involves *cyr1*⁻, *cdc25*⁻, and *ras1*⁻ *ras2*⁻ mutations that are defective in the activation of adenylyl cyclase and block activity of A-kinase. In the absence of A-kinase, cells cease to grow and arrest in G1.[9,22,26] A second group involves *RAS2*^val19 and *bcy1*⁻ mutations that yield elevated or unrestricted A-kinase activity. They prevent cells from arresting in G1 following nutrient starvation or heat shock. Therefore, such mutations cause a sporulation deficiency and sensitivities to heat shock and nutrient starvation.[9,27]

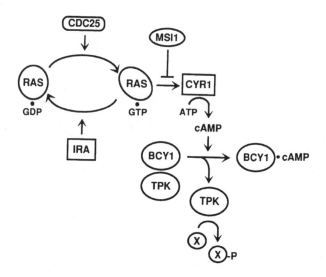

FIGURE 2. The RAS-cAMP pathway in *S. cerevisiae*. The blunt head arrow indicates inhibition and X indicates the substrate for A-kinase.

The *ira1⁻* mutation was identified in the process of isolating mutations affecting the RAS-cAMP pathway as a suppressor of the *cyr2⁻* mutation which is allelic to *cdc25⁻*.[28] Disruption of the *IRA1* gene results in sensitivities to heat shock and nitrogen starvation. Diploids homozygous for the disrupted *IRA1* gene are deficient to sporulation.[29] These phenotypes are typical of the *RAS2^{val19}* and *bcy1⁻* mutations. The *ira1⁻* mutation belongs to the second group of mutations in the RAS-cAMP pathway. In an attempt to further elucidate the components operating in the RAS-cAMP pathway, we have isolated multicopy suppressor genes of the heat shock sensitive phenotype caused by the *ira1* mutation. One is the *MSI1* gene, which encodes a protein homologous to the β-subunit of mammalian heterotrimeric G protein.[30] Another gene is the *IRA2* gene, which is a homolog of *IRA1*. Disruption of the *IRA2* gene also results in increased sensitivity to heat shock, nitrogen starvation, and sporulation defects as in the case of disruption of the *IRA1* gene.[31]

B. GENETIC ANALYSIS OF *IRA*

Disruption of either the *IRA1* gene or the *IRA2* gene suppresses the lethality of the *cdc25⁻* mutation, but not the lethality of a *ras1⁻ ras2⁻* double mutation or a *cyr1⁻* mutation. Moreover, the *ira1⁻* and *ira2⁻* mutants show increased levels of cAMP.[29,31] These results suggest that IRA proteins function at or upstream of RAS as negative regulators in the RAS-cAMP pathway. The *cdc25⁻* and *ira⁻* mutations are mutual suppressors; *ira⁻* suppresses the growth defect of *cdc25⁻* and *cdc25⁻* suppresses the heat shock sensitivity phenotype of *ira⁻*. This finding suggests that CDC25 and IRA act on RAS antagonistically.

Sensitivity to heat shock and to nitrogen starvation is more severe in the *ira2*⁻ mutant than in the *ira1*⁻ mutant, and the *ira1*⁻ *ira2*⁻ double mutant is the most sensitive to these stresses. These results indicate that the *IRA1* and *IRA2* gene products additively regulate the RAS-cAMP pathway. Overexpression of *IRA2* suppresses the phenotypes associated with the disruption of *IRA1*, but overexpression of *IRA1* does not suppress the phenotypes of the disruption of *IRA2*, suggesting their functions are similar but not the same.

IRA1 and *IRA2* genes encode proteins of 2938 and 3079 amino acids, respectively. The IRA2 protein is 45% identical to the IRA1 protein. When IRA1 and IRA2 proteins are divided into three parts, amino terminal, middle, and carboxyl terminal, the identities between IRA1 and IRA2 proteins are 31, 52, and 53%, respectively, indicating that IRA1 and IRA2 proteins are less homologous in their amino terminal region. In addition, the IRA2 protein has an extension of 180 amino acid residues at the amino terminus of the amino acid sequences of IRA1 and IRA2 proteins are aligned for maximum matching (Figure 3A). In the middle of the IRA1 and IRA2 proteins, there is a region showing a weak but significant homology with the carboxyl terminus of bovine GAP (Figure 3B). When these regions of IRA1 and IRA2 proteins (amino acid residues 1550 to 1880 of IRA1 and amino acid residues 1650 to 2025 of IRA2) are aligned with bovine GAP (amino acid residues 700 to 1047) for maximum matching, they are 22% identical and 45% homologous to GAP. The carboxyl terminal domain of bovine GAP (amino acid residues 702 to 1044) is sufficient to bind ras and to catalytically stimulate ras GTPase activity.[32] The region of GAP homologous to IRA proteins is within this carboxyl terminal catalytic domain. To define the essential domain of IRA proteins, *IRA2* was examined by deletion analysis for the ability to complement the *ira2*⁻ mutation. The amino acid sequence from 1609 to 1998 of the IRA2 protein is sufficient for suppression of the *ira2*⁻ mutation. This region corresponds to the IRA domain homologous to mammalian GAP. The homology between IRA proteins and GAP, thus, seems to be significant, suggesting that IRA proteins bind RAS and have a GAP-like activity. As described below, this possibility was supported by the biochemical analysis of IRA proteins.

The essential domain for IRA2 function resides in the amino acid sequences from 1609 to 1998, suggesting that the domains flanking the essential region may have regulatory functions. IRA proteins may function as detectors of nutritional limitation, since IRA1 and IRA2 act in an upstream position in the RAS-cAMP pathway. In this case, when a nutrient such as a nitrogen source becomes scarce in the medium, the regulatory domain(s) of IRA proteins will transfer the signal to the catalytic domain, which in turn will decrease the population of GTP-bound RAS proteins to reduce cAMP production. The net result of a nutrient limitation would, thus, be inhibition of entry into the next mitotic cycle. The amino acid sequence divergence at the amino terminal regions between IRA1 and IRA2 proteins may indicate that these regions respond to different nutrient signals.

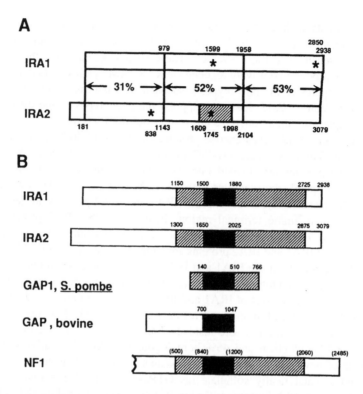

FIGURE 3. Comparison of protein structures. (A) Schematic diagram of the structure of IRA1 and IRA2 proteins. Residue numbers are indicated outside the boxes. Percent homologies refer to identities in protein sequences. The cross-hatched section indicates an essential domain for the IRA2 function. * indicates the potential site for A-kinase phosphorylation. (B) Schematic representation of the structure of GAP homologs. Solid sections represent core sequences conserved among these proteins. The cross-hatched sections represent extended homology among IRA1, IRA2, *S. pombe* GAP1, and NF-1. Relevant amino acid residues are numbered for each protein and the numbering of NF-1 is tentative.

In yeast, cAMP production is under feedback control.[33] Wild-type cells show a rapid biphasic change in cAMP levels in response to glucose. However, in *ira⁻* mutant cells, cAMP levels rise and remain elevated in response to glucose.[29,31] Thus, IRA proteins may be components of a feedback pathway that controls RAS activity. IRA1 and IRA2 proteins have two potential sites for A-kinase phosphorylation. Arg-Arg-X-Ser or Thr, and one of the two sites (IRA1, residue 1599; IRA2, residue 1745) is located at the same position in both proteins. These sites may be involved in feedback control of cAMP formation conducted by A-kinase.

C. BIOCHEMICAL ANALYSIS OF IRA

It is widely believed that *ras* proteins function as elements of signal transduction pathways and are capable of activating their effectors when bound to

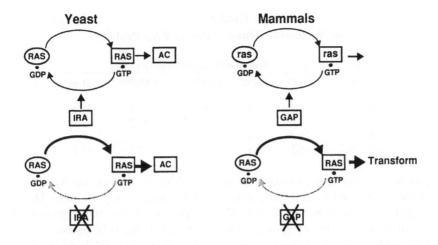

FIGURE 4. Comparison between IRA and GAP functions. AC -- adenylyl cyclase.

GTP but not when bound to GDP.[34,35] Field et al.[10] demonstrated that purified adenylyl cyclase could be activated by RAS2 protein only in the presence of GTP. The guanine nucleotide content in the *ras* protein is thought to be regulated by *ras* protein itself through its intrinsic GTPase activity[1-4] or possibly by other regulatory proteins such as GAP[19,20] and GDP/GTP exchange protein (Figure 1).[36,37] The *RAS2*[val19] mutation, analogous to the oncogenic variation in mammalian H-*ras*[val12], reduces intrinsic GTPase activity and causes hyperactivation of adenylyl cyclase.[38] The CDC25 protein is proposed to act as a GDP/GTP exchange protein.[15-18] The phenotypic similarity of the *ira⁻* mutants to the *RAS2*[val19] mutant suggests that the IRA proteins function to reduce the levels of GTP-bound RAS proteins. In the absence of IRA function, the RAS proteins are thought to accumulate in the GTP-bound state and constitutively activate adenylyl cyclase (Figure 4). From the fact that IRA proteins inhibit the function of RAS proteins in a fashion antagonistic to the function of the CDC25 protein, it can be assumed that IRA proteins stimulate the formation of GDP-bound RAS proteins by acting as a GAP protein.

To examine the possibility that IRA proteins function to decrease the population of GTP-bound RAS proteins, Tanaka and Nakafuku et al.[34] undertook a quantitative analysis of the guanine nucleotides bound to RAS proteins in *ira* mutants as well as in the wild-type strains (Table 1). The percent of RAS1 or RAS2 protein bound with GTP in the wild-type strain is less than 1%. In the *ira1⁻* or *ira2⁻* mutant, by contrast, the amount of the GTP-bound form of the RAS1 or RAS2 protein is significantly increased. These results indicate that both IRA1 and IRA2 proteins function to decrease the population of GTP-bound RAS proteins. In the *ira1⁻ ira2⁻* double mutant, the level of GTP-bound form of both RAS1 and RAS2 proteins increased dramatically. The IRA1 and IRA2 proteins, thus, cooperate to regulate the guanine nucleotide-bound state of RAS

TABLE 1
The Nucleotide-Bound States of RAS Proteins

Genotype	Protein with GTP-bound form (%)				
	RAS1	RAS2	RAS2^{val19}	RAS2^{thr66}	c-H-*ras*
IRA1 IRA2	0.9	9.4	19.2	74.1	43.9
ira1⁻ IRA2	3.2	3.4	—	80.2	—
IRA1 ira2⁻	24.2	3.8	—	—	—
ira1⁻ ira2⁻	44.3	11.1	27.6	83.3	47.2

proteins. The oncogenic mutant RAS protein, RAS2^{val19}, accumulates much more as the GTP-bound form than do wild-type RAS proteins in yeast cells. This is explained by the impaired intrinsic GTPase activity of the mutant protein. In mammalian cells, the cytoplasmic factor GAP binds the GTP-bound form of both normal and oncogenic *ras* and stimulates the intrinsic GTPase activity of normal *ras* protein but not that of oncogenic *ras* proteins, ras^{val12}, and ras^{thr59}, which have impaired GTPase activity.[39] In agreement with this, the corresponding mutant RAS2 proteins, RAS2^{val19} and RAS2^{thr66}, show a high percentage of GTP-bound form in wild-type as well as in *ira⁻* mutant cells. This property of *ira⁻* mutants suggests that the IRA1 and IRA2 proteins function in a manner similar to mammalian GAP.

Recently, Tanaka and Tamanoi et al.[35] have presented the following evidence that the IRA2 protein indeed possesses a GAP activity for yeast RAS proteins:

1. Extracts prepared from yeast cells overexpressing the IRA2 protein (amino acid residues 527 to 2255) stimulated the release of γ-phosphate from the GTP-bound RAS2 protein, indicating that the IRA2 protein stimulates the GTPase activity of the RAS2 protein, on the other hand, the IRA2 protein does not stimulate the GTPase activity of the RAS2^{val19} protein.

2. To demonstrate that the IRA2 protein is responsible for the GAP activity detected in crude yeast extract, Tanaka et al. constructed a fusion protein of IRA2 whose amino terminus was added to the HA1 epitope. This peptide is recognized by the monoclonal antibody 12CA5. The 12CA5 antibody precipitates a significant GAP activity from extracts prepared from cells expressing the HA1-IRA2 fusion protein. Thus, it seems that the IRA2 protein itself has GAP activity.

3. A 1082 bp DNA fragment encoding amino acid residues 1665 to 2027 (362 amino acids) of the IRA2 protein was fused to the 3' end of the glutathione-*S*-transferase. This region of the IRA2 protein is essential for its function and is homologous to the catalytic domain of mammalian GAP. The fusion protein partially purified from *Escherichia coli* cells stimulates the GTPase activity of the RAS2 protein but not that of the RAS2^{val19} mutant protein. These results provide evidence that IRA proteins are yeast homologs of mammalian GAP proteins.

D. IRA1 vs. IRA2

As previously mentioned, the functions of IRA1 and IRA2 proteins are similar but not the same. What is the functional difference between IRA1 and IRA2? From the fact that yeast has two RAS proteins, RAS1 and RAS2, a simple prediction is that each IRA protein has specificity for each RAS protein. The $ras2^-$ mutation clearly suppresses the phenotypes of both $ira1^-$ and $ira2^-$ mutations, but the $ras1^-$ mutation cannot suppress the phenotypes of $ira1^-$ and weakly suppresses those of $ira2^-$.[31] These results suggest that the phenotypes of $ira2^-$ are caused by the activation of RAS1 and RAS2 proteins, although the contribution by RAS1 activation is lower than that of RAS2 activation, possibly because of weaker expression of RAS1 than RAS2.[40] In contrast, the phenotypes of $ira1^-$ are caused by the activation of only the RAS2 protein. Thus, some preference appears to exist between RAS and IRA proteins, a speculation supported by the guanine nucleotide analyses (Table 1). When RAS1 or RAS2 protein is overexpressed in the $ira1^-$ mutant, both RAS proteins seem to be equally activated, whereas, in the $ira2^-$ mutant, the RAS1 protein is more activated than the RAS2 protein.[34] Therefore, the IRA2 protein appears to act on the RAS1 protein more efficiently than on the RAS2 protein.

III. COMPARISON OF YEAST IRA WITH MAMMALIAN GAP AND NF-1

A. EXPRESSION OF MAMMALIAN GAP IN YEAST

Normal mammalian H-*ras* protein expressed in yeast remains constitutively in the GTP-bound form and appears to be refractory to IRA proteins (Table 1).[34] Moreover, the IRA2 protein partially purified from *E. coli* cannot stimulate the GTPase activity of the H-*ras* protein.[35] These results indicate that IRA proteins recognize yeast RAS proteins but not mammalian *ras* proteins, suggesting that the domain(s) of yeast RAS proteins required for productive interaction with IRA proteins reside with amino acid sequence different from that of mammalian *ras*. One possible site is the unique carboxyl terminal stretch of about 150 amino acids in the RAS1 and RAS2 proteins. However, a mutant protein of RAS2 that lacks this stretch (amino acid residues 175 to 300[41]) binds guanine nucleotides in an IRA-dependent manner,[34] suggesting that the carboxyl terminal region of yeast RAS proteins is not involved in interaction with IRA proteins. This possibility is supported by GAP assays *in vitro*.[35] Crude extracts of yeast overexpressing IRA2 and IRA2 protein purified from *E. coli* extracts stimulated the GTPase activity of the truncated RAS2 protein which lacks the unique carboxyl terminal 120 amino acid stretch of RAS. Thus, the domain of RAS proteins required for the interaction with IRA proteins seems to reside in the conserved amino terminal region.

Gibbs et al.[20] have shown that purified bovine GAP can stimulate the GTPase activity of yeast RAS1 protein, indicating that mammalian GAP recognizes mammalian *ras* and yeast RAS proteins. In contrast, yeast IRA cannot stimulate

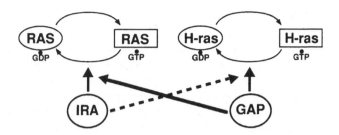

FIGURE 5. Action of IRA and GAP on yeast RAS and mammalian H-*ras* proteins. Dashed lines indicate no stimulation.

the GTPase activity of mammalian H-*ras* (Figure 5). Moreover, mammalian GAP can interact with the *ras*-related GTP-binding protein K*rev*-1, but is incapable of stimulating its GTPase activity.[42] Thus, GAP and IRA have specificity toward substrate proteins. Since the putative effector region in H-*ras* (amino acid residues 32 to 40), K*rev*-1, and yeast RAS proteins is identical, additional regions are important for the interaction between *ras*/K*rev*-1/yeast RAS and GAP/ IRA. Consistent with this possibility, a peptide having amino acid residues 17 to 32 of H-*ras* competes with H-*ras* in the interaction with GAP.[43]

The structural and functional similarity between IRA proteins and mammalian GAP raises the possibility that GAP replaces IRA function in yeast. Expression of truncated GAP, amino acid residues 664 to 1044, suppresses the phenotypes of *ira*⁻ mutants, but not those of the RAS2^val19 mutant.[34] Expression of mammalian GAP causes a decrease in the amount of the GTP-bound form of yeast RAS1 protein in *ira2*⁻ mutant cells. On the other hand, GAP expression does not affect the level of GTP-bound RAS2^val19. These results indicate that expression of mammalian GAP can functionally complement an IRA deficiency by induction of the hydrolysis of GTP bound to yeast RAS proteins. Thus, mammalian GAP can function as a negative regulator of wild-type RAS proteins. Mammalian GAP also negatively regulates mammalian H-*ras* in yeast. When H-*ras* is expressed in yeast, there are high levels of the GTP-bound form. Expression of GAP causes a decrease in the amount of the GTP-bound form of H-*ras* in yeast, indicating that mammalian GAP stimulates the GTPase activity of H-*ras* in yeast. The CDC25 protein is thought to be required for the activation of yeast RAS proteins by catalyzing nucleotide exchange.[15-18] The *cdc25* temperature-sensitive *(ts)* mutation is suppressed by the mammalian H-*ras* or H-*ras*^val12 genes[41] because the mammalian *ras* is insensitive to the action of yeast IRA proteins so that it does not require catalytic activation by the CDC25 protein. However, Wigler et al.[44] have shown that *cdc25-ts* cells expressing GAP and carrying wild-type H-*ras* become temperature-sensitive for growth. By contrast, *cdc25-ts* cells carrying the mutant H-*ras*^val12 are not affected by the presence of GAP.[44] These results suggest that mammalian GAP can downregulate wild-type H-*ras* but cannot activate H-*ras*^val12 protein when coexpressed in yeast.

B. FUNCTION OF GAP IN MAMMALIAN CELLS

Two models for the role of GAP in mammalian cells have been proposed. In the first model, GAP serves as a negative regulator of mammalian *ras* function. This negative regulatory role is consistent with the following observations:

1. As discussed in a previous section, mammalian GAP stimulates the GTPase activity of mammalian *ras* and decreases *ras* stimulatory function on yeast adenylyl cyclase when both proteins are expressed in yeast.[34,44]
2. Overexpression of GAP in mammalian cells decreases the percentage of *ras* in the GTP-bound state.[45]
3. Overexpression of GAP reverses transformation induced by overexpression of the normal H-*ras* gene but not by the activated mutant v-*ras* gene.[46]
4. Activation of the T-cell antigen receptor increases the levels of GTP-bound *ras*. This activation seems to occur through inactivation of GAP by protein kinase C activation.[47]

In the second model, GAP is a downstream target of mammalian *ras*. Evidence which supports this possibility also exists: (1) mutations in the effector region of *ras* that reduce GAP binding are incapable of transforming cells[43,48,49] and (2) GAP is required for *ras* to inhibit coupling between muscarinic receptors and atrial potassium channels.[50] Thus, the precise role of GAP in the function of the *ras* proteins is not yet fully understood. A possible explanation of these contradictory findings is that GAP functions both as a target and a negative regulator of *ras*. As originally proposed by McCormick,[39] the interaction of GTP-bound *ras* and GAP could simultaneously transmit the *ras* signal downstream and promote its termination by stimulating hydrolysis of GTP bound to GDP. This model is based on an analogy with EF-Tu where interaction with its target, ribosomes, stimulates its GTPase activity.

Some approaches could be proposed to distinguish between the possibility that GAP is solely a negative regulator, and the alternative model that GAP functions both as a target and a negative regulator of *ras*. First, if *ras* mutations that fail to interact with GAP would cause transformation of cells, this would support the possibility that GAP does not have a positive function. Second, if GAP would act as a negative regulator of *ras*, it would be a recessive oncogene, since loss of its function would increase the level of GTP-bound *ras* protein and thereby, stimulate proliferation (Figure 4). This point will be discussed in the section on NF-1.

If GAP functions both as a target and a negative regulator of *ras*, this possibility leads to several interesting questions with regard to the yeast system. Do yeast IRA proteins have additional functions? Although we could find only a negative regulatory function for IRA proteins, it still remains that IRA proteins might be a target of yeast RAS stimulating unknown signaling pathway(s) different from that stimulating adenylyl cyclase. Wigler[51] has already proposed that yeast RAS regulates at least one pathway in addition to stimulation of adenylyl cyclase. Can yeast adenylyl cyclase stimulate GTPase activity of yeast RAS? Since this enzyme appears to be a target of yeast RAS, it is interesting to test the possibility that adenylyl cyclase might have a GAP function.

C. NF-1 IS HOMOLOGOUS TO YEAST IRA

Recently, a gene that is involved in the pathogenesis of the NF-1 disease was identified.[52-54] The *NF-1* gene encodes a protein of at least 2485 amino acids. Sequence analysis of this gene demonstrated that the NF-1 protein has a homology with the catalytic domains of yeast IRA and mammalian GAP proteins.[21] While the similarity of NF-1 to GAP is restricted to the catalytic domain, similarity to yeast IRA proteins extends well outside this region (Figure 3B). Thus, NF-1 protein is more similar to IRA proteins than to GAP. This evidence raises the prospect that NF-1 protein interacts with *ras* and acts as a GAP for *ras*. Three groups have presented clear evidence supporting this possibility.[55-57] The NF-1 catalytic domain which is homologous to IRA and GAP stimulates the GTPase activity of wild-type yeast RAS2 and mammalian *ras* proteins, but not oncogenic mutant *ras* proteins, RAS2^{val19} and H-*ras*^{val12}. Thus, the GTPase-stimulating activity of the NF1 peptide is similar to that of both yeast IRA and mammalian GAP. Expression of the NF-1 catalytic domain can complement the loss of yeast IRA function.

Several mutations, point mutations,[52] deletions,[53] and an insertion,[54] have been identified in NF-1 patients. Thus, it appears that inactivation of the *NF-1* gene leads to observations of loss of growth control in cells of NF-1 patients. Based on the model that NF-1 acts as an upstream negative regulator of *ras*, inactivation of NF-1 could result in accumulation of GTP-bound *ras* and thus, contribute to abnormal cellular proliferation. This model implies that the *NF-1* gene is an antioncogene. It is also possible to explain the observations by a model based on the idea that NF-1 is an effector of *ras* function. A signal generated by GTP-bound *ras* would inhibit cellular proliferation but stimulate differentiation. Inactivation of this signal transduction could block differentiation, resulting in uncontrolled cellular proliferation. In fact, activated *ras* induces differentiation and causes the cessation of growth in some neural cells such as rat PC12.[58-60]

There are at least two GTPase-activating proteins (GAP and NF-1) acting on *ras* in mammalian cells. Since both proteins appear to be ubiquitously expressed,[54,61] most cells presumably contain GAP and NF-1. One possibility is that NF-1 is a negative regulator of *ras*, while GAP is the effector. The more extensive similarity of NF-1 to yeast IRA suggests that NF-1 may be a mammalian counterpart of yeast IRA. Does yeast have a counterpart of mammalian GAP?

D. A GENE ENCODING A GAP-LIKE PROTEIN IN FISSION YEAST

The fission yeast *S. pombe* has a single gene, *RAS1*.[62] Disruption of the gene blocks mating function and inhibits sporulation but has no effect on growth or adenylyl cyclase. The mating defect phenotype can be complemented by mammalian *ras*. An activating mutation $RAS1^{val17}$ causes hypersensitivity to mating pheromone and blocks mating functions.[63,64] These results suggest that the *S. pombe* RAS1 protein is involved in the signal transduction pathway that responds to mating pheromone. Genetic analysis has revealed regulators, *STE6* and *GAP1*, that modulate the activity of RAS1 protein. STE6 is homologous to *S. cerevisiae* CDC25, suggesting that STE6 acts as a nucleotide exchange factor.[65] Sequence analysis of *GAP1* indicates that it encodes a protein of 766 amino acids that is homologous to mammalian GAP.[66] Compared with IRA and NF-1, the GAP1 protein has relatively short amino and carboxyl terminal regions flanking the catalytic domain of the mammalian GAP (Figure 3B). Disruption of the *GAP1* gene results in the same phenotype caused by the activated *RAS1* mutation, $RAS1^{val17}$. Moreover, the phenotypes of the *STE6* disruption are analogous to those of the *RAS1* disruption and are suppressed by the *GAP1* disruption and the $RAS1^{val17}$ mutation. These results suggest that *S. pombe* GAP1 primarily functions as a negative regulator of RAS1.

IV. CONCLUDING REMARKS

Yeast has been utilized to analyze RAS-mediated signal tranduction. The use of yeast as a model makes it possible to use powerful genetic and biochemical analyses that are not available in mammalian cell systems. Genetic analysis has allowed identification of yeast IRA proteins which regulate the activity of RAS. On the other hand, mammalian GAP was identified biochemically at first. Based on the results obtained with mammalian GAP, it has been confirmed that yeast IRA stimulates GTPase activity of RAS biochemically. We anticipate that the concepts learned from a combination of yeast genetics and mammalian cell biology and biochemistry will elucidate the role of GAP in *ras*-mediated cellular transformation.

ACKNOWLEDGMENTS

We thank Fuyuhiko Tamanoi, Masato Nakafuku, Yoshito Kaziro, and Masayuki Yamamoto for sharing results and manuscripts before publication. We also thank Rosamaria Ruggieri for a critical reading of this paper.

REFERENCES

1. **Gibbs, J. B., Sigal, I. S., Poe, M., and Scolnick, E. M.,** Intrinsic GTPase activity distinguishes normal and oncogenic *ras* p21 molecules, *Proc. Natl. Acad. Sci. U.S.A.*, 81, 5704, 1984.
2. **McGrath, J. P., Capon, D. J., Goeddel, D. V., and Levinson, A. D.,** Comparative biochemical properties of normal and activated human *ras* protein, *Nature*, 310, 644, 1984.

3. **Sweet, R. W., Yokoyama, S., Kamata, T., Feramisco, J. R., Rosenberg, M., and Gross, M.,** The product of *ras* is a GTPase and the T24 oncogenic mutant is deficient in this activity, *Nature*, 311, 273, 1984.

4. **Walter, M., Clark, S. G., and Levinson, A. D.,** The oncogenic activation of human p21 ras by a novel mechanism, *Science*, 233, 649, 1986.

5. **DeFeo-Jones, D., Scolnick, E. M., Koller, R., and Dhar, R.,** *ras*-related gene sequences identified and isolated from *Saccharomyces cerevisiae*, *Nature*, 306, 707, 1983.

6. **Powers, S., Kataoka, T., Fasano, O., Goldfarb, M., Strathern, J., Broach, J., and Wigler, M.,** Genes in *S. cerevisiae* encoding proteins with domains homologous to the mammalian *ras* proteins, *Cell*, 36, 607, 1984.

7. **Matsumoto, K., Uno, I., and Ishikawa, T.,** Genetic analysis of the role of cAMP in yeast, *Yeast*, 1, 15, 1985.

8. **Tatchell, K.,** *RAS* genes and growth control in *Saccharomyces cerevisiae*, *J. Bacteriol.*, 166, 364, 1986.

9. **Toda, T., Uno, I., Ishikawa, T., Powers, S., Kataoka, T., Broek, D., Cameron, S., Broach, J. B., Matsumoto, K., and Wigler, M.,** In yeast *RAS* proteins are controlling elements of adenylate cyclase, *Cell*, 40, 27, 1985.

10. **Field, J., Nikawa, J., Broek, D., MacDonald, B., Rodgers, L., Wilson, I. A., Lerner, R. A., and Wigler, M.,** Purification of a *RAS*-responsive adenylyl cyclase complex from *Saccharomyces cerevisiae* by use of an epitope addition method, *Mol. Cell. Biol.*, 8, 2159, 1988.

11. **DeFeo-Jones, D., Tatchell, K., Robinson, L. C., Sigal, I. S., Vass, W. C., Lowy, D. R., and Scolnick, E. M.,** Mammalian and yeast *ras* gene products: biological function in their heterologous systems, *Science*, 288, 179, 1985.

12. **Kataoka, T., Powers, S., Cameron, S., Fasano, O., Goldfarb, M., Broach, J., and Wigler, M.,** Functional homology of mammalian and yeast *ras* genes, *Cell*, 40, 19, 1985.

13. **Beckner, S. K., Hattori, S., and Shih, T. Y.,** the ras oncogene product p21 is not a regulatory component of adenylate cyclase, *Nature*, 317, 71, 1985.

14. **Birchmeier, C., Broek, D., and Wigler, M.,** *RAS* proteins can induce meiosis in Xenopus oocytes, *Cell*, 43, 615, 1985.

15. **Camonis, J. H., Kalekine, M., Gondre, B., Garreau, H., Boy-Marcotte, E., and Jacquet, M.,** Characterization, cloning and sequence analysis of the *CDC25* gene which controls the cyclic AMP level of *Saccharomyces cerevisiae*, *EMBO J.*, 5, 375, 1986.

16. **Broek, D., Toda, T., Michaeli, T., Levin, L., Birchmeier, C., Zoller, M., Powers, S., and Wigler, M.,** The S. cerevisiae *CDC25* gene product regulates the RAS/adenylate cyclase pathway, *Cell*, 48, 789, 1987.

17. **Daniel, J., Becker, J. M., Enari, E., and Levitzki, A.,** The activation of adenylate cyclase by guanyl nucleotides in *Saccharomyces cerevisiae* is controlled by the *CDC25* start gene product, *Mol. Cell. Biol.*, 7, 3857, 1987.

18. **Robinson, L. C., Gibbs, J. B., Marshall, M. S., Sigal, I. S., and Tatchell, K.,** *CDC25*: a component of the ras-adenylate cyclase pathway in *Saccharomyces cerevisiae*, *Science*, 235, 1218, 1987.

19. **Trahey, M. and McCormick, F.,** A cytoplasmic protein stimulates normal N-*ras* p21 GTPase, but does not affect oncogenic mutants, *Science*, 238, 542, 1987.

20. **Gibbs, J. B., Schaber, M. D., Allard, W. J., Sigal, I. S., and Scolnick, E. M.,** Purification of ras GTPase activating protein from bovine brain, *Proc. Natl. Acad. Sci. U.S.A.*, 85, 5026, 1988.

21. **Xu, G., O'Connell, P., Viskochil, D., Cawthon, R., Robertson, M., Culver, M., Dunn, D., Stevens, J., Gesteland, R., White, R., and Weiss, R.,** The neurofibromatosis type 1 gene encodes a protein related to GAP, *Cell*, 62, 599, 1990.

22. **Matsumoto, K., Uno, I., Oshima, Y., and Ishikawa, T.,** Isolation and characterization of yeast *Saccharomyces cerevisiae* mutants deficient in adenylate cyclase and cyclic AMP dependent protein kinase, *Proc. Natl. Acad. Sci. U.S.A.*, 79, 2355, 1982.

23. Kataoka, T., Broek, D., and Wigler, M., DNA sequence and characterization of the *S. cerevisiae* gene encoding adenylate cyclase, *Cell*, 43, 493, 1985.

24. Toda, T., Cameron, S., Sass, P., Zoller, M., Scott, J. D., McMullen, B., Hurwitz, M., Krebs, E. G., and Wigler, M., Cloning and characterization of *BCY1*, a locus encoding a regulatory subunit of the cyclic AMP-dependent protein kinase in *Saccharomyces cerevisiae*, *Mol. Cell. Biol.*, 7, 1371, 1987.

25. Toda, T., Cameron, S., Sass, P., Zoller, M., and Wigler, M., Three different genes in *S. cerevisiae* encode the catalytic subunit of the cyclic AMP-dependent protein kinase, *Cell*, 50, 277, 1987.

26. Pringle, J. R. and Hartwell, L. H., The *Saccharomyces cerevisiae* cell cycle, in *The Molecular Biology of the Yeast Saccharomyces cerevisiae: Life Cycle and Inheritance*, Strathern, J. N., Jones, E. W., and Broach, J. R., Eds., Cold Spring Harbor, New York, 1981, 97.

27. Matsumoto, K., Uno, I., and Ishikawa, T., Initiation of meiosis in yeast mutants defective in adenylate cyclase and cyclic AMP dependent protein kinase, *Cell*, 32, 417, 1983.

28. Matsumoto, K., Uno, I., Kato, K., and Ishikawa, T., 1985, Isolation and characterization of a phosphoprotein phosphatase-deficient mutant in yeast, *Yeast*, 1, 25, 1985.

29. Tanaka, K., Matsumoto, K., and Toh-e, A., *IRA1*, an inhibitory regulator of the *RAS/* cyclic AMP pathway in *Saccharomyces cerevisiae*, *Mol. Cell. Biol.*, 9, 757, 1989.

30. Ruggieri, R., Tanaka, K., Nakafuku, M., Kaziro, Y., Toh-e, A., and Matsumoto, K., *MSI1*, a negative regulator of the RAS-cAMP pathway in *Saccharomyces cerevisiae*, *Proc. Natl. Acad. Sci. U.S.A.*, 86, 8778, 1989.

31. Tanaka, K., Nakafuku, M., Tamanoi, F., Kaziro, Y., Matsumoto, K., and Toh-e, A., *IRA2*, a second gene of *Saccharomyces cerevisiae* that encodes a protein with a domain homologous to mammalian *ras* GTPase-activating protein, *Mol. Cell. Biol.*, 10, 4303, 1990.

32. Marshall, M. S., Hill, W. S., Ng, A. S., Vogel, U. S., Schaber, M. D., Scolnick, E. M., Dixon, R. A. F., Sigal, I. S., and Gibbs, J. B., A C-terminal domain of GAP is sufficient to stimulate *ras* p21 GTPase activity, *EMBO J.*, 8, 1105, 1989.

33. Nikawa, J., Cameron, S., Toda, T., Ferguson, K. M., and Wigler, M., Rigorous feedback control of cAMP levels in *Saccharomyces cerevisiae*, *Genes Dev.*, 1, 931, 1987.

34. Tanaka, K., Nakafuku, M., Satoh, T., Marshall, M. S., Gibbs, J. B., Matsumoto, K., Kaziro, Y., and Toh-e, A., *S. cerevisiae* genes *IRA1* and *IRA2* encode proteins that may be functionally equivalent to mammalian *ras* GTPase activating protein, *Cell*, 60, 803, 1990.

35. Tanaka, K., Lin, B. K., Wood, D. R., and Tamanoi, F., IRA2, an upstream negative regulator of RAS in yeast, is a RAS GTPase activating protein (GAP), *Proc. Natl. Acad. Sci. U.S.A.*, 88, 468, 1991.

36. Downward, J., Riehl, E., Wu, L., and Weinberg, R. A., Identification of a nucleotide exchange-promoting activity for p21^{ras}, *Proc. Natl. Acad. Sci. U.S.A.*, 87, 5998, 1990.

37. Wolfman, A. and Macara, I. G., A cytosolic protein catalyzes the release of GDP from p21, *Science*, 248, 247, 1990.

38. Broek, D., Samily, N., Fasano, O., Fujiyama, A., Tamanoi, F., Northup, J., and Wigler, M., Differentiation activation of yeast adenylate cyclase by wild-type and mutant *ras* proteins, *Cell*, 41, 763, 1985.

39. McCormick, F., *ras* GTPase activating protein: signal transmitters and signal terminator, *Cell*, 56, 5, 1989.

40. Breviario, D., Hinnebusch, A. G., and Dhar, R., Multiple regulatory mechanisms control the expression of the *RAS1* and *RAS2* genes of *Saccharomyces cerevisiae*, *EMBO J.*, 7, 1805, 1988.

41. Marshall, M. S., Gibbs, J. B., Scolnick, E. M., and Sigal, I. S., Regulatory function of the *Saccharomyces cerevisiae* RAS C-terminus, *Mol. Cell. Biol.*, 7, 2309, 1987.

42. Hata, Y., Kikuchi, A., Sasaki, T., Schaber, M. D., Gibbs, J. B., and Takai, Y., Inhibition of the *ras* p21 GTPase-activating protein-stimulated GTPase activity of c-Ha-*ras* p21 by *smg* p21 having the same putative effector domain as *ras* p21s, *J. Biol. Chem.*, 265, 7104, 1990.

43. Schaber, M. D., Garsky, V. M., Boylan, D., Hill, W. S., Scolnick, E. M., Marshall, M. S., Sigal, I. S., and Gibbs, J. B., Ras interactions with the GTPase-activating protein (GAP), *Proteins Struct. Funct. Genet.*, 6, 306, 1989.

44. Ballester, R., Michaeli, T., Ferguson, K., Xu, H.-P., McCormick, F., and Wigler, M., Genetic analysis of mammalian GAP expressed in yeast, *Cell*, 59, 681, 1989.

45. Gibbs, J. B., Marshall, M. S., Scolnick, E. M., Dickson, R. A. F., and Vogel, U. S., Mutation of guanine nucleotides bound to Ras in NIH 3T3 cells by oncogenes, and the GTPase activating protein GAP, *J. Biol. Chem.*, 265, 20437, 1990.

46. Zhang, K., DeClue, J. E., Vass, W. C., Papageorge, A. G., McCormick, F., and Lowy, D. R., Suppression of c-*ras* transformation by GTPase-activating protein, *Nature*, 346, 754, 1990.

47. Downward, J., Graves, J. D., Warne, P. H., Rayter, S., and Cantrell, D. A., Stimulation of p21 Ras upon T-cell activation, *Nature*, 346, 719, 1990.

48. Adari, H., Lowy, D. R., Willumsen, B. M., Der, C. J., and McCormick, F., Guanosine triphosphate activating protein (GAP) interacts with the p21 effector binding domain, *Science*, 240, 518, 1988.

49. Cales, C., Hancock, J. F., Marshall, C. J., and Hall, A., The cytoplasmic protein GAP is implicated as the target for regulation by the *ras* gene product, *Nature*, 332, 548, 1988.

50. Yatani, A., Okabe, K., Polakis, P., Halenbeck, R., McCormick, F., and Brown, A. M., *ras* p21 and GAP inhibit coupling of muscarinic receptors to atrial K^+ channels, *Cell*, 61, 769, 1990.

51. Wigler, M., Field, J., Powers, S., Broek, D., Toda, T., Cameron, S., Nikawa, J., Michaeli, T., Coliceli, J., and Ferguson, K., Studies of *RAS* function in the yeast *S. cerevisiae*, *Cold Spring Harbor Symp. Quant. Biol.*, 53, 649, 1988.

52. Cawthon, R. M., Weiss, R., Xu, G., Viskochil, D., Culver, M., Stevens, J., Robertson, M., Dunn, D., Gesteland, R., O'Connell, P., and White, R., A major segment of the neurofibromatosis type 1 gene: cDNA sequence, genomic structure and point mutations, *Cell*, 62, 193, 1990.

53. Viskochil, D., Buchberg, A. M., Xu, G., Cawthon, R. M., Stevens, J., Wolff, R. K., Culver, M., Carey, J. C., Copeland, N. G., Jenkins, N. A., White, R., and O'Connell, P., Deletions and a translocation interrupt a cloned gene at the neurofibromatosis type 1 locus, *Cell*, 62, 187, 1990.

54. Wallace, M. R., Marchuk, D. A., Andersen, L. B., Letcher, R., Odeh, H. M., Saulino, A. M., Fountain, J. W., Brereton, A., Nicholson, J., Mitchell, A. L., Brownstein, B. H., and Collins, F. S., Type 1 neurofibromatosis gene: identification of a large transcript disrupted in three NF1 patients, *Science*, 249, 181, 1990.

55. Ballester, R., Marchuk, D., Boguski, M., Saulino, A., Letcher, R., Wigler, M., and Collins, F., The *NF1* locus encodes a protein functionally related to mammalian GAP and yeast *IRA* proteins, *Cell*, 63, 851, 1990.

56. Martin, G. A., Viskochil, D., Bollag, G., McCabe, P. C., Crosier, W. J., Haubruck, H., Conroy, L., Clark, R., O'Connell, P., Cawthon, R. M., Innis, M. A., and McCormick, F., The GAP-related domain of the neurofibromatosis type 1 gene product interacts with *ras* p21, *Cell*, 63, 843, 1990.

57. Xu, G., Lin, B., Tanaka, K., Dunn, D., Wood, D., Gesteland, R., White, R., Weiss, R., and Tamanoi, F., The catalytic domain of the neurofibromatosis type 1 gene product stimulates *ras* GTPase and complements *ira* mutants of *S. cerevisiae*, *Cell*, 63, 835, 1990.

58. Bar-Sagi, D. and Feramisco, J. R., Microinjection of the *ras* oncogene protein into PC12 cells induces morphological differentiation, *Cell*, 42, 841, 1985.

59. Noda, M., Ko, M., Ogura, A., Liu, D., Amano, T., Takano, T., Ikawa, Y., Sarcoma viruses carrying *ras* oncogenes induce differentiation-associated properties in a neuronal cell line, *Nature*, 318, 73, 1985.

60. Satoh, T., Nakamura, S., and Kaziro, Y., Induction of neurite formation in PC12 cells by microinjection of proto-oncogenic Ha-*ras* protein pre-incubated guanosine-5'-*O*-(3-thio-triphosphate), *Mol. Cell. Biol.*, 7, 4553, 1987.

61. Trahey, M., Wong, G., Halenbeck, R., Rubinfeld, B., Martin, G. A., Ladner, M., Long, C. M., Crosier, W. J., Watt, K., Koths, K., and McCormick, F., Molecular cloning of two types of GAP complementary DNA from human placenta, *Science*, 242, 1697, 1988.

62. **Fukui, Y. and Kaziro, Y.**, Molecular cloning and sequence analysis of a *ras* gene from *Schizosaccharomyces pombe*, *EMBO J.*, 4, 687, 1985.

63. **Fukui, Y., Kozasa, T., Kaziro, Y., Takeda, T., and Yamamoto, M.**, Role of a *ras* homolog in the life cycle of *Schizosaccharomyces pombe*, *Cell*, 44, 329, 1986.

Chapter 24

GAPs SPECIFIC FOR THE rap1/Krev-1 PROTEIN

Paul Polakis

TABLE OF CONTENTS

0-8493-5214-2/93/$0.00 + $.50

445

I. INTRODUCTION

Among the myriad of recently discovered low molecular mass GTP-binding proteins[1,2] the rap1/Krev-1 protein exhibits the highest degree of identity to p21[ras], particularly in the so-called effector region where the two proteins are virtually identical.[3] While they have apparently different, possibly even opposing activities *in vivo,* the sequence homology between the ras and rap1 proteins suggests they are regulated by similar mechanisms. It is clear from the cellular transforming activities of various p21[ras] mutants that the GTP-bound form of the ras protein is required for its activity.[4] Likewise, the GTP-bound form of the rap1 protein is probably responsible for whatever function it performs in the cell. The activities of these proteins are terminated by hydrolysis of GTP to GDP. In support of this is the finding that an *in vitro* activity of p21[rap1], the inhibition of rasGAP activity, is GTP-dependent.[5,6] Moreover, the ability of p21[rap1] to suppress transformation by v-*ras* also appears to be GTP-dependent.[7] However, p21[rap1] possesses only a very weak intrinsic GTPase activity[6] and therefore, once activated, would be expected to remain active for a considerable duration. As for p21[ras], though, there exist GTPase-activating proteins (GAPs) that specifically stimulate GTP hydrolysis by p21[rap1].[8,9] The discovery of GAPs specific for p21[rap1], as well as for other ras-related proteins, uncovers a common theme inherent in the regulation of the ras-related family of GTP-binding proteins. However, on closer examination of rap1-GAP we find that the analogy to the ras system breaks down, revealing an entirely new class of GAP; one that seems related to ras-GAP only by virtue of being a GAP.

II. PURIFICATION AND MOLECULAR CLONING OF rap1-GAP

The identification of a GAP specific for p21[rap1] was originally reported by Takai and colleagues.[8] These authors showed that cytosolic preparations form bovine brain contained GAP activities for both p21[ras] and p21[rap1] and that these two activities could be partially resolved by size exclusion chromatography. Further resolution of the rap1-GAP activity on matrix gel orange chromatography revealed two distinct peaks, designated *smg* p21 GAP1 and -2. The two partially purified forms of rap1-GAP specifically stimulated the GTPase activity of the rap1 protein but neither was active toward p21[ras], p20[rho], or *smg* 25A. Both GAPs were also ineffective as guanine nucleotide exchange factors. The molecular basis for the multiple forms of rap1-GAP was not determined in this study, but in addition to being clearly resolved chromatographically, the partially purified GAP1 and GAP2 activities were further distinguished by molecular size analysis. These results suggest that more than one GAP specific for the rap1 protein exists in brain. This was also shown to be the case for platelets in a similar study reported by the same laboratory.[10]

A GAP activity specific for p21[rap1] was also identified in the human pro-myelocytic HL-60 cell line.[9] The rap1-GAP activity was found both in the cytosolic form and membrane fractions of these cells. The membrane-bound activity represented about 13% of the total activity recovered from these cells and remained stably associated with the membrane unless treated with detergent. Size-exclusion chromatography of the cytosolic and solubilized membrane rap1-GAP activities revealed that the cytosolic form was considerably larger than the membrane rap1-GAP. The reason for this difference was not clarified, but again it suggests that the presence of multiple GAPs specific for p21[rap1]. The membrane associated rap1-GAP was solubilized in an active form with detergents and purified to near homogeneity by four successive steps of column chromatography. An 88-kDa polypeptide was found to comigrate with the rap1-GAP activity on the final step of purification. This purified preparation migrated with a mobility nearly identical to that of the crude, freshly solubilized membrane-associated rap1-GAP on size exclusion HPLC, suggesting that the protein had been purified in an intact form. The 88-kDa rap1-GAP exhibited a high specific activity for p21[rap1] and did not stimulate the GTPase activity of several other ras-related GTP-binding proteins, including p21[ras], rhoB, G25K, and rac 1.

The identification of the 88-kDa rap1-specific GAP clearly distinguished this protein from the previously identified 120-kDa ras-GAP[11,12] and demonstrated that the different GAP activities were likely due to distinct gene products. Additionally, the lack of reactivity of purified rap1-GAP with antibodies specific to ras-GAP suggested that the proteins were not highly related. This was confirmed following the purification and amino acid sequencing of the 88-kDa rap1-GAP purified from bovine brain membranes. Amino acid sequence generated from peptide fragments of this protein were used to design oligonucleotides which led to the isolation of a full-length cDNA coding for rap1-GAP.[13] The deduced amino acid sequence contains no significant homology to ras-GAP nor to any known protein sequences contained in the available databases. Although the protein was originally identified in a membrane fraction and required detergent for its solubilization, the amino acid sequence does not contain any obvious hydrophobic regions. This suggests that the attachment of rap1-GAP to the membrane is not likely due to direct insertion of the polypeptide into the lipid bilayer, but rather may require its interaction with other membrane proteins. Alternatively, a posttranslational addition of a hydrophobic moiety could account for the membrane binding although the characteristic amino acid sequences required for isoprenylation, myristoylation, or a carboxyl terminal glycophosphatidylinositol modification are not apparent in the rap1-GAP sequence.

III. EXPRESSION AND CHARACTERIZATION OF rap1-GAP

Expression of rap1-GAP cDNA in insect Sf9 cells infected with recombinant baculovirus resulted in the high level production of a diffusely staining 85- to

95-kDa polypeptide detected on SDS-polyacrylamide gels.[13] This protein was detected in both the membrane and cytosolic fractions of the recombinant cell and exhibited high specific activity toward p21^{rap1}. The expressed protein appeared to consist of several molecular mass forms suggesting a posttranslational modification event. This was confirmed by pulse-chase labeling experiments showing that the translated rap1-GAP polypeptide increased in molecular mass following its synthesis in the Sf9 cell. The nature of this modification is not clear, however, recent experiments suggest that phosphorylation is at least in part responsible for the appearance of the multiple bands detected on SDS-gels.[14] The carboxyl terminal region of rap1-GAP contains good consensus sites for both cAMP-dependent kinase and the p34^{cdc2} cell cycle kinase.[15] Interestingly, a 26 amino acid segment of rap1-GAP containing these consensus sites is duplicated in a rap1-GAP encoded by an mRNA splicing variant. Polymerase chain reaction (PCR) analysis of mRNAs derived from several tissue sources indicated that the splicing variant containing this repeated motif was exclusive to brian and placenta. The recombinant protein expressed by this variant exhibits an apparently normal specific activity toward p21^{rap1}. In addition, preliminary results indicate that mutants of rap1-GAP completely lacking the phosphorylation consensus sites are still active toward p21^{rap1}.[14] These observations suggest that the carboxyl terminal region of rap1-GAP may harbor regulatory elements but is not essential to GAP activity.

p21^{rap1} is expressed in a wide variety of cell lines and tissues.[16] It would follow that rap1-GAP should also exhibit a similar ubiquitous distribution. This appears to be the case based on measurements of rap1-GAP activity in crude cell lysates and tissue homogenates.[13] However, Western and Northern analysis of the rap1-GAP cloned in our laboratory indicated a rather limited tissue distribution for this particular gene product.[13] The 85- to 95-kDa protein could be detected in membrane fractions prepared from adult rat brain, but heart, liver, lung, spleen, and testes did not contain detectable levels. Homogenates of adult human tissues also contained undetectable or very low levels, however, the corresponding fetal organs contained considerably higher amounts. rap1-GAP was also identified in the promyelocytic HL-60 cell and in human erythroleukemia cells, but the reactivity was dramatically reduced following their induced differentiation to mature cell lines. The rap1-GAP mRNA was also abundant in certain cancer cell lines, particularly the Wilms' tumor SK-NEP1 and the SK-MEL-3 melanoma cells, but was either undetectable or very low in numerous other sources tested. The limited expression of rap1-GAP suggests a specialized function for this molecule and its preferential expression in dividing cells suggests that this function may be relevant to cell growth and division.

The observation that rap1-GAP is found predominantly in dividing cells is consistent with the proposal that p21^{rap1} itself may have an activity that is antagonistic to p21^{ras}. The rap1 gene, Krev-1, was identified as a suppressor of ras-transformation in fibroblasts.[16] Although it is not clear how the rap1 protein exerts its suppressor activity, it is conceivable that it binds competitively to the

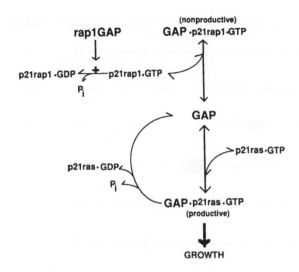

FIGURE 1. Representation of the proposed inhibition of p21[ras] signaling by p21[rap1]. In this model GAP is considered an effector target for p21[ras]. rap1-GAP is proposed to limit the competitive binding of p21[rap1] to GAP by eliminating the GTP-bound form of p21[rap1].

p21[ras] effector target. p21[rap1] contains an effector sequence identical to that found in p21[ras] and *in vitro* it binds with high affinity to ras-GAP without affect on its GTPase activity.[5,6] Because the binding to ras-GAP, as well as the suppressor activity of p21[rap1], both appear to be GTP-dependent processes, rap1-GAP would be expected to reduce the effectiveness of p21[rap1] by maintaining it in the GDP-bound state. High levels of rap1-GAP would, thus, favor ras activity by limiting the amount of p21[rap1] present in the GTP-bound state. This proposed relationship between the function of the ras and rap1 proteins (Figure 1) is consistent with the high levels of rap1-GAP detected in dividing cells. The effector target in this model may be ras-GAP itself, as previously proposed,[17,18] or an as yet unidentified protein with the capacity to bind both p21[rap1] and p21[ras].

IV. MUTATIONAL ANALYSIS OF p21[rap1] SENSITIVITY TO rap1-GAP

p21[ras] and p21[rap1] are structurally quite similar, in fact, as noted above, p21[rap1] will bind to ras-GAP. In addition to this, certain mutations appear to have functionally analogous consequences for the two proteins. Substitution of valine for glycine at position 12 results in a p21[ras] with potent transforming activity[4] and a p21[rap1] with stronger transformation-suppressor activity.[7] This mutation also abrogates the ability of GAP to stimulate p21[ras] GTPase activity.[12] Similarly, we have found that p21[rap1-val12] is insensitive to stimulation by rap1-GAP[19] (Figure 2). Similarities in the structural arrangement of these two proteins can also be drawn from the functional outcome of effector site mutations. The

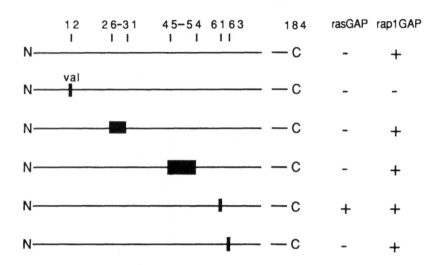

FIGURE 2. Mutational analysis of p21^rap1^. Mutants were generated containing either a G12V substitution or substitutions corresponding to amino acids in the analogous position of p21^ras^ as indicated by the blocked areas. The purified mutant proteins were tested for stimulation of GTPase activity by ras-GAP and rap1-GAP.

ras and rap1 proteins contain virtually identical sequence in the so-called effector region (residues 32 and 44). Mutational analysis indicates that this region is important to both proteins for biological activity and responsiveness to their GAPs. Certain mutations in the effector region of p21^ras^ result in a well-documented loss of transforming activity as well as a resistance to stimulation by GAP.[20] A p21^rap1^ with an effector site mutation has reduced suppressor activity[6] and is also unresponsive to stimulation by rap1-GAP.[21] Considering their extensive homology, it is not surprising that p21^ras^ and p21^rap1^ are impaired in a similar fashion by introduction of common point mutations. However, in spite of their obvious similarities, p21^ras^ will not associate with rap1-GAP and neither of the GAPs will stimulate the GTPase activities of the heterologous GTP-binding protein. Even more striking is the discovery that the two GAPs have entirely different primary structures.[13] This suggests that within the nonconserved sequences of these two proteins there exist amino acid residues that are critical to maintaining exclusive interactions with their respective GAPs. Candidate sequences would obviously not include the so-called effector region because p21^rap1^ and p21^ras^ are identical here. However, sequences flanking both sides of the effector site show considerable divergence. Lowy and colleagues showed that these flanking sequences are important in determining whether ras-rap chimeric proteins induced cellular transformation or suppressed ras transformation.[22] This demonstrated that the sequences adjacent to the effector site are important for the distinct cellular activities of the ras and rap1 proteins. However, other evidence suggests these regions are not important to p21^ras^ and p21^rap1^ for specific recognition of their GAPs. Our laboratory has found that substituting the effector flanking sequences of 21^{rap1} with those of p21^{ras} did not confer ras-GAP sensitivity to p21^{rap1} (Figure 2).[19]

Another area of sequence divergence is located in the loop 4 region positioned between amino acid residues 59 and 65 of p21^{ras}. This region exhibits a high temperature factor in the crystal structure of p21^{ras} and has been proposed to interact with ras-GAP.[23] Substituting p21^{ras} residues 61 to 65 into p21^{rap1} resulted in a chimeric protein that was sensitive to ras-GAP as well as to rap1-GAP (Figure 2).[19] A similar result was obtained on substituting only the glutamine 61 residue found in p21^{ras} (Figure 2).[6,19] This demonstrates that position 61 is a critical residue in determining reactivity to ras-GAP. Furthermore, because the T61Q mutant of p21^{rap1} is sensitive to rap1-GAP, but p21^{ras} is not, there must be residues in addition to position 61 important for an interaction with rap1-GAP. It appears that Threonine 61 in p21^{rap1} is important for maintaining a resistance to stimulation by ras-GAP but is not critical for activation by rap1-GAP. We have also found that the rap1 homolog in yeast, RSR, is also sensitive to rap1-GAP[24] but contains an isoleucine residue at position 61. These observations underscore an important difference in the mechanisms by which p21^{rap1} and p21^{ras} undergo GTPase stimulation by their GAPs. There is not a strict requirement for a specific residue at position 61 for rap1-GAP stimulation, yet substitutions at this position in p21^{ras} render the protein insensitive to stimulation by its GAP.[12]

V. PERSPECTIVES ON rap1-GAP FUNCTION

What does rap1-GAP do? Thus far we can only offer a rather cryptic reply; it stimulates the GTPase activity of p21^{rap1}. However, in this capacity it is likely that rap1-GAP is interfaced to other signaling events in the cell. For ras-GAP, only one third of the entire molecule is required for GAP activity[25] while the remainder of the protein includes regulatory domains, hydrophobic sequences and additional uncharacterized primary structure. Similarly, preliminary analysis suggests that only a portion of the rap1-GAP sequence is required for its activity toward p21^{rap1}, with the remainder of the protein containing regulatory elements.[14] Perhaps rather than asking what does rap1-GAP do, we should be asking what is doing rap1-GAP? What are the kinases that phosphorylate rap1-GAP *in vivo* and does this phosphorylation affect its activity or subcellular localization? What are the signals that lead to the downregulation of rap1-GAP expression levels following cellular differentiation? Are there any proteins that associate with rap1-GAP *in vivo*? Forthcoming answers to these questions will help define the signaling pathways in which rap1-GAP participates.

It is perplexing that two such functionally homologous proteins, ras-GAP and rap1-GAP, having reactivities toward two such structurally homologous substrates, p21^{ras} and p21^{rap1}, do not bear any resemblance to each other at the level of primary structure. Conserved sequences thought to be required for ras-GAP catalytic activity[26] cannot be found in rap1-GAP. The src homology domains, known to be important for the association of ras-GAP with growth factor receptors[27] and other tyrosine-phosphorylated proteins,[28] are also absent in rap1-GAP. Considering this, the two GAPs appear to have little in common outside of their ability to act as signaling terminators for their respective substrates. Yet the apparently opposing biological activities of the ras and rap1 proteins, along with the ability of p21[rap1] to bind ras-GAP, implies an intracellular network involving the coordinated regulation of these two GAPs. Expression studies indicate that rap1-GAP may be involved in the programming of cell growth and division.[13] ras-GAP is certainly involved in these processes. As the signaling networks become more clearly delineated we may ultimately discover the intersection connecting the GAPs.

REFERENCES

1. **Bourne, H. R., Sanders, D. A., and McCormick, F.,** The GTPase superfamily: a conserved switch for diverse cell functions, *Nature,* 348, 125, 1990.
2. **Downward, J.,** The ras superfamily of small GTP-binding proteins, *Trends Biochem. Sci.,* 15, 469, 1990.
3. **Pizon, V., Chardin, P., Lerosey, I., Olofsson, B., and Tavitian, A.,** Human cDNAs rap1 and rap2 homologous to the Drosophila gene Dras3 encode proteins closely related to ras in the "effector" region, *Oncogene,* 3, 201, 1988.
4. **Barbacid, M.,** *ras* Genes, *Annu. Rev. Biochem.,* 56, 779, 1987.
5. **Hata, Y., Kikuchi, A., Sasaki, T., Schaber, M. D., Gibbs, J. B., and Takai, Y.,** Inhibition of the ras p21 GTPase-activating protein-stimulated GTPase activity of c-Ha-ras p21 by smg p21 having the same putative effector domain as ras p21s, *J. Biol. Chem.,* 265, 7104, 1990.
6. **Frech, M., John, J., Pizon, V., Chardin, P., Tavitian, A., Clark, R., McCormick, F., and Wittinghofer, A.,** Inhibition of GTPase activating protein stimulation of Ras-p21 GTPase by the Krev-1 gene product, *Science,* 249, 169, 1990.
7. **Kitayama, H., Matsuzaki, T., Ikawa, Y., and Noda, M.,** Genetic analysis of the Kirsten-*ras*-revertant 1 gene: potentiation of its tumor suppressor activity by a specific point mutations, *Proc. Natl. Acad. Sci. U.S.A.,* 87, 4284, 1990.
8. **Kikuchi, A., Sasaki, T., Araki, S., Hata, Y., and Takai, Y.,** Purification and characterization from bovine brain cytosol of two GTPase-activating proteins specific for smg p21, a GTP-binding protein having the same effector domain as c-ras p21s, *J. Biol. Chem.,* 264, 9133, 1989.
9. **Polakis, P. G., Rubinfeld, B., Evans, T., and McCormick, F.,** Purification of a plasma membrane-associated GTPase-activating protein specific for rap1/K-rev-1 from HL60 cells, *Proc. Natl. Acad. Sci. U.S.A.,* 88, 239, 1991.
10. **Ueda, T., Kikuchi, A., Ohga, N., Yamamoto, J., and Takai, Y.,** GTPase activating proteins for the smg-21 GTP-binding protein having the same effector domain as the ras proteins in human platelets, *Biochem. Biophys. Res. Commun.,* 159, 1411, 1989.
11. **Trahey, M., Wong, G., Halenbeck, R., Rubinfeld, B., Martin, G. A., Ladner, M., Long, C. M., Crosier, W. J., Watt, K., Koths, K., and McCormick, F.,** Molecular cloning of two types of GAP complementary DNA from human placenta, *Science,* 242, 1697, 1988.
12. **Vogel, U. S., Dixon, R. A., Schaber, M. D., Diehl, R. E., Marshall, M. S., Scolnick, E. M., Sigal, I. S., and Gibbs, J. B.,** Cloning of bovine GAP and its interaction with oncogenic ras p21, *Nature,* 335, 90, 1988.

13. **Rubinfeld, B., Munemitsu, S., Clark, R., Conroy, L., Watt, K., Crosier, W., McCormick, F., and Polakis, P.,** Molecular cloning of a GTPase activating protein specific for the Krev-1 protein p21^{rap1}, *Cell*, 65, 1033, 1991.

14. **Polakis, P., Rubinfeld, B., and McCormick, F.,** unpublished data.

15. **Kemp, B. E. and Pearson, R. B.,** Protein kinase recognition sequence motifs, *Trends Biochem. Sci.*, 15, 342, 1991.

16. **Kitayama, H., Sugimoto, Y., Matsuzaki, T., Ikawa, Y., and Noda, M.** A ras-related gene with transformation suppressor activity, *Cell*, 56, 77, 1989.

17. **McCormick, F.,** ras GTPase activating protein: signal transmitter and signal terminator, *Cell*, 56, 5, 1989.

18. **Hall, A.,** *ras* and GAP — Who's controlling whom?, *Cell*, 61, 921, 1990.

19. **Haubruck, H., Polakis, P., McCabe, P., Conroy, L., Clark, R., Innis, M., and McCormick, F.,** Mutational analysis of rap1/Krev-1 protein; sensitivity to GTPase activating proteins and suppression of the yeast cdc24 budding defect, *J. Cell. Biochem.*, 15B, 138, 1991.

20. **Adari, H., Lowy, D. R., Willumsen, B. M., Der, C. J., and McCormick, F.,** Guanosine triphosphatase activating protein (GAP) interacts with the p21 ras effector binding domain, *Science*, 240, 518, 1988.

21. **Quiliam, L. A., Der, C. J., Clark, R., O'Rouke, E. C., Zhang, K., McCormick, F., and Bokoch, G.,** Biochemical characterization of Baculovirus-expressed rap1A/Krev-1 protein and its regulation by GTPase activating proteins, *Mol. Cell. Biol.*, 10, 2901, 1990.

22. **Zhang, K., Noda, M., Vass, W. C., Papageorge, A. G., and Lowy, D. R.,** Identification of small clusters of divergent amino acids that mediate the opposing effects of *ras* and Krev-1, *Science*, 249, 162, 1990.

23. **Krengel, U., Schlichting, L., Scherer, A., Schumann, R., Frech, M., John, J., Kabsch, W., Pai, E. F., and Wittinghofer, A.,** Three-dimensional structures of H-ras p21 mutants: molecular basis for their inability to function as signal switch molecules, *Cell*, 62, 539, 1990.

24. **Haubruck, H., McCabe, P., and Polakis, P.,** unpublished data.

25. **Marshall, M. S., Hill, W. S., Ng, A. S., Vogel, U. S., Schaber, M. D., Scolnick, E. M., Dixon, R. A., Sigal, I. S., and Gibbs, J. B.,** A C-terminal domain of GAP is sufficient to stimulate ras p21 GTPase activity, *EMBO J.*, 8, 1105, 1989.

26. **Ballester, R., Marchuk, D., Boguski, M., Saulino, A., Lechter, R., Wigler, M., and Collins, F.,** The NF1 locus encodes a protein functionally related to mammalian GAP and yeast IRA proteins, *Cell*, 63, 851, 1990.

27. **Anderson, D., Koch, C. A., Grey, L., Ellis, C., Moran, M. F., and Pawson, T.,** Binding of SH2 domains of phospholipase C gamma 1, GAP, and Src to activated growth factor receptors, *Science*, 250, 979, 1990.

28. **Moran, M. F., Koch, C. A., Anderson, D., Ellis, C., England, L., Martin, G. S., and Pawson, T.,** Src homology region 2 domains direct protein-protein interactions in signal transduction, *Proc. Natl. Acad. Sci. U.S.A.*, 87, 8622, 1990.

Chapter 25

EXCHANGE FACTORS

Julian Downward

TABLE OF CONTENTS

0-8493-5214-2/93/$0.00 + $.50

I. INTRODUCTION

The ras superfamily of GTPases bind guanosine triphosphate and hydrolyse it to guanosine diphosphate. The rate of hydrolysis of GTP on p21^{ras} is greatly increased by GTPase-activating proteins such as GAP and NF-1 in mammalian cells (see Chapter 22) and IRA proteins in yeast (see Chapter 23). The intrinsic rate at which ras proteins will exchange bound nucleotide for nucleotide in solution is extremely slow in the presence of physiological levels of magnesium ions:[1] the half-life for exchange of bound nucleotide is in the order of one hour. Without the intervention of some system for stimulating nucleotide exchange, ras proteins would be expected to be almost exclusively GDP-bound in the cell. Several lines of evidence indicate that the GTP-bound form of the protein is biologically active while the GDP-bound form is inactive (see Chapter 6). Without exchange factors, normal ras would therefore be continuously locked in the "off" state. *A priori,* it therefore, seems likely that exchange factors should exist for ras proteins in order to allow the accumulation of biologically effective levels of the active p21-GTP form. It should be noted that since more than 95% of the guanine nucleotide in the cytosol is GTP rather than GDP,[2] the exchange of nucleotide need not necessarily be a directed replacement of GDP by GTP: the circumstances of p21^{ras} in the cell ensure that this will come about anyway.

The potential importance of stimulating the nucleotide exchange rate on ras proteins is clear from the fact that a whole class of transforming mutations in p21^{ras} are in regions of interaction of the protein with the guanine ring of the nucleotide: residues 116, 119, and 146.[3] These mutations decrease the affinity of p21^{ras} for guanine nucleotide, often by many orders of magnitude. This results in a greatly increased basal nucleotide exchange rate of the protein: indeed, for several of these mutant proteins GTP binding has not been directly demonstrated. The affinity of wild-type p21^{ras} for GTP is about 20 nM and for GDP is about 40 nM,[4] while the concentration of GTP in the cytosol approaches 1 mM and GDP about 10 to 50 μM. The affinity of p21^{ras} for nucleotide could therefore be decreased by more than 10,000-fold before the amount of unoccupied ras protein in the cell became significant. These mutant proteins could, thus, have exchange rates increased by orders of magnitude and would, therefore, presumably be extensively GTP-bound in the cell since the elevated exchange rate would overwhelm the effects of GAP-like proteins. It is, thus, clear that elevating the guanine nucleotide exchange rate of p21^{ras}, either by mutation or by interaction with an exchange factor, could have a significant biological effect in the cell.

Magnesium ions play an important part in the interaction of guanine nucleotides with p21^{ras}. X-ray crystallography shows that nucleotide-bound p21^{ras} contains a magnesium ion coordinated between two oxygen atoms in the β- and γ-phosphate groups of the GTP, two molecules of water, and residues Threonine 35 and Serine 17.[5] In the presence of high levels of Mg^{2+} ions, as are found in mammalian cells, nucleotide normally exchanges very slowly. Removal of Mg^{2+} ions with chelators to give free magnesium concentrations below 1 μM results in a greatly reduced affinity of p21^{ras} for GTP and GDP and consequently, a much more rapid exchange rate. Several researchers have suggested that exchange factors for ras proteins might act by mimicking the removal of Mg^{2+} ions from the nucleotide binding site.[1]

In addition to these theoretical considerations, it also seems likely that exchange factors should exist for ras proteins by comparison with other GTP-binding proteins. The G protein family of large heterotrimeric GTP-binding proteins are known to have regulated guanine nucleotide exchange.[6] Their activity is controlled by membrane receptors that contain seven transmembrane spans. When the receptor becomes activated, usually by binding to a soluble ligand, it undergoes a conformational change and interacts with the inactive GDP-bound form of the G protein causing stimulation of nucleotide exchange and, thus, binding of GTP. Thus, for the G proteins the activated receptors act as exchange factors.

Another class of GTP-binding proteins that are well-characterized are the elongation and initiation factors involved in protein synthesis. The bacterial protein EF-Tu binds and hydrolyzes GTP. Once it is GDP-bound, it has an absolute requirement for interaction with the exchange factor EF-Ts to allow exchange of GDP for GTP.[7] Once again an exchange protein is essential for the function of this GTP-binding protein.

In addition to proteins that enhance the rate of exchange of nucleotide on ras-like proteins, it is clearly possible that other proteins might exist that have other effects on nucleotide exchange. In the Chapter 27 by Takai a family of proteins that inhibit guanine nucleotide exchange on ras-related proteins are described: these have been termed GDI for guanine dissociation inhibitors. Whether these proteins are important in controlling the activation state of small GTPases in whole cells is not yet clear.

II. EXCHANGE FACTORS FOR YEAST ras PROTEINS

The first clear indication for the existence of an exchange promoting factor for p21^{ras} was found in the yeast *Saccharomyces cerevisiae*. The cell division cycle gene CDC25 was shown to regulate the RAS/adenylate cyclase pathway in this organism.[8-10] The CDC25 gene product is dispensable in cells containing an activated RAS allele such as RAS2 Gly^{12}→Val or a RAS mutant with an increased rate of nucleotide exchange such as RAS2 Thr^{152}→Ile. Also, mutationally activated alleles of CDC25 cause a phenotype similar to yeast with a genetically activated RAS/adenylate cyclase pathway. In *S. cerevisiae* this pathway is under the control of fermentable sugars such as glucose in the medium: addition of glucose causes cAMP levels in the cells to rise rapidly. Mutations in CDC25 can prevent this regulation. It, therefore, appears that CDC25 encodes an upstream regulator of RAS proteins.

Experiments on adenylate cyclase activity in membrane preparations from CDC25-disrupted mutants showed a reduction in the normal response to guanine nucleotides.[10] These data are consistent with CDC25 encoding a guanine nucleotide exchange factor for *S. cerevisiae* RAS proteins. However, it has not been possible to demonstrate biochemically that the CDC25 gene product directly stimulates the exchange of guanine nucleotide on the RAS proteins.

Recently, a gene has been discovered in *S. cerevisiae* that is functionally and structurally related to CDC25: this has been named SDC25.[11] While this gene is not essential, the 3′ terminal part of it can suppress the requirement for CDC25. Furthermore, partially purified preparations of the protein encoded by the carboxyl terminal portion of the SDC25 gene can directly stimulate guanine nucleotide exchange on both *S. cerevisiae* RAS2 protein and human p21[c-Ha-*ras*].[12] In addition, overexpression of this protein fragment in mammalian cells can cause a considerable increase in the GTP/GDP ratio on p21[ras], indicating that the SDC25 protein is causing activation of p21[ras], presumably through stimulation of guanine nucleotide exchange.[13] SDC25 encodes the only exchange factor that is active on mammalian p21[ras] whose primary structure is known at this time. The SDC25 and CDC25 gene products are discussed in more detail in the Chapter 26.

A homolog of CDC25 and SDC25 has been found in the fission yeast *Schizosaccharomyces pombe*. The ste6 gene product shows sequence similarity to both CDC25 and SDC25 proteins and appears to function upstream of the ras1 protein in this organism.[14] It seems likely that it is also a nucleotide exchange factor for ras proteins.

III. EXCHANGE FACTORS FOR MAMMALIAN p21[ras]

Guanine nucleotide exchange promoting factors for p21[ras] have long been sought in the mammalian system with little success until very recently. Three different groups have now reported the identification and characterization of exchange factors for p21[ras] using different approaches.

A. rGEF

In the laboratory of Tohru Kamata an exchange factor has been found in detergent extracts of bovine brain membranes.[15,16] This was initially identified when it was found that Triton® X-100 extracts of membrane preparations from bovine brain, but not cytosolic extracts, were capable of increasing the rate of loss of [3H]GDP from purified wild-type p21[ras] in the presence of unlabeled nucleotide. Such activity was also found in similar preparations from a wide range of tissues and also from NIH 3T3 fibroblasts and *Xenopus laevis* oocytes. Upon purification, the exchange activity ran on a Superose 12 FPLC column with an apparent relative molecular weight of 100,000. Further purification to homogeneity revealed that the activity was associated with a protein that ran as a single band on an SDS-polyacrylamide gel with a molecular weight of 35,000. This protein was termed "rGEF" for ras guanine nucleotide exchange factor.

This exchange activity is sensitive to heat and proteases and is blocked by the anti-p21^{ras} monoclonal antibody Y13-259. It acts on both wild-type and Valine 12 mutant p21^{ras}. rGEF is effective at high concentrations of magnesium ions (2 mM), similar to those found in the cell cytosol, under which conditions the exchange rate is normally very slow. Pure rGEF stimulates the exchange rate by 30- to 40-fold, displaying first order kinetics. Analysis of the mechanism of action of rGEF indicates that it decreases the affinity of p21^{ras} for GTP by about seven-fold and for GDP by about two-fold. rGEF also stimulates the binding of labeled nucleotide to p21^{ras}, confirming that the reaction is a true exchange. Reversibility of the effects of rGEF have not yet been reported.

rGEF, a 35,000 molecular weight membrane protein, is the best characterized of the mammalian exchange factors for p21^{ras}. An unexpected characteristic of it is that, unlike the other mammalian exchange factors for p21^{ras}, rGEF acts equally well on a number of ras-like proteins, both those from the ras subfamily such as p21^{rap1A} and p23^{R-ras} and also rho and rab proteins. While it is clear that rGEF could act in the cell to maintain a certain level of activated p21^{ras}-GTP, it seems unlikely that it would be responsible for the regulated control of the activation state of p21^{ras} owing to its lack of specificity. Presumably, the signals that control the activation state of p21^{ras} are very different from those that control the activation states of rho and rab proteins.

b. ras-GRF

Wolfman and Macara[17] have characterized another exchange factor for p21^{ras} found in rat brain, but in this case in the cytosolic rather than the membrane fraction. High concentrations of cytosolic extracts from the rat brain could catalyze the release of [α-^{32}P]GDP from p21^{ras} in the presence of cold GDP and high concentrations of magnesium ions. A dose response curve was sharply sigmoidal, suggesting that the interaction of the exchange factor with p21^{ras} was highly cooperative. The exchange activity was heat- and protease-sensitive and ran on a Superose 12 FPLC sizing column with an apparent relative molecular weight of 100,000 to 160,000. This protein was named "ras-GRF" for RAS-guanine nucleotide releasing factor.

ras-GRF is able to stimulate both the release of [α-^{32}P]GDP from p21^{ras} in exchange for cold GDP and the uptake of [α-^{32}P]GDP onto p21^{ras} from solution. Investigation of the effects of detergents on the activity of ras-GRF showed no inhibition that might hide the presence of the exchange factor in membrane extracts. The cytosolic location of ras-GRF, therefore, seems to be a real difference from rGEF. Models of ras action based on the heterotrimeric G proteins have generally assumed that any guanine nucleotide exchange factor for p21^{ras} should be membrane-associated. p21^{ras} itself is membrane bound so one might perhaps expect that molecules that control its activity in response ultimately to extracellular stimuli would also be membrane-bound. It should be noted, however, that there is no *a priori* reason for a ras exchange factor to be membrane-bound. Several guanine nucleotide exchange factors for vectorial processes such as EF-Tu and eIF-2 are soluble proteins.

Another interesting feature of ras-GRF is that its activity in crude cytosolic extracts appears to be stable only in the presence of phosphatase inhibitors. This suggests that phosphorylation is required to activate the enzyme and that its activity may be regulated in the cell. Importantly, while ras-GRF was active on both wild-type and viral p21^{ras}, it was inactive toward the ras-related protein p25^{rab3A}. ras-GRF is, thus, clearly distinct from rGEF, both in its different subcellular localization and its greater specificity: it would appear to be a better candidate for a physiological regulator of p21^{ras} in mammalian cells.

C. REP

From the laboratory of Robert Weinberg, Downward et al.[18] have also reported the characterization of a mammalian exchange factor for p21^{ras}. Cytosolic extracts of human placenta were found to contain an activity that promoted the exchange of nucleotide from [α-^{32}P]GTP-loaded p21^{ras} in the presence of unlabeled GTP or GDP and high magnesium ion concentrations. This activity, which was heat- and protease-sensitive, was partially purified on a number of chromatography columns. On an AcA 34 sizing column the activity showed an apparent molecular weight of about 60,000. The protein responsible for this activity was named "REP" for ras exchange promoting factor.

REP was shown to act reversibly on p21^{ras} and not to contain any degradative activity. It stimulated the rate of exchange of nucleotide on p21^{ras} by more than 20-fold. A number of mutant ras proteins were analyzed for their sensitivity to REP. All were found to be sensitive including point mutants in the effector domain and at the activating positions 12, 59, 61, 116, and 119, although some mutations at codons 59 and 61 caused a considerable reduction in sensitivity to REP relative to wild-type p21^{ras}. Some deletion mutants ($\triangle$165 to 184 and $\triangle$101 to 109) also had reduced sensitivity to REP. This exchange factor was, however, found not to be a substrate for REP. Subsequent work has also shown it to be ineffective on p21^{rap1A} (Downward, unpublished observations).

REP, therefore, appears to be much more similar to ras-GRF than to rGEF. Indeed, recent work has shown that the apparent size of REP on a Superose 12 FPLC sizing column is close to 100,000, similar to that of ras-GRF (Downward, unpublished observations). Although the preparation of REP did not involve the intentional use of phosphatase inhibitors as was the case for ras-GRF, it was found that good yields of REP activity were only obtained when phosphate buffers were used for the placental homogenization and early purification steps: phosphate buffers are known to inhibit the action of phosphatases. As is the case for ras-GRF, REP is a cytosolic protein although there is some evidence that a latent form of the enzyme can be released from membrane preparations by gentle proteolysis. It is obviously possible that both ras-GRF and REP could be proteolytic fragments of larger membrane-bound proteins. It could also be the case that proteolysis might be required to activate the enzyme *in vitro*.

Neither REP nor ras-GRF have yet been purified to homogeneity: this, along with cDNA cloning, will be required before it is clear whether or not they are the same enzyme. The structure of these proteins might be expected to be related to CDC25, SDC25, and ste6 if they are genuinely regulators of p21^{ras} function. Unfortunately, complementation assays for mammalian cDNAs that can replace the function of yeast RAS nucleotide exchange factors have not yet been successful: it is possible that the degree of evolutionary conservation in this part of the RAS system is relatively low.

D. FUNCTION OF EXCHANGE FACTORS IN CONTROLLING MAMMALIAN p21^{ras}

There are two simple ways in which the activity of p21^{ras} might be regulated in the mammalian cell: one is through the GTPase-activating proteins like GAP and NF-1 and the other is through the guanine nucleotide exchange proteins, particularly the specific ones such as ras-GRF and REP. Other possibilities include alterations in subcellular localization through changes in post translational modification and perhaps, direct interaction of exchange and GTPase inhibitors with p21^{ras}. GAP appears likely to be regulated within the cell in some way (see Chapter 22) but it is not yet known whether the exchange activities can be regulated.

A recent report has gone some way toward addressing these problems: Downward et al.[19] showed that in intact ^{32}P-labeled T cells, stimulation of the T-cell antigen receptor, or any treatment that causes protein kinase C activation, leads to a very rapid and extensive activation of p21^{ras} as measured by the conversion of p21-GDP to p21-GTP. This activation was accompanied by a decrease in the assayable "GAP" activity in extracts of the cells. It was, therefore, possible that protein kinase C activation was leading to suppression of the activity of GAP-like proteins and consequent activation of p21^{ras}. However, since the activation was complete within about one minute, the rate of nucleotide exchange on p21^{ras} in the activated cells must be very high: if it was not, it would take an hour or so to achieve a major accumulation of GTP on p21 after GAP proteins had been inactivated.

In order to study this further a permeabilized cell system was used: this involved making membrane pores in peripheral blood T lymphoblasts with the bacterial toxin streptolysin O. [α-^{32}P]GTP could then be added to the cells and allowed to bind to the endogenous p21^{ras}. At the end of the incubation the cells were lysed with detergent and p21^{ras} bound to labeled nucleotide immunoprecipitated with antibody 259. It was found that the increase in the ratio of GTP to GDP on p21^{ras} in response to protein kinase C activation could be mimicked in this system. Interestingly, the rate of binding of nucleotide to p21^{ras} was very rapid in these cells, not just in the presence of protein kinase C activators but also in untreated quiescent cells. The half-life for exchange of nucleotide on p21^{ras} in this system was less than one minute in both stimulated and quiescent T cells at 37°C, as compared with about an hour in membrane preparations or for purified p21^{ras}. This did not appear to be a direct effect of streptolysin O itself. This technique has revealed that the rate of exchange of nucleotide on

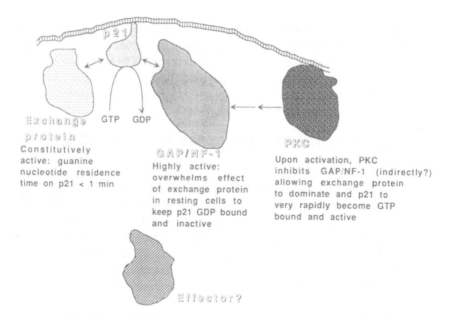

FIGURE 1. A model for the control of the activation state of p21^{ras} in T cells.

p21^{ras} is also elevated in permeabilized fibroblasts, although not as greatly as in T cells (J. D. Graves, D. A. Cantrell, and J. Downward, unpublished observations). For the T cells, the activation of p21^{ras} in response to protein kinase C agonists can be explained by a model in which exchange factors are constitutively activated and continuously stimulating the uptake of fresh GTP onto p21^{ras}. In quiescent cells the GAP proteins are also highly active and cause p21^{ras} to be predominantly GDP bound: they overwhelm the action of the exchange factors. Upon stimulation of the cell by the appropriate extracellular signal, in the case of T cells, activators of protein kinase C, the activity of the GAP proteins immediately drops dramatically. The exchange factors remain highly active so GTP now accumulates rapidly on p21^{ras} and is hydrolyzed only slowly. p21^{ras}, thus, becomes activated (see Figure 1).

This system would clearly involve hydrolysis of large amounts of GTP, but it gives a very highly regulatable and rapid response. It is not yet clear whether the suppression of GAP-like activity is due to control of GAP, NF-1 protein, or other GAP-like factors. Each could be regulated by different stimuli. In this system, the T cell, it appears that the exchange activity is always highly active, although in other systems it may well also be regulated. There is, thus, the possibility that p21^{ras} could respond to a great many different stimuli.

IV. EXCHANGE FACTORS FOR ras-LIKE PROTEINS

A number of exchange promoting factors for ras-related proteins have been characterized in the laboratory of Yoshimi Takai. The most extensively studied is specific for p21^{rap1A} and p21^{rap1B}, also known as *smg* p21A and *smg* p21B. The protein has been termed "GDS" for GDP dissociation stimulator. smg p21 GDS has been purified to homogeneity from bovine brain cytosol.[20] The protein stimulates the dissociation of [³H]GDP and [³⁵S]GTPγS from, and the association of GTPγS to, p21^{rap1}, but is without activity toward p21^{ras}, p21^{rhoB}, and p25^{rab3A}. It has no GAP-like activity and is distinct from the GDP dissociation inhibitors (GDIs) previously characterized in brain cytosol (see Chapter 27). The GDS is not active on p21^{rap1} whose carboxyl terminal five amino acids have been removed by limited proteolysis with trypsin: this implies that the modified C terminus of the protein is important for the interaction with the exchange factor and could be the actual site of contact.[21] There is also evidence that p21^{rap1} could form a relatively stable stoichiometric complex with its exchange factor *in vitro*.[22]

The GDS activity runs as two peaks on Mono Q FPLC columns. Both peaks (GDS-1 and -2) have a molecular weight of 53,000 on SDS-polyacrylamide gel electrophoresis and are indistinguishable by peptide map analysis. Partial amino acid sequence has now been obtained from purified *smg* p21 GDS and used to isolate cDNA clones.[23] These reveal an open reading frame encoding a protein of 558 amino acids with a molecular weight 61,000. Expression of this protein in *Escherichia coli* confirms that it possesses guanine nucleotide exchange activity specific for *smg* p21/p21^{rap1}. The primary structure of this GDS shows some low levels of homology to CDC25 and SDC25 protein and, in a different region, rather poorer homology to IRA1 and NF-1 proteins, but not to GAP. The homology to CDC25/SDC25 is in the region of these proteins thought to possess the catalytic exchange activity toward yeast RAS proteins, but the homology to IRA1 and NF-1 is outside the GTPase-activating catalytic domain. The significance of these homologies, particularly to IRA1 and NF-1, has not yet been determined.

Recent reports from Takai's laboratory have revealed some interesting clues as to the regulation of *smg* p21 GDS and some other activities that it possesses. p21^{rap1A} and p21^{rap1B} are both known to be phosphorylated by the cAMP-dependent protein kinase.[24] The phosphorylated residue in p21^{rap1B} has been shown to be serine[179], which is very close to the processed carboxyl terminal cysteine[181] which is isoprenylated with a geranylgeranyl group.[25] p21^{rap1B} purified from human platelets will bind to synaptic plasma membranes: this binding is reduced when the p21^{rap1B} is phosphorylated by protein kinase A.

In addition to stimulation of guanine nucleotide exchange activity, another activity that *smg* p21 GDS has been found to display is the ability to inhibit binding of p21^{rap1B} to plasma membranes and to induce the dissociation of prebound protein from membranes.[22] Furthermore, both the catalytic activities of *smg* p21 GDS are markedly more effective upon protein kinase A phosphorylated p21^{rap1B} than on the unphosphorylated protein.[25] Basal exchange rates and basal and rap-GAP-stimulated GTPase rates are not effected by the phosphorylation.

Treatment of intact platelets with agents that raise cAMP levels cause a translocation of p21^rap1B from membranes to the cytosol. Presumably, this is caused in a large part by increased sensitivity of phosphorylated p21^rap1B to the membrane-dissociation activity of GDS. It is not yet known whether p21^rap1B also becomes activated (more GTP-bound) at the same time, although this has been proposed by Takai: p21^rap1B-GTP would then be able to interact with its effector, which could be cytoplasmic. Control of *smg* p21 GDS, a dual function protein, thus, seems to be through posttranslational modification of its substrate rather than of the exchange factor itself.

In addition to *smg* p21 GDS, Takai's laboratory have also reported the identification and characterization of exchange factors for rho proteins.[26] These were found in bovine brain cytosol: fractionation on a Mono-Q FPLC column revealed at least two peaks of activity which have been named rho-GDS-1 and -2. It is not known whether they are related to each other. These factors have exchange activity toward p20^rhoA and p20^rhoB, but not toward p21^ras, p21^rap1B, or p25^rab3A. They do not appear to be related to the GDI or GAP proteins specific for rho proteins. The rho-GDS activities have not yet been purified to homogeneity.

An exchange factor specific for p25^rab3A has also been recently discovered in the laboratory of Ian Macara (Macara, personal communication).

V. NOMENCLATURE

The nomenclature of this rapidly expanding field is currently very confused: in addition to GEF, GRF, REP, and GDS, the term GNRP, for guanine nucleotide releasing protein has also been used in some review articles recently. Since the term GDS has already been used in a larger number of publications than all the others together and because it complements the term GDI, for GDP dissociation inhibitor, which has already been accepted by most in the field, I would like to propose that all the highly specific exchange factors for ras-like proteins are henceforth named GDS, preceded by the name of the gene for whose product they are specific (e.g., ras-GDS, rap1-GDS, rho-GDS).

VI. CONCLUSION

The last two years have seen the discovery of a number of guanine nucleotide exchange factors for ras-like proteins in mammalian systems. Most of these appear to be cytosolic proteins and are highly specific in their actions. One only, rGEF, is membrane-bound and effective on all members of the ras superfamily assayed. Two groups have reported specific exchange factors for p21^ras from different sources: these may well be related or identical. Specific exchange factors for the rap1 and rho proteins have also been reported. The exchange factor for p21^rap1 is the only one to have been molecularly cloned: it is unusual in that it also possesses an activity that promotes dissociation of p21^rap1 from membranes.

The important question of regulation of these exchange factors is only just beginning to be addressed. In addition to biochemical approaches, genetics is likely to play a significant role in understanding this system. Much has already been learnt from yeast about the functioning of CDC25, SDC25, and ste6, but the study of ras function is also well advanced in more complex organisms such as *Dictyostelium discoideum, Caenorhabditis elegans* and *Drosophila melanogaster.* There is particular reason to believe that genetic study of the control of ras protein may be fruitful in the multicellular organism *C. elegans*, the nematode worm. Several genes that function upstream of a ras gene, *let-60*, have been identified due to abnormalities in vulval formation shown by mutants.[27,28] One in particular, *lin-3*, could perhaps be a candidate exchange factor in this system. The discovery of an ever increasing number of activities toward ras proteins indicates that the regulation of the ras superfamily could be very complex indeed.

REFERENCES

1. **Hall, A. and Self, A. J.**, The effect of Mg^{2+} on the guanine nucleotide exchange rate of p21N-ras, *J. Biol. Chem.*, 261, 10963, 1986.
2. **Schramm, M. and Selinger, Z.**, Message transmission: receptor controlled adenylate cyclase system, *Science*, 225, 1350, 1984.
3. **Clanton, D. J., Hattori, S., and Shih, T. Y.**, Mutations of the ras gene product p21 that abolish guanine nucleotide binding, *Proc. Natl. Acad. Sci. U.S.A.*, 83, 5076, 1986.
4. **Stein, R. B., Robinson, P. S., and Scolnick, E. M.**, Guanine nucleotide binding to ras proteins, *J. Virol.*, 50, 343, 1984.
5. **Pai, E. F., Krengel, U., Petsko, G. A., Goody, R. S., Kabsch, W., and Wittinghofer, A.**, Refined crystal structure of the triphosphate conformation of H-ras p21 at 1.35 Å resolution: implications for the mechanism of GTP hydrolysis, *EMBO J.*, 9, 2351, 1990.
6. **Gilman, A. G.**, G proteins: transducers of receptor-generated signals, *Annu. Rev. Biochem.*, 56, 615, 1987.
7. **Thompson, R. C.**, Elongation factors, *Trends Biochem. Sci.*, 13, 91, 1988.
8. **Robinson, L. C., Gibbs, J. B., Marshall, M. S., Sigal, I. S., and Tatchell, K.**, CDC25: a component of the RAS-adenylate cyclase pathway in *Saccharomyces cerevisiae*, *Science*, 235, 1218, 1987.
9. **Broek, D., Toda, T., Michaeli, T., Levin, L., Birchmeier, C., Zoller, M., Powers, S., and Wigler, M.**, The *S. cerevisiae* CDC25 gene product regulates the RAS/adenylate cyclase pathway, *Cell*, 48, 789, 1987.
10. **Daniel, J., Becker, J. M., Enari, E., and Levitski, A.**, The activation of adenylate cyclase by guanyl nucleotides in *S. cerevisiae* is controlled by the CDC25 start gene product, *Mol. Cell. Biol.*, 7, 3857, 1987.
11. **Boy-Marcotte, E., Damak, F., Camonis, J., Garreau, H., and Jacquet, M.**, The C-terminal part of a gene partially homologous to CDC25 gene suppresses the cdc25-5 mutation in *S. cerevisiae*, *Gene*, 77, 21, 1989.
12. **Crechet, J.-B., Poullet, P., Mistou, M.-Y., Parmeggiani, A., Camonis, J., Boy-Marcotte, E., Damak, F., and Jacquet, M.**, Enhancement of the GDP-GTP exchange of RAS proteins by the carboxyl-terminal domain of SCD25, *Science*, 248, 866, 1990.

13. **Rey, I., Schweighoffer, F., Barlat, I., Camonis, J., Boy-Marcotte, E., Guilbaud, R., Jacquet, M., and Tocque, B.,** The COOH-domain of the product of the SCD25 gene elicits activation of p21-ras proteins in mammalian cells, *Oncogene,* 6, 347, 1991.

14. **Hughes, D. A., Fukui, Y., and Yamamoto, M.,** Homologous activators of ras in fission and budding yeast, *Nature,* 344, 355, 1990.

15. **West, M., Kung, H., and Kamata, T.,** A novel membrane factor stimulates guanine nucleotide exchange reaction of ras proteins, *FEBS Lett.,* 259, 245, 1990.

16. **Huang, Y. K., Kung, H.-F., and Kamata, T.,** Purification of a factor capable of stimulating the guanine nucleotide exchange reaction of ras proteins and its effect on ras-related small molecular mass G proteins, *Proc. Natl. Acad. Sci. U.S.A.,* 87, 8008, 1990.

17. **Wolfman, A. and Macara, I.,** A cytosolic protein catalyzes the release of GDP from p21ras, *Science,* 248, 67, 1990.

18. **Downward, J., Riehl, R., Wu, L., and Weinberg, R. A.,** Identification of a nucleotide exchange-promoting factor for p21ras, *Proc. Natl. Acad. Sci. U.S.A.,* 87, 5998, 1990.

19. **Downward, J., Graves, J. D., Warne, P. H., Rayter, S., and Cantrell, D. A.,** Stimulation of p21ras upon T-cell activation, *Nature,* 346, 719, 1990.

20. **Yamamoto, T., Kaibuchi, K., Mizuno, T., Hiroyoshi, M., Shirataki, H., and Takai, Y.,** Purification and characterization from bovine brain cytosol of proteins that regulate the GDP/GTP exchange reaction of smg p21s, ras p21-like GTP-binding proteins, *J. Biol. Chem.,* 265, 16626, 1990.

21. **Hiroyoshi, M., Kaibuchi, K., Kawamura, S., Hata, Y., and Takai, Y.,** Role of the C-terminal region of smg p21, a ras p21 like small GTP binding protein, in membrane and smg p21 GDP/GTP exchange protein interactions, *J. Biol. Chem.,* 266, 2962, 1991.

22. **Kawamura, S., Kaibuchi, K., Hiroyoshi, M., Hata, Y., and Takai, Y.,** Stoichiometric interaction of smg p21 with its GDP/GTP exchange protein and its novel action to regulate the translocation of smg p21 between membrane and cytoplasm, *Biochem. Biophys. Res. Commun.,* 174, 1095, 1991.

23. **Kaibuchi, K., Mizuno, T., Fujioka, H., Yamamoto, T., Kishi, K., Fukumoto, Y., Hori, Y., and Takai, Y.,** Molecular cloning and characterization of the stimulatory GDP/GTP exchange protein (GDS) for smg p21s, ras p21-like small GTP-binding proteins, *Mol. Cell. Biol.,* 11, 2873, 1991.

24. **Kawata, M., Kikuchi, A., Hoshijima, M., Yamamoto, K., Hashimoto, E., Yamamura, H., and Takai, Y.,** Phosphorylation of smg p21, a ras p21-like GTP-binding protein, by cyclic AMP-dependent protein kinase in a cell-free system and in response to prostaglandin E1 in intact human platelets, *J. Biol. Chem.,* 264, 15,688, 1989.

25. **Hata, Y., Kaibuchi, K., Kawamura, S., Hiroyoshi, M., Shirataki, H., and Takai, Y.,** Enhancement of the actions of smg p21 GDP/GTP exchange protein by the protein kinase A-catalysed phosphorylation of smg p21, *J. Biol. Chem.,* 266, 6571, 1991.

26. **Isomura, M., Kaibuchi, K., Yamamoto, T., Kawamura, S., Katayama, M., and Takai, Y.,** Partial purification and characterization of GDP dissociation stimulator (GDS) for the rho proteins from bovine brain cytosol, *Biochem. Biophys. Res. Commun.,* 169, 652, 1990.

27. **Beitel, G. J., Clark, S. G., and Horvitz, H. R.,** C.elegans ras gene let-60 acts as a switch in the pathway of vulval development, *Nature,* 348, 503, 1990.

28. **Han, M. and Sternberg, P. W.,** let-60, a gene that specifies cell fates during C. elegans vulval induction, encodes a ras protein, *Cell,* 63, 921, 1990.

Chapter 26

PROPERTIES OF CDC25-LIKE PROTEINS

**Andrea Parmeggiani, Michel-Yves Mistou, Eric Jacquet,
Patrick Poullet, and Jean-Bernard Créchet**

TABLE OF CONTENTS

I. INTRODUCTION

In *Saccharomyces cerevisiae* the cell division cycle gene *CDC25* is a fundamental element of the RAS/adenylate cyclase pathway which controls the production of cAMP, an internal signal for the onset of cell division.[1-5] Temperature-sensitive mutations of this gene lead to a growth arrest in the G1 phase. Several lines of evidence indicate that the *CDC25* gene product, which is probably involved in monitoring environmental signals, acts upstream of the RAS proteins which regulate adenylate cyclase activity.[6,7] The two homologous proteins RAS1 and RAS2, which share considerable similarities with mammalian ras proteins, are GTPases and, as for all members of this class of proteins, their function is controlled by GTP and GDP, which respectively, induce the active and the inactive conformation.[8-11] Because of the tight binding of GDP and GTP, the key steps of the RAS-guanine nucleotide cycle are the dissociation of the RAS-GDP complex, which conditions the regeneration of the active complex RAS-GTP, and the hydrolysis of GTP which turns off the active state of RAS. The very low rates of the intrinsic GDP to GTP exchange and GTPase of RAS imply the existence of effectors to assure a rapid transient response to the extracellular stimuli which lead to cAMP production.[12-14] In *S. cerevisiae* the product of *IRA2* (and probably also that of *IRA1*) corresponds functionally to mammalian GTPase-activating protein (GAP).[15] Concerning the GDP to GTP exchange of RAS, there is enough evidence that the *CDC25* gene product is a regulator of this reaction. This was first suggested by the observation that dominant mutations of RAS2 increasing the level of the RAS2-GTP complex can bypass the growth arrest caused by thermosensitive *cdc25* mutations.[4,5,16] The use of reconstituted systems for cAMP production *in vitro* has further supported the involvement of the *CDC25* gene product in the nucleotide exchange of RAS proteins.[17] It has been reported that partially purified CDC25-β-galactosidase fusion products or extracts from yeast cells with overexpressed *CDC25* gene can accelerate the RAS2-GDP dissociation *in vitro*.[18]

In 1989, the group of M. Jacquet observed that overexpression of the 3′ terminal portion of a newly discovered *S. cerevisiae* gene was capable of suppressing only the thermosensitive phenotype of the *cdc25-5* mutation and not the phenotypes dependent on other mutated components of the RAS/adenylate cyclase pathway.[19] This gene, first named *SCD25* (for suppressor of *cdc25*) and then renamed *SDC25* for reasons of international nomenclature, encodes a product with a significant sequence similarity to the *CDC25* gene product, particularly in the C terminal domain.[20] Surprisingly, the full-length *SDC25* gene was unable to compensate for the cdc25-5 mutation. Genetic analysis has shown that the suppressing activity of the *SDC25* 3′ terminal region occurs upstream of RAS proteins and adenylate cyclase, as known for the action of the *CDC25* gene. The production in *Escherichia coli* of the encoded protein, the SDC25 C terminal domain (SDC25 C-domain), has made possible its isolation and purification.[21] The SDC25 C domain is a very active guanine nucleotide dissociation stimulator (GDS) of RAS2 *in vitro*, leading to a fast exchange of RAS2-bound GDP with

ACTIVATION   INACTIVATION

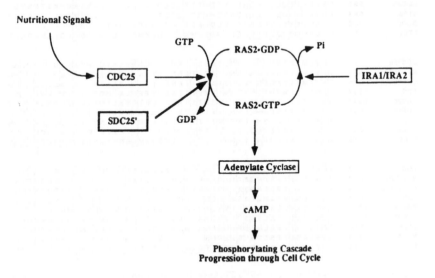

FIGURE 1. Schematic representation of the cycle of RAS proteins and action of the known effectors in the RAS-adenylate cyclase pathway of *S. cerevisiae*. The step influenced by the SDC25 C domain (SDC25') is indicated.

free GTP. Its action provides the biochemical explanation for the suppression of *cdc25-5* mutation by the *SDC25* 3' terminal region. In Figure 1, we schematically represent the CDC25/RAS/adenylate cyclase pathway in *S. cerevisiae* and indicate the step controlled by this GDS.

In recent years, other genes have been isolated that contain sequence similarities with the C terminal moiety of the *CDC25* gene. They are: *ste6* in *Schizosaccharomyces pombe, BUD5,* and *LTE1* in *S. cerevisiae, SOS* in *Drosophila melanogaster* and *CDC25^{Mm}* in mouse.[22-27] These observations indicate the existence of a rapidly growing number of proteins which act in different organisms and pathways, and are structurally and very probably functionally related to the *CDC25* gene product. In Figure 2, an overview of the sequence similarities of the conserved domain of these proteins is reported (c.f., Note Added in Proof at end of chapter).

II. GENERAL PROPERTIES OF THE CDC25 GENE PRODUCT

The *CDC25* gene contains an open reading frame encoding 1589 amino acid residues and in agreement with this the molecular weight of the product has been found to be 180 kDa.[1,4,28,29] Studies using polyclonal antibodies have shown that

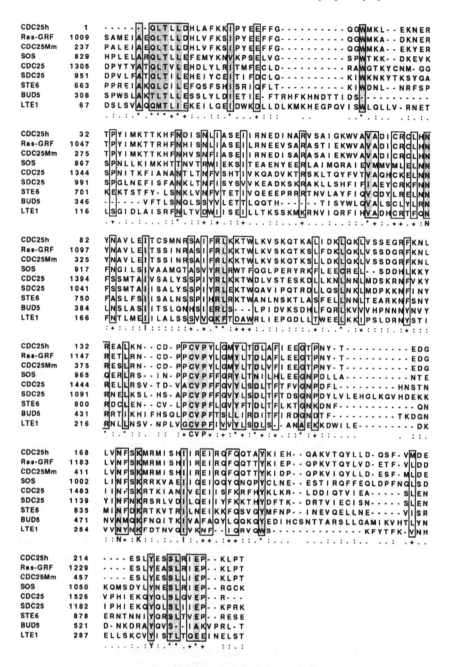

FIGURE 2. Primary sequence similarities in the most conserved domain of CDC25-like proteins. The alignment was obtained by using the CLUSTAL V program and the computer facilities of CITI2 between the sequences of human CDC25h, rat Ras-GRF, mouse CDC25Mm,[27] *Drosophila melanogaster* SOS,[25] *S. cerevisiae* CDC25,[19] SDC25,[20] BUD5,[23] LTE1,[24] and *S. pombe* STE6.[22] (For the references of CDC25h and Ras-GRF, see Note Added in Proof at end of chapter.) Shaded boxes indicate identical amino acids in at least 8 of 9 sequences; open boxes enlight the most conserved motifs.

in *S. cerevisiae* the CDC25 protein is present in a very low amount, even when high copy number plasmids are expressed in protease-deficient strains.[28,29] Discrepancies concerning the intracellular distribution of the product of overexpressed *CDC25* gene, may be due to the different yeast strains used and growth conditions. Jones et al.[18] report that CDC25 is equally distributed in cell membrane and cytoplasm, whereas Garreau et al.[28] and Vanoni et al.[29] emphasize its tight association with the particulate fraction. How this association takes place is not yet clear. In the N terminal domain of CDC25, an SH3 motif can be recognized within residues 65 and 129.[29] This consensus sequence is found in proteins associated with the membrane as part of the site responsible for the anchorage to the membrane.[30] Only a small portion of the *CDC25* gene product can be solubilized even with high concentrations of guanidinium-HCl or urea, suggesting that strong hydrophobic interactions are involved in its binding to the membrane. A transmembrane domain has been proposed in the hydrophobic region 1452 to 1473.[20,31,32] Interestingly, deletion of residues 1255 to 1550 renders the CDC25 protein soluble.[29] Many potential sites for *N*-glycosylation and one for *O*-glycosylation at the Thr/Ser rich amino terminus can be identified in the CDC25 sequence.[33] This has suggested that the N terminal domain of CDC25 is, in part, secreted through the cell membrane and lies in the extracellular space.[4,32] However, in a recent report no experimental evidence for glycosylation was found.[29] The CDC25 sequence predicts seven to nine phosphorylation sites for cAMP-dependent protein kinase.[20,33] Figure 3 resumes some structural properties of the *CDC25* gene product.

The phenotypic effects following gene deletions have been interpreted as the existence of functional domains in the CDC25 protein.[32] The N terminal moiety is apparently needed for sporulation, gluconeogenic function, and growth on nonfermentable carbohydrate sources, but not for viability. The C domain is essential for viability except for the last 38 amino acid residues and is involved in the glucose-induced cAMP production.[1,2,13,14,32,34,35] A 3' region of the *CDC25* gene encoding at most 330 C terminal amino acids is sufficient for complementing *cdc25* mutations.[36] Therefore, the catalytic domain of CD25 ought to be located in this region. From genetic data, it can be inferred that the minimally active CDC25 C domain is comprised within 286 amino acid residues (residues 1258 to 1543) of the most conserved part of the C domain.[3,36]

Thus, the difficulties in isolating and purifying an active *CDC25* gene product have not yet permitted a satisfactory investigation of its biochemical properties. Properties of its activity were indirectly derived from the investigation of two mutated RAS2 proteins suppressing the *CDC25* requirement, RAS2G19V and RAS2T152I, whose interaction with guanine nucleotides was studied in detail *in vitro*.[37] Whereas RAS2G19V shows a low intrinsic GTPase associated with an increased stability of the complex with GTP, RAS2T1521 is characterized by a strongly accelerated GDP to GTP exchange reaction. Even though the specific modifications induced by these two mutations are different, in both cases they lead to increased levels of the active RAS2-GTP complex. These

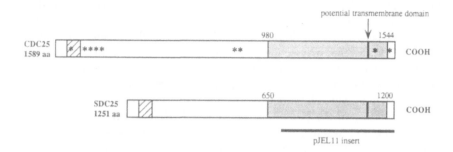

FIGURE 3. Schematic representation of some properties of the CDC25 and SDC25 gene products. The shaded area indicates the region of most conserved homology (47% identity); the hatched bar, the region containing the SH3 consensus sequence; the stars, the potential phosphorylation sites; the vertical solid bar, the predicted transmembrane domain; and the horizontal solid bar, the 3′ terminal region of SDC25 contained in pJEL11.

results are in line with conclusions concerning the CDC25 function as derived from genetic experiments.[4,5,16] A direct confirmation for the GDS activity of CDC25 has been recently obtained using immunopurified CDC25 fragments fused with β-galactosidase or extracts of yeast cells containing overexpressed CDC25 gene.[18] These preparations were able to enhance to some extent the GDP dissociation of RAS2. The low activity may be an inherent property of these preparations or of the assay system, but may also mean that *in vivo* additional as yet unidentified elements are required for the activation of the *CDC25* gene product. In fact, in the yeast cell, in the conditions tested, the overexpression of the *CDC25* gene product or its 3′ terminal region has been reported not to induce a phenotype of permanently activated cAMP production (lack of glycogen accumulation, heat shock sensitivity, and sporulation defect in diploids strains), unlike the activated phenotype described to be associated with RAS2G19V or the overexpression of the *SDC25* 3′ terminal region (see below).[4,20,38]

Concerning the interaction between CDC25 and RAS2, Powers et al.[39] have described mutated *RAS2* genes that cause a temperature-sensitive loss of RAS functions. This dominant-negative phenotype associated with mutant RAS2G22A and RAS2A25P could be reversed by the overexpression of the *CDC25* gene in the presence of the wild-type allele of *RAS2*, a phenomenon that was interpreted as a consequence of the formation of a tight complex between the mutated RAS2 and the *CDC25* gene product. These observations support a direct interaction between RAS2 and CDC25.

III. PROPERTIES OF *CDC25*-LIKE GENE PRODUCTS

The discovery of the other members of the CDC25-like family has pointed to the importance of the CDC25-like functions in the cell. Like the *ras1* gene, the *ste6* gene is essential for mating in *S. pombe*. It acts upstream of *ras1* and encodes a product of 911 amino acid residues.[22] A region of the ste6 C domain spanning 270 amino acid residues shows a 34% identity with the C domain of the *CDC25* gene product and also a significant similarity with the *SDC25* gene product. The similarity between this and the budding yeast system has been emphasized by the observation that ras1G17V (corresponding to the activating mutant RAS2G19V in *S. cerevisiae*) compensates for the inactivation of the *ste6* gene. This leads to the conclusion that the ste6 product is involved in the GDP to GTP exchange of ras1.

The *S. cerevisiae* gene *BUD5* is required for both the axial and bipolar pattern of the bud sites.[23] It encodes a protein of 538 amino acid residues which shares a significant similarity with the C domain of CDC25 (19% identity in the last 400 C terminal residues). This similarity suggests that BUD5 may also be a GDS of RAS-like proteins such as the *BUD1* and *CDC42* gene products.[40,41] Of remarkable interest is the observation of Powers et al.[42] that high copy number plasmids carrying *BUD5* gene can partially suppress the growth defect associated with RAS2G22A mutation, suggesting a direct interaction between these two proteins as in the case of CDC25 and RAS2 (see above).[39] However, BUD5 cannot substitute for CDC25 in the activation of RAS2. Therefore, whether the interaction between BUD5 and RAS2 has some physiological significance remains an open question.

Neither connections with ras (or ras-like) pathway(s) nor any other function has yet been attributed to the *S. cerevisiae LTE1* gene but its deletion influences the growth at low temperatures.[24] It codes for a product of 468 amino acid residues, of which the 308 C terminal residues show 22% identity with the most conserved part of the CDC25 C domain.[20]

In *Drosophila* the *SOS* gene encodes a product of 1596 amino acid residues with a middle domain (residues 641 to 1020) sharing a pronounced identity (28%) with the C domain of CDC25 and SDC25.[25,26] Genetic studies imply that this protein acts in the tyrosine kinase pathway, probably as a GDS of ras1.

It should be stressed that the action of these proteins as putative GDP to GTP exchange factors of ras-like proteins is up to now exclusively based on their sequence similarity to CDC25; no biochemical work has proved this putative role.

Recently, a 3' terminal region of a mouse gene has been identified by complementation with *CDC25* that acts as suppressor of *cdc25* gene mutations in *S. cerevisiae*.[27] Experiments have been carried out with the partially purified GST-fushed product containing the last 287 C-terminal amino acids. This protein displays a GDS activity on the GDP-RAS2(p21) complex.[43]

Since the SDC25 C domain, the first found and so far the best characterized CDC25-like protein, represents a model GDS for studying the basic mechanism for nucleotide release of ras proteins, the following sections report its biochemical properties in detail.

IV. THE *SDC25* GENE PRODUCT AND ITS C DOMAIN

The *SDC25*-encoded product is approximately 20% shorter than the *CDC25* gene product: 1251 vs. 1589 amino acid residues.[19,20] The sequences of these two proteins share a significant similarity that is particularly evident in the C domain (47% identity obtained by an optimal alignment of the regions 650 to 1200 of SDC25 and 980 to 1544 of CDC25). As deduced from the *SDC25* open reading frame, the encoded protein should have a molecular weight of 145 kDa. The SDC25 protein also contains an SH3 motif related to residues 33 to 95 in the N terminal end (vs. residues 65 to 129 in CDC25) and a putative hydrophobic transmembrane domain in the C domain (residues 1102 to 1116 vs. residues 1455 to 1469 in CDC25).[20] The potential phosphorylation sites for cAMP-dependent protein kinase present in CDC25 are not conserved in SDC25, whose unique potential phosphorylation site (residue 439) is situated in a nonconserved region.[19] The location of the *SDC25* gene product within the *S. cerevisiae* cell has yet to be established but the sequence similarities to CDC25 support its association with the membrane. Figure 3 resumes some structural properties of the *SDC25* gene product.

As already mentioned, overexpression of the *SDC25* 3′ terminal region (encoding in the original work the 584 C terminal residues), but not of the intact gene, suppresses the *cdc25-5* mutation.[19] Genetic evidence indicates that the minimal suppressor fragment is comprised in the last 335 C terminal residues.[20] Deletion of residues 263 to 608 in the intact *SDC25* gene product can restore the capacity of suppressing *cdc25*, whereas a larger deletion including the same region (Δ 140 to 874) does not. The lack of effect with the intact gene is not due to deficient expression and may depend on elements of the N terminal domain of SDC25, that regulate the activity of the protein and act as negative effectors on the catalytic domain.[20] Alternatively, a specific activation mechanism may be required. Overexpression of the *SDC25* 3′ terminal region induces in *S. cerevisiae* a phenotype of permanently activated cAMP production, similar to that obtained with RAS2G19V. Thus far, neither deletion nor overexpression of the intact *SDC25* gene could be associated with a specific phenotype. The existence of redundancy between the *CDC25* and *SDC25* genes has been excluded by genetic evidence.[20] As a possible interpretation of these results, the *SDC25* gene product may not act on RAS, at least as a component of the adenylate cyclase pathway, and may be a GDS of one of the many ras-like proteins. However, the high specificity and activity of the SDC25 C-domain with respect to RAS2 and Ha-*ras* p21 (see later) make it more probable that the intact SDC25 protein could monitor as yet unknown selective signals to RAS proteins.

A recombinant SDC25 C domain with GDS activity *in vitro* can be produced in *E. coli*, as recently has also been obtained with C terminal fragments of CDC25 and chimeric CDC25/5DC25 constructions.[18,44,45]

To obtain the SDC25 C domain as a recombinant protein in *E. coli*, the 3′ terminal part of the *SDC25* gene, cloned in pUC19 (pJEL11) or pTTQ19, respectively, under the control of the *Lac* and *Tac* promotor, has been expressed as a chimeric protein including the first 13 N terminal residues of the β-galactosidase and the last 550 C terminal residues of SDC25.[20,46] Also in this system, the production of the SDC25 C domain is low; its presence in the total cell extract is hardly detectable by Coomassie-blue staining of SDS-PAGE. Immunoblot experiments, using rabbit polyclonal antibodies (see below), have shown that in the sonicated cell extract of *E. coli,* about 70% of the SDC25 C domain, is associated with the insoluble cell fraction. After a fractionation of the cell extract supernatant with ammonium sulfate, the purification procedure includes three chromatographic steps in the order: Q-Sepharose (Pharmacia), Ultrogel AcA54 (IBF) and Mono-Q (Pharmacia). The first and third anion exchange chromatographies are carried out with the Pharmacia FPLC system. In all three purification steps the SDC25 C domain is isolated as a single peak. The active fractions are detected either by measuring the dissociation rate of the RAS2-GDP complex using the filtration method on nitrocellulose disks or by immunoblots of SDS-PAGE. For the tests, a pure full-length or truncated RAS2 (proteolytically cleaved at His210) is used.[37,43,47]

The major difficulty in the purification of SDC25 C domain is encountered after the Ultrogel AcA54 filtration; aggregation phenomena, in which hydrophobic interactions may play a major role, lead to progressive precipitation and inactivation of the protein.[46] These precipitates, containing more than 80% pure SDC25 C domain, have successfully been used for the production of polyclonal antibodies. The tendency to aggregate increases with an increasing degree of purification. Attempts to stabilize the purified SDC25 C domain have thus far been unsuccessful. Therefore, the last chromatographic step on Mono-Q, though yielding a virtually pure protein, has not been routinely performed due to the minimal amount of soluble protein recovered. In most of our studies, a partially purified SDC25 C domain has been used that is stable and virtually devoid of contaminating guanine nucleotide-binding proteins or GTPases.

As deduced from the gene sequence, the SDC25 C domain should have a molecular weight of 64 kDa. The immunoblot of SDS-PAGE shows that a protein of this length is, indeed, present when the *E. coli* cells are directly extracted with cracking buffer; however, if the cells are sonicated, the active component of the extract mainly coincides with a 54 kDa protein.[46] This suggests the occurrence of proteolytic phenomena.

V. THE SDC25 C DOMAIN ENHANCES *in vitro* THE GUANINE NUCLEOTIDE DISSOCIATION OF RAS2 AND Ha-*ras* p21

The effect of the SDC25 C domain on the dissociation of the RAS2 complexes with guanine nucleotides is assayed *in vitro* by measuring the following reaction:

$$RAS2 \cdot {}^*GDP({}^*GTP) + GDP(GTP) \rightarrow RAS2 \cdot GDP(GTP) + {}^*GDP({}^*GTP)$$

where the star indicates a radioactive label.

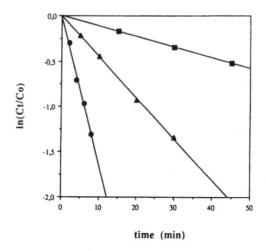

FIGURE 4. Effect of the SDC25 C domain on the dissociation rate of the RAS2·GDP complex. Intrinsic dissociation rate (■); in the presence of 2 nM (▲), and 10 nM SDC25 C domain (●). For more details of the experimental conditions see Créchet et al.[21]

The reversed reaction, the formation of the RAS2 · *GDP(*GTP) complex, can be neglected if a large excess (>500 times) of cold nucleotide is added. The RAS2 · *guanine nucleotide complex is obtained by incubating nucleotide-free RAS2 with a *guanine nucleotide or alternatively by the addition of EDTA that accelerates the RAS2-bound nucleotide exchange with a free *nucleotide.[37,46] As shown in Figure 4, the addition of two different amounts of the SDC25 C domain proportionally stimulates the dissociation rate of RAS2·GDP. Its effect on RAS2·GDP is stronger than on RAS2·GTP (for details, see following section).[37] The SDC25 C domain also enhances the association rate between RAS2 and GDP or GTP, an effect which is about one fourth that observed on the dissociation rate.

The SDC25 C-domain is also active on the human Ha-*ras* p21 protein. Its stimulation of the release of the p21-bound GDP is comparable to that on RAS2.[21,46] By using filtration on nitrocellulose disks we have observed stimulations of several hundred-fold of the dissociation rate of the wild-type p21 · GDP complex, i.e., the maximum effect that can be evaluated by this technique. Since the action of the SDC25 C domain on RAS2 and p21 is similar, the results described in the following paragraphs are virtually valid for both these two proteins, if not otherwise mentioned.

As a consequence of the fast release of GDP from ras proteins induced by the SDC25 C domain, the exchange between protein-bound GDP and free GTP is strongly enhanced and follows a dose-response effect that is linear in the tested concentration range (Figure 5 and inset therein).

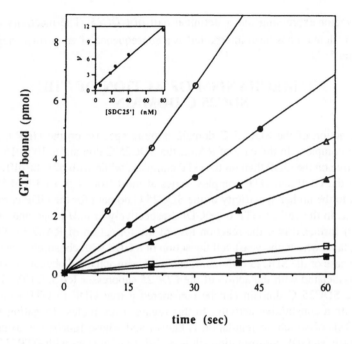

FIGURE 5. Stimulation by the SDC25 C domain of the GDP to GTP exchange of Ha-*ras* p21. Intrinsic exchange rate (■); in the presence of 6 (□), 18 (▲), 24 (△), 40 (●), and 80 n*M* (○) SDC25 C domain. The inset indicates the linearity of the dose-response effect; ν = pmol GTP-bound · min^{-1}. The SDC25 C domain has been indicated as SDC25'. For more details of the experimental conditions, see Créchet et al.[21]

We have observed that the SDC25 C domain acts catalytically and that the reaction follows a Michealis-Menten saturation kinetics in the presence of increasing concentrations of either the p21·GDP complex or the nucleotide.

The action of the SDC25 C domain during the exchange reaction has been followed by measuring the shift of the intrinsic fluorescence related to the GDP and GTP complexes of two mutated p21 proteins in which a tryptophane residue was introduced either in position 56 or 64 by site-directed mutagenesis.[48]

The regeneration of the active GTP-bound form of ras proteins induced by the SDC25 C domain has also been shown *in vivo,* by transfection of CHO cells with expression vectors encoding the SDC25 C domain.[49] Under these conditions, the GTP-bound form of ras proteins was strongly increased, as determined by immunoprecipitation analysis, indicating that the SDC25 C domain can overcome the negative regulation by GAP of the level of the ras·GTP complex. As found for mutated ras proteins with transforming activity as p21G12K, the SDC25 C domain can also enhance the transcription of the *CAT* reporter gene under control of the LTR region of HIV-1. These effects could not be achieved by transfection with a vector expressing the 3′ terminal part of *CDC25*. The inactivity of the CDC25 C domain has been interpreted as a consequence of an intrinsic negative regulation.[49]

VI. MECHANISM OF ACTION OF THE SDC25 C DOMAIN

The action of the SDC25 C domain is more specific on the GDP than on the GTP complex. In the case of RAS2, the SDC25 C domain is 10 to 15 times more active on the GDP than on the GTP complex, while with p21 the difference between the GDP and GTP complex varies at most from 1.5- to 3-fold.[45,46] In addition to the higher specificity of the SDC25 C domain for the GDP complex, the excess in the cell of GTP over GDP (respectively in mM range and around 0.1 mM) further drives the reaction toward the formation of RAS2 · GTP.[50,51] This explains why in the yeast cell the action of a GDS with constitutive activity, as in the case of the SDC25 C domain, can induce an activated phenotype similar to that associated with the action of the *CDC25* suppressor RAS2G19V. Therefore, the SDC25 C domain can be considered a true GDP to GTP exchange factor with a constitutive activity. In this regard it resembles elongation factor Ts (EF-Ts), of which no regulation is known and whose function is to unlock the tight EF-Tu·GDP conformation allowing EF-Tu to interact with GTP.[52,53] EF-Ts acts as a ligand-exchange catalyst in a two-substrate system.[54-56] Since the dissociation of EF-Tu·GDP represents the rate-limiting step for the formation of the active EF-Tu·GTP complex, the function of EF-Ts is directed to stabilize the reactive intermediate, the nucleotide-free EF-Tu, as the crucial step in a double displacement mechanism.[55] Although in the case of the SDC25 C domain a thorough kinetic study has not yet been provided, our results strongly suggest that the action of the SDC25 C domain in the GDP to GTP exchange of RAS proteins also follows a substitution mechanism, whose events are outlined in Figure 6. This mechanism predicts the formation of a complex between the SDC25 C domain and guanine nucleotide-free ras proteins. We have indeed identified on gel filtration the existence of a nucleotide-free, stable RAS2 · SDC25 C domain complex, as is since long known for EF-Tu and EF-Ts.[57,58] Addition of GDP causes the dissociation of this complex.

The ability of the SDC25 C domain to act equally well on the GDP complexes of RAS2 and p21 has allowed us, in collaboration with the group of Dr. A. Wittinghofer (Max-Planck-Institut für Biochemie, Heidelberg), to study the GDS activity of the SDC25 C domain on a number of p21 mutants.[58] Substitutions affecting the coordination of Mg^{2+} nucleotide or the interaction of the guanine ring were found to mimick the action of GDS; those in the phosphate-binding loop L1 or deletion of the last 23 C terminal residues of p21 were not relevant for the GDS effect. Substitutions in the switch region 1 (loop L2) and 2 (loop L4) strongly impair the response to the SDC25 C domain, without however,

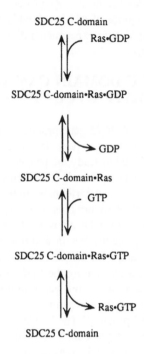

SDC25 C-domain

Ras•GDP

SDC25 C-domain•Ras•GDP

GDP

SDC25 C-domain•Ras

GTP

SDC25 C-domain•Ras•GTP

Ras•GTP

SDC25 C-domain

FIGURE 6. Proposed mechanism of the GDP to GTP exchange of ras proteins induced by the SDC25 C domain.

decreasing its affinity for the mutated p21. These results suggest that GDS does not bind to these two regions of the p21. We have suggested that the inhibitory effect of mutations in switch regions 1 and 2 very likely depends on the specific functions of these two regions, inducing selective conformations during the GDP to GTP exchange reaction.

The irrelevance of the C domain of ras proteins for the action of SDC25 C domain differs from the observations of Downward et al.,[59] concerning the action of a GDS activity from human placenta on urea-renaturated p21 proteins. To explain this discrepancy, one should emphasize that the SDC25 C domain which was identified as a suppressor of a regulated yeast gene *(CDC25)* can express its constitutive activity, independently of a specific activation mechanism which may be required for other GDS. Another GDS acting on *smg* p21 (rap1), c-Ki-*ras* p21 and *rhoA* p21 has been reported to need the presence of a post-translationally processed C domain.[60,60a]

Our observations suggest that exposed elements of the p21 molecule, other than the loops L1 and L2, may be involved in the binding of a GDS. A number of genetic studies suggest as possible candidates the helix α4, residues 75 or 103 to 108, or in RAS2 residues 80 and 81 (73 and 74 in p21).[61-63]

The definition of the regions of p21 involved in the action of the SDC25 C domain and GAP will help design experiments directed to clarify the action of the effectors of ras proteins.

VII. THE SDC25 C DOMAIN CAN STIMULATE THE ADENYLATE CYCLASE ACTIVITY

A crucial function of the *CDC25* gene product in *S. cerevisiae* is the positive regulation of the adenylate cyclase activity. Reconstituted systems for adenylate cyclase activity have been used to study the properties of the CDC25-dependent RAS stimulation on adenylate cyclase.[17,64] Evidence has been presented indicating that the *CDC25* gene product participates in the formation of a membrane-bound ternary complex with RAS proteins and adenylate cyclase. The control of adenylate cyclase activity by CDC25 is mediated by the GTP- and GDP-bound forms of the RAS proteins, but in addition to this a direct interaction between adenylate cyclase and CDC25 has been postulated.[17]

We have measured the activation by the SDC25 C domain of the adenylate cyclase using (1) yeast membranes from yeast strains carrying genes encoding active and inactive CDC25 and RAS proteins and (B) the GDP- and GTP-bound forms of recombinant RAS2 protein.[65] The SDC25 C domain can restore *in vitro* the production of cAMP from *RAS2 cdc25*⁻ membranes, but can little influence that of *RAS2 CDC25* membranes. Increasing concentrations of the SDC25 C domain lead to the saturation of the system; the cAMP production becomes comparable to that obtained in the presence of an amount of $MnCl_2$ inducing optimum adenylate cyclase activity. All evidence indicates that SDC25 C domain activates adenylate cyclase by regenerating the RAS2·GTP complex from RAS2·GDP. No direct action on adenylate cyclase in the absence of RAS2 could be detected. We could show *in vitro* that adenylate cyclase activity is modulated as a function of the ratio between the SDC25 C domain, as a positive effector, and the catalytic domain of GAP, as a negative effector of the active RAS2·GTP complex. The use of this GDS provides a new system to define the dynamics of the ras-guanine nucleotide cycle and of the reconstituted adenylate cyclase reaction. Recently, the CDC25^Mm has also been found to stimulate the *S. cerevisiae* adenylate cyclase in a RAS-dependent manner.[27,43]

VIII. SPECIFICITY OF THE ACTION OF THE SDC25 C TERMINAL DOMAIN

The specificity of the SDC25 C domain has been tested in our Laboratory on small ras-like proteins from mammals (rap2, ralA, rap1A), *S. cerevisiae* (YPT1) and *S. pombe* (ryh1). These proteins are phylogenetically related to the *ras* gene products, displaying a structural similarity between 30 and 50%.[66,67]

No guanine nucleotide-releasing effect by the SDC25 C domain has been observed on these proteins, nor does the SDC25 C domain affect the guanine nucleotide-binding of the elongation factor Tu and its eukaryotic counterpart, EF-1α from *S. cerevisiae* and calf brain. Interestingly, Huang et al.[68] describes a p21 guanine nucleotide exchange factor from bovine brain that also acts on small ras-like proteins such as R-*ras*, rap1A, rab 1B, and rho. Another GDP to GTP exchange factor from brain acting on the ras-like proteins rap1 (= *smg*p21), have been reported to be inactive on Ha-*ras* p21, but is stimulating Ki-*ras*, rho, rab3, rac1, and p21.[69] Therefore, it appears that the action of the GDP to GTP exchange factors can, in some cases, display highly selective properties and in other cases affect diverse ras-like proteins. Evidently, further work is required for a comprehensive picture of the structural requirements for the specificity of these proteins.

IX. CONCLUSIONS AND PERSPECTIVES

Our understanding of the activity of the CDC25-like proteins is still very incomplete and in a large part supported by indirect observations. For CDC25 and probably also for SDC25, this is mainly due to the association with membrane structures and to the low concentration in the cell, both features that make the isolation and purification of these proteins in the native state very difficult. The *CDC25* gene product acts in close functional relationships with RAS proteins and adenylate cyclase. Whether the *CDC25* gene is the most upstream element of the RAS/adenylate cyclase pathway, has not yet been defined; other ligands may mediate the reception of extracellular stimuli to CDC25. As indicated by the glucose-dependent stimulation of the cAMP production in the cell, the regulation of the CDC25 activity appears to be under control of metabolic signals. Thus, CDC25 may be part of a complex system which transduces nutritional stimuli across the cell membrane, but at the present state of our knowledge it is not yet possible to speculate on effectors that could modulate its activity. Recent genetic work has enlighted a sequence of elements participating in a signal transduction pathway in the R7 cells of *Drosophila* eye.[25,26] In this succession of events, membrane receptors transmit the signal for the activation of a CDC25-like protein that controls the activity of the ras1 pathway.

The isolation of a SDC25 fragment with constitutive nucleotide-releasing activity on RAS2 or p21 has supplied a valuable tool to investigate the biochemical characteristics of the CDC25-like protein family and of the GDSs in general. The detailed investigation of the structural elements of the molecule regulating the activity of the CDC25 and SDC25 C terminal domains may be crucial for understanding the activation mechanisms of these proteins.

A major role for the activity of CDC25 and maybe of SDC25 is likely played by their integration in the cell membrane as part of a complex in which RAS proteins and possibly adenylate cyclase and/or other components participate. The posttranslational modifications of ras proteins, needed for their attachment to the cell membrane, may be pivotal for their activity. It is known that for the ras-like protein *smg* p21, purified from blood platelets or brain, the presence of the C domain is essential for the activity of its specific GDS.[60]

The isolation of a GDS with a strong constitutive activity allows it to set up *in vitro* systems for the kinetic evaluation of the ras protein cycle and the regulation by its effectors. The identification of the SDC25 C domain, active on both yeast and mammalian ras proteins, will help identify specific exchange factors from higher eukaryotes, which may have some sequence similarity with CDC25, or novel proteins involved in the regulation of the ras or ras-like activities. This approach has already proved its value by the isolation of a mammalian mouse GDS (CDC25^{Mm}) identified as a suppressor of *cdc25*.[27,43]

Protein factors influencing the regeneration cycle of GTP-binding proteins are ubiquitous and have long since been known and well-characterized for quite a few systems such as protein biosynthesis, transport across membranes, and hormonal and sensorial response.[10,11] Recent articles from different laboratories have described a number of GDSs for mammalian ras proteins.[59,68,71,72] The preliminary characterization of their action shows commune features with that of the SDC25 C domain. It is not yet known but it is probable that these factors also share sequence similarities with the CDC25-like family. On the basis of these recent developments, the discovery of many more CDC25-like proteins controlling the most diverse pathways in the cell, can be expected in the time to come. Therefore, elucidation of the mechanisms regulating the functions of these GDS will represent the next major goal.

NOTE ADDED IN PROOF

A considerable number of reports dealing with mammalian and yeast GDSs of Ras or ras-like proteins have appeared after the submission of this manuscript. It is important to mention a few of them. Concerning new GDSs, a gene fragment from *Saccharomyces kluyveri* has been identified, sharing in the 863 C terminal amino acids 50% identity with CDC25 and capable of complementing the loss of the CDC25 function in *S. cerevisiae*.[73] Albright et al.[74] used sequences derived from CDC25 and ste6 to identify a cDNA from a mouse cell line, encoding a specific GDS of the ras-related proteins ral1A and ral1B (ralGDS). RalGDS can be phosphorylated on serine residues, but no evidence was found that this phosphorylation affected its activity toward the ralA and ralB GTPase. Cen et al.[75] isolated 6 full-length clones of cDNAs, which appear to be derived from the CDC25^{Mm} gene,[27] encoding products from 666 to 1260 amino acids. As an explanation, the authors propose that *CDC25^{Mm}* is a complex gene, whose transcription may be controlled by several promotors or by differential splicing. The various products could be tissue specific and have other functions in addition to GDS activity. A 241 amino acid region near the N terminus of the largest products was found to show homology to a domain shared by bcr, vav, dbl, and CDC24. The C terminal part (last 287 amino acid residues) of CDC25^{Mm} cleaved from

GST has been isolated in pure form and characterized biochemically (E. Jacquet and A. Parmeggiani, unpublished results). Shou et al.[76] cloned from a rat brain library a cDNA that encodes a 140-kDa exchange factor of ras p21 (Ras-GRF), of which the last 310 C terminal amino acids shows a 28% identity with CDC25 and 85% with CDC25^{Mn}. The C domain of this GDS, that was only detected in rat brain and not in other tissues, accelerated the nucleotide release of H- and N-ras p21 *in vitro* but not of the related GTP binding proteins RalA and CDC42Hs. As reported above for CDC25^{Mm}, the N terminal domain of Ras-GRF shows similarity to bcr and dbl (a GDS of CDC42Hs). Wei et al.[77] using degenerate oligonucleotides, encoding conserved sequences of CDC25-related gene products, identified fragments of human (*CDC25h*) and mouse cDNA (*CDC25m*) that were amplified via polymerase chain reaction. A chimeric molecule, part mouse and part yeast CDC25, can suppress the loss of the CDC25 function in *S. cerevisiae*. The known sequences of these two GDSs are virtually identical to H-GRF55 and CDC25^{Mm} (see also in the text). Schweighoffer et al.[78] have isolated a gene encoding a human GDS (H-GRF55) by screening a human brain cDNA library with oligonucleotide primers derived from the cDNA sequence of *CDC25^{Mm}*. Its cloning has led to the isolation of a 2.8 kb cDNA predicted to encode a protein of 488 amino acids, whose fusion product with glutathione-S-transferase acts *in vitro* as a specific GDS. *In situ* hybridization on human chromosomes revealed a localization on band 15 q2.4. Chardin and collaborators (personal communication) have isolated a human cDNA encoding two human proteins (hSOS1 and hSOS2) that are closely related to the drosophila ras exchange factor SOS and share a domain of homology with other ras exchange factor from yeast (CDC25 and SDC25) and rodents (CDC25^{Mm} or Ras-GRF), and can complement CDC25 deficiency in *S. cerevisiae*. These hSOS proteins are ubiquitously expressed in mammalian cells and may form a complex with the receptor tyrosine kinase associated protein GRB2. Two murine homologs of the drosophila *SOS* genes (mSOS-1 and mSOS-2) have been isolated and also found to be widely expressed.[79]

Concerning the properties of the *CDC25* and *SDC25* gene products, P. Poullet, J. B. Créchet, A. Bernardi, and A. Parmeggiani (unpublished results) have purified the SDC25 C domain to a high extent, despite its tendency to form inactive aggregates. The stable nucleotide-free complex between the SDC25 C domain and RAS2 or Ha-ras p21, a likely intermediate of the GDP to GTP exchange reaction, has been isolated on gel filtration and shown to be dissociated by GDP or GTP. Barlat et al.[80] observed that the SDC25 C domain functions as an oncoprotein in NIH3T3 cells. Schweighoffer et al.[81] reported that the SDC25 C domain overcomes the dominant inhibitory activity of mutant Ha-ras S17N. Yeast strains disrupted at the *RAS1* and *RAS2* loci, expressing H-ras p21 and the catalytic domain of GAP, and containing the *cdc25-2* mutation, were used to show that CDC25 stimulates the GDP-to-GTP exchange of H-ras p21.[82] Gross et al.[83] could raise antibodies against a β-galactosidase-CDC25 fusion protein inhibiting the GTP-dependent adenylate cyclase activity of *S. cerevisiae* and strongly cross-reacting with the SDC25 C domain and also with mammalian proteins of ca. 150 kDa from various tissues. A particularly interesting report by Gross et al.[84] described the important regulatory role of the phosphorylation of CDC25 in *S. cerevisiae*. By using highly selective anti-CDC25 antibodies they observed that the cell CDC25 is hyperphosphorylated within seconds in response to glucose by the cAMP-dependent kinase. The hyperphosphorylation relocalizes part of the CDC25 from the cell membrane to the cytoplasm.

ACKNOWLEDGMENTS

Cited work done in the laboratory of A. Parmeggiani was supported by grants of the Association pour la Recherche sur le Cancer, Ligue Nationale Française contre le Cancer, Institut National de la Santé et de la Recherche Médicale. We are indebted to Drs. M. Jacquet and E. Boy-Marcotte (Université Paris XI, Orsay), E. Martegani (Universitá di Milano), and P. H. Anborgh for fruitful discussion. We are grateful to Drs. M. Vanoni, M. Merola, and C. Lallemand for valuable help in carrying out the alignment in Figure 2.

REFERENCES

1. **Camonis, J., Kalékine, M., Gondré, B., Garreau, H., Boy-Marcotte, E., and Jacquet, M.,** Characterization, cloning and sequence analysis of the CDC25 gene which controls the cyclic AMP level of *Saccharomyces cerevisiae*, *EMBO J.*, 5, 375, 1986.
2. **Daniel, J. and Simchen, C.,** Clones from two different genomic regions complement the *cdc25* start mutation of *Saccharomyces cerevisiae*, *Curr. Genet.*, 10, 643, 1986.
3. **Martegani, E., Baroni, M. D., Frascotti, G., and Alberghina, L.,** Molecular cloning and transcriptional analysis of the start gene *CDC25* of *Saccharomyces cerevisiae*, *EMBO J.*, 5, 2363, 1986.
4. **Broek, D., Toda, T., Michaeli, T., Levin, L., Birchmeier, C., Zoller, M., Powers, S., and Wigler, M.,** The *Saccharomyces cerevisiae CDC25* gene product regulates the RAS/ adenylate cyclase pathway, *Cell*, 48, 789, 1987.
5. **Robinson, L. C., Gibbs, J. B., Marshall, M. S., Sigal, I. S., and Tatchell, K.,** *CDC25*: a component of the ras-adenylate cyclase pathway in *Saccharomyces cerevisiae*, *Science*, 235, 1218, 1987.
6. **Gibbs, J. B. and Marshall, M. S.,** The ras oncogenes — an important regulatory element in lower eukaryotic organisms, *Microbiol. Rev.*, 53, 171, 1989.
7. **Jacquet, M., Camonis, J., Boy-Marcotte, E., Damak, F., and Garreau, H.,** The cyclic AMP producing pathway in *Saccharomyces cerevisiae* involves *CDC25* and *RAS* genes products, in *The Guanine Nucleotide Binding Proteins: Common Structural and Functional Properties*, Bosch, L., Kraal, B., and Parmeggiani, A., Eds., NATO ASI series Vol. 165, Plenum Publishing, New York, 241, 1989.

8. **Barbacid, M.,** *ras* genes, *Annu. Rev. Biochem.,* 56, 779, 1987.

9. **Broach, J. R. and Deschenes, R. J.,** The function of RAS genes in *Saccharomyces cerevisiae, Adv. Cancer Res.,* 54, 79, 1990.

10. **Bourne, H. R., Sanders, D. A., and McCormick, F.,** The GTPase superfamily: a conserved switch for diverse cell functions, *Nature,* 238, 125, 1990.

11. **Bourne, H. R., Sanders, D. A., and McCormick, F.,** The GTPase superfamily: conserved structure and molecular mechanism, *Nature,* 239, 117, 1991.

12. **Mazon, M. J., Gancedo, J. M., and Gancedo, C.,** Phosphorylation and inactivation of yeast fructose-biphosphatase *in vivo* by glucose and by proton ionophoresis, *Eur. J. Biochem.,* 127, 605, 1982.

13. **Thevelein, J. M.,** Cyclic-AMP content and trehalase activation in vegetative cells and ascopores of yeast, *Arch. Microbiol.,* 138, 64, 1984.

14. **Van Aelst, L., Boy-Marcotte, E., Camonis, J., Thevelein, J. M., and Jacquet, M.,** The C-terminal Part of the *CDC25* gene product plays a key role in signal transduction in the glucose-induced modulation of cAMP level in *Saccharomyces cerevisiae, Eur. J. Biochem.,* 193, 675, 1990.

15. **Tanaka, K., Boris, K. L., Wood, D. R., and Tamanoi, F.,** IRA2, An upstream negative regulator of RAS in yeast, is a RAS GTPase-activating protein, *Proc. Natl. Acad. Sci. U.S.A.,* 88, 468, 1991.

16. **Camonis, J. and Jacquet, M.,** A new RAS mutation that suppresses the *CDC25* gene requirement for growth of *Saccharomyces cerevisiae, Mol. Cell. Biol.,* 8, 2980, 1988.

17. **Engelberg, D., Simchen, G., and Levitzki, A.,** *In vitro* reconstitution of CDC25 regulated *S. cerevisiae* adenylyl cyclase and its kinetic properties, *EMBO J.,* 9, 641, 1990.

18. **Jones, S., Vignais, M. L., and Broach, J. R.,** The CDC25 protein of *S. cerevisiae* promotes exchange of guanine nucleotides bound to Ras, *Mol. Cell. Biol.,* 11, 2641, 1991.

19. **Boy-Marcotte, E., Damak, F., Camonis, J., Garreau, H., and Jacquet, M.,** The C-terminal part of a gene partially homologous to CDC25 gene suppresses the *cdc25-5* mutation in *Saccharomyces cerevisiae, Gene,* 77, 21, 1989.

20. **Damak, F., Boy-Marcotte, E., Le-Roscquet, D., Guilbaud, R., and Jacquet, M.,** *SDC25,* a *CDC25*-like gene which contains a RAS-activating domain and is a dispensable gene of *Saccharomyces cerevisiae, Mol. Cell. Biol.,* 11, 202, 1991.

21. **Créchet, J.-B., Poullet, P., Mistou, M. Y., Parmeggiani, A., Camonis, J., Boy-Marcotte, E., Damak, F., and Jacquet, M.,** Enhancement of the GDP-GTP exchange reaction of RAS proteins by the carboxyl-terminal domain of SDC25, *Science,* 248, 866, 1990.

22. **Hughes, D. A., Fukui, Y., and Yamamoto, M.,** Homologous activators of ras in fission and budding yeast, *Nature,* 344, 355, 1990.

23. **Chant, J., Corrado, K., Pringle, J. R., and Herskowitz, I.,** The yeast *BUD5* gene, which encodes a putative GDP-GTP exchange factor, is necessary for bud-sites selection and interacts with bud-formation gene *BEM1, Cell,* 65, 1213, 1991.

24. **Wickner, W. J., Koh, T. J., Crowley, J. C., O'Neil, J., and Kaback, D. B.,** Molecular cloning of chromosome I DNA from *Saccharomyces cerevisiae:* isolation of the *MAK16* gene and analysis of an adjacent gene essential for growth at low temperatures, *Yeast,* 3, 51, 1987.

25. **Simon, M. A., Bowtell, D. D. L., Dodson, G. S., Laverty, T. R., and Rubin, G. M.,** Ras1 and putative guanine nucleotide exchange factor perform crucial steps in signaling by the sevenless protein tyrosine kinase, *Cell,* 67, 701, 1992.

26. **Bonfini, L., Karlovich, C. A., Dasgupta, C., and Banerjet, U.,** The Son of sevenless gene product: a putative activator of Ras, *Science,* 255, 603, 1992.

27. **Martegani, E., Vanoni, M., Zippel, R., Coccetti, P., Brambilla, R., Ferrari, C., Sturani, E., and Alberghina, L.,** Cloning by functional complementation of a mouse cDNA encoding a homolog of CDC25, a *Saccharomyces cerevisiae,* RAS activator, *EMBO J.,* 11, 2151, 1992.

28. **Garreau, H., Camonis, J., Guitton, C., and Jacquet, M.,** The *Saccharomyces cerevisiae* CDC25 gene product is a 180 kDa polypeptide and is associated with a membrane fraction, *FEBS Lett.,* 269, 53, 1990.

29. **Vanoni, M., Vavassori, M., Frascotti, G., Martegani, E., and Alberghina, L.,** Over-expression of the CDC25 gene, an upstream element of the RAS'Adenylyl cyclase pathway in *Saccharomyces cerevisiae,* allows immunological identification and characterization of its gene product, *Biochem. Biophys. Res. Commun.,* 172, 61, 1990.

30. **Roadeway, A. R. F., Sternberg, M. J. E., and Bentley, D. L.,** Similarity in membrane proteins, *Nature,* 342, 624, 1989.

31. **Daniel, J.,** The *CDC25* "start" gene of *Saccharomyces cerevisiae:* sequencing of the active C-terminal fragment and regional homologies with rhodopsin and cytochrome P450, *Curr. Genet.,* 10, 879, 1986.

32. **Munder, Th., Mink, M., and Küntzel, H.,** Domains of the *Saccharomyces cerevisiae CDC25* gene controlling mitosis and meiosis, *Mol. Gen. Genet.,* 214, 271, 1988.

33. **Shamor, C., Munder, Th., and Küntzel, H.,** Site-directed mutagenesis of the *Saccharomyces cerevisiae CDC25* gene: effects on mitotic growth and cAMP signalling, *Mol. Gen. Genet.,* 223, 426, 1990.

34. **Lisziewicz, J., Godany, A., Förster, R. H., and Kuntzel, H.,** Isolation and nucleotide sequence of a *Saccharomyces cerevisiae* protein kinase gene suppressing the cell cycle start mutation *cdc25, J. Biol. Chem.,* 262, 2549, 1987.

35. **Munder, Th. and Küntzel, H.,** Glucose-induced cAMP signaling in *Saccharomyces cerevisiae* is mediated by the CDC25 protein, *FEBS Lett.,* 242, 341, 1989.

36. **Petitjean, A., Hilger, F., and Tatchell, K.,** Comparison of thermosensitive alleles of the CDC25 gene involved in the cAMP metabolism of *S. cerevisiae, Genetics,* 124, 797, 1990.

37. **Créchet, J. B., Poullet, P., Camonis, J., Jacquet, M., and Parmeggiani, A.,** Different kinetic properties of the two mutants RAS2Ile152 and RAS2Val19, that suppress the CDC25 requirement in RAS/adenylate cyclase pathway in *Saccharomyces cerevisiae, J. Biol. Chem.,* 265, 1563, 1990.

38. **Frascotti, G., Coccetti, P., Vanoni, M. A., Alberghina, L., and Martegani, E.,** The Overexpression of the 3' terminal region of the *CDC25* gene of *S. cerevisiae* causes growth inhibition and alteration of purine nucleotides pools, *Biochim. Biophys. Acta,* 1089, 206, 1991.

39. **Powers, S. K., O'Neill, K., and Wigler, M.,** Dominant yeast and mammalian RAS mutants that interfere with the CDC25-dependent activation of wild-type Ras in *Saccharomyces cerevisiae, Mol. Cell Biol.,* 9, 390, 1989.

40. **Bender, A. and Pringle, J. R.,** Use of a screen for synthetic lethal and multicopy suppressor mutants to identify two new genes involved in morphogenesis in *Saccharomyces cerevisiae, Mol. Cell. Biol.,* 11, 1295, 1991.

41. **Johnson, D. and Pringle, J. R.,** Molecular characterization of *CDC42,* a gene involved in the development of cell polarity, *J. Cell Biol.,* 111, 143, 1990.

42. **Powers, S., Gonzales, E., Christensen, T., Cubert, J., and Broek, D.,** Functional cloning of *BUD5,* a *CDC25*-related gene from *S. cerevisiae* that can suppress a dominant-negative RAS2 mutant, *Cell,* 65, 1225, 1991.

43. **Jacquet, E., Vanoni, M., Ferrari, C., Alberghina, L., Martegani, E., and Parmeggiani, A.,** A mouse CDC25-like product enhances the formation of the active GTP complex of human ras p21 and *Saccharomyces cerevisiae* RAS2 proteins, *J. Biol. Chem.,* 267, 24181, 1992.

44. **Jacquet, E. and Parrini, M. C.,** unpublished results.

45. **Boy-Marcotte, E., Buu, A., Soustelle, Ch., Poullet, P., Parmeggiani, A., and Jacquet, M.,** The C-terminal part of the *CDC25* gene product has a Ras-nucleotide exchange activity when present in a chimeric SDC25-CDC25 protein, *Current Genetics,* in press.

46. **Poullet, P.,** Ph.D. thesis, University of Paris XI, 1992.

47. **Jonák, J.,** personal communication.

48. **Antonny, B., Chardin, P., Roux, M., and Chabre, M.,** GTP hydrolysis mechanisms in ras p21 and in the ras-GAP complex studied by fluorescense measurements on tryptophan mutants, *Biochemistry,* 30, 8287, 1991.

49. **Rey, I., Schweigehoffer, F., Barlat, I., Camonis, J., Boy-Marcotte, E., Guilbaud, R., Jacquet, M., and Tocquet, B.,** The C-domain of the product of the *Saccharomyces cerevisiae SDC25* gene elicits activation of p21 ras proteins in mammalian cells, *Oncogene,* 6, 347, 1991.

50. **Goodrich, G. A. and Burrel, H. R.,** Micromeasurement of nucleoside 5′-triphosphates using coupled bioluminescence, *Anal. Biochem.,* 127, 395, 1982.

51. **Proud, C. G.,** Guanine nucleotides, protein phosphorylation and the control of translation, *Trends Biochem. Sci.,* 11, 73, 1986.

52. **Miller, D. L. and Weissbach, H.,** Factors involved in the transfer of aminoacyl-tRNA to the ribosome, in *Molecular Mechanisms in Protein Biosynthesis,* Weissbach, H. and Petska, S., Eds., Academic Press, New York, 323, 1977.

53. **Kaziro, Y.,** The role of guanosine 5′-triphosphate in polypeptide chain elongation, *Biochem. Biophys. Acta,* 505, 95, 1978.

54. **Chau, V., Romero, G., and Biltonen, R. L.,** Kinetic studies on the interaction of *Escherichia coli* K12 elongation factor Tu with GDP and elongation factor Ts, *J. Biol. Chem.,* 256, 5591, 1981.

55. **Hwang, Y. W. and Miller, D. L.,** A study of the kinetic mechanism of elongation factor Ts, *J. Biol. Chem.,* 260, 11498, 1985.

56. **Wessling-Resnick, M., Kelleher, D. J., Weiss, E. R., and Johnson, G. L.,** Enzymatic model for receptor activation of GTP-binding regulatory proteins, *Trends, Biol. Sci.,* 12, 473, 1987.

57. **Lucas-Lenard, J. and Lipmann, F.,** Protein biosynthesis, *Annu. Rev. Biochem.,* 40, 409, 1971.

58. **Mistou, M. Y., Jacquet, E., Poullet, P., Rensland, H., Gideon, P., Schlichting, I., Wittinghofer, A., and Parmeggiani, A.,** Mutations of Ha-ras p21, that define important regions for the molecular mechanism of the SDC25 C-domain, a guanine nucleotide dissociation stimulator, *EMBO J.,* 11, 2391, 1992.

59. **Downward, J., Riehl, R., Wu, L., and Weinberg, R. A.,** Identification of a nucleotide exchange-promoting activity for p21^{ras}, *Proc. Natl. Acad. Sci. U.S.A.,* 87, 5998, 1990.

60. **Hiroyoshi, M., Kaibuchi, K., Kjawamura, S., Hata, Y., and Takai, Y.,** Role of the C-terminal region of smg p21, a ras-like small GTP-binding protein, in membrane and smg p21 GDP/GTP exchange protein interactions, *J. Biol. Chem.,* 266, 2962, 1991.

60a. **Mizuno, T., Kaibuchi, K., Yamamoto, T., Kawamura, M., Sakoda, T., Fujioka, H., Matsuura, Y., and Takai, Y.,** A stimulatory GDP/GTP exchange protein for smg p21 is active on the post-translationally processed form of c-Ki-ras p21 and rhoA p21, *J. Biol. Chem.,* 88, 6442, 1991.

61. **Beitel, G. J., Clark, S. G., and Horvitz, H. R.,** *Caenorhabditis elegans ras* gene *let-60* acts as switch in the pathway of vulval induction, *Nature,* 348, 503, 1991.

62. **Willumsen, B. M., Vass, W. C., Velu, T. J., Papageorge, A. G., Schiller, J. T., and Lowy, D. R.,** The bovine papillomavirus E5 oncogene can cooperate with *ras:* identification of p21 amino acids critical for transformation by c-ras^H but not v-ras^H, *Mol. Cell. Biol.,* 11, 6026, 1991.

63. **Verrotti, A., Créchet, J.-B., Di Blasi, F., Mirisola, M., Seidita, G., Kavounis, C., Nastopoulos, V., Burderi, E., De Vendittis, E., Parmeggiani, A., and Fasano, O.,** Ras residues that are distant from the GDP binding site play a critical role in exchange factor-stimulated release of GDP, *EMBO J.,* 11, 2855, 1992.

64. **Daniel, J., Becker, J. M., Enari, E., and Lewitzky, A.,** The activation of adenylate cyclase by guanyl nucleotides in *Saccharomyces cerevisiae* is controlled by *CDC25* start gene product, *Mol. Cell. Biol.,* 7, 3857, 1987.

65. **Créchet, J. B., Poullet, P., Bernardi, A., Fasano, O., and Parmeggiani, A.,** Properties of the SDC25 C-domain, a GDP to GTP exchange factor of *S. cerevisiae* Ras proteins, and modulation of adenylyl cyclase activity, *J. Biol. Chem.,* in press.

66. **Chardin, P.,** The *ras* superfamily of proteins, *Biochimie,* 70, 865, 1988.

67. **Hall, A.,** The cellular functions of small GTP-binding proteins, *Science,* 249, 635, 1990.
68. **Huang, Y. K., Kung, H. F., and Kamata, T.,** Purification of a factor capable of stimulating the guanine nucleotide exchange reaction of ras proteins and its effect on ras-related small molecular mass G proteins, *Proc. Natl. Acad. Sci. U.S.A.,* 87, 8008, 1990.
69. **Hiraoka, K., Kaibuchi, K., Ando, S., Musha, T., Takaishi, K., Mizuno, T., Asada, M., Ménard, L., Tomhave, E., Didsburg, J., Snyderman, R., and Takai, Y.,** Both stimulatory and inhibitory GDP/GTP exchange protein, smg GDS and rho GDI, are active on multiple small GTP-binding proteins, *Biochem. Biophys. Res. Commun.,* 182, 921, 1992.
70. **Yamamoto, T., Kaibuchi, K., Mizuno, T., Hiroyoshi, M., Shirataki, H., and Takai, Y.,** Purification and characterization from bovine brain cytosol of proteins that regulate the GDP/GTP exchange reaction of *smg* p21s, *ras* p21-like GTP-binding proteins, *J. Biol. Chem.,* 265, 16,626, 1990.
71. **West, M., Kung, H. F., and Kamata, T.,** A novel membrane factor stimulates guanine nucleotide exchange reaction of ras proteins, *FEBS Lett.,* 259, 245, 1990.
72. **Wolfman, A. and Macara, I. G.,** A cytosolic protein catalyzes the release of GDP from p21^{ras}, *Science,* 248, 67, 1990.
73. **Prigozy, T., Gonzales, E., and Broek, D.,** *Gene,* 117, 67, 1992.
74. **Albright, C. F., Giddings, B. W., Liu, J., Vito, M., and Weinberg, R. A.,** *EMBO J.,* 12, 339, 1993.
75. **Cen, H., Papageorge, A. G., Zippel, R., Lowy, D. R., and Zhang, K.,** *EMBO J.,* 11, 4007, 1992.
76. **Shou, C., Farnsworth, C. L., Neel, B. G., and Feig, L. A.,** *Nature,* 358, 351, 1992.
77. **Wei, W., Mosteller, R. D., Sanyal, P., Gonzales, E., McKinney, D., Dasgupta, C., Li, P., Liu, B.-X., and Broek, D.,** *Proc. Natl. Acad. Sci. U.S.A.,* 89, 7100, 1992.
78. **Schweighoffer, F., Faure, M., Fath, I., Chevallier-Multon, M. C., Apiou, F., Dutrillaux, B., Martegani, E., Jacquet, M., and Tocque, B.,** *Oncogene,* in press, 1993.
79. **Bowtell, D., Fu, P., Simon, M., and Senior, P.,** *Proc. Natl. Acad. Sci. U.S.A.,* 89, 6511, 1992.
80. **Barlat, I., Schweighoffer, F., Chevallier-Multon, M. C., Duchesne, M., Fath, I., Landais, D., Jacquet, M., and Tocque, B.,** *Oncogene,* in press.
81. **Schweighoffer, F., Cai, H., Chevallier-Multon, M. C., Fath, I., Cooper, G., and Tocque, B.,** *Mol. Cell Biol.,* 13, 39, 1993.
82. **Segal, M., Marbach, I., Engelberg, D., Simchen, G., and Levitzki, A.,** *J. Biol. Chem.,* 267, 22747, 1992.
83. **Gross, E., Marbach, I., Engelberg, D., Segal, M., Simchen, G., and Levitzki, A.,** *Mol. Cell Biol.,* 12, 2653, 1992.
84. **Gross, E., Goldberg, D., and Levitzki, A.,** *Nature,* 360, 762, 1992.

Chapter 27

GDP/GTP EXCHANGE PROTEINS FOR SMALL GTP-BINDING PROTEINS

Yoshimi Takai, Kozo Kaibuchi, Akira Kikuchi, Masahito Kawata, Takuya Sasaki, and Takeshi Yamamoto

TABLE OF CONTENTS

489

I. INTRODUCTION

Small GTP-binding proteins (G proteins) have GDP-bound inactive and GTP-bound active forms which are interconvertible by GDP/GTP exchange and GTPase reactions. The former reaction is regulated by GDP/GTP exchange proteins (GEPs), and the latter reaction is catalyzed by small G proteins by themselves and is regulated by GTPase-activating and -inhibiting proteins (GAPs and GIPs, respectively; Figure 1). The rate-limiting step for the GDP/GTP exchange reaction is the dissociation of GDP from small G proteins. Since the intracellular concentration of GTP is much higher than that of GDP, once GDP dissociates from small G proteins, GTP is associated with the guanine nucleotide-free form of small G proteins. Two types of GEPs have been identified: one is a stimulatory type, named GDP dissociation stimulator (GDS), and the other is an inhibitory type, named GDP dissociation inhibitor (GDI). GDS stimulates the dissociation of GDP from and thereby the subsequent binding of GTP to each small G protein, whereas GDI inhibits the dissociation of GDP from and thereby, the subsequent binding of GTP to each small G protein. GDS has been found for *ras* p21, *smg* p21 (identical to the *rap1* protein), and *rho* p21, and GDI has been found for *smg* p25A (identical to the *rab3A* protein) and *rho* p21. Among these regulatory proteins, *smg* p21 GDS and *smg* p25A GDI have been most extensively investigated. In this article, we mainly review these two regulatory proteins and briefly *ras* p21 GDS, *rho* p21 GDS, and *rho* p21 GDI.

II. *smg* p21 GDS

A. PHYSICAL PROPERTIES

We have purified *smg* p21 GDS to near homogeneity from bovine brain cytosol.[1] The molecular weight estimated by sodium dodecylsulfate polyacrylamide gel electrophoresis (SDS-PAGE) and the S value are nearly the same at about 53,000. We have also cloned its cDNA from a bovine brain cDNA library and determined its primary structure.[2] The protein is composed of 558 amino acids with a calculated molecular weight of about 61,000. Homology search has revealed that *smg* p21 GDS shares a low amino acid sequence homology with the yeast *CDC25* and *SCD25* proteins, which have been suggested to serve as GEPs for the yeast *RAS2* protein[3,4] and the *IRA1* and *NF 1* proteins, which belong to the family of *ras* p21 GAP,[5,6] but not with *ras* p21 GAP,[7,8] *smg* p25A GDI,[9] *rho* p21 GDI,[10] and the βγ-subunits of heterotrimeric G proteins such as G_s and G_i.[11]

B. THE ACTIONS

smg p21 GDS forms a complex with both the GDP- and GTP-bound forms of *smg* p21 at a molar ratio of 1:1 as tested with *smg* p21B.[12] *smg* p21 GDS stimulates the dissociation of both GDP and GTP to *smg* p21A and -B and the binding of both nucleotides to each *smg* p21 to the same degrees in a cell-free

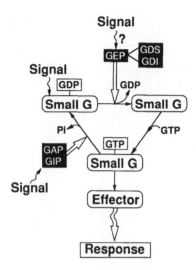

FIGURE 1. Mode of activation of small G proteins.

system.[1] However, since the intracellular concentration of GTP is much higher than that of GDP, *smg* p21 GDS may stimulate the GDP/GTP exchange reaction of *smg* p21 in intact cells.

smg p21 binds to membranes through a hydrophobic and polybasic C terminal region as described below when examined with *smg* p21B.[13] *smg* p21 GDS forms a complex with both the GDP- and GTP-bound forms of *smg* p21B and thereby inhibits its binding to membranes and induces its dissociation from the membranes in a cell-free system using the purified *smg* p21B, *smg* p21 GDS, and synaptic plasma membranes.

C. THE C TERMINAL REGION OF *smg* p21 AND THE *smg* p21 GDS ACTIONS

Both *smg* p21A and -B have a CAAX C terminal structure[14,15] where A is an aliphatic amino acid and X is any amino acid. *smg* p21B is posttranslationally modified: it is first geranylgeranylated followed by removal of the AAX portion and the subsequent carboxymethylation of the exposed cysteine residue (Figure 2).[16] *smg* p21B has a polybasic region upstream of the prenylated cysteine residue but lacks the second cysteine residue in the C terminal region and is not palmitoylated. The posttranslational modifications of *smg* p21B are similar to those of Ki-*ras* p21 that also has a polybasic region and is not palmitoylated.[17-19] In the case of Ha- and N-*ras* p21s, they lack the polybasic region but have the second cysteine residue which is palmitoylated.[19]

When *smg* p21B is digested with trypsin to a limited extent, it is cleaved at the site of Arg^{176}-Lys^{177}, resulting in the formation of the N terminal fragment with a molecular weight of 20,500 and the C terminal fragment with a molecular weight of about 1,000.[13] The N terminal fragment has the consensus amino acid

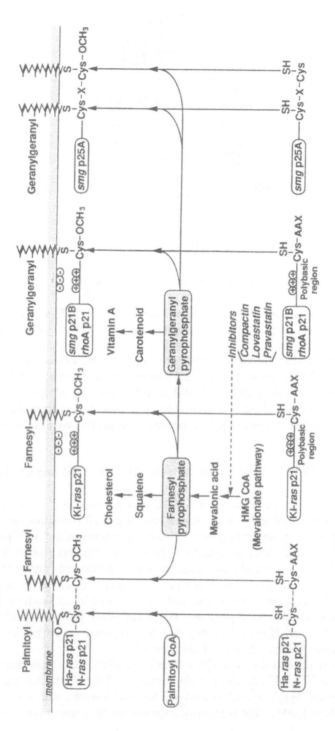

FIGURE 2. Posttranslational modifications of the C terminal region of small G proteins.

sequences for GDP/GTP-binding and GTPase activities and indeed exhibits these activities but does not bind to membranes anymore. In contrast, the C terminal fragment lacks the GDP/GTP-binding and GTPase activities but binds to membranes. Moreover, *smg* p21 GDS is inactive for the N terminal fragment, indicating that the C terminal region, most presumably either or both of the polybasic and/or hydrophobic regions, is essential for the *smg* p21 GDS action.[13] Consistently, *smg* p21 GDS is inactive for the posttranslationally unmodified *smg* p21B, which is purified from the soluble fraction of the insect cells overproducing *smg* p21B by use of a baculovirus expression system, under the conditions where *smg* p21 GDS is active for the posttranslationally modified *smg* p21B, which is purified from the membrane fraction of the insect cells.[20] In contrast, *smg* p21 GAP is active for both the N terminal fragment and the posttranslationally unmodified protein, indicating that the C terminal region is not necessary for the *smg* p21 GAP action.[13]

D. PHOSPHORYLATION OF *smg* p21 AND THE *smg* p21 GDS ACTION

Both *smg* p21A and -B are phosphorylated by A-kinase.[21-23] The site of this phosphorylation of *smg* p21B is Ser[179] in the C terminal region.[24] There are clustered basic amino acids just upstream of the phosphorylated serine residue. The phosphorylation of this site reduces the membrane-binding activity of *smg* p21B. Presumably, the polybasic region may interact with the acidic polar head groups of membrane phospholipids and thereby facilitate the membrane-binding activity of the geranylgeranylated *smg* p21B. The phosphorylation of the serine residue may reduce this ionic interaction.

In addition, the A-kinase-catalyzed phosphorylation of *smg* p21 makes the *smg* p21 GDS action sensitive to inhibiting the binding of *smg* p21B to membranes.[24] In intact resting platelets, *smg* p21B is mostly found in the membrane fraction. Upon stimulation of platelets by cAMP-elevating prostaglandin E_1 and dibutyryl cAMP, the phosphorylated form of *smg* p21B is recovered in the cytosol fraction, suggesting that the A-kinase-catalyzed phosphorylation of *smg* p21B induces its translocation from the membranes to the cytoplasm in intact platelets.[24,25] Moreover, the A-kinase-catalyzed phosphorylation of *smg* p21B enhances the *smg* p21 GDS action to stimulate the GDP/GTP exchange reaction. The phosphorylation of *smg* p21B by itself does not affect the dissociation of GDP from and the binding of GTP to *smg* p21B.

E. MODE OF ACTIVATION OF *smg* p21 BY *smg* p21 GDS

A possible mode of activation of *smg* p21 in intact cells is schematically shown in Figure 3. In resting cells, the GDP-bound inactive form of *smg* p21 binds to membrane. When A-kinase is activated, *smg* p21 is phosphorylated and becomes sensitive to the *smg* p21 GDS action, resulting in the complex formation with *smg* p21 GDS. The complex then dissociates from the membrane to the cytoplasm. The GDP/GTP exchange reaction is induced by *smg* p21 GDS, and

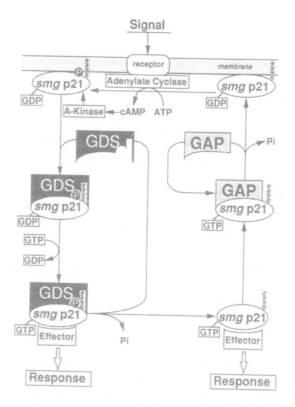

FIGURE 3. Possible mode of activation of *smg* p21.

the GDP-bound inactive form is converted to the GTP-bound active form. This form can still be complexed with *smg* p21 GDS and this complex recognizes its specific effector protein and exerts its biological actions. *smg* p21 is then dephosphorylated resulting in the dissociation of *smg* p21 GDS from *smg* p21. The GTP-bound active form free of *smg* p21 GDS can, of course, also recognize its specific effector protein. This form is then complexed with *smg* p21 GAP and is converted to the GDP-bound inactive form which finally binds to the membrane. Thus, the *smg* p21 activity is regulated in this way by these two types of regulatory protein and the A-kinase-catalyzed phosphorylation of *smg* p21. It has recently been reported that *smg* p21B tightly binds to the membrane skeleton after aggregation of human platelets by thrombin in the presence of Ca^{2+} in the medium,[26] but we have not confirmed this result.

F. POSSIBLE FUNCTIONS OF *smg* p21

It is well known that thrombin and collagen induce platelet activation and cAMP-elevating prostaglandins antagonize this activation.[27] Both the C-kinase and Ca^{2+} pathways are involved in this platelet activation and A-kinase antagonizes this activation. This biological response system is generally called,

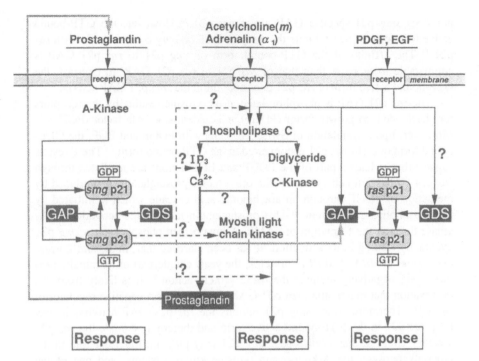

FIGURE 4. Possible crosstalk among the A-kinase, C kinase, Ca^{2+}, and tyrosine kinase systems through *smg* p21.

"bidirectional system", and is present not only in platelets but also in many other cell types such as smooth muscle cells and lymphocytes. At least four sites of the A-kinase action are possible: the receptor-linked phospholipase C activation, the IP_3-regulated Ca^{2+} mobilization, calmodulin-dependent myosin light chain kinase activation, and the downstream of the C-kinase and Ca^{2+} pathways. Since *smg* p21 is at the downstream of A-kinase, it is possible that *smg* p21 plays a role in the crosstalk between the A-kinase system and the C-kinase and Ca^{2+} system as schematically shown in Figure 4.

In contrast to the biological response system called the bidirectional system, there is another type of system called the "mono-directional system", and this system is present in many cell types such as pancreatic acinar cells, parotid gland cells, and hepatocytes.[27] In this system, A-kinase does not antagonize the C-kinase and Ca^{2+} pathways and by itself exerts its specific actions. In these cell types, *smg* p21 may mediate at least a part of the A-kinase actions in a manner independent of C-kinase and Ca^{2+}.

On the other hand, *smg* p21B and the *rap1A* proteins (identical to *smg* p21A) inhibit the *ras* p21 GAP activity in a cell-free system.[28,29] *ras* p21 GAP interacts with at least three regions of *ras* p21 and one of them is the effector domain.[30] Although both *smg* p21A and -B have the same amino acid sequence as that of the effector domain of *ras* p21, *ras* p21-specific GAP does not affect *smg* p21 nor does *smg* p21-specific GAP affect *ras* p21.[28] However, the GTP-bound active form of *smg* p21 inhibits the *ras* p21 GAP activity competitively with *ras* p21.[28] The affinity of the GTP-bound form of *smg* p21 to *ras* p21 GAP is comparable to or more than that of *ras* p21 to *ras* p21 GAP. In contrast, the GTP-bound active form of *ras* p21 does not inhibit the *smg* p21 GAP activity.[28,31]

The *ras* p21 GAP is phosphorylated at its tyrosine residue by the receptors for platelet-derived growth factor (PDGF) and epidermal growth factor (EGF).[32-34] Moreover, upon stimulation of Swiss 3T3 cells with PDGF and EGF, the GDP-bound form of c-Ha-*ras* p21 is converted to the GTP-bound form.[35] These results suggest that at least a part of the PDGF and EGF actions are mediated through *ras* p21. In certain cell types, on the other hand, prostaglandin is produced by the action of PDGF through an arachidonic acid cascade which is initiated by the phospholipase C system.[36] This prostaglandin induces the activation of A-kinase through the formation of cAMP. A-kinase then phosphorylates *smg* p21 eventually leading to its activation. It has been clarified that, in *Saccharomyces cerevisiae*, the *IRA1* and *IRA2* proteins, the yeast counterpart of mammalian *ras* p21 GAP, negatively regulate the *RAS2* protein action.[37] It is likely from this observation that mammalian *ras* p21 GAP also serves as a negative regulator of *ras* p21. Therefore, since *smg* p21 inhibits the *ras* p21 GAP activity, it may keep *ras* p21 in the GTP-bound active form and thereby potentiate the *ras* p21 actions. Thus, in these cell types, *smg* p21 may play an important role in the crosstalk between the A-kinase and tyrosine kinase systems, and one of the potential effector proteins of *smg* p21 may be *ras* p21 GAP as schematically shown in Figure 4. Of course, it is possible that *smg* p21 antagonizes the *ras* p21 actions as described for the K*rev*-1 protein (identical to *smg* p21A).[38] These apparently different actions of *smg* p21 may depend on cell types. Moreover, it is possible that *smg* p21 recognizes its specific effector proteins and exerts its specific actions in a manner independent of C-kinase, Ca^{2+}, and tyrosine kinase in certain cell types.

III. *smg* p25A GDI

A. PHYSICAL PROPERTIES

We have purified *smg* p25A GDI to near homogeneity from bovine brain cytosol.[39] The molecular weight estimated by SDS-PAGE and the S value are about 54,000 and 65,000, respectively. We have also cloned its cDNA from a bovine brain cDNA library and determined its primary structure.[9] The protein is composed of 447 amino acids with a calculated molecular weight of about 51,000. Homology search has revealed that *smg* p25A GDI shares a low amino acid sequence homology with *rho* p21 GDI[10] and the *CDC25* and *SCD25* proteins,[3,4] but not with the *IRA1* and *NF 1* proteins,[5,6] *ras* p21 GAP,[7,8] and the βγ-subunits of heterotrimeric G proteins.[11]

B. THE ACTIONS

smg p25A GDI forms a complex with only the GDP-bound form of *smg* p25A at a molar ratio of 1:1.[40] *smg* p25A GDI inhibits the dissociation of GDP from and the subsequent-binding of GTP to *smg* p25A. Thus, *smg* p25A GDI inhibits the GDP/GTP exchange reaction of *smg* p25A.

smg p25A binds to membranes through a hydrophobic C terminal region as described below. *smg* p25A GDI forms a complex with the GDP-bound form of *smg* p25A and thereby inhibits its binding to membranes and moreover induces its dissociation from the membranes.[40]

C. THE C TERMINAL REGION OF *smg* p25A AND THE *smg* p25A GDI ACTIONS

smg p25A has a CXC C terminal structure which is posttranslationally modified: both the cysteine residues are geranylgeranylated and the C terminal cysteine residue is carboxymethylated (Figure 2).[41] When purified *smg* p25A is digested with *Achromobacter* protease I to a limited extent, *smg* p25A is cleaved at the site of Lys[186]-Met[187], and the N terminal fragment with a molecular weight of 20,000 and the C terminal fragment with a molecular weight of about 2000 are produced.[42] The N terminal fragment has the consensus amino acid sequences for GDP/GTP-binding and GTPase activities and indeed exhibits these activities but does not bind to membranes anymore. In contrast, the C terminal fragment lacks the GDP/GTP-binding and GTPase activities and membrane-binding activity, suggesting that both the N and C terminal regions are essential for the binding of *smg* p25A to membranes. Moreover, *smg* p25A GDI is inactive for the N terminal fragment, suggesting that the C terminal region, most presumably the hydrophobic region, is essential for the *smg* p25A GDI action. This conclusion is also supported by the fact that *smg* p25A synthesized in *Escherichia coli* lacking the posttranslational modifications of the C terminal region neither binds to membranes nor is sensitive to the *smg* p25A GDI action.[42]

D. MODE OF ACTIVATION OF *smg* p25A BY *smg* p25A GDI

A possible mode of activation of *smg* p25A in intact cells is schematically shown in Figure 5. *smg* p25A is present in cytoplasm in the GDP-bound inactive form which is complexed with *smg* p25A GDI. When a signal comes to either *smg* p25A or *smg* p25A GDI, *smg* p25A dissociates from *smg* p25A GDI. Subsequently, *smg* p25A is first converted to the GTP-bound active form and then binds to membrane. The GTP-bound active form interacts with the putative effector protein which may be present in the membrane. The GTP-bound active form of smg p25A is complexed with GAP[43] and is converted to the GDP-bound inactive form. This form of *smg* p25A then becomes complexed with *smg* p25A GDI, and the complex dissociates from the membrane and is translocated to the cytoplasm. Thus, the *smg* p25A activity may be regulated by its GDI and GAP in a cyclical manner. Consistent with our hypothesis, it has recently been reported that in intact synaptosomes the *rab3A* protein (identical to *smg* p25A) dissociates from the synaptic vesicles during the neurotransmitter release.[44]

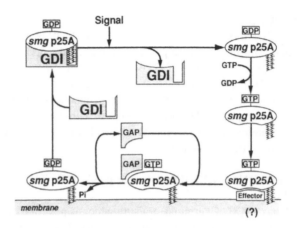

FIGURE 5.   Possible mode of activation of *smg* p25A.

E. POSSIBLE FUNCTIONS OF *smg* p25A AND *smg* p25A GDI

In mammalian cells, there are two types of secretion: one is a secretagogue-regulated secretion type and another is a constitutive secretion type.[45] Neurons, exocrine and endocrine cells are regulated secretory cells whereas hepatocytes and lymphocytes are constitutive secretory cells. *smg* p25A is mainly found in secretory cells with a regulated secretion type.[46] Particularly, in neurons, it is mainly localized in presynaptic membranes and synaptic vesicles and partly in the presynaptic cytosol.[47] The similar observation has been reported for the *rab3A* protein (identical to *smg* p25A).[48]

Ca^{2+} is involved in the last stage of secretory processes, that is the fusion of the secretory vesicles with plasma membranes.[49] C-kinase enhances this process.[27] On the other hand, it has recently been reported that GTP and its stable GTP analogs stimulate secretion, presumably the fusion of secretory vesicles with the plasma membrane, in permeabilized cells, whereas the fusion processes among the Golgi membranes are mostly inhibited by the stable GTP analog.[50-52] Endocytosis is also affected by GTP and its stable analogs.[48] The synthetic peptides designed from the putative effector domain of the *rab* proteins inhibit the fusion processes of intracellular vesicles of CHO cells.[48] Accumulating evidence suggests that G proteins, presumably small G proteins, may be involved in the exoendocytotic processes. It is likely that *smg* p25A as well as the *rab* proteins may be the most probable candidates that may regulate exoendocytotic processes as schematically shown in Figure 6.

We have recently found that *smg* p25A GDI is present in the secretory cells with both regulated and constitutive secretion types[53] and that *smg* p25A GDI recognizes not only *smg* p25A, but also a rat liver small G protein similar to but different from *smg* p25A and the yeast *SEC4* protein in a cell-free purified system.[54,55] These results suggest that *smg* p25A GDI is also involved in secretory processes.

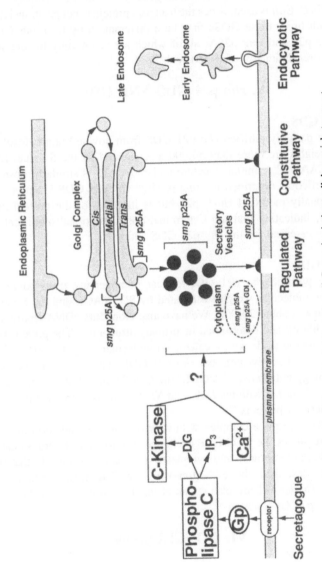

FIGURE 6. Possible mode of action of small G proteins in intracellular vesicle transport.

IV. *ras* p21 GDS

There are four papers reporting the existence of *ras* 21 GDS, and in one of them the protein is purified to near homogeneity.[56-59] In these studies, *ras* p21 synthesized in *E. Coli* is used. Since the bacterial protein is not posttranslationally modified with lipid, these GDSs may be a different entity from our GDS and GDI for *smg* p21 and *smg* p25A, all of which are active only for the small G proteins posttranslationally modified with lipid.

V. *rho* p21 GDS AND GDI

A. *rho* p21 GDS

We have partially purified *rho* p21 GDS from bovine brain cytosol.[60] The molecular weights are more than 50,000 as estimated by the S value. *rho* p21 has also a CAAX C terminal structure and *rho*A p21 is similarly posttranslationally modified as described for *smg* p21B.[61] *rho* p21 GDS is active for this posttranslationally modified *rho*A p21 but is inactive for the protein purified from *E. coli,* indicating that the C terminal posttranslational modifications of *rho*A p21 are essential for the *rho* p21 GDS action.[62]

B. *rho* p21 GDI

We have purified *rho* p21 GDI to near homogeneity from bovine brain cytosol.[63] The molecular weights estimated by SDS-PAGE and the S value are nearly the same at about 27,000. We have also cloned its cDNA from a bovine brain cDNA library and determined its primary structure.[10] The protein is composed of 204 amino acids with a calculated molecular weight of about 23,000. Homology search has revealed that *rho* p21 GDI shares a low amino acid sequence homology with *smg* p25A GDI,[9] the *CDC25* and *SCD25* proteins,[3,4] and *ras* p21 GAP,[7,8] but not with the *IRA1* and *NF 1* proteins,[5,6] and the βγ-subunits of heterotrimeric G proteins.[11]

rho p21 GDI forms a complex with only the GDP-bound form of *rho* p21 and thereby regulates the GDP/GTP exchange reaction and the translocation of *rho* p21 between membranes and cytoplasm in a manner similar to that of *smg* p25A GDI as described above.[64] Moreover, the C terminal posttranslational modifications of *rho* p21 are also essential for the *rho* p21 GDI action as well as for the *rho* p21 GDS action.[62]

VI. CONCLUSION

There are at least two types of GDP/GTP exchange proteins for each small G protein named GDS and GDI, in addition to GAP and GIP. The posttranslational modifications of the C terminal region of each small G protein are essential not only for its membrane-binding activity but also for the action of each GEP. While the *ras* p21 activity is regulated by tyrosine kinases of the PDGF and

EGF receptors, the *smg* p21 activity is regulated by A-kinase. Accumulating evidence suggests that either small G proteins by themselves or their regulatory proteins are regulated by membrane receptors or their intracellular signal transduction systems and thus constitute a huge network with them for intracellular regulatory mechanism. It is important to clarify the direct linkage of each small G protein with receptors and other intracellular messenger systems.

ACKNOWLEDGMENTS

We are grateful to Junko Yamaguchi for her skillful secretarial assistance. We apologize in advance for the incompleteness of the referencing because of space constraints.

REFERENCES

1. **Yamamoto, T., Kaibuchi, K., Mizuno, T., Hiroyoshi, M., Shirataki, H., and Takai, Y.**, Purification and characterization from bovine brain cytosol of proteins that regulate the GDP/GTP exchange reaction of *smg* p21s, *ras* p21-like GTP-binding proteins, *J. Biol. Chem.*, 265, 16626, 1990.

2. **Kaibuchi, K., Mizuno, T., Fujioka, H., Yamamoto, T., Kishi, K., Fukumoto, Y., Hori, Y., and Takai, Y.**, Molecular cloning of the cDNA for stimulatory GDP/GTP exchange protein for *smg* p21s (*ras* p21-like small GTP-binding proteins) and its characterization of stimulatory GDP/GTP exchange protein, *Mol. Cell. Biol.*, 11, 2873, 1991.

3. **Broek, D., Toda, T., Michaeli, T., Levin, L., Birchmeier, C., Zoller, M., Powers, S., and Wigler, M.**, The *S. cerevisiae CDC25* gene product regulates the RAS/adenylate cyclase pathway, *Cell*, 48, 789, 1987.

4. **Créchet, J.-B., Poullet, P., Mistou, M.-Y., Parmeggiani, A., Camonis, J., Boy-Marcotte, E., Damak, F., and Jacquet, M.**, Enhancement of the GDP/GTP exchange of *RAS* proteins by the carboxyl-terminal domain of *SCD25*, *Science*, 248, 866, 1990.

5. **Tanaka, K., Matsumoto, K., and Toh-e, A.**, *IRA1*, an inhibitory regulator of the *ras*-cyclic AMP pathway in *Saccharomyces cerevisiae*, *Mol. Cell. Biol.*, 9, 757, 1989.

6. **Cawthon, R., Weiss, R., Xu, G., Viskochil, D., Culver, M., Stevens, J., Robertson, M., Dunn, D., Gesteland, R., O'Connell, P., and White, R.**, A major segment of the neurofibromatosis type 1 gene: cDNA sequence, genomic structure, and point mutations, *Cell*, 62, 193, 1990.

7. **Vogel, U. S., Dixon, R. A. F., Schaber, M. D., Diehl, R. E., Marshall, M. S., Scolnick, E. M., Sigal, I. S., and Gibbs, J. B.**, Cloning of bovine GAP and its interaction with oncogenic *ras* p21, *Nature*, 335, 90, 1988.

8. **Trahey, M., Wong, G., Halenbeck, R., Rubinfeld, B., Martin, G. A., Ladner, M., Long, C. M., Crosier, W. J., Watt, K., Koths, K., and McCormick, F.**, Molecular cloning of two types of GAP complementary DNA from human placenta, *Science*, 242, 1697, 1988.

9. **Matsui, Y., Kikuchi, A., Araki, S., Hata, Y., Kondo, J., Teranishi, Y., and Takai, Y.**, Molecular cloning and characterization of a novel type of regulatory protein (GDI) for *smg* p25A, a *ras* p21-like GTP-binding protein, *Mol. Cell. Biol.*, 10, 4116, 1990.

10. **Fukumoto, Y., Kaibuchi, K., Hori, Y., Fujioka, H., Araki, S., Ueda, T., Kikuchi, A., and Takai, Y.,** Molecular cloning and characterization of a novel type of regulatory protein (GDI) for the *rho* proteins, *ras* p21-like small GTP-binding proteins, *Oncogene,* 5, 1321, 1990.

11. **Fong, H. K. W., Hurley, J. B., Hopkins, R. S., Miake-Lye, R., Johnson, M. S., Doolittle, R. F., and Simon, M. I.,** Repetitive segmental structure of the transducin β subunit: homology with the *CDC4* gene and identification of related mRNAs, *Proc. Natl. Acad. Sci. U.S.A.,* 83, 2162, 1986.

12. **Kawamura, S., Kaibuchi, K., Hiroyoshi, M., Hata, Y., and Takai, Y.,** Stoichiometric interaction of *smg* p21 with its GDP/GTP exchange protein and its novel action to regulate the translocation of *smg* p21 between membrane and cytoplasm, *Biochem. Biophys. Res. Commun.,* 174, 1095, 1991.

13. **Hiroyoshi, M., Kaibuchi, K., Kawamura, S., Hata, Y., and Takai, Y.,** Role of the C-terminal region of *smg* p21, a *ras* p21-like small GTP-binding protein, in membrane and *smg* p21 GDP/GTP exchange protein interactions, *J. Biol. Chem.,* 266, 2962, 1991.

14. **Kawata, M., Matsui, Y., Kondo, J., Hishida, T., Teranishi, Y., and Takai, Y.,** A novel small molecular weight GTP-binding protein with the same putative effector domain as the *ras* proteins in bovine brain membranes, purification, determination of primary structure, and characterization, *J. Biol. Chem.,* 263, 18965, 1988.

15. **Matsui, Y., Kikuchi, A., Kawata, M., Kondo, J., Teranishi, Y., and Takai, Y.,** Molecular cloning of *smg* p21B and identification of *smg* p21 purified from bovine brain and human platelets as *smg* p21B, *Biochem. Biophys. Res. Commun.,* 166, 1010, 1990.

16. **Kawata, M., Farnsworth, C. C., Yoshida, Y., Gelb, M. H., Glomset, J. A., and Takai, Y.,** Posttranslationally processed structure of the human platelet protein *smg* p21B: Evidence for geranylgeranylation and carboxyl methylation of the C-terminal cysteine, *Proc. Natl. Acad. Sci. U.S.A.,* 87, 8960, 1990.

17. **Rine, J. and Kim, S. H.,** A role for isoprenoid lipids in the localization and function of an oncoprotein, *New Biol.,* 2, 219, 1990.

18. **Glomset, J. A., Gelb, M. H., and Farnsworth, C. C.,** Prenyl proteins in eukaryotic cells: a new type of membrane anchor., *Trends Biochem. Sci.,* 15, 139, 1990.

19. **Hancock, J. F., Paterson, H., and Marshall, C. J.,** A polybasic domain or palmitoylation is required in addition to the CAAX motif to localize p21^{ras} to the plasma membrane, *Cell,* 63, 133, 1990.

20. **Mizuno, T., Kaibuchi, K., Yamamoto, T., Kawamura, M., Sakoda, T., Fujioka, H., Matsuura, Y., and Takai, Y.,** A stimulatory GDP/GTP exchange protein for *smg* p21 is active on the post-translationally processed form c-Ki-*ras* p21 and *rho*A p21, *Proc. Natl. Acad. Sci. U.S.A.,* 88, 6442, 1991.

21. **Hoshijima, M., Kikuchi, A., Kawata, M., Ohmori, T., Hashimoto, E., Yamamura, H., and Takai, Y.,** Phosphorylation by cyclic AMP-dependent protein kinase of a human platelet Mr 22,000 GTP-binding protein (*smg* p21) having the same putative effector domain as the *ras* gene products, *Biochem. Biophys. Res. Commun.,* 157, 851, 1988.

22. **Kawata, M., Kikuchi, A., Hoshijima, M., Yamamoto, K., Hashimoto, E., Yamaura, H., and Takai, Y.,** Phosphorylation of *smg* p21, a *ras* p21-like GTP-binding protein, by cyclic AMP-dependent protein kinase in a cell-free system and in response to prostaglandin E$_1$ in intact human platelets, *J. Biol. Chem.,* 264, 15688, 1989.

23. **Kawata, M., Kawahara, Y., Sunako, M., Araki, S., Koide, M., Tsuda, T., Fukuzaki, H., and Takai, Y.,** The molecular heterogeneity of the *smg*-21/*Krev*-1/*rap*1 proteins, a GTP-binding protein having the same effector domain as *ras* p21s, in bovine aortic smooth muscle membranes, *Oncogene,* 6, 841, 1991.

24. **Hata, Y., Kaibuchi, K., Kawamura, S., Hiroyoshi, M., Shirataki, H., and Takai, Y.,** Enhancement of the actions of *smg* p21 GDP/GTP exchange protein by the protein kinase A-catalyzed phosphorylation of *smg* p21, *J. Biol. Chem.,* 266, 6571, 1991.

25. **Lapetina, E. G., Lacal, J. C., Reep, B. R., and Vedia, L. M.,** A *ras*-related protein is phosphorylated and translocated by agonists that increase cAMP levels in human platelets, *Proc. Natl. Acad. Sci. U.S.A.,* 86, 3131, 1989.

26. **Fischer, T. H., Gatling, M. N., Lacal, J. C., and White, II, G. C.,** *rap*1B, a cAMP-dependent protein kinase substrate, associates with the platelet cytoskeleton, *J. Biol. Chem.,* 265, 19405, 1990.

27. **Takai, Y., Kikkawa, U., Kaibuchi, K., and Nishizuka, Y.,** Membrane phospholipid metabolism and signal transduction for protein phosphorylation, *Adv. Cyclic Nucleotide Protein Phosphorylation Res.,* 18, 119, 1984.

28. **Hata, Y., Kikuchi, A., Sasaki, T., Schaber, M. D., Gibbs, J. B., and Takai, Y.,** Inhibition of the *ras* p21 GTPase-activating protein-stimulated GTPase activity of c-Ha-*ras* p21 by *smg* p21 having the same putative effector domain as *ras* p21s, *J. Biol. Chem.,* 265, 7104, 1990.

29. **Frech, M., John, J., Pizon, V., Chardin, P., Tavitian, A., Clark, R., McCormick, F., and Wittinghofer, A.,** Inhibition of GTPase activating protein stimulation of *Ras* p21 GTPase by the K*rev*-1 gene product, *Science,* 249, 169, 1990.

30. **Schaber, M. D., Garsky, V. M., Boylan, D., Hill, W. S., Scolnick, E. M., Marshall, M. S., Sigal, I. S., and Gibbs, J. B.,** *Ras*-interaction with the GTPase-activating protein (GAP), *Proteins,* 6, 306, 1989.

31. **Kikuchi, A., Sasaki, T., Araki, S., Hata, Y., and Takai, Y.,** Purification and characterization from bovine brain cytosol of two GTPase-activating proteins specific for *smg* p21, a GTP-binding protein having the same effector domain as c-*ras* p21s, *J. Biol. Chem.,* 264, 9133, 1989.

32. **Molloy, C. J., Bottaro, D. P., Fleming, T. P., Marshall, M. S., Gibbs, J. B., and Aaronson, S. A.,** PDGF induction of tyrosine phosphorylation of GTPase activating protein, *Nature,* 342, 711, 1989.

33. **Ellis, C., Moran, M., McCormick, F., and Pawson, T.,** Phosphorylation of GAP and GAP-associated proteins by transforming and mitogenic tyrosine kinases, *Nature,* 343, 377, 1990.

34. **Kaplan, D. R., Morrison, D. K., Wong, G., McCormick, F., and Williams, L. T.,** PDGF β-receptor stimulates tyrosine phosphorylation of GAP and association of GAP with a signaling complex, *Cell,* 61, 125, 1990.

35. **Saton, T., Endo, M., Nakafuku, M., Akiyama, T., Yamamoto, T., and Kaziro, Y.,** Accumulation of p21^{ras}-GTP in response to stimulation with epidermal growth factor and oncogene products with tyrosine kinase activity, *Proc. Natl. Acad. Sci. U.S.A.,* 87, 7926, 1990.

36. **Williams, L. T.,** Signal transduction by the platelet-derived growth factor receptor, *Science,* 243, 1564, 1989.

37. **Tanaka, K., Nakafuku, M., Satoh, T., Marshall, M. S., Gibbs, J. B., Matsumoto, K., Kaziro, Y., and Toh-e, A.,** *S. cerevisiae* genes *IRA1* and *IRA2* encode proteins that may be functionally equivalent to mammalian *ras* GTPase activating protein, *Cell,* 60, 803, 1990.

38. **Kitayama, H., Matsuzaki, T., Ikawa, Y., and Noda, M.,** Genetic analysis of the Kirsten-*ras*-revertant 1 gene: potentiation of its tumor suppressor activity by specific point mutations, *Proc. Natl. Acad. Sci. U.S.A.,* 87, 4284, 1990.

39. **Sasaki, T., Kikuchi, A., Araki, S., Hata, Y., Isomura, M., Kuroda, S., and Takai, Y.,** Purification and characterization from bovine brain cytosol of a protein that inhibits the dissociation of GDP from and the subsequent binding of GTP to *smg* p25A, a *ras* p21-like GTP-binding protein, *J. Biol. Chem.,* 265, 2333, 1990.

40. **Araki, S., Kikuchi, A., Hata, Y., Isomura, M., and Takai, Y.,** Regulation of reversible binding of *smg* p25A, a *ras* p21-like GTP-binding protein, to synaptic plasma membranes and vesicles by its specific regulatory protein, GDP dissociation inhibitor, *J. Biol. Chem.,* 265, 13,007, 1990.

41. **Farnsworth, C. C., Kawata, M., Yoshida, Y., Takai, Y., Gelb, M. H., and Glomset, J. A.,** The carboxyl-terminal domain of the small GTP-binding protein smg p25A contains two geranylgeranylated cysteine residues and a methyl ester, *Proc. Natl. Acad. Sci. U.S.A.,* 88, 6196, 1991.

42. **Araki, S., Kaibuchi, K., Sasaki, T., Hata, Y., and Takai, Y.,** Role of the C-terminal region of *smg* p25A in its interaction with membranes and the GDP/GTP exchange protein, *Mol. Cell. Biol.,* 11, 1438, 1991.

43. **Burstein, E. S., Linko-Stentz, K., Lu, Z., and Macara, I. G.**, Regulation of the GTPase activity of the *ras*-like protein p25^{rab3A}, *J. Biol. Chem.*, 266, 2689, 1991.

44. **Mollard, G. F., Südhof, T. C., and Jahn, R.**, A small GTP-binding protein dissociates from synaptic vesicles during exocytosis, *Nature*, 349, 79, 1991.

45. **Wattenberg, B. W.**, The molecular control of transport vesicle fusion, *New Biol.*, 2, 505, 1990.

46. **Mizoguchi, A., Kim, S., Ueda, T., and Takai, Y.**, Tissue distribution of *smg* p25A, a *ras* p21-like GTP-binding protein, studied by use of a specific monoclonal antibody, *Biochem. Biophys. Res. Commun.*, 162, 1438, 1989.

47. **Mizoguchi, A., Kim, S., Ueda, T., Kikuchi, A., Yorifuji, H., Hirokawa, N., and Takai, Y.**, Localization and subcellular distribution of *smg* p25A, a *ras* p21-like GTP-binding protein, in rat brain, *J. Biol. Chem.*, 265, 11872, 1990.

48. **Balch, W. B.**, Small GTP-binding proteins in vesicular transport, *Trends, Biochem. Sci.*, 15, 473, 1990.

49. **De Lisle, R. C. and Williams, J. A.**, Regulation of membrane fusion in secretory exocytosis, *Annu. Rev. Physiol.*, 48, 225, 1986.

50. **Gomperts, B. D.**, Involvement of guanine nucleotide-binding protein in the gating of Ca^{2+} by receptors, *Nature*, 306, 64, 1983.

51. **Wollheim, C. B., Ullrich, S., Meda, P., and Vallar, L.**, Regulation of exocytosis in electrically permeabilized insulin-secreting cells — evidence for Ca^{2+} dependent and independent secretion, *Biosci. Rep.*, 7, 443, 1987.

52. **Melancon, P., Glick, B. S., Malhotra, V., Weidman, P. J., Serafini, T., Gleason, M. L., Orci, L., and Rothman, J. E.**, Involvement of GTP-binding "G" proteins in transport through the Golgi stack, *Cell*, 51, 1053, 1987.

53. **Nonaka, H., Kaibuchi, K., Shimizu, K., Yamamoto, J., and Takai, Y.**, Tissue and subcellular distributions of an inhibitory GDP/GTP exchange protein (GDI) for *smg* p25A by use of its antibody, *Biochem. Biophys. Res. Commun.*, 174, 556, 1991.

54. **Ueda, T., Takeyama, Y., Ohmori, T., Ohyanagi, H., Saitoh, Y., and Takai, Y.**, Purification and characterization from rat liver cytosol of a GDP dissociation inhibitor (GDI) for liver 24KG, a *ras* p21-like GTP-binding protein, with properties similar to those of *smg* p25A GDI, *Biochemistry*, 30, 909, 1991.

55. **Sasaki, T., Kaibuchi, K., Kabcenelle, A. K., Novick, P. J., and Takai, Y.**, A mammalian inhibitory GDP/GTP exchange protein (GDP dissociation inhibitor) for *smg* p25A is active on the yeast *SEC4* protein, *Mol. Cell. Biol.*, 11, 2909, 1991.

56. **West, M., Kung, H., and Kamata, T.**, A novel membrane factor stimulates guanine nucleotide exchange reaction of *ras* proteins, *FEBS Lett.*, 259, 245, 1990.

57. **Wolfman, A. and Macara, I. G.**, A cytosolic protein catalyzes the release of GDP from p21^{ras}, *Science*, 248, 67, 1990.

58. **Downward, J., Riehl, R., Wu, L., and Weinberg, R. A.**, Identification of a nucleotide exchange-promoting activity for p21^{ras}, *Proc. Natl. Acad. Sci. U.S.A.*, 87, 5998, 1990.

59. **Huang, Y. K., Kung, H.-F., and Kamata, T.**, Purification of a factor capable of stimulating the guanine nucleotide exchange reaction of ras proteins and its effect on ras-related small molecular mass G proteins, *Proc. Natl. Acad. Sci. U.S.A.*, 87, 8008, 1990.

60. **Isomura, M., Kaibuchi, K., Yamamoto, T., Kawamura, S., Katayama, M., and Takai, Y.**, Partial purification and characterization of GDP dissociation stimulator (GDS) for the *rho* proteins from bovine brain cytosol, *Biochem. Biophys. Res. Commun.*, 169, 652, 1990.

61. **Katayama, M., Kawata, M., Yoshida, Y., Horiuchi, H., Yamamoto, T., Matsuura, Y., and Takai, Y.**, The post translationally modified C-terminal structure of bovine aortic smooth muscle *rho*A p21, *J. Biol. Chem.*, 266, 12639, 1991.

62. **Hori, Y., Kikuchi, A., Isomura, M., Katayama, M., Miura, Y., Fujioka, H., Kaibuchi, K., and Takai, Y.**, Post-translational modifications of the C-terminal region of the *rho* protein are important for its interaction with membranes and the stimulatory and inhibitory GDP/GTP exchange proteins, *Oncogene*, 6, 515, 1991.

63. **Ueda, T., Kikuchi, A., Ohga, N., Yamamoto, J., and Takai, Y.,** Purification and characterization from bovine brain cytosol of a novel regulatory protein inhibiting the dissociation of GDP from and the subsequent binding of GTP to *rho*B p20, a *ras* p21-like GTP-binding protein, *J. Biol. Chem.,* 265, 9373, 1990.
64. **Isomura, M., Kikuchi, A., Ohga, N., and Takai, Y.,** Regulation of binding of *rho*B p20 to membranes by its specific regulatory protein, GDP dissociation inhibitor, *Oncogene,* 6, 119, 1991.

INDEX

INDEX

INDEX

A

abl genes, 105
Actin, 204, 274, 292, 298
Actinomycin D, 261
Actin-β-tubulin, 368
Activating mutations, 428, see also specific types
Activation of *ras* genes, 5–14
  carcinogen specificity and, 18
  detection and, 6–7
  as insufficient for carcinogenesis, 22–27
  liver tumors and, 12
  lung tumors and, 12–14
  lymphomas and, 10–11
  mammary tumors and, 9–10
  skin tumors and, 7–9
  tissue specificity and, 14–17
Acute leukemias, 15, 86
Acylation, 66
Adenocarcinomas, 86, see also Cancer; specific
  types
Adenomas, 21, 22, 23, 86, see also Cancer;
  specific types
Adenosine diphosphate (ADP)-ribosylation, 274,
  275, 288, 291–293, 335
Adenosine diphosphate (ADP)-ribosylation factors
  (ARFs), 189, 194, 218, 219, 225, 271,
  335
Adenosine diphosphate (ADP)-ribosylation
  substrates, 284, 290
Adenosine diphosphate (ADP)-ribosyltransferase,
  271, 284, 288, 289–293
S-Adenosylhomocysteine, 71
Adenylate cyclase, 122, 184, 457, 468, 474, 480
  regulation of, 127
Adenylate kinase, 48
Adenylyl cyclase, 158, 162–166, 428, 433
Adipocytes, 108
ADP, see Adenosine diphosphate
Aflatoxin B, 12, 19
Agglutinability, 166
A-kinase, see Cyclic AMP-dependent protein
  kinase
Alanine, 378
Albumin, 26
Albumin promoters, 20
Aliphatic amino acids, 66, 157, 163, 267, see
  also specific types
Alkaline phosphatase reporter gene, 90
Allele-specific oligonucleotide hybridization, 7

Amino acids, 4, 71, 205, see also specific types
  aliphatic, 66, 157, 163, 267
  in *Dictyostelium*, 176, 177
  K*rev* gene and, 235
  rac gene products and, 284, 289
  rho genes and, 260, 267
  substitutions of, 235
  in transformation suppressor activities, 238–241
  in yeasts, 163
  YPT genes and, 378, 379
Ampicillin, 232
Angiogenesis, 78
Antibodies, 138, 416, see also specific types
  anti-beta-galactosidase, 483
  anti-phosphotyrosine, 416
  anti-*ras*, 87
  CDC25 gene and, 469, 475, 483
  in *Dictyostelium*, 177, 180
  IRA genes and, 434
  K*rev* gene and, 233, 234
  monoclonal, 103, 131, 177, 233, 434
  neutralizing, 122, 415–416
  polyclonal, 469, 475
  rab genes and, 359
  ras-specific, 274
  rho genes and, 274
  SAS genes and, 394
Antigens, 20, 27, 253, see also specific types
Antioncogenes, 438
Anti-phosphotyrosine antibodies, 416
Anti-*ras* antibody, 87
Aplysia
  californica, 260, 261
  spp., 121, 204, 260, 275, 276
Arachidonic acid, 135, 417
ARFs, see Adenosine diphosphate (ADP)-
  ribosylation factors
arl genes, 189
Asparagine, 163, 290, 291, 378
Atrial potassium channels, 437
Autocrine, 129
Autophosphorylation, 105, 315, 378, 416
awd genes, 194–195

B

Bacterial elongation factor, 377
Bacterial toxins, 271, see also specific types
Baculovirus, 322, 493
Baker's yeast, see *Schizosaccharomyces pombe*
Basement membrane, 87

T - #0449 - 101024 - C0 - 234/156/29 - PB - 9781138562301 - Gloss Lamination